Communications and Control Engineering

Published titles include:

Stability and Stabilization of Infinite Dimensional Systems with Applications
Zheng-Hua Luo, Bao-Zhu Guo and Omer Morgul

Nonsmooth Mechanics (Second edition)
Bernard Brogliato

Nonlinear Control Systems II
Alberto Isidori

L_2-Gain and Passivity Techniques in Nonlinear Control
Arjan van der Schaft

Control of Linear Systems with Regulation and Input Constraints
Ali Saberi, Anton A. Stoorvogel and Peddapullaiah Sannuti

Robust and H_∞ Control
Ben M. Chen

Computer Controlled Systems
Efim N. Rosenwasser and Bernhard P. Lampe

Dissipative Systems Analysis and Control
Rogelio Lozano, Bernard Brogliato, Olav Egeland and Bernhard Maschke

Control of Complex and Uncertain Systems
Stanislav V. Emelyanov and Sergey K. Korovin

Robust Control Design Using H_∞ Methods
Ian R. Petersen, Valery A. Ugrinovski and Andrey V. Savkin

Model Reduction for Control System Design
Goro Obinata and Brian D.O. Anderson

Control Theory for Linear Systems
Harry L. Trentelman, Anton Stoorvogel and Malo Hautus

Functional Adaptive Control
Simon G. Fabri and Visakan Kadirkamanathan

Positive 1D and 2D Systems
Tadeusz Kaczorek

Identification and Control Using Volterra Models
Francis J. Doyle III, Ronald K. Pearson and Bobatunde A. Ogunnaike

Non-linear Control for Underactuated Mechanical Systems
Isabelle Fantoni and Rogelio Lozano

Robust Control (Second edition)
Jürgen Ackermann

Flow Control by Feedback
Ole Morten Aamo and Miroslav Krstić

Learning and Generalization (Second edition)
Mathukumalli Vidyasagar

Constrained Control and Estimation
Graham C. Goodwin, María M. Seron and José A. De Doná

Randomized Algorithms for Analysis and Control of Uncertain Systems
Roberto Tempo, Giuseppe Calafiore and Fabrizio Dabbene

Switched Linear Systems
Zhendong Sun and Shuzhi S. Ge

Subspace Methods for System Identification
Tohru Katayama

Digital Control Systems
Ioan D. Landau and Gianluca Zito

Efim N. Rosenwasser and Bernhard P. Lampe

Multivariable Computer-controlled Systems

A Transfer Function Approach

With 27 Figures

Springer

Efim N. Rosenwasser, Dr. rer. nat. Dr. Eng.
State Marine Technical University
Lozmanskaya str. 3
190008 Saint Petersburg
Russia

Bernhard P. Lampe, Dr. rer. nat. Dr. Eng.
University of Rostock
Institute of Automation
18051 Rostock
Germany

Series Editors
E.D. Sontag · M. Thoma · A. Isidori · J.H. van Schuppen

British Library Cataloguing in Publication Data
Rosenwasser, Efim
Multivariable computer-controlled systems : a transfer function approach. - (Communications and control engineering)
1.Automatic control
I.Title II.Lampe, Bernhard P.
629.8
ISBN-13: 9781846284311
ISBN-10: 1846284317

Library of Congress Control Number: 2006926886

Communications and Control Engineering Series ISSN 0178-5354
ISBN-10: 1-84628-431-7 e-ISBN 1-84628-432-5 Printed on acid-free paper
ISBN-13: 978-1-84628-431-1

Printed in Germany

9 8 7 6 5 4 3 2 1

Springer Science+Business Media
springer.com

To Elena and Bärbel

Preface

Classical control theory comprehends two principal approaches for continuous-time and discrete-time linear time-invariant (LTI) systems. The first, constituted of frequency-domain methods, is based on the concepts of the transfer function and the frequency response. The second approach arises from the state space concept and uses either differential or difference equations for describing dynamical systems.

Although these approaches were originally separate, it was finally accepted that rather than hindering each other, they are, in fact, complementary; therefore more constructive and comprehensive methods of investigation could be developed by applying and combining frequency-domain techniques with state-space ones [68, 55, 53, 40, 49].

A different situation exists in the theory of linear computer-controlled systems, which are a subclass of sampled-data (SD) systems, because they are built of both continuous- and discrete-time components. Traditionally, approximation methods, where the problem is reduced to a complete investigation of either continuous- or discrete-time LTI models, predominate in this theory. This assertion can easily be corroborated by studying the leading monograph in this field [14]. However, to obtain rigorous results, a unified and accurate description of discrete- as well as continuous-time elements in continuous time is needed. Unfortunately, as a consequence of this approach, the models become variable in time. Over the last few years, a series of methods has been developed for this more complicated problem; a lot of them are cited in [30, 158]. An analysis of those references, however, shows that there are no frequency-domain methods for the analysis and design of SD systems that could be applied analogously to those used in the theory of LTI systems.

The reason for this deficiency seems to be the lack of a transfer function concept for this wider class of systems that would parallel the classical transfer function for LTI systems [7]. Difficulties in introducing such a concept are caused by the fact that linear computer-controlled systems are non-stationary and have periodically varying coefficients.

In [148] the authors demonstrated that these difficulties could be conquered by the concept of the parametric transfer function (PTF) $w(s,t)$, which, in contrast with the ordinary transfer function for LTI systems, depends on an additional parameter: the time t. Applying the PTF permits the development of frequency methods for the analysis and design of SD systems after the pattern of classical methods and by doing so provides important additional results, practical methods and solutions for a number of new problems. Last but not least, the PTF yields deeper insight into the structure and nature of SD systems.

Though, for the most part, [148] handles single-input, single-output (SISO) systems, it pays attention to practical constraints and makes clear the broad potential of the PTF approach. Since its publication, the authors have taken forward a number of investigations which extend these methods to multi-input, multi-output (MIMO) systems. The results of these investigations are summarized in the present monograph. In place of the PTF, we now make use of the parametric transfer matrix (PTM) $w(s,t)$. In making this extension, obstacles arise because, in contrast with the transfer matrix for LTI systems, the PTM $w(s,t)$ is not a rational function of the argument s. Fortunately, these obstacles turn out to be surmountable and we have developed investigation methods that use only polynomial and rational matrices. Though the final results are stated in a fairly general form, they open new possibilities for solving important classes of applied multivariable problems for which other methods fail.

The theory presented in *Multivariable Computer-controlled Systems* is conceptually based on the work of J.S. Tsypkin [177], who proposed a general frequency-domain description of SD systems in the complex s-plane, and L.A. Zadeh [198, 199], who introduced the PTF concept into automatic control theory. Other significant results in this field are due to J.R. Raggazini, J.T. Tou and S.S.L. Chang, [136, 175, 29]. A version of the well-known Wiener-Hopf method by D. Youla et al. [196] was a useful tool.

The main body of the book consists of ten chapters and is divided into three parts. Part I (Chapters 1–3) contains preliminary algebraic material. Chapters 1 and 2 handle the fundamentals of polynomial and rational matrices that are necessary for understanding ideas explained later. Chapter 3 describes a class of rational matrices that are termed "normal" in the text. At first sight, these matrices seem to have a number of exotic properties because their entries are bounded by a multitude of algebraic conditions; however, it follows from the results of Chapters 1, 2 and 3 that, in practical applications, it is with these matrices that we mostly have to deal.

Chapters 4 and 5 form Part II of the book, dedicated to some control problems which are also necessary to further investigations but which are of additional, independent, importance. Chapter 4 handles the eigenvalue assignment problem and the structure of the characteristic matrix of the closed systems, where the processes are given by polynomial pairs or by polynomial matrix description (PMD), from a general standpoint.

Chapter 5 looks more deeply into the question whether the z- and ζ-transforms are applicable for the investigation of normal and anomalous discrete systems. In this connection it considers the construction of controllable forward and backward models of such systems.

We would emphasize that Chapters 1, 2, 4 and 5 use many results that are known from the fundamental literature: see, for instance, [51, 133, 69, 114, 27, 80, 68, 206, 111, 167, 164] and others. We have decided to include this material in the main body of the current book because of the following considerations:

1. Their inclusion makes the book more readable because it reduces the reader's need for additional literature to a minimum.
2. The representation of the information is adapted to be more suited to achieving the objectives of this particular book.
3. Chapters 1, 2, 4 and 5 contain a number of new results that, in our opinion, will be interesting for readers who are not directly engaged with SD systems.

Among the latter results are the concept of the simple polynomial matrix, its property of structural stability and the analysis of rational matrices on basis of their dominance and subordination, all of which appear in Chapters 1 and 2. Chapter 2 also details investigations into the reducibility of rational transfer matrices. Chapter 4 covers the theorems on eigenvalue and eigenstructure assignment for control systems with PMD processes. In Chapter 5, the investigation of the applicability of the z- and ζ- (or Taylor-) transformations to the mathematical description of anomalous discrete systems is obviously new, as is the generation of controllable forward and backward models for such systems.

Part III (Chapters 6–10) is mainly concerned with frequency methods for the investigation of MIMO SD systems. Chapter 6 presents a frequency approach for parametrically discretized continuous MIMO processes and makes clear the mutual algebraic properties of the continuous process and the discrete model.

Chapter 7 is dedicated to the mathematical description of the standard SD system. It is here that we introduce the PTM, among other substantial methods of description, and make careful investigation of its properties. Stability and stabilizing problems for closed-loop systems, in which the polynomial solution of the stabilizing problem obtained has very general character, are studied. Particularly, cases with pathological sampling periods are included.

Chapter 8 deals with the analysis of the response of the standard SD system to stationary stochastic excitation and with the solution of the $\mathcal{H}_2$ optimization problem on the basis of the PTM concept and the Wiener-Hopf method. The method presented is extremely general and, in addition to finding the optimal control program, it permits us to state a number of fundamental properties of the optimal system: its structure and the set of its poles, for instance.

Chapter 9 describes the methods of the preceding three chapters in greater detail for the special case of single-loop MIMO SD systems. This is done with the supposition that the transfer matrices of all continuous parts are normal and that the sampling period is non-pathological. When theses suppositions hold, important special cancellations take place; thus, the critical case, in which the transfer matrices of continuous elements contain poles on the imaginary axis, is considered. In this way, the fact, important for applications, that the solvability of the connected $\mathcal{H}_2$ problem in the critical case depends on the location of the critical elements inside the control loop with respect to the input and output of the system is stated. In this case there may be situations in which the $\mathcal{H}_2$ problem has no solution.

Chapter 10 is devoted to the $\mathcal{L}_2$ problem for the standard SD system; it contains, as special cases, the design of optimal tracking systems and the redesign problem. In our opinion, this case constitutes a splendid example for demonstrating the possibilities of frequency methods. This chapter demonstrates that in the multidimensional case the solution of the $\mathcal{L}_2$ problem always leads to a singular quadratic functional for that a set of minimizing control programs exists. Applying Laplace transforms during the evaluation by the Wiener-Hopf method allows us to find the complete set of optimal solutions; by doing this, input signals of finite duration and constant signals are included. We know of no alternative methods for constructing the general solution to this problem.

The book closes with four appendices. Appendix A gives a short introduction to the ζ-transformation (Taylor transformation), and its relationship to other operator transformations for discrete sequences. In Appendix B some auxiliary formulae are derived.Appendix C, written by Dr. K. Polyakov, presents the MATLAB® *DirectSDM* Toolbox. Using this toolbox, various $\mathcal{H}_2$ and $\mathcal{L}_2$ problems for single-loop MIMO systems can be solved numerically.Appendix D, composed by Dr. V. Rybinskii, describes a design method for control with guaranteed performance. These controllers guarantee a required performance for arbitrary members of certain classes of stochastic disturbances. The MATLAB® *GarSD* Toolbox, used for the numerical solution of such problems, is also presented.

In our opinion, the best way to get well acquainted with the content of the book is, of course, the thorough reading of all the chapters in sequence, starting with Chapter 1. We recognize, however, that this requires effort and staying-power of the reader and an expert, interested only in SD systems, can start directly with Chapter 6, looking into the preceding chapters only when necessary.

The book is written in a mathematical style. We do not include elementary introductory material on the functioning or the physical and technological characteristics of computer-controlled systems; likewise, there is no relation of the theory and practice of such systems in an historical context. For material of this sort, we refer the reader to the extensive literature in those fields, the above mentioned reference [14] by Åström and Wittenmark, for example. From this viewpoint, our book and its predecessor [148] can be seen as extensions of [14] the titles of both being inspired by it.

Multivariable Computer-controlled Systems is addressed to engineers and scientific workers involved in the investigation and design of computer-controlled systems. It can also be used as a complementary textbook on process-oriented methods in computer-controlled systems by students on courses in control theory, communications engineering and related fields. Practically oriented mathematicians and engineers working in systems theory will find interesting insight in the following pages. The mathematical tools used in this book are, in general, included in basic mathematics syllabuses for engineers at technical universities. Necessary additional material is given directly in the text. The References section is by no means a complete bibliography as it contains only those works we used directly in the preparation of the book.

The authors gratefully acknowledge the financial support by the German science foundation (Deutsche Forschungsgemeinschaft), especially Dr. Andreas Engelke for his engagement and helpful hints. Due to Mrs. Hannelore Gellert from the University of Rostock and Mrs. Ludmila Patrashewa from the Saint Petersburg University of Ocean Technology, additional support by the Euler program of the German academic exchange service (Deutscher Akademischer Austauschdienst) was possible, which is thankfully mentioned. We are especially indebted to the professors B.D.O. Anderson, K.J. Åström, P.M. Frank, G.C. Goodwin, M. Grimble, T. Kaczorek, V. Kučera, J. Lunze, B. Lohmann, M. Šebek, A. Weinmann and a great number of unnamed colleagues for many helpful discussions and valuable remarks.

The engaged work of Oliver Jackson from Springer helped us to overcome various editorial problems - we appreciate his careful work. We thank Sri Ramoju Ravi for comments after reading the draft version.

The MATLAB®-Toolbox *DirectSDM* by K.Y. Polyakov is available as free download from

`http://www.iat.uni-rostock.de/blampe/matlab_toolbox.html`

In case of any problems, please contact `bernhard.lampe@uni-rostock.de`.

Rostock,
May 17, 2006

Efim Rosenwasser
Bernhard Lampe

Contents

Part I Algebraic Preliminaries

Part III Frequency Methods for MIMO SD Systems

Appendices

Part I

Algebraic Preliminaries

1

Polynomial Matrices

1.1 Basic Concepts of Algebra

1. Let a certain set A with elements $a, b, c, d, \ldots$ be given. Assume that over the set A an algebraic operation is defined which relates every pair of elements (a, b) to a third element $c \in A$ that is called the result of the operation.

If the named operation is designated by the symbol '$*$', then the result is symbolically written as

$$a * b = c\,.$$

In general, we have $a * b \neq b * a$. However, if for any two elements a, b in A the equality $a * b = b * a$ holds, then the operation '$*$' is called *commutative*. The operation '$*$' is named *associative*, if for any $a, b, c \in A$ the relation

$$(a * b) * c = a * (b * c)$$

is true.

The set A is called a *semigroup*, if an associative operation '$*$' is defined in it. A semigroup A is called a *group* , if it contains a neutral element e, such that for every $a \in A$

$$a * e = e * a = a$$

is correct, and furthermore, for any $a \in A$ there exists a uniquely determined element $a^{-1} \in A$, such that

$$a * a^{-1} = a^{-1} * a = e\,. \tag{1.1}$$

The element a^{-1} is called the *inverse* element of a. A group, where the operation '$*$' is commutative, is called a commutative group or *Abelian group*.

In many cases the operation '$*$' in an Abelian group is called *addition*, and it is designated by the symbol '$+$'. This notation is called *additive*. In additive notation the neutral element is called the zero element , and it is denoted by the symbol '0' (zero).

In other cases the operation '$*$' is called *multiplication*, and it is written in the same way as the ordinary multiplication of numbers. This notation is named *multiplicative*. The neutral element in the multiplicative notation is designated by the symbol '1' (one). For the inverse element in multiplicative notation is used a^{-1}, and in additive notation we write $-a$. In the last case the inverse element $-a$ is also named the *opposite* element to a.

2. The set A is called an (associative) ring, if the two operations 'addition' and 'multiplication' are defined on A. Hereby, the set A forms an Abelian group with respect to the 'addition', and a semigroup with respect to the 'multiplication'. From the membership to an Abelian group it follows

$$(a+b)+c = a+(b+c)$$

and

$$a+b = b+a\,.$$

Moreover, there exists a zero element 0, such that for an arbitrary $a \in A$

$$a+0 = 0+a = a\,.$$

The element 0 is always uniquely determined. Between the operations 'addition' and 'multiplication' of a ring the relations

$$(a+b)c = ac+bc\,, \qquad c(a+b) = ca+cb$$

(left and right distributivity) are valid.

In many cases rings are considered, which possess a number of further properties. If for any two a, b always $ab = ba$ is true, then the ring is called *commutative*. If a unit element exists with $1a = a1$ for all $a \in A$, then the ring is named as a *ring with unit element*. The element 1 in such a ring is always uniquely determined.

The non-zero elements a, b of a ring, satisfying $ab = 0$, are named *(left resp. right) zero divisor*. A ring is called *integrity region* , if it has no zero divisor.

3. A commutative associative ring with unit element, where every non-zero element a has an inverse a^{-1} that satisfies Equation (1.1), is called a *field*. In others words, a field is a ring, where all elements different from zero with respect to multiplication form a commutative group. It can be shown that an arbitrary field is an integrity region. The set of complex or the set of real numbers with the ordinary addition and multiplication as operations are important examples for fields. In the following, these fields will be designated by $\mathbb{C}$ and $\mathbb{R}$, respectively.

1.2 Polynomials

1. Let $\mathcal{N}$ be a certain commutative associative ring with unit element, especially it can be a field. Let us consider the infinite sequence $(a_0, a_1, \ldots, a_k; 0, \ldots)$, where $a_k \neq 0$, and all elements starting from a_{k+1} are equal to zero. Furthermore, we write

$$(a_0, a_1, \ldots, a_k; 0, \ldots) = (b_0, b_1, \ldots, b_k; 0, \ldots),$$

if and only if $a_i = b_i$ $(i = 0, \ldots, k)$. Over the set of elements of the above form, the operations addition and multiplication are introduced in the following way. The sum is defined by the relation

$$(a_0, a_1, \ldots, a_k; 0, \ldots) + (b_0, b_1, \ldots, b_k; 0, \ldots) = (a_0 + b_0, a_1 + b_1, \ldots, a_k + b_k; 0, \ldots)$$

and the product of the sequences is given by

$$\begin{aligned}(a_0, a_1, \ldots, a_k; 0, \ldots)&(b_0, b_1, \ldots, b_k; 0, \ldots)\\ &= (a_0 b_0, a_0 b_1 + a_1 b_0, \ldots, a_0 b_k + a_1 b_{k-1} + \ldots + a_k b_0, \ldots, a_k b_k; 0, \ldots).\end{aligned} \tag{1.2}$$

It is easily proven that the above explained operations addition and multiplication are commutative and associative. Moreover, these operations are distributive too. Any element $a \in \mathcal{N}$ is identified with the sequence $(a; 0, \ldots)$. Furthermore, let λ be the sequence

$$\lambda = (0, 1; 0, \ldots).$$

Then using (1.2), we get

$$\lambda^2 = (0, 0, 1; 0, \ldots), \quad \lambda^3 = (0, 0, 0, 1; 0, \ldots), \quad etc.$$

Herewith, we can write

$$\begin{aligned}(a_0, a_1, \ldots, a_k; 0, \ldots) &= \\ &= (a_0; 0, \ldots) + (0, a_1; 0, \ldots) + \ldots + (0, \ldots, 0, a_k; 0, \ldots)\\ &= a_0 + a_1(0, 1; 0, \ldots) + \ldots + a_k(0, \ldots, 0, 1; 0, \ldots)\\ &= a_0 + a_1\lambda + a_2\lambda^2 + \ldots + a_k\lambda^k.\end{aligned}$$

The expression on the right side of the last equation is called a *polynomial* in λ with coefficients in $\mathcal{N}$. It is easily shown that this definition of a polynomial is equivalent to other definitions in elementary algebra. For $a_k \neq 0$ the polynomial $a_k\lambda^k$ is called the term of the polynomial

$$f(\lambda) = a_0 + a_1\lambda + \ldots + a_k\lambda^k \tag{1.3}$$

with the highest power. The number k is called the *degree* of the polynomial (1.3), and it is designed by $\deg f(\lambda)$. If we have in (1.3) $a_0 = a_1 = \ldots = a_k = 0$, then the polynomial (1.3) is named the *zero polynomial* . A polynomial with $a_k = 1$ is called *monic.* If for two polynomials $f_1(\lambda)$, $f_2(\lambda)$ the relation $f_1(\lambda) = af_2(\lambda)$ with $a \in \mathcal{N}$ is valid, then these polynomials are called *equivalent.* In what follows, we will use the notation $f_1\lambda) \approx f_2(\lambda)$ for the fact that the polynomials $f_1(\lambda)$ and $f_2(\lambda)$ are equivalent.

Inside this book we only consider polynomials with coefficients from the real number field $\mathbb{R}$ or the complex number field $\mathbb{C}$. Following [206] we use the notation $\mathbb{F}$ for a field that is either $\mathbb{R}$ or $\mathbb{C}$. The set of polynomials over these fields are designated by $\mathbb{R}[\lambda]$, $\mathbb{C}[\lambda]$ or $\mathbb{F}[\lambda]$ respectively. The sets $\mathbb{R}[\lambda]$ and $\mathbb{C}[\lambda]$ are commutative rings without zero divisor. In what follows, the elements in $\mathbb{R}[\lambda]$ are called real polynomials.

2. Some general properties of polynomials are listed below:

1. Any polynomial $f(\lambda) \in \mathbb{C}[\lambda]$ with $\deg f(\lambda) = n$ can be written in the form
$$f(\lambda) = a_n(\lambda - \lambda_1)\cdots(\lambda - \lambda_n)\,. \tag{1.4}$$
This representation is unique up to permutation of the factors. Some of the numbers $\lambda_1, \ldots, \lambda_n$ that are the roots of the polynomial $f(\lambda)$, could be equal. In that case the product (1.4) is represented by
$$f(\lambda) = a_n(\lambda - \lambda_1)^{\mu_1}\cdots(\lambda - \lambda_q)^{\mu_q}, \quad \mu_1 + \ldots + \mu_q = n, \tag{1.5}$$
where all λ_i, $(i = 1, \ldots, q)$ are different. The number μ_i, $(i = 1, \ldots, q)$ is called the multiplicity of the root λ_i. If $f(\lambda) \in \mathbb{R}[\lambda]$ then a_n is a real number, and in the products (1.4), (1.5) for every complex root λ_i there exists the conjugate complex root with equal multiplicity.
2. For given polynomials $f(\lambda)$, $d(\lambda) \in \mathbb{F}[\lambda]$ there exists a uniquely determined pair of polynomials $q(\lambda)$, $r(\lambda) \in \mathbb{F}[\lambda]$, such that
$$f(\lambda) = q(\lambda)d(\lambda) + r(\lambda)\,, \tag{1.6}$$
where
$$\deg r(\lambda) < \deg d(\lambda)\,.$$
Hereby, the polynomial $q(\lambda)$ is called the *entire part*, and the polynomial $r(\lambda)$ is the *remainder* from the division of $f(\lambda)$ by $d(\lambda)$.
3. Let us have $f(\lambda)$, $g(\lambda) \in \mathbb{F}[\lambda]$. It is said, that the polynomial $g(\lambda)$ is a *divisor* of $f(\lambda)$, and we write $g(\lambda)|f(\lambda)$, if
$$f(\lambda) = q(\lambda)g(\lambda)$$
is true, where $q(\lambda)$ is a certain polynomial.
The greatest common divisor (GCD) of the polynomials $f_1(\lambda)$ and $f_2(\lambda)$ should be designated by $p(\lambda)$. At the same time the GCD is a divisor of $f_1(\lambda)$ and $f_2(\lambda)$, and it possesses the greatest possible degree . Up to

equivalence, the GCD is uniquely determined. Any GCD $p(\lambda)$ permits a representation of the form

$$p(\lambda) = f_1(\lambda)m_1(\lambda) + f_2(\lambda)m_2(\lambda),$$

where $m_1(\lambda)$, $m_2(\lambda)$ are certain polynomials in $\mathbb{F}[\lambda]$.

4. The two polynomials $f_1(\lambda)$ and $f_2(\lambda)$ are called *coprime* if their monic GCD is equal to one, that means, up to constants, these polynomials possess no common divisors. For the polynomials $f_1(\lambda)$ and $f_2(\lambda)$ to be coprime, it is necessary and sufficient that there exist polynomials $m_1(\lambda)$ and $m_2(\lambda)$ with
$$f_1(\lambda)m_1(\lambda) + f_2(\lambda)m_2(\lambda) = 1.$$
5. If
$$f_1(\lambda) = p(\lambda)\tilde{f}_1(\lambda), \qquad f_2(\lambda) = p(\lambda)\tilde{f}_2(\lambda),$$
where $p(\lambda)$ is a GCD of $f_1(\lambda)$ and $f_2(\lambda)$, then the polynomials $\tilde{f}_1(\lambda)$ and $\tilde{f}_2(\lambda)$ are coprime.

1.3 Matrices over Rings

1. Let $\mathcal{N}$ be a commutative ring with unit element forming an integrity region, such that $ab = 0$ implies a or b equal to zero, where 0 is the zero element of the ring $\mathcal{N}$. Then from $ab = 0$, $a \neq 0$ it always follows $b = 0$.

2. The rectangular scheme

$$A = \begin{bmatrix} a_{11} & \dots & a_{1m} \\ \vdots & \vdots & \vdots \\ a_{n1} & \dots & a_{nm} \end{bmatrix} \tag{1.7}$$

is named a rectangular *matrix over the ring* $\mathcal{N}$, where the a_{ik}, $(i = 1,\dots,n;\ k = 1,\dots,m)$ are elements of the ring $\mathcal{N}$. In what follows, the set of matrices is designated by $\mathcal{N}_{nm}$. The integers n and m are called the *dimension* of the matrix. In case of $m = n$ we speak of a *quadratic* matrix A, for $m < n$ of a *vertical* and for $m > n$ of a *horizontal* matrix A. For matrices over rings the operations addition, (scalar) multiplication with elements of the ring $\mathcal{N}$, multiplication of matrices by matrices and transposition are defined. All these operations are defined in the same way as for matrices over numbers [51, 44].

3. Every quadratic matrix $A \in \mathcal{N}_{nn}$ is related to its determinant $\det A$ which is calculated in the same way as for number matrices. However, in the given case the value of $\det A$ is an element of the ring $\mathcal{N}$. A matrix A with $\det A \neq 0_{\mathcal{N}}$ is called *regular* or *non-singular*, for $\det A = 0_{\mathcal{N}}$ it is called *singular*.

4. For any matrix $A \in \mathcal{N}_{nn}$ there uniquely exists a matrix $\operatorname{adj} A$ of the form

$$\operatorname{adj} A = \begin{bmatrix} A_{11} & \dots & A_{n1} \\ \vdots & \vdots & \vdots \\ A_{1n} & \dots & A_{nn} \end{bmatrix}, \tag{1.8}$$

where A_{ik} is the algebraic complement (the adjoint) of the element a_{ik} of the matrix A, which is received as the determinant of those matrix that remains by cutting the $i-$th row and $k-$th column multiplied by the sign-factor $(-1)^{i+k}$. The matrix $\operatorname{adj} A$ is called the *adjoint of the matrix* A. The matrices A and $\operatorname{adj} A$ are connected by the relation

$$A(\operatorname{adj} A) = (\operatorname{adj} A)A = (\det A)I_n , \tag{1.9}$$

where the identity matrix I_n is defined by

$$I_n = \begin{bmatrix} 1_{\mathcal{N}} & 0_{\mathcal{N}} & \dots & 0_{\mathcal{N}} \\ 0_{\mathcal{N}} & 1_{\mathcal{N}} & \dots & 0_{\mathcal{N}} \\ \vdots & \vdots & \ddots & \vdots \\ 0_{\mathcal{N}} & 0_{\mathcal{N}} & \dots & 1_{\mathcal{N}} \end{bmatrix} = \operatorname{diag}\{1_{\mathcal{N}}, \ \dots \ , 1_{\mathcal{N}}\}$$

with the unit element $1_{\mathcal{N}}$ of the ring $\mathcal{N}$, and diag means the diagonal matrix.

5. In the following, matrices of dimension $n \times 1$ are called as *columns* and matrices of dimension $1 \times m$ as *rows*, and both are referred to as *vectors* . The number n is named the *height* of the column, and the number m the *width* of the row, and both are the *length* of the vector.

Let $u_1, u_2, \dots, u_k$ be rows of $\mathcal{N}_{1m}$. As a *linear combination* of the rows $u_1, \dots, u_k$, we term the row

$$\tilde{u} = c_1 u_1 + \dots + c_k u_k ,$$

where the c_i, $(i = 1, \dots, k)$ are elements of the ring $\mathcal{N}$. The set of rows $\{u_1, \dots, u_k\}$ is named *linear dependent*, if there exist coefficients $c_1, \dots, c_k$, that are not all equal to zero, such that $\tilde{u} = O_{1m}$. Here and furthermore, O_{nm} designates the *zero matrix, i.e.* that matrix in $\mathcal{N}_{nm}$ having all its elements equal to the zero element $0_{\mathcal{N}}$.

If the equation

$$cu = c_1 u_1 + \dots + c_k u_k$$

is valid with a $c \neq 0_{\mathcal{N}}$, then we say that the column u depends linearly on the columns $u_1, \dots, u_k$.

For the set $\{u_1, \dots, u_k\}$ of columns to be linear dependent, it is necessary and sufficient that one column depends linearly on the others in the sense of the above definition.

For rows over the ring $\mathcal{N}$ the important statement is true: Any set of rows of the width m with more than m elements is linear dependent. In analogy, any set of columns of height n with more than n elements is also linear dependent.

6. Let a finite or infinite set $\mathcal{U}$ of rows of width m be given. Furthermore, let r be the maximal number of linear independent elements of $\mathcal{U}$, where due to the above statement $r \leq m$ is valid. An arbitrary subset of r linear independent rows of $\mathcal{U}$ is called a *basis* of the set $\mathcal{U}$, , the number r itself is called the *normal rank* of $\mathcal{U}$. All that is said above can be directly transferred to sets of columns.

7. Let a matrix $A \in \mathcal{N}_{nm}$ be given, and $\mathcal{U}$ should be the set of rows of A, and $\mathcal{V}$ the set of its columns. Then the following important statements take place:

1. The normal rank of the set $\mathcal{U}$ of the rows of the matrix A is equal to the normal rank of the set $\mathcal{V}$ of its columns. The common value of these ranks is called the *normal rank of the matrix* A, and it is designated by $\operatorname{rank} A$.
2. The normal rank of the matrix A is equal to the highest order of its subdeterminants (*minors*) different from zero. (Here zero again means the zero element of the ring $\mathcal{N}$.)
3. For the linear independence of all rows (columns) of a quadratic matrix, it is necessary and sufficient that it is non-singular.

For arbitrary matrices $A \in \mathcal{N}_{nm}$, the above statements imply

$$\operatorname{rank} \mathcal{V} = \operatorname{rank} \mathcal{U} \leq \min(n, m) \overset{\triangle}{=} \gamma_A \,.$$

Hereinafter, the symbol '$\overset{\triangle}{=}$' stands for equality by definition. In the following, we say that the matrix A has *maximal or full normal* rank, if $\operatorname{rank} A = \gamma_A$, or that it is *non-degenerated.* In the following the symbol 'rank' also denotes the rank of an ordinary number matrix. This notation does not lead to contradictions, because for matrices over the fields of real or complex numbers the normal rank coincides with the ordinary rank.

8. For Matrix (1.7), the expression

$$A\begin{pmatrix} i_1 & i_2 & \dots & i_p \\ k_1 & k_2 & \dots & k_p \end{pmatrix} = \det \begin{bmatrix} a_{i_1 k_1} & \dots & a_{i_1 k_p} \\ \vdots & \vdots & \vdots \\ a_{i_p k_1} & \dots & a_{i_p k_p} \end{bmatrix}$$

denotes the minor of the matrix A, which is calculated by the elements, that are at the same time members of the rows with the numbers $i_1, \dots, i_p$, and of the columns with the numbers $k_1, \dots, k_p$. Let

$$C = AB$$

be given with $C \in \mathcal{N}_{nm}$, $A \in \mathcal{N}_{n\ell}$, $B \in \mathcal{N}_{\ell m}$. Then if $n = m$, the matrix C is quadratic, and for $n \leq \ell$ we have

$$\det C = \sum_{1\le k_1<\cdots<k_n\le\ell} \cdots \sum A\begin{pmatrix} 1 & 2 & \dots & n \\ k_1 & k_2 & \dots & k_n \end{pmatrix} B\begin{pmatrix} k_1 & k_2 & \dots & k_n \\ 1 & 2 & \dots & n \end{pmatrix} .$$

This relation is called *Binet-Cauchy-formula* [51]. For $n > \ell$ we obtain $\det C = 0$.

The formula of Binet-Cauchy permits to express an arbitrary minor of a product by the corresponding minors of its factors. For $p \le \ell$ this formula takes the form

$$C\begin{pmatrix} i_1 & \dots & i_p \\ j_1 & \dots & j_p \end{pmatrix} = \sum_{1\le k_1<\cdots<k_p\le\ell} \cdots \sum A\begin{pmatrix} i_1 & i_2 & \dots & i_p \\ k_1 & k_2 & \dots & k_p \end{pmatrix} B\begin{pmatrix} k_1 & k_2 & \dots & k_p \\ j_1 & j_2 & \dots & j_p \end{pmatrix} .$$

For $p > \ell$, all minors in this relation will be equal to zero.

1.4 Polynomial Matrices

1. By a *polynomial matrix* $A(\lambda)$ we mean a matrix of the form

$$A(\lambda) = \begin{bmatrix} a_{11}(\lambda) & \dots & a_{1m}(\lambda) \\ \vdots & \vdots & \vdots \\ a_{n1}(\lambda) & \dots & a_{nm}(\lambda) \end{bmatrix},$$

where all elements are polynomials in $\mathbb{F}[\lambda]$, especially also in $\mathbb{R}[\lambda]$ or $\mathbb{C}[\lambda]$. The set of these matrices will be designated by $\mathbb{F}_{nm}[\lambda]$, or symbolised directly by $\mathbb{R}_{nm}[\lambda]$ resp. $\mathbb{C}[\lambda]$, and their subsets containing the constant matrices, are denoted by $\mathbb{F}_{nm}$, $\mathbb{R}_{nm}$ or $\mathbb{C}_{nm}$, respectively. The matrices in $\mathbb{R}_{nm}$ and $\mathbb{R}_{nm}[\lambda]$ are called real.

2. Let, especially

$$u_i(\lambda) = \begin{bmatrix} a_{i1}(\lambda) & \dots & a_{im}(\lambda) \end{bmatrix}, \quad (i = 1, 2, \dots, p)$$

be a certain set of rows with width m. The above defined rows will be called *linear dependent* in $\mathbb{F}_{1m}[\lambda]$, if and only if there exist polynomials $c_i(\lambda) \in \mathbb{F}[\lambda]$ that are not all zero at the same time, such that

$$\sum_{i=1}^{p} c_i(\lambda)u_i(\lambda) = O_{1m} .$$

Here $O_{\ell m}$ is the matrix of dimension $\ell \times m$ with all elements equal to the zero polynomial.

Based on this fundamental understanding of these definitions, all derived concepts and insight can be transferred to polynomial matrices, especially the normal rank of matrices over rings and also the formula of Binet-Cauchy.

3. Any polynomial matrix $A(\lambda) \in \mathbb{F}_{nm}[\lambda]$ can be written in the form

$$A(\lambda) = A_0\lambda^q + A_1\lambda^{q-1} + \ldots + A_q\,, \tag{1.10}$$

where A_i, $(i = 1, \ldots, q)$ are constant matrices in $\mathbb{F}_{nm}$. The matrix A_0 is named the *highest coefficient* of the polynomial matrix $A(\lambda)$. If $A_0 \neq O_{nm}$ is true, then the number q is called the *degree of the polynomial matrix* $A(\lambda)$, and it is designated by $\deg A(\lambda)$. If $n \neq m$, or $\det A(\lambda) \equiv 0$ in case of $n = m$, the matrix $A(\lambda)$ is called *singular*. For $\det A(\lambda) \not\equiv 0$ the matrix $A(\lambda)$ is called *non-singular*. A non-singular matrix (1.10) is called *regular*, if $\det A_0 \neq 0$, and *anomalous*, if $\det A_0 = 0$ is true.

In the general case we have

$$\deg[A(\lambda)B(\lambda)] \leq \deg A(\lambda) + \deg B(\lambda)\,. \tag{1.11}$$

However, if one of the factors is regular, then

$$\deg[A(\lambda)B(\lambda)] = \deg A(\lambda) + \deg B(\lambda)\,. \tag{1.12}$$

4. For $A(\lambda) \in \mathbb{F}_{nn}[\lambda]$, Matrix (1.10) is related to its determinant $\det A(\lambda)$, which itself is a polynomial in $\mathbb{F}[\lambda]$. In accordance with the above statements the matrix is non-singular if its determinant is different from the zero polynomial. A non-singular matrix $A(\lambda)$ is related to the non-negative number

$$\operatorname{ord} A(\lambda) = \deg \det A(\lambda)$$

that is called the *order* of the matrix $A(\lambda)$. The degree and order of a matrix $A(\lambda)$ are connected by the inequalities

$$\operatorname{ord} A(\lambda) \leq n \deg A(\lambda) \tag{1.13}$$

or

$$\deg A(\lambda) \geq \frac{1}{n} \operatorname{ord} A(\lambda)\,. \tag{1.14}$$

For a regular matrix $A(\lambda)$ the inequalities (1.13), (1.14) become equalities. In general for a given order $\operatorname{ord} A(\lambda)$, the degree of a matrix $A(\lambda)$ can be an arbitrary large number. A non-singular quadratic polynomial matrix $A(\lambda)$ with $\operatorname{ord} A(\lambda) = 0$, *i.e.* $\det A(\lambda) = \text{const.} \neq 0$ is called *unimodular*.

Example 1.1. The matrix

$$A(\lambda) = \begin{bmatrix} \lambda - 2 & 1 \\ \lambda^4 - 5\lambda^3 + 6\lambda^2 - 5\lambda + 6 & \lambda^3 - 3\lambda^2 + 4\lambda - 4 \end{bmatrix}$$

can be written in the form (1.10) as

$$A(\lambda) = A_0\lambda^4 + A_1\lambda^3 + A_2\lambda^2 + A_3\lambda + A_4\,,$$

where

$$A_0 = \begin{bmatrix} 0 & 0 \\ 1 & 0 \end{bmatrix}, \quad A_1 = \begin{bmatrix} 0 & 0 \\ -5 & 1 \end{bmatrix},$$

$$A_2 = \begin{bmatrix} 0 & 0 \\ 6 & -3 \end{bmatrix}, \quad A_3 = \begin{bmatrix} 1 & 0 \\ -5 & 4 \end{bmatrix}, \quad A_4 = \begin{bmatrix} -2 & 1 \\ 6 & -4 \end{bmatrix}.$$

In the present case we have $\deg A(\lambda) = 4$. The matrix $A(\lambda)$ is non-singular, because of $n = m = 2$ and $\det A(\lambda) \not\equiv 0$. At the same time $\operatorname{ord} A(\lambda) = 2$ due to

$$\det A(\lambda) = 4\lambda^2 - 7\lambda + 2 .$$

Moreover, the matrix $A(\lambda)$ is anomalous, because $\det A_0 = 0$. □

Example 1.2. Let

$$A(\lambda) = \begin{bmatrix} 1 & \lambda^2 + 4 \\ \lambda^3 - 3\lambda + 5 & \lambda^5 + \lambda^3 + 5\lambda^2 - 12\lambda + 21 \end{bmatrix}$$

be given. In this case we have $\deg A(\lambda) = 5$. At the same time $\det A(\lambda) = 1$ and $\operatorname{ord} A(\lambda) = 0$, thus the matrix $A(\lambda)$ is unimodular. □

1.5 Left and Right Equivalence of Polynomial Matrices

1. Let two polynomial matrices $A_1(\lambda), A_2(\lambda) \in \mathbb{F}_{nm}[\lambda]$ be given. The matrices $A_1(\lambda)$ and $A_2(\lambda)$ are called *left-equivalent*, if one of them can be generated from the other by applying the following operations, which are called *left elementary operations*:

1. Exchange of two rows
2. Multiplying the elements of any row with one and the same non-zero number in $\mathbb{F}$
3. Adding to the elements of any row, the by one and the same polynomial in $\mathbb{F}[\lambda]$ multiplied corresponding elements of another row.

It is known that matrices $A_1(\lambda), A_2(\lambda)$ are left-equivalent if and only if there exists a unimodular matrix $p(\lambda)$, such that

$$A_1(\lambda) = p(\lambda)A_2(\lambda) .$$

2. By applying left elementary operations, any matrix $A(\lambda)$ can be given a special form that later on is named the left canonical form of Hermite.

Theorem 1.3 (following [113]). *Let the matrix $A(\lambda) \in \mathbb{F}_{nm}[\lambda]$ have maximal rank γ_A, and the first γ_A columns of $A(\lambda)$ should have a minor of non-vanishing order γ_A. Then in dependence of its dimension, the matrix $A(\lambda)$ can be transformed by left elementary operations into one of the three forms:*

$\gamma_A = m = n$:

$$p(\lambda)A(\lambda) = \begin{bmatrix} g_{11}(\lambda) & g_{12}(\lambda) & \dots & g_{1n}(\lambda) \\ 0 & g_{22}(\lambda) & \dots & g_{2n}(\lambda) \\ \vdots & \vdots & \ddots & \vdots \\ 0 & 0 & \dots & g_{nn}(\lambda) \end{bmatrix} \triangleq \tilde{A}_l(\lambda)\,, \tag{1.15}$$

$\gamma_A = n < m$:

$$p(\lambda)A(\lambda) = \begin{bmatrix} g_{11}(\lambda) & g_{12}(\lambda) & \dots & g_{1n}(\lambda) & g_{1,n+1}(\lambda) & \dots & g_{1m}(\lambda) \\ 0 & g_{22}(\lambda) & \dots & g_{2n}(\lambda) & g_{2,n+1}(\lambda) & \dots & g_{2m}(\lambda) \\ \vdots & \vdots & \ddots & \vdots & \vdots & \dots & \vdots \\ 0 & 0 & \dots & g_{nn}(\lambda) & g_{n,n+1}(\lambda) & \dots & g_{nm}(\lambda) \end{bmatrix} \triangleq \tilde{A}_l(\lambda)\,, \tag{1.16}$$

$\gamma_A = m < n$:

$$p(\lambda)A(\lambda) = \begin{bmatrix} g_{11}(\lambda) & g_{12}(\lambda) & \dots & g_{1m}(\lambda) \\ 0 & g_{22}(\lambda) & \dots & g_{2m}(\lambda) \\ \vdots & \vdots & \ddots & \vdots \\ 0 & 0 & \dots & g_{mm}(\lambda) \\ 0 & 0 & \dots & 0 \\ \vdots & \vdots & \dots & \vdots \\ 0 & 0 & \dots & 0 \end{bmatrix} \triangleq \tilde{A}_l(\lambda)\,. \tag{1.17}$$

In (1.15)–(1.17) the matrix $p(\lambda)$ is unimodular, and the $g_{ii}(\lambda)$ are monic polynomials, where every $g_{ii}(\lambda)$ is of highest degree in its column. Doing so, the matrix $\tilde{A}_l(\lambda)$ is uniquely determined by $A(\lambda)$. Moreover, in Formulae (1.15) and (1.16) the matrix $p(\lambda)$ is also uniquely committed. ■

In the following, the suitable matrix $\tilde{A}_l(\lambda)$ is said to be the *left canonical form* of the corresponding matrix $A(\lambda)$ or also its *left Hermitian form.*

3. By *right elementary operations* , we understand the above declared operations for columns instead of rows. Two matrices are called *right-equivalent*, if we can generate any one from the other by applying right elementary operations. Two polynomial matrices $A_1(\lambda)$, $A_2(\lambda)$ are right-equivalent if and only if there exists a unimodular matrix $q(\lambda)$ with

$$A_1(\lambda) = A_2(\lambda)q(\lambda)\,.$$

In analogy to Theorem 1.3 the following theorem holds.

Theorem 1.4. *Let the matrix $A(\lambda)$ have the maximal rank γ_A, and the first γ_A rows of $A(\lambda)$ should possess a non-zero minor of order γ_A. Then according to its dimension, by applying right elementary operations, the matrix $A(\lambda)$ can be transformed into one of the three forms:*

$\boldsymbol{\gamma_A = m = n:}$

$$A(\lambda)q(\lambda) = \begin{bmatrix} g_{11}(\lambda) & 0 & \dots & 0 \\ g_{21}(\lambda) & g_{22}(\lambda) & \dots & 0 \\ \vdots & \vdots & \ddots & \vdots \\ g_{n1}(\lambda) & g_{n2}(\lambda) & \dots & g_{nn}(\lambda) \end{bmatrix} \triangleq \tilde{A}_r(\lambda), \tag{1.18}$$

$\boldsymbol{\gamma_A = n < m:}$

$$A(\lambda)q(\lambda) = \begin{bmatrix} g_{11}(\lambda) & 0 & & \dots & 0 \dots 0 \\ g_{21}(\lambda) & g_{22}(\lambda) & 0 & \dots & 0 \dots 0 \\ \vdots & \vdots & \ddots & \vdots & \vdots \dots \vdots \\ g_{n1}(\lambda) & g_{n2}(\lambda) & \dots & g_{nn}(\lambda) & 0 \dots 0 \end{bmatrix} \triangleq \tilde{A}_r(\lambda), \tag{1.19}$$

$\boldsymbol{\gamma_A = m < n:}$

$$A(\lambda)q(\lambda) = \begin{bmatrix} g_{11}(\lambda) & 0 & \dots & 0 \\ g_{21}(\lambda) & g_{22}(\lambda) & \dots & 0 \\ \vdots & \vdots & \ddots & \vdots \\ g_{m1}(\lambda) & g_{m2}(\lambda) & \dots & g_{mm}(\lambda) \\ g_{m+1,1}(\lambda) & g_{m+1,2}(\lambda) & \dots & g_{m+1,m}(\lambda) \\ \vdots & \vdots & \dots & \vdots \\ g_{n1}(\lambda) & g_{n2}(\lambda) & \dots & g_{nm}(\lambda) \end{bmatrix} \triangleq \tilde{A}_r(\lambda). \tag{1.20}$$

In (1.18)–(1.20) the matrix $q(\lambda)$ is unimodular, and the $g_{ii}(\lambda)$ are monic polynomials, where every $g_{ii}(\lambda)$ has the highest degree in its row. Doing so, the matrix $\tilde{A}_r(\lambda)$ is uniquely determined by $A(\lambda)$. Moreover, the matrix $q(\lambda)$ in (1.18) and (1.20) is also uniquely committed. ■

The suitable matrix $\tilde{A}_r(\lambda)$ in (1.18)–(1.20) is said to be the *right canonical form* of the polynomial matrix $A(\lambda)$, or its *right Hermitian form.*

Example 1.5. Let

$$A(\lambda) = \begin{bmatrix} \lambda^4 + 1 & \lambda^4 + 3 \\ \lambda^6 + 2\lambda^2 + 1 & \lambda^6 + 4\lambda^2 + 2 \end{bmatrix}.$$

In this case we have $\det A(\lambda) = \lambda^4 - 2\lambda^2 - 1$. Hence $\deg A(\lambda) = 6$ and $\operatorname{ord} A(\lambda) = 4$. The matrix

$$p(\lambda) = \begin{bmatrix} 0.5 & 0.5(1 - \lambda^2) \\ -(\lambda^2 + 1) & \lambda^4 + 1 \end{bmatrix} \begin{bmatrix} 1 & 0 \\ -\lambda^2 & 1 \end{bmatrix}$$

is unimodular. By direct calculation we confirm

$$p(\lambda)A(\lambda) = \begin{bmatrix} 1 & 2.5 - 0.5\lambda^2 \\ 0 & \lambda^4 - 2\lambda^2 - 1 \end{bmatrix},$$

which has degree 4. The matrix on the right side of the last equation is a Hermitian canonical form. As a conclusion of Theorem 1.3, it follows that this Hermitian form $\tilde{A}_l(\lambda)$ and its transformation matrix $p(\lambda)$ are uniquely determined. It should be remarked that the matrix $\tilde{A}_l(\lambda)$ possesses not the smallest possible degree of all matrices that are left-equivalent to $A(\lambda)$. Indeed, consider the product

$$A_1(\lambda) = \begin{bmatrix} 1 & -\lambda^2 \\ 0 & 1 \end{bmatrix} \begin{bmatrix} 1 & 0 \\ -\lambda^2 & 1 \end{bmatrix} A(\lambda) = \begin{bmatrix} 1-\lambda^2 & 3-\lambda^2 \\ \lambda^2+1 & \lambda^2+2 \end{bmatrix}.$$

Then obviously $\deg A_1(\lambda) = 2$. This is the minimal degree, and this result confirms Inequality (1.14). □

1.6 Row and Column Reduced Matrices

1. Let the non-singular quadratic matrix $A(\lambda) \in \mathbb{F}_{nn}[\lambda]$ be given, and $a_1(\lambda), \ldots, a_n(\lambda)$ be the rows of $A(\lambda)$. With the notation

$$\alpha_i = \deg a_i(\lambda)\,, \quad (i = 1, \ldots, n)\,,$$

the matrix $A(\lambda)$ can be written in the form

$$A(\lambda) = \operatorname{diag}\{\lambda^{\alpha_1}, \ldots, \lambda^{\alpha_n}\}A_0 + A_1(\lambda)\,. \tag{1.21}$$

Herein $A_1(\lambda)$ is a matrix, where the degree of its i-th row is smaller than α_i, and A_0 is a constant matrix. Formula (1.21) could be transformed into

$$A(\lambda) = \operatorname{diag}\{\lambda^{\alpha_1}, \ldots, \lambda^{\alpha_n}\}\left(A_0 + A_1\lambda^{-1} + \ldots + A_p\lambda^{-p}\right), \tag{1.22}$$

where $p \geq 0$ is an integer, and the A_i are constant matrices. The number

$$\alpha_l = \alpha_1 + \ldots + \alpha_n$$

is named the *left order* of the matrix $A(\lambda)$. Denote

$$\alpha_{\max} = \max_{1 \leq i \leq n} \{\alpha_i\}\,.$$

Then obviously

$$\deg A(\lambda) = \alpha_{\max}\,. \tag{1.23}$$

In analogy, assuming that $b_1(\lambda), \ldots, b_n(\lambda)$ are the columns of $A(\lambda)$ and

$$\beta_i = \deg b_i(\lambda)\,,$$

we generate the representation

$$A(\lambda) = \left(B_0 + B_1\lambda^{-1} + \ldots + B_q\lambda^{-q}\right) \operatorname{diag}\{\lambda^{\beta_1}, \ldots, \lambda^{\beta_n}\}\,. \tag{1.24}$$

The number

$$\beta_r = \beta_1 + \ldots + \beta_n$$

is called the *right order* of the matrix $A(\lambda)$. Introduce the notation

$$\beta_{\max} = \max_{1 \le i \le n} \{\beta_i\}\,.$$

Then we obtain

$$\deg A(\lambda) = \beta_{\max}\,.$$

Example 1.6. [68]: Consider the matrix

$$A(\lambda) = \begin{bmatrix} \lambda^2 - 1 & \lambda & -3\lambda \\ \lambda^2 & \lambda - 1 & 2\lambda - 1 \\ \lambda + 2 & -\lambda & 2 \end{bmatrix}\,. \tag{1.25}$$

In this case we have $\alpha_1 = 2$, $\alpha_2 = 2$, $\alpha_3 = 1$, and the left order of the matrix $A(\lambda)$ becomes $\alpha_l = 5$. In the representation (1.21) we get

$$A_0 = \begin{bmatrix} 1 & 0 & 0 \\ 1 & 0 & 0 \\ 1 & -1 & 0 \end{bmatrix}, \quad A_1(\lambda) = \begin{bmatrix} -1 & \lambda & -3\lambda \\ 0 & \lambda - 1 & 2\lambda - 1 \\ 2 & 0 & 2 \end{bmatrix}$$

and therefore, (1.22) yields

$$A(\lambda) = \operatorname{diag}\{\lambda^2,\ \lambda^2,\ \lambda\}\left(A_0 + A_1\lambda^{-1} + A_2\lambda^{-2}\right) \tag{1.26}$$

with

$$A_0 = \begin{bmatrix} 1 & 0 & 0 \\ 1 & 0 & 0 \\ 1 & -1 & 0 \end{bmatrix}, \quad A_1 = \begin{bmatrix} 0 & 1 & -3 \\ 0 & 1 & 2 \\ 2 & 0 & 2 \end{bmatrix}, \quad A_2 = \begin{bmatrix} -1 & 0 & 0 \\ 0 & -1 & -1 \\ 0 & 0 & 0 \end{bmatrix}\,. \tag{1.27}$$

At the same time we have $\beta_1 = 2$, $\beta_2 = 1$, $\beta_3 = 1$, and the right order of $A(\lambda)$ becomes $\beta_r = 4$. The representation (1.24) takes the form

$$A(\lambda) = \left(B_0 + B_1\lambda^{-1} + B_2\lambda^{-2}\right) \operatorname{diag}\{\lambda^2,\ \lambda,\ \lambda\}\,,$$

where

$$B_0 = \begin{bmatrix} 1 & 1 & -3 \\ 1 & 1 & 2 \\ 0 & -1 & 0 \end{bmatrix}, \quad B_1 = \begin{bmatrix} 0 & 0 & 0 \\ 0 & -1 & -1 \\ 1 & 0 & 2 \end{bmatrix}, \quad B_2 = \begin{bmatrix} -1 & 0 & 0 \\ 0 & 0 & 0 \\ 2 & 0 & 0 \end{bmatrix}\,.$$

□

2. The matrix $A(\lambda)$ is said to be *row reduced*, if we have in the representation (1.21)

$$\det A_0 \neq 0 \tag{1.28}$$

and it said to be *column reduced*, if in the representation (1.24)

$$\det B_0 \neq 0$$

is true.

Column-reduced matrices can be generated from row-reduced matrices simply by transposition. Therefore, in the following only row-reduced matrices will be considered.

Lemma 1.7. *For Matrix (1.21) to be row reduced, a necessary and sufficient condition is the validity of the equation*

$$\operatorname{ord} A(\lambda) = \deg \det A(\lambda) = \alpha_l\,. \tag{1.29}$$

Proof. From (1.21) we get

$$\det A(\lambda) = \lambda^{\alpha_l} \det A_0 + a_1(\lambda)$$

with $\deg a_1(\lambda) < \alpha_l$. For (1.29) to be valid, (1.28) is necessary. If, conversely, (1.29) is fulfilled, then $\det A_0 \neq 0$ is true, and the matrix $A(\lambda)$ is row reduced. ∎

Example 1.8. For Matrix (1.25) we get $\det A_0 = 0$, $\det B_0 = 5$, therefore the matrix $A(\lambda)$ is column reduced but not row reduced. Hereby, we obtain $\operatorname{ord} A(\lambda) = \beta_r = 4$. □

Theorem 1.9 ([133]). *Any non-singular matrix $A(\lambda)$ can be made row-reduced by left-equivalent transforms.*

Proof. Assume the matrix $A(\lambda)$ be given in form of Representation (1.22). Then for $\det A_0 \neq 0$ the matrix $A(\lambda)$ is already row reduced. Therefore, take a singular matrix A_0, *i.e.* $\det A_0 = 0$. Then there exists a non-zero row vector $\nu = (\nu_1, \dots, \nu_n)$ such that

$$\nu A_0 = O_{1n} \tag{1.30}$$

is fulfilled. Let $\nu_{i_1}, \dots, \nu_{i_q}$, $(1 \leq q \leq n)$ be the non-zero components of ν, and $\alpha_{i_1}, \dots, \alpha_{i_q}$ are the corresponding exponents α_i. Denote

$$\psi = \max_{1 \leq j \leq q} \{\alpha_{i_j}\}\,,$$

and let γ be a value of the index j, for which $\alpha_{i_\gamma} = \psi$ is valid. Then the row

$$\nu(\lambda) = \begin{bmatrix} \nu_1 \lambda^{\psi - \alpha_1} & \nu_2 \lambda^{\psi - \alpha_2} & \dots & \nu_\gamma & \dots & \nu_{n-1} \lambda^{\psi - \alpha_{n-1}} & \nu_n \lambda^{\psi - \alpha_n} \end{bmatrix} \tag{1.31}$$

comes out as a polynomial. Now consider the matrix $P(\lambda)$ that is generated from the identity matrix I_n by exchanging the γ-th row by the row (1.31):

$$P(\lambda) = \begin{bmatrix} 1 & 0 & \dots 0 \dots & 0 & 0 \\ \vdots & \ddots & \dots \vdots \dots & \vdots & \vdots \\ \nu_1\lambda^{\psi-\alpha_1} & \nu_2\lambda^{\psi-\alpha_2} & \dots \nu_\gamma \dots & \nu_{n-1}\lambda^{\psi-\alpha_{n-1}} & \nu_n\lambda^{\psi-\alpha_n} \\ \vdots & \vdots & \dots \vdots \dots & \ddots & \vdots \\ 0 & 0 & \dots 0 \dots & 0 & 1 \end{bmatrix}. \tag{1.32}$$

Due to $\det P(\lambda) = \nu_\gamma \neq 0$, the matrix $P(\lambda)$ is unimodular. Therefore, as shown in [133], the equation

$$P(\lambda)A(\lambda) = \operatorname{diag}\{\lambda^{\alpha_1}, \dots, \lambda^{\alpha_n}\}\left(\tilde{A}_0 + \tilde{A}_1\lambda^{-1} + \dots + \tilde{A}_p\lambda^{-p}\right) \tag{1.33}$$

holds with

$$\tilde{A}_i = DA_i\,, \tag{1.34}$$

and the matrix D is generated from the identity matrix I_n by exchanging the γ-th row by the row ν:

$$D = \begin{bmatrix} 1 & 0 & \dots & 0 & \dots & 0 & 0 \\ \vdots & \ddots & \dots & \vdots & \dots & \vdots & \vdots \\ \nu_1 & \nu_2 & \dots & \nu_\gamma & \dots & \nu_{n-1} & \nu_n \\ \vdots & \vdots & \dots & \vdots & \dots & \ddots & \vdots \\ 0 & 0 & \dots & 0 & \dots & 0 & 1 \end{bmatrix}. \tag{1.35}$$

Obviously, $\det D = \nu_\gamma \neq 0$. From (1.30) and (1.33) it follows that the γ-th row of the matrix $\tilde{A}_0$ is identical to zero. That means, Equation (1.33) can be written in the form

$$\begin{aligned} P(\lambda)A(\lambda) = \operatorname{diag}\{\lambda^{\alpha_1}, \dots, \lambda^{\alpha_{\gamma-1}}, \lambda^{\alpha_\gamma - 1}, \lambda^{\alpha_{\gamma+1}}, \dots, \lambda^{\alpha_n}\}\,\cdot \\ \left(\tilde{\tilde{A}}_0 + \tilde{\tilde{A}}_1\lambda^{-1} + \dots + \tilde{\tilde{A}}_p\lambda^{-p}\right), \end{aligned} \tag{1.36}$$

where the matrices $\tilde{\tilde{A}}_i$ $(i = 0, \dots, p-1)$ are built from the matrices $\tilde{A}_i$ by substituting their γ-th rows by the γ-th row of $\tilde{A}_{i+1}$. Hereby, the γ-th row of $\tilde{A}_p$ is substituted by the zero row. If in Relation (1.36) the matrix $\tilde{\tilde{A}}_0$ is regular, then the matrix $P(\lambda)A(\lambda)$ is row reduced, and the transformation procedure finishes. If, however, the matrix $\tilde{\tilde{A}}_0$ is already singular, we have to repeat the transformation procedure again. It was shown in [133] that for a non-singular matrix $A(\lambda)$ this algorithm after a finite number of steps yields a row-reduced matrix. ∎

Example 1.10. Generate a row-reduced form of the matrix $A(\lambda)$ in (1.26), (1.27). Equation (1.30) leads to the system of linear equations

$$\nu_1 + \nu_2 + \nu_3 = 0\,, \qquad -\nu_3 = 0\,,$$

so we choose $\nu = \begin{bmatrix} 1 & -1 & 0 \end{bmatrix}$, $\gamma = 1$ and $\psi = 2$. Applying (1.31), (1.32) and (1.34) yields

$$P(\lambda) = D = \begin{bmatrix} 1 & -1 & 0 \\ 0 & 1 & 0 \\ 0 & 0 & 1 \end{bmatrix}.$$

Using this result and (1.34), we find

$$\tilde{A}_0 = \begin{bmatrix} 0 & 0 & 0 \\ 1 & 0 & 0 \\ 1 & -1 & 0 \end{bmatrix}, \quad \tilde{A}_1 = \begin{bmatrix} 0 & 0 & -5 \\ 0 & 1 & 2 \\ 2 & 0 & 2 \end{bmatrix}, \quad \tilde{A}_2 = \begin{bmatrix} -1 & 1 & 1 \\ 0 & -1 & -1 \\ 0 & 0 & 0 \end{bmatrix}.$$

Now, exchange the first row of $\tilde{A}_0$ by the first row of $\tilde{A}_1$, the first row of $\tilde{A}_1$ by the first row of $\tilde{A}_2$, and the first row of $\tilde{A}_2$ by the zero row. As result we get

$$\tilde{\tilde{A}}_0 = \begin{bmatrix} 0 & 0 & -5 \\ 1 & 0 & 0 \\ 1 & -1 & 0 \end{bmatrix}, \quad \tilde{\tilde{A}}_1 = \begin{bmatrix} -1 & 1 & 1 \\ 0 & 1 & 2 \\ 2 & 0 & 2 \end{bmatrix}, \quad \tilde{\tilde{A}}_2 = \begin{bmatrix} 0 & 0 & 0 \\ 0 & -1 & -1 \\ 0 & 0 & 0 \end{bmatrix}.$$

The matrix $\tilde{\tilde{A}}_0$ is regular. Therefore, the procedure stops, and with the help of (1.36) we get

$$\begin{aligned} P(\lambda)A(\lambda) &= \operatorname{diag}\{\lambda, \lambda^2, \lambda\} \left(\tilde{\tilde{A}}_0 + \tilde{\tilde{A}}_1 \lambda^{-1} + \tilde{\tilde{A}}_2 \lambda^{-2} \right) \\ &= \begin{bmatrix} -1 & 1 & -5\lambda + 1 \\ \lambda^2 & \lambda - 1 & 2\lambda - 1 \\ \lambda + 2 & -\lambda & 2 \end{bmatrix}. \end{aligned}$$

□

3. A number of useful properties of row-reduced matrices follows from the above explanations.

Theorem 1.11. (see [69]) *Let the matrices*

$$\begin{aligned} A(\lambda) &= \operatorname{diag}\{\lambda^{\alpha_1}, \ldots, \lambda^{\alpha_n}\} \left(\tilde{A}_0 + \tilde{A}_1 \lambda^{-1} + \ldots \right), \\ B(\lambda) &= \operatorname{diag}\{\lambda^{\phi_1}, \ldots, \lambda^{\phi_n}\} \left(\tilde{B}_0 + \tilde{B}_1 \lambda^{-1} + \ldots \right) \end{aligned} \tag{1.37}$$

be given. Then, if the matrices $A(\lambda)$ and $B(\lambda)$ are row reduced and left equivalent, then the sets of numbers $\{\alpha_1, \ldots, \alpha_n\}$ and $\{\phi_1, \ldots, \phi_n\}$ coincide. ∎

Corollary 1.12. *If the matrices $A(\lambda)$ and $B(\lambda)$ are left equivalent, and the matrix $A(\lambda)$ is row reduced, then*

$$\sum_{i=1}^{n} \alpha_i \leq \sum_{i=1}^{n} \phi_i$$

is true, where the equality takes place if and only if the matrix $B(\lambda)$ is also row reduced.

Proof. Because the matrices (1.37) are left equivalent, they possess the same order. Therefore, by Lemma 1.7 it follows

$$\phi_1 + \ldots + \phi_n \geq \operatorname{ord} B(\lambda) = \operatorname{ord} A(\lambda) = \alpha_1 + \ldots + \alpha_n \,,$$

where the equality in the left part exactly takes place, when the matrix $B(\lambda)$ is row reduced. ∎

Corollary 1.13. *Under the conditions of Theorem 1.11,*

$$\deg A(\lambda) = \deg B(\lambda) \,. \tag{1.38}$$

Proof. From (1.23) and (1.37), we get

$$\deg A(\lambda) = \max_{1 \leq i \leq n} \{\alpha_i\} = \max_{1 \leq i \leq n} \{\phi_i\} = \deg B(\lambda) \,,$$

because the sets of numbers α_i and ϕ_i coincide. ∎

Corollary 1.14. *Let the matrices $A(\lambda)$ and $B(\lambda)$ in (1.37) be left equivalent, where the matrix $A(\lambda)$ is row reduced, but the matrix $B(\lambda)$ is not row reduced. Then we have*

$$\deg A(\lambda) \leq \deg B(\lambda) \,. \tag{1.39}$$

Proof. In contrary to the claim, we assume

$$\chi = \deg B(\lambda) < \deg A(\lambda) = \alpha_{\max} \,.$$

Then $B(\lambda)$ is given in a row-reduced form by applying Relation (1.36). In this manner, we get

$$Q(\lambda)B(\lambda) = \operatorname{diag}\{\lambda^{\tilde{\phi}_1}, \ldots \lambda^{\tilde{\phi}_n}\} \left(\tilde{B}_0 + \tilde{B}_1 \lambda^{-1} + \ldots\right)$$

with a unimodular matrix $Q(\lambda)$ and $\det \tilde{B}_0 \neq 0$. From (1.36), it is seen that

$$\deg[Q(\lambda)B(\lambda)] \leq \deg B(\lambda) = \chi < \alpha_{\max} = \deg A(\lambda) \,,$$

which contradicts Equation (1.38). Hence it follows (1.39). ∎

1.7 Equivalence of Polynomial Matrices

1. The matrices $A_1(\lambda)$, $A_2(\lambda) \in \mathbb{F}_{nm}[\lambda]$ are called *equivalent*, if

$$A_1(\lambda) = p(\lambda) A_2(\lambda) q(\lambda) \tag{1.40}$$

is true with unimodular matrices $p(\lambda)$, $q(\lambda)$. Obviously, left-equivalent or right-equivalent matrices are also equivalent. Formula (1.40) says that the matrices $A_1(\lambda)$ and $A_2(\lambda)$ are equivalent if and only if they could be generated of each other by left or right elementary operations.

2.

Theorem 1.15 ([51]). *Any $n \times m$ matrix $A(\lambda)$ with the normal rank ρ is equivalent to the matrix*

$$S_A(\lambda) = \begin{bmatrix} S_\rho(\lambda) & O_{\rho,m-\rho} \\ O_{n-\rho,\rho} & O_{n-\rho,m-\rho} \end{bmatrix}, \tag{1.41}$$

where the matrix $S_\rho(\lambda)$ has the form

$$S_\rho(\lambda) = \operatorname{diag}\{a_1(\lambda), \ldots, a_\rho(\lambda)\}, \tag{1.42}$$

and the $a_i(\lambda)$ are monic polynomials, where every polynomial $a_{i+1}(\lambda)$ is divisible by $a_i(\lambda)$. ■

Matrix (1.41) is uniquely determined by the matrix $A(\lambda)$, and it is named as the *Smith-canonical form* of the matrix $A(\lambda)$.

Corollary 1.16. *It follows immediately from Relations (1.40)–(1.42), that under the condition* $\operatorname{rank} A(\lambda) = \rho$*, there exists only a finite set of numbers $\tilde{\lambda}_i$, $(i = 1, \ldots, q)$ such that the number matrix $A(\tilde{\lambda}_i)$ satisfies the inequality* $\operatorname{rank} A(\tilde{\lambda}_i) < \rho$. ■

Corollary 1.17. *Let a finite number of matrices $A_1(\lambda), \ldots, A_p(\lambda)$ with* $\operatorname{rank} A_i(\lambda) = \rho_i$, *$(i = 1, \ldots, p)$ be given. Then for all fixed values $\tilde{\lambda}$, excluding a certain finite set, the condition* $\operatorname{rank} A_i(\tilde{\lambda}) = \rho_i$ *is fulfilled.* ■

1.8 Normal Rank of Polynomial Matrices

1. Utilising the results from the preceding section, we are able to transfer known results over the rank of number matrices to the normal rank of polynomial matrices.

Theorem 1.18. *Assume*

$$D(\lambda) = A(\lambda)B(\lambda)$$

with polynomial matrices $A(\lambda)$, $B(\lambda)$, $D(\lambda)$ of sizes $n \times \ell$, $\ell \times m$ and $n \times m$, respectively. Then the relations

$$\operatorname{rank} D(\lambda) \leq \min\{\operatorname{rank} A(\lambda), \operatorname{rank} B(\lambda)\} \tag{1.43}$$

and

$$\operatorname{rank} D(\lambda) \geq \operatorname{rank} A(\lambda) + \operatorname{rank} B(\lambda) - \ell \tag{1.44}$$

are true.

Relations (1.43), (1.44) are named *inequalities of Sylvester.*

Proof. Assume

$$\operatorname{rank} A(\lambda) = \rho_A, \ \operatorname{rank} B(\lambda) = \rho_B, \ \operatorname{rank} D(\lambda) = \rho_D$$

with

$$\rho_D > \min\{\rho_A, \rho_B\}\,. \tag{1.45}$$

Then due to Corollary 1.17, there exists a value $\lambda = \tilde{\lambda}$ with

$$\operatorname{rank} A(\tilde{\lambda}) = \rho_A, \quad \operatorname{rank} B(\tilde{\lambda}) = \rho_B, \quad \operatorname{rank} D(\tilde{\lambda}) = \rho_D\,.$$

It is possible to apply the Sylvester inequalities to the number matrices

$$D(\tilde{\lambda}) = A(\tilde{\lambda})B(\tilde{\lambda})\,,$$

which gives

$$\operatorname{rank} D(\tilde{\lambda}) \le \min\{\operatorname{rank} A(\tilde{\lambda}) \operatorname{rank} B(\tilde{\lambda})\}\,.$$

But this contradicts (1.45). This contradiction proves the validity of Inequality (1.43).

Inequality (1.44) could be proved analogously because a corresponding inequality holds for constant matrices. ∎

2. From the inequalities of Sylvester (1.43), (1.44) ensue the following relations

$$\operatorname{rank}[A(\lambda)B(\lambda)] = \operatorname{rank} B(\lambda) \ \text{ for } \operatorname{rank} A(\lambda) = \ell \tag{1.46}$$

and

$$\operatorname{rank}[A(\lambda)B(\lambda)] = \operatorname{rank} A(\lambda) \ \text{ for } \operatorname{rank} B(\lambda) = \ell\,. \tag{1.47}$$

Herein, $\operatorname{rank} A(\lambda) = \ell$ can only be fulfilled for $n \ge \ell$, and $\operatorname{rank} B(\lambda) = \ell$ only for $m \ge \ell$. Especially, Equation (1.46) is valid if the matrix $A(\lambda)$ is non-singular, and Equation (1.47) holds if matrix $B(\lambda)$ is non-singular.

3. For arbitrary matrices $A(\lambda) \in \mathbb{F}_{nm}[\lambda]$ we introduce the notation

$$\operatorname{def} A(\lambda) \stackrel{\triangle}{=} \min\{n, m\} - \operatorname{rank} A(\lambda) = \gamma_A - \operatorname{rank} A(\lambda)$$

and call it as the *normal defect* of $A(\lambda)$. Obviously, we always have $\operatorname{def} A(\lambda) \ge 0$.

Arising from this fact and the above considerations, we conclude that the rank of any matrix does not decrease if it is multiplied from the left by a vertical or square matrix with defect zero. Analogously, the multiplication from right by an horizontal or square matrix with defect zero also would not change the rank.

4. Applying the above thoughts used in the proofs of Theorem 1.18, the known statements for number matrices [51] can also be proved for polynomial matrices.

Theorem 1.19. *The matrices $A(\lambda)$, $B(\lambda)$ and the matrix $D(\lambda)$ should be connected by*

$$D(\lambda) = \begin{bmatrix} A(\lambda) & B(\lambda) \end{bmatrix}.$$

Then,

$$\operatorname{rank} D(\lambda) \leq \operatorname{rank} A(\lambda) + \operatorname{rank} B(\lambda). \tag{1.48}$$

■

Theorem 1.20. *For any polynomial matrices $A(\lambda)$ and $B(\lambda)$ of equal dimension,*

$$\operatorname{rank}[A(\lambda) + B(\lambda)] \leq \operatorname{rank} A(\lambda) + \operatorname{rank} B(\lambda).$$

■

Remark 1.21. A corresponding relation to the last one was proven for number matrices in [147].

1.9 Invariant Polynomials and Elementary Divisors

1. Applying (1.41) and (1.42), the Smith-canonical form of a polynomial matrix $A(\lambda)$ can be written in the form

$$S_A(\lambda) = \left[\begin{array}{cccc|c} h_1(\lambda) & 0 & \ldots & 0 & \\ 0 & h_1(\lambda)h_2(\lambda) & \ldots & 0 & \\ \vdots & \vdots & \ddots & \vdots & O_{\rho,m-\rho} \\ 0 & 0 & \ldots & h_1(\lambda)h_2(\lambda)\cdots h_\rho(\lambda) & \\ \hline & & O_{n-\rho,\rho} & & O_{n-\rho,m-\rho} \end{array}\right], \tag{1.49}$$

where $h_1(\lambda), \ldots, h_\rho(\lambda)$ are scalar monic polynomials given by the relations

$$h_1(\lambda) = a_1(\lambda), \quad h_1(\lambda)h_2(\lambda) = a_2(\lambda), \quad \ldots \quad , h_1(\lambda)h_2(\lambda)\cdots h_\rho(\lambda) = a_\rho(\lambda). \tag{1.50}$$

2. The polynomials $a_i(\lambda)$, $(i = 1, \ldots, \rho)$ configured by (1.42) are called the *invariant polynomials* of the matrix $A(\lambda)$. It was shown that the coincidence of the sets of their invariant polynomials is not only a necessary but a sufficient condition for two polynomial matrices $A_1(\lambda)$ and $A_2(\lambda)$ to be equivalent, [51].

3. The monic greatest common divisor of all minors of i-th order for the matrix $A(\lambda)$ is named its i-th *determinantal divisor*. If rank $A(\lambda) = \rho$ is true, then there exist ρ determinant divisors $D_1(\lambda)$, $D_2(\lambda)$, ... , $D_\rho(\lambda)$.

$D_\rho(\lambda)$ is named the *greatest determinantal divisor*. It can be shown that the set of determinantal divisors is invariant against equivalence transformations on the matrix $A(\lambda)$.

4. The invariant polynomials $a_i(\lambda)$ are connected with the polynomials $D_i(\lambda)$ by the relation

$$a_i(\lambda) = \frac{D_i(\lambda)}{D_{i-1}(\lambda)}, \quad D_0(\lambda) = 1, \quad (i = 1, \ldots, \rho). \tag{1.51}$$

Therefore, it follows from (1.51)

$$D_1(\lambda) = a_1(\lambda),\ D_2(\lambda) = a_1(\lambda)a_2(\lambda), \quad \ldots \quad , D_\rho(\lambda) = a_1(\lambda)a_2(\lambda)\cdots a_\rho(\lambda). \tag{1.52}$$

If Representations (1.49), (1.50) are used, then these relations can be written in the form

$$\begin{aligned} &D_1(\lambda) = h_1(\lambda),\ D_2(\lambda) = h_1^2(\lambda)h_2(\lambda), \quad \ldots \\ &D_\rho(\lambda) = h_1^\rho(\lambda)h_2^{\rho-1}(\lambda)\cdots h_\rho(\lambda). \end{aligned}$$

5. Suppose the greatest common determinantal divisor $D_\rho(\lambda)$ be given by the linear factors

$$D_\rho(\lambda) = (\lambda - \lambda_1)^{\nu_{1\rho}} \cdots (\lambda - \lambda_q)^{\nu_{q\rho}}, \tag{1.53}$$

where all numbers λ_i are different. We take from (1.51) that every invariant polynomial $a_i(\lambda)$ permits a factorisation of the form

$$a_i(\lambda) = (\lambda - \lambda_1)^{\mu_{1i}} \cdots (\lambda - \lambda_q)^{\mu_{qi}}, \quad (i = 1, \ldots, \rho) \tag{1.54}$$

with

$$0 \le \mu_{pi} \le \mu_{p,i+1} \le \nu_{p\rho}, \quad (p = 1, \ldots, q).$$

The factors different from one in the expression (1.54) are called *elementary divisors* of the polynomial matrix $A(\lambda)$ in the field $\mathbb{C}$. In general, every root λ_i is configured to several elementary divisors. It follows from the above said, that the set of invariant polynomials uniquely determines the set of elementary divisors. The reverse is also true if the rank of the matrix $A(\lambda)$ is known.

Example 1.22. Assume the rank of the matrix $A(\lambda)$ to be equal to four, and the whole of its elementary divisors to be

$$(\lambda - 2)^2,\ (\lambda - 2)^2,\ \lambda - 2,\ \lambda - 3,\ \lambda - 3,\ \lambda - 4.$$

Then as the set of invariant polynomials, we obtain

$$a_4(\lambda) = (\lambda-2)^2(\lambda-3)(\lambda-4),\ a_3(\lambda) = (\lambda-2)^2(\lambda-3),\ a_2(\lambda) = \lambda-2,\ a_1(\lambda) = 1.$$

Using the set of invariant polynomials, we are able to specify immediately the Smith-canonical form of the matrix $A(\lambda)$. In the present case we get

$$S_A(\lambda) = \begin{bmatrix} 1 & 0 & 0 & 0 \\ 0 & \lambda-2 & 0 & 0 \\ 0 & 0 & (\lambda-2)^2(\lambda-3) & 0 \\ 0 & 0 & 0 & (\lambda-2)^2(\lambda-3)(\lambda-4) \end{bmatrix}.$$ □

6. For diagonal and block-diagonal matrices the system of elementary divisor can be constructed by the elementary divisors of its elements.

Lemma 1.23 ([51]). *The system of elementary divisors of any diagonal matrix is the unification of the elementary divisors of its elements.* ■

Example 1.24. Let the diagonal matrix

$$A(\lambda) = \begin{bmatrix} \lambda^2 & 0 & 0 & 0 \\ 0 & \lambda(\lambda-1) & 0 & 0 \\ 0 & 0 & (\lambda-1)^2 & 0 \\ 0 & 0 & 0 & \lambda(\lambda-1) \end{bmatrix}$$

be given. By decomposition of all diagonal elements into factors (1.54), we obtain the totality of elementary divisors

$$\lambda^2,\ \lambda,\ \lambda-1,\ (\lambda-1)^2,\ \lambda,\ \lambda-1$$

and, finally, we find the Smith-canonical form

$$S_A(\lambda) = \begin{bmatrix} 1 & 0 & 0 & 0 \\ 0 & \lambda(\lambda-1) & 0 & 0 \\ 0 & 0 & \lambda(\lambda-1) & 0 \\ 0 & 0 & 0 & \lambda^2(\lambda-1)^2 \end{bmatrix}.$$ □

Lemma 1.25 ([51]). *The system of elementary divisors of the block-diagonal matrix*

$$A_d(\lambda) = \begin{bmatrix} A_1(\lambda) & O & \dots & O \\ O & A_2(\lambda) & \dots & O \\ \vdots & \vdots & \ddots & \vdots \\ O & O & \dots & A_n(\lambda) \end{bmatrix} = \operatorname{diag}\{A_1(\lambda), \dots, A_n(\lambda)\},$$

where the $A_i(\lambda)$, $(i = 1, \dots, n)$ are rectangular matrices of any dimension, is built by the unification of the elementary divisors of their block elements. ■

Example 1.26. We choose $n = 2$ and

$$A_1(\lambda) = \begin{bmatrix} \lambda & 1 & 0 \\ 0 & \lambda & 1 \\ 0 & 0 & \lambda \end{bmatrix}, \quad A_2(\lambda) = \begin{bmatrix} \lambda & 0 & 0 \\ 0 & 1 & 0 \\ 0 & 0 & \lambda - a \end{bmatrix}.$$

We realise immediately that the matrix $A_1(\lambda)$ possesses the one and only elementary divisor λ^3. The matrix $A_2(\lambda)$ for $a = 0$ has the two equal elementary divisors λ and λ. In case of $a \neq 0$, we find for $A_2(\lambda)$ the two different elementary divisors λ and $\lambda - a$. That's why for $a \neq 0$ the totality of elementary divisors of the matrix $A_d(\lambda) = \operatorname{diag}\{A_1(\lambda),\, A_2(\lambda)\}$ consists of $\lambda^3,\ \lambda,\ \lambda - a$, and the Smith-canonical form comes out as

$$S_{A_d}(\lambda) = \operatorname{diag}\{1,\, 1,\, 1,\, 1,\, \lambda,\, \lambda^3(\lambda - a)\}.$$

However, in case of $a = 0$, we find

$$S_{A_d}(\lambda) = \operatorname{diag}\{1,\, 1,\, 1,\, \lambda,\, \lambda,\, \lambda^3\}. \qquad \square$$

Remark 1.27. The above example illustrates the fact that the dependence of the Smith-canonical form (or the totality of its elementary divisors) on the coefficients of the polynomial matrix is numerically unstable.

1.10 Latent Equations and Latent Numbers

1. Let the non-singular matrix $A(\lambda) \in \mathbb{F}_{nn}[\lambda]$ be given. The polynomial

$$d_A(\lambda) = \det A(\lambda)$$

is said to be the *characteristic polynomial* of the matrix $A(\lambda)$, and the equation

$$d_A(\lambda) = 0 \tag{1.55}$$

is its *characteristic equation.* The roots of the characteristic equation are called the *eigenvalues* of the matrix $A(\lambda)$. For $A(\lambda) \in \mathbb{F}_{nn}[\lambda]$, the characteristic polynomial $d_A(\lambda)$ is equivalent to the greatest determinantal divisor $D_n(\lambda)$. Therefore, the characteristic equation (1.55) is equivalent to

$$D_\rho(\lambda) = 0\,, \tag{1.56}$$

where $\rho = n$ is the normal rank of the matrix $A(\lambda)$.

2. Let us consider an arbitrary matrix $A(\lambda) \in \mathbb{F}_{nm}[\lambda]$ having full rank $\rho = \gamma_A$. For it, Equation (1.56) always has a sense, and for $n \neq m$ this will be called its *latent equation.* The roots of Equation (1.56) are named *latent roots (numbers)* of the matrix $A(\lambda)$. Obviously, the latent numbers are equal to the numbers λ_i, that are configured by the factorisation (1.53). The latent roots of square matrices coincide with its eigenvalues.

Owing to (1.52), the latent equation can be written in the form

$$a_1(\lambda)a_2(\lambda)\cdots a_\rho(\lambda) = 0\,.$$

Hence it follows that every latent number is the root of at least one invariant polynomial.

3. In the following we investigate the question on the important relation between the rank of the polynomial matrix $A(\lambda)$ and the rank of the number matrix $A(\tilde\lambda)$ that is generated from $A(\lambda)$ by substituting $\lambda = \tilde\lambda$, where $\tilde\lambda$ is a given complex number.

Theorem 1.28. *Suppose that the matrix $A(\lambda)$ possesses the rank $\rho = \gamma_A$. Then, if $\tilde\lambda$ does not coincide with one of the latent roots, the relation*

$$\operatorname{rank} A(\tilde\lambda) = \rho$$

is true. However, if $\tilde\lambda = \lambda_i$ takes place with a certain latent number λ_i, then we have

$$\operatorname{rank} A(\tilde\lambda) = \rho - d_i\,, \tag{1.57}$$

where d_i is the number of different elementary divisors, that is connected with the latent number λ_i.

Proof. If $\lambda = \tilde\lambda$ is not a latent number, then it follows from (1.41), (1.42)

$$\operatorname{rank} S_A(\tilde\lambda) = \rho\,.$$

Due to (1.40), we obtain

$$A(\tilde\lambda) = p(\tilde\lambda)S_A(\lambda)q(\tilde\lambda)\,,$$

where we read $\operatorname{rank} A(\tilde\lambda) = \operatorname{rank} S_A(\tilde\lambda) = \rho$, because the multiplication with the non-singular matrices $p(\tilde\lambda)$, $q(\tilde\lambda)$ does not change the rank. Now, let $\tilde\lambda = \lambda_i$, where λ_i is a latent number. Then there exists a number $d_i \geq 1$, such that

$$a_\rho(\lambda_i) = 0, \ldots,\; a_{\rho-d_i+1}(\lambda_i) = 0,\; a_{\rho-d_i}(\lambda_i) \neq 0, \ldots,\; a_1(\lambda_i) \neq 0\,.$$

Obviously, d_i is equal to the number of different elementary divisors of the latent root λ_i. Hence it follows with the help of (1.40)–(1.42)

$$\operatorname{rank} A(\lambda_i) = \operatorname{rank} S_A(\lambda_i) = \rho - d_i\,,$$

which is equivalent to (1.57). ∎

Corollary 1.29. *The equation for the defect*

$$\operatorname{def} A(\lambda_i) = d_i\,, \quad (i = 1, \ldots, q)$$

is true. ∎

Corollary 1.30. *It follows from Theorem 1.28, that the latent numbers* λ_i *of a non-degenerated matrix* $A(\lambda)$ *are exactly those numbers* λ_i*, for which*

$$\operatorname{rank} A(\lambda_i) < \gamma_A$$

becomes valid. ■

4. For a non-degenerated matrix $A(\lambda)$, let the monic greatest common divisor of the minors of γ_A-th order be equal to 1. In that case, the latent equation (1.56) has no roots, thus the matrix $A(\lambda)$ also does not possess latent roots. Such polynomial matrices are said to be *alatent.* All invariant polynomials of an alatent matrix are equal to 1.

Alatent square matrices turn out to be unimodular. For an alatent matrix $A(\lambda)$, the number matrix $A(\tilde{\lambda})$ for all $\tilde{\lambda}$ possesses its maximal rank.

Theorem 1.31. *The non-degenerated* $n \times m$ *matrix* $A(\lambda)$ *with* $n < m$ *proves to be alatent, if and only if there exists a unimodular matrix* $\psi(\lambda)$*, that meets*

$$A(\lambda) = \begin{bmatrix} I_n & O_{n,m-n} \end{bmatrix} \psi(\lambda) .$$

Proof. Under the made suppositions, due to Theorem 1.4, the Hermitian form $\tilde{A}_r(\lambda)$ has the shape

$$\tilde{A}_r(\lambda) = \begin{bmatrix} I_n & O_{n,m-n} \end{bmatrix} ,$$

and the claimed relation emerges from (1.19) for $\psi(\lambda) = q^{-1}(\lambda)$. ■

Analogously, we conclude from (1.17) that for $n > m$ the vertical $n \times m$ matrix $A(\lambda)$ is alatent if and only if

$$A(\lambda) = \varphi(\lambda) \begin{bmatrix} I_m \\ O_{n-m,m} \end{bmatrix}$$

becomes true with a certain unimodular matrix $\varphi(\lambda)$.

5. A non-degenerated matrix $A(\lambda) \in \mathbb{F}_{nm}[\lambda]$ is said to be *latent,* if it has latent roots. Due to (1.40)–(1.42) it is clear that for $n < m$ a latent matrix $A(\lambda)$ allows the representation

$$A(\lambda) = a(\lambda) b(\lambda) , \tag{1.58}$$

where

$$a(\lambda) = p(\lambda) \operatorname{diag}\{a_1(\lambda), \ldots, a_n(\lambda)\} ,$$

$$b(\lambda) = \begin{bmatrix} I_n & O_{n,m-n} \end{bmatrix} q(\lambda)$$

and $p(\lambda)$, $q(\lambda)$ are unimodular matrices.

Obviously, $\det a(\lambda) \approx a_1(\lambda) \cdots a_n(\lambda)$ is valid. The matrix $b(\lambda)$ proves to be alatent, *i.e.* its rank is equal to $n = \rho$ for all $\lambda = \tilde{\lambda}$. A corresponding representation for $n > m$ is also possible.

6.

Theorem 1.32. *Suppose the $n \times m$ matrix $A(\lambda)$ to be alatent. Then every submatrix generated from any of its rows is also alatent.*

Proof. Take a positive integer $p < n$ and present the matrix $A(\lambda)$ in the form

$$A(\lambda) = \left[\begin{array}{ccc} a_{11}(\lambda) & \dots & a_{1m}(\lambda) \\ \vdots & \dots & \vdots \\ a_{p1}(\lambda) & \dots & a_{pm}(\lambda) \\ \hline a_{p+1,1}(\lambda) & \dots & a_{p+1,m}(\lambda) \\ \vdots & \dots & \vdots \\ a_{n1}(\lambda) & \dots & a_{nm}(\lambda) \end{array}\right] = \left[\begin{array}{c} A_p(\lambda) \\ \hline A_1(\lambda) \end{array}\right]. \tag{1.59}$$

It is indirectly shown that the submatrix $A_p(\lambda)$ over the line turns out to be alatent. Suppose the contrary. Then owing to (1.58), we get

$$A_p(\lambda) = a_p(\lambda) b_p(\lambda),$$

where the matrix $a_p(\lambda)$ is latent, and $\operatorname{ord} a_p(\lambda) > 0$. Applying this result from (1.59),

$$A(\lambda) = \begin{bmatrix} a_p(\lambda) & O_{p,n-p} \\ O_{n-p,p} & I_{n-p} \end{bmatrix} \begin{bmatrix} b_p(\lambda) \\ A_1(\lambda) \end{bmatrix}$$

is acquired. Let $\tilde{\lambda}$ be an eigenvalue of the matrix $a_p(\lambda)$, so

$$A(\tilde{\lambda}) = \begin{bmatrix} a_p(\tilde{\lambda}) & O_{p,n-p} \\ O_{n-p,p} & I_{n-p} \end{bmatrix} \begin{bmatrix} b_p(\tilde{\lambda}) \\ A_1(\tilde{\lambda}) \end{bmatrix}$$

is valid.Because the rank of the first factors on the right side is smaller than n, this implies $\operatorname{rank} A(\tilde{\lambda}) < n$, which is in contradiction to the supposed alatency of $A(\lambda)$. ∎

Remark 1.33. In the same way, it is shown that any submatrix of an alatent matrix $A(\lambda)$ built from any of its columns also becomes alatent.

Corollary 1.34. *Every submatrix built from any rows or columns of a unimodular matrix is alatent.* ∎

1.11 Simple Matrices

1. A non-degenerated latent $n \times m$ matrix $A(\lambda)$ of full rank $\rho = \gamma_A$ is called *simple*, if

$$D_\rho(\lambda) = a_\rho(\lambda), \quad D_{\rho-1}(\lambda) = D_{\rho-2}(\lambda) = \ldots = D_1(\lambda) = 1.$$

In dependence on the dimension for a simple matrix $A(\lambda)$ from (1.40)–(1.42), we derive the representations

$$\gamma_A = n = m: \; A(\lambda) = \varphi(\lambda)\operatorname{diag}\{1,\dots,1,a_n(\lambda)\}\psi(\lambda)\,,$$

$$\gamma_A = n < m: \; A(\lambda) = \varphi(\lambda)\begin{bmatrix} 1 \dots 0 & 0 & 0 \dots 0 \\ \vdots \;\ddots\; \vdots & \vdots & \vdots \;\dots\; \vdots \\ 0 \dots 1 & 0 & 0 \dots 0 \\ 0 \dots 0 & a_n(\lambda) & 0 \dots 0 \end{bmatrix}\psi(\lambda)\,,$$

$$\gamma_A = m < n: \; A(\lambda) = \varphi(\lambda)\begin{bmatrix} 1 \dots 0 & 0 \\ \vdots \;\ddots\; \vdots & \vdots \\ 0 \dots 1 & 0 \\ 0 \dots 0 & a_m(\lambda) \\ 0 \dots 0 & 0 \\ \vdots \;\dots\; \vdots & \vdots \\ 0 \dots 0 & 0 \end{bmatrix}\psi(\lambda)\,,$$

where $\varphi(\lambda)$ and $\psi(\lambda)$ are unimodular matrices.

2. From the last relations, we directly deduce the following statements:

a) For a non-degenerated matrix $A(\lambda)$ to be simple, it is necessary and sufficient that every latent root λ_i is configured to only one elementary divisor.

b) Let the non-degenerated $n \times m$ matrix $A(\lambda)$ of rank γ_A have the latent roots $\lambda_1, \dots, \lambda_q$. Then for the simplicity of $A(\lambda)$, the relation

$$\operatorname{rank} A(\lambda_i) = \gamma_A - 1, \quad (i = 1, \dots, q)$$

or, equivalently the condition

$$\operatorname{def} A(\lambda_i) = 1, \quad (i = 1, \dots, q) \tag{1.60}$$

is necessary and sufficient.

c) Another criterion for the membership of a matrix $A(\lambda)$ to the class of simple matrices yields the following theorem.

Theorem 1.35. *A necessary and sufficient condition for the simplicity of the $n \times n$ matrix $A(\lambda)$ is, that there exists a $n \times 1$ column $B(\lambda)$, such that the matrix $L(\lambda) = \begin{bmatrix} A(\lambda) & B(\lambda) \end{bmatrix}$ becomes alatent.*

Proof. Sufficiency: Let the matrix $\begin{bmatrix} A(\lambda) & B(\lambda) \end{bmatrix}$ be alatent and λ_i, $(i = 1, \dots, q)$ are the eigenvalues of $A(\lambda)$. Hence it follows

$$\operatorname{rank}\begin{bmatrix} A(\lambda_i) & B(\lambda_i) \end{bmatrix} = n, \quad (i = 1, \dots, q)\,.$$

Hereby, we deduce from Theorem 1.28, that we need Condition (1.60) to be satisfied if the last conditions should be fulfilled, *i.e.* the matrix $A(\lambda)$ has to be simple.
Necessity: It is shown that for a simple matrix $A(\lambda)$, there exists a column $B(\lambda)$, such that the matrix $\begin{bmatrix} A(\lambda) & B(\lambda) \end{bmatrix}$ becomes alatent. Let us have $\det A(\lambda) = d(\lambda)$ and $\Delta(\lambda) \approx d(\lambda)$ as the equivalent monic polynomial. Then the matrix $A(\lambda)$ can be written in the form

$$A(\lambda) = \varphi(\lambda)\operatorname{diag}\{1, \ldots, 1, \Delta(\lambda)\}\psi(\lambda),$$

where $\varphi(\lambda)$, $\psi(\lambda)$ are unimodular $n \times n$ matrices. The matrix $Q(\lambda)$ of the shape

$$Q(\lambda) = \begin{bmatrix} I_{n-1} & O_{n-1,1} & O_{n-1,1} \\ O_{1,n-1} & \Delta(\lambda) & 1 \end{bmatrix}$$

is obviously alatent, because it has a minor of n-th order that is equal to one. The matrix $\tilde{\psi}(\lambda)$ with

$$\tilde{\psi}(\lambda) = \begin{bmatrix} \psi(\lambda) & O_{n1} \\ O_{1n} & 1 \end{bmatrix} = \operatorname{diag}\{\psi(\lambda), 1\}$$

is unimodular. Applying the last two equations, we get

$$\varphi(\lambda)Q(\lambda)\tilde{\psi}(\lambda) = \begin{bmatrix} A(\lambda) & B(\lambda) \end{bmatrix} = L(\lambda)$$

with

$$B(\lambda) = \varphi(\lambda)\begin{bmatrix} 0 \\ \vdots \\ 0 \\ 1 \end{bmatrix}. \tag{1.61}$$

The matrix $L(\lambda)$ is alatent per construction. ∎

Remark 1.36. If the matrix $\varphi(\lambda)$ is written in the form

$$\varphi(\lambda) = \begin{bmatrix} \varphi_1(\lambda) & \ldots & \varphi_n(\lambda) \end{bmatrix},$$

where $\varphi_1(\lambda), \ldots, \varphi_n(\lambda)$ are the corresponding columns, then from (1.61), we gain

$$B(\lambda) = \varphi_n(\lambda).$$

3. Square simple matrices possess the property of structural stability, which will be explained by the next theorem.

Theorem 1.37. *Let the matrices* $A(\lambda) \in \mathbb{F}_{nn}(\lambda)$, $B(\lambda) \in \mathbb{F}_{nn}[\lambda]$ *be given, where the matrix* $A(\lambda)$ *is simple, but the matrix* $B(\lambda)$ *is of any structure. Furthermore, let us have* $\det A(\lambda) = d(\lambda)$ *and*

$$\det[A(\lambda) + \epsilon B(\lambda)] = d(\lambda) + \epsilon d_1(\lambda, \epsilon)\,, \tag{1.62}$$

where $d_1(\lambda, \epsilon)$ *is a polynomial, satisfying the condition*

$$\deg d_1(\lambda, \epsilon) < \deg d(\lambda)\,. \tag{1.63}$$

Then there exists a positive number ϵ_0*, such that for* $|\epsilon| < \epsilon_0$ *all matrices* $A(\lambda) + \epsilon B(\lambda)$ *are simple.*

Proof. The proof splits into several stages.

Lemma 1.38. *Let* $\|\cdot\|$ *be a certain norm for finite-dimensional number matrices. Then for any matrix* $B = [\,b_{ik}\,] \in \mathbb{F}_{nn}$ *the estimation*

$$\max_{1\le i,k\le n} |b_{ik}| \le \beta\|B\| \tag{1.64}$$

is true, where $\beta > 0$ *is a constant, independent of* B.

Proof. Let $\|\cdot\|_1$ and $\|\cdot\|_2$ be any two norms in the space $\mathbb{C}_{nn}$. Due to the finite dimension of $\mathbb{C}_{nn}$, any two norms are equivalent, that means, for an arbitrary matrix B, we have

$$\alpha_1\|B\|_1 \le \|B\|_2 \le \alpha_2\|B\|_1\,,$$

where α_1, α_2 are positive constants not depending on the choice of B.

Take

$$\|B\|_1 = \max_{1\le i\le n} \sum_{k=1}^{n} |b_{ik}|\,,$$

then under the assumption $\|\cdot\|_2 = \|\cdot\|$, we win

$$|b_{ik}| \le \|B\|_1 \le \alpha_1^{-1}\|B\|\,,$$

which is adequate to (1.64) with $\beta = \alpha_1^{-1}$. ■

Lemma 1.39. *Let the matrix* $A \in \mathbb{F}_{nn}$ *be non-singular and* $\|\cdot\|$ *be a certain norm in* $\mathbb{F}_{nn}$*. Then, there exists a positive constant* α_0*, such that for* $\|B\| < \alpha_0$*, all matrices* $A + B$ *become non-singular.*

Proof. Assume $|b_{ik}| \le \beta\|B\|$, where $\beta > 0$ is the constant configured in (1.64). Then we expand

$$\det(A + B) = \det A + \varphi(A, B)\,,$$

where $\varphi(A, B)$ is a scalar function of the elements in A and B. For it, an estimation

$$|\varphi(A, B)| < \mu_1\beta\|B\| + \mu_2\beta^2\|B\|^2 + \ldots + \mu_n\beta^n\|B\|^n$$

is true, where μ_i, $(i = 1, \ldots, n)$ are constants, that do not depend on B. Hence there exists a number $\alpha_0 > 0$, such that $\|B\| < \alpha_0$ always implies

$$|\varphi(A, B)| < |\det A|\,.$$

That's why for $\|B\| < \alpha_0$, the desired relation $\det(A + B) \ne 0$ holds. ■

Lemma 1.40. *For the matrix $A \in \mathbb{F}_{nn}$, we assume* $\operatorname{rank} A = \rho$, *and let* $\|\cdot\|$ *be a certain norm in* $\mathbb{F}_{nn}$. *Then there exists a positive constant* α_0, *such that for* $B \in \mathbb{F}_{nn}$ *with* $\|B\| < \alpha_0$ *always*

$$\operatorname{rank}(A+B) \geq \rho\,.$$

Proof. Let A_ρ be a non-zero minor of order ρ of A. Lemma 1.39 delivers the existence of a number $\alpha_0 > 0$, such that for $\|B\| < \alpha_0$, the minor of the matrix $A+B$ corresponding to A_ρ is different from zero. However, this means that the rank will not reduce after addition of B, but that was claimed by the lemma. ∎

Proof of Theorem 1.37 Let

$$d(\lambda) = d_0\lambda^k + \ldots + d_k = 0, \quad d_0 \neq 0$$

be the characteristic polynomial of the matrix $A(\lambda)$. Then we obtain from (1.62) and (1.63)

$$\det[A(\lambda) + \epsilon B(\lambda)] \stackrel{\triangle}{=} d(\lambda, \epsilon) = d_0\lambda^k + d_1(\epsilon)\lambda^{k-1} + \ldots + d_k(\epsilon)\,,$$

where

$$d_i(\epsilon) = d_i + d_{i1}\epsilon + d_{i2}\epsilon^2 + \ldots \quad , \; (i = 1, \ldots, k)$$

are polynomials in the variable ϵ with $d_i(0) = d_i$. Let $\tilde{\lambda}$ be a root of the equation

$$d(\lambda, 0) = d(\lambda) = 0$$

with multiplicity ν, *i.e.* an eigenvalue of the matrix $A(\lambda)$ with multiplicity ν. Since the matrix $A(\lambda)$ is simple, we obtain

$$\operatorname{rank} A(\tilde{\lambda}) = n - 1\,.$$

Hereby, due to Lemma 1.40 it follows the existence of a constant $\tilde{\alpha}$, such that for every matrix $G \in \mathbb{C}_{nn}$ with $\|G\| < \tilde{\alpha}$ the relation

$$\operatorname{rank}[A(\tilde{\lambda}) + G] \geq n - 1 \tag{1.65}$$

is fulfilled. Now consider the equation

$$d(\lambda, \epsilon) = 0\,.$$

As known from [188], for $|\epsilon| < \delta$, where $\delta > 0$ is sufficiently small, there exist ν continuous functions $\tilde{\lambda}_i(\epsilon)$, $(i = 1, \ldots, \nu)$, such that

$$d(\tilde{\lambda}_i(\epsilon), \epsilon) = \det[A(\tilde{\lambda}_i(\epsilon)) + \epsilon B(\tilde{\lambda}_i(\epsilon))] = 0\,, \tag{1.66}$$

where some of the functions $\tilde{\lambda}_i(\epsilon)$ may coincide. Thereby, the limits

$$\lim_{\epsilon \to 0} \tilde{\lambda}_i(\epsilon) = \tilde{\lambda}_i \,, \quad (i = 1, \ldots, \nu)$$

exist, and we can write

$$\tilde{\lambda}_i(\epsilon) = \tilde{\lambda}_i + \tilde{\psi}_i(\epsilon) \,,$$

where $\tilde{\psi}_i(\epsilon)$ are continuous functions with $\tilde{\psi}(0) = 0$. Consequently, we get

$$\begin{aligned} A(\tilde{\lambda}_i(\epsilon)) + \epsilon B(\tilde{\lambda}_i(\epsilon)) &= A(\tilde{\lambda}_i + \tilde{\psi}_i(\epsilon)) + \epsilon B(\tilde{\lambda}_i + \tilde{\psi}_i(\epsilon)) \\ &= A(\tilde{\lambda}_i) + G_i(\epsilon) \end{aligned}$$

with

$$G_i(\epsilon) = \epsilon B(\tilde{\lambda}_i) + \tilde{L}_i(\epsilon)$$

and the matrices $\tilde{L}_i(\epsilon)$ for $|\epsilon| < \delta$ depend continuously on ϵ, and $\tilde{L}_i(0) = O_{nn}$ holds. Next choose a constant $\tilde{\epsilon} > 0$ with the property that for $|\epsilon| < \tilde{\epsilon}$ and all $i = 1, \ldots, \nu$, the relation

$$\|G_i(\epsilon)\| = \|\epsilon B(\tilde{\lambda}_i) + \tilde{L}_i(\epsilon)\| < \tilde{\alpha}$$

is true. Therefore, we receive for $|\epsilon| < \tilde{\epsilon}$ from (1.65)

$$\operatorname{rank}[A(\tilde{\lambda}_i(\epsilon)) + \epsilon B(\tilde{\lambda}_i(\epsilon))] \geq n - 1 \,.$$

On the other side, it follows from (1.66), that for $|\epsilon| < \delta$, we have

$$\operatorname{rank}[A(\tilde{\lambda}_i(\epsilon)) + \epsilon B(\tilde{\lambda}_i(\epsilon))] \leq n - 1 \,.$$

Comparing the last two inequalities, we find for $|\epsilon| < \min\{\tilde{\epsilon}, \delta\}$

$$\operatorname{rank}[A(\tilde{\lambda}_i(\epsilon)) + \epsilon B(\tilde{\lambda}_i(\epsilon))] = n - 1 \,.$$

The above considerations can be made for all eigenvalues of the matrix $A(\lambda)$, therefore, Theorem 1.37 is proved by (1.60). ■

1.12 Pairs of Polynomial Matrices

1. Let us have $a(\lambda) \in \mathbb{F}_{nn}[\lambda]$, $b(\lambda) \in \mathbb{F}_{nm}[\lambda]$. The entirety of both matrices is called a *horizontal pair*, and it is designated by $(a(\lambda), b(\lambda))$. On the other side, if we have $a(\lambda) \in \mathbb{F}_{mm}[\lambda]$ and $c(\lambda) \in \mathbb{F}_{nm}[\lambda]$, then we speak about a *vertical pair* and we write $[a(\lambda), c(\lambda)]$. The pairs $(a(\lambda), b(\lambda))$ and $[a(\lambda), c(\lambda)]$ may be configured to the rectangular matrices

$$R_h(\lambda) = \begin{bmatrix} a(\lambda) & b(\lambda) \end{bmatrix} \,, \quad R_v(\lambda) = \begin{bmatrix} a(\lambda) \\ c(\lambda) \end{bmatrix} \,, \tag{1.67}$$

where the first one is horizontal, and the second one is vertical. Due to

$$R'_v(\lambda) = \begin{bmatrix} a'(\lambda) \; c'(\lambda) \end{bmatrix},$$

the properties of vertical pairs can immediately deduced from the properties of horizontal pairs. Therefore, we will consider now only horizontal pairs. The pairs $(a(\lambda), b(\lambda))$, $[a(\lambda), c(\lambda)]$ are called *non-degenerated* if the matrices (1.67) are non-degenerated. If not supposed explicitly otherwise, we will always consider non-degenerated pairs.

2. Let for the pair $(a(\lambda), b(\lambda))$ exist a polynomial matrix $g(\lambda)$, such that

$$a(\lambda) = g(\lambda)a_1(\lambda), \quad b(\lambda) = g(\lambda)b_1(\lambda) \tag{1.68}$$

with polynomial matrices $a_1(\lambda)$, $b_1(\lambda)$. Then the matrix $g(\lambda)$ is called a *common left divisor* of the pair $(a(\lambda), b(\lambda))$. The common left divisor $g(\lambda)$ is named as a *greatest common left divisor* (GCLD) of the pair $(a(\lambda), b(\lambda))$, if for any left common divisor $g_1(\lambda)$

$$g(\lambda) = g_1(\lambda)\alpha(\lambda)$$

with a polynomial matrix $\alpha(\lambda)$ is true. As known any two GCLD are right-equivalent [69].

3. If the pair $(a(\lambda), b(\lambda))$ is non-degenerated, then from Theorem 1.4, it follows the existence of a unimodular matrix

$$r(\lambda) = \overset{\begin{matrix} n & \quad\quad m \end{matrix}}{\begin{bmatrix} r_{11}(\lambda) & r_{12}(\lambda) \\ r_{21}(\lambda) & r_{22}(\lambda) \end{bmatrix}} \begin{matrix} n \\ m \end{matrix} \tag{1.69}$$

for which

$$\begin{bmatrix} a(\lambda) \; b(\lambda) \end{bmatrix} \begin{bmatrix} r_{11}(\lambda) \; r_{12}(\lambda) \\ r_{21}(\lambda) \; r_{22}(\lambda) \end{bmatrix} = \overset{\begin{matrix} n & \; m \end{matrix}}{\begin{bmatrix} N(\lambda) \; O \end{bmatrix}} n \tag{1.70}$$

holds. As known [69], the matrix $N(\lambda)$ is a GCLD of the pair $(a(\lambda), b(\lambda))$.

4. The pair $(a(\lambda), b(\lambda))$ is called *irreducible*, if the matrix $R_h(\lambda)$ in (1.67) is alatent. From the above considerations, it follows that the pair $(a(\lambda), b(\lambda))$ is irreducible, if and only if there exists a unimodular matrix $r(\lambda)$ according to (1.69) with

$$\begin{bmatrix} a(\lambda) \; b(\lambda) \end{bmatrix} r(\lambda) = \begin{bmatrix} I_n \; O_{nm} \end{bmatrix}.$$

5. Let

$$s(\lambda) = r^{-1}(\lambda) = \begin{bmatrix} s_{11}(\lambda) \; s_{12}(\lambda) \\ s_{21}(\lambda) \; s_{22}(\lambda) \end{bmatrix}$$

be a unimodular polynomial matrix. Then we get from (1.70)

$$\begin{bmatrix} a(\lambda) \; b(\lambda) \end{bmatrix} = \begin{bmatrix} N(\lambda) \; O_{nm} \end{bmatrix} \begin{bmatrix} s_{11}(\lambda) & s_{12}(\lambda) \\ s_{21}(\lambda) & s_{22}(\lambda) \end{bmatrix}.$$

Hence it follows immediately

$$a(\lambda) = N(\lambda)s_{11}(\lambda), \quad b(\lambda) = N(\lambda)s_{12}(\lambda),$$

that can be written in the form

$$\begin{bmatrix} a(\lambda) \; b(\lambda) \end{bmatrix} = N(\lambda) \begin{bmatrix} s_{11}(\lambda) \; s_{12}(\lambda) \end{bmatrix}.$$

Due to Corollary 1.34, the pair $(s_{11}(\lambda), s_{12}(\lambda))$ is irreducible. Therefore, the next statement is true:

If Relation (1.68) is true, and $g(\lambda)$ is a GCLD of the pair $(a(\lambda), b(\lambda))$, then the pair $(a_1(\lambda), b_1(\lambda))$ is irreducible.

The reverse statement is also true:

If Relation (1.68) is valid, and the pair $(a_1(\lambda), b_1(\lambda))$ is irreducible, then the matrix $g(\lambda)$ is a GCLD of the pair $(a(\lambda), b(\lambda))$.

6. A necessary and sufficient condition for the irreducibility of the pair $(a(\lambda), b(\lambda))$ with the $n \times n$ polynomial matrix $a(\lambda)$ and the $n \times m$ polynomial matrix $b(\lambda)$ is the existence of an $n \times n$ polynomial matrix $X(\lambda)$ and an $m \times n$ polynomial matrix $Y(\lambda)$, such that the relation

$$a(\lambda)X(\lambda) + b(\lambda)Y(\lambda) = I_n \tag{1.71}$$

becomes true [69].

7. All what is said up to now, can be transferred practically without change to vertical pairs $[a(\lambda), c(\lambda)]$. In this case, instead of the concepts common left divisor and GCLD we introduce the concepts *common right divisor* and *greatest common right divisor* (GCRD) . Hereby, if

$$p(\lambda) \begin{bmatrix} a(\lambda) \\ c(\lambda) \end{bmatrix} = \overset{m}{\begin{bmatrix} L(\lambda) \\ O_{nm} \end{bmatrix}} \begin{matrix} m \\ n \end{matrix}$$

is valid with a unimodular matrix $p(\lambda)$, then $L(\lambda)$ is a GCRD of the corresponding pair $[a(\lambda), c(\lambda)]$. If $L(\lambda)$ and $L_1(\lambda)$ are two GCRD, then they are related by

$$L(\lambda) = f(\lambda)L_1(\lambda)$$

where $f(\lambda)$ is a unimodular matrix.

The vertical pair $[a(\lambda), c(\lambda)]$ is called *irreducible*, if the matrix $R_v(\lambda)$ in (1.67) is alatent. The pair $[a(\lambda), c(\lambda)]$ turns out to be irreducible, if and only if, there exists a unimodular matrix $p(\lambda)$ with

$$p(\lambda)\begin{bmatrix} a(\lambda) \\ c(\lambda) \end{bmatrix} = \begin{bmatrix} I_m \\ O_{nm} \end{bmatrix}.$$

Immediately, it is seen that the pair $[a(\lambda), c(\lambda)]$ is exactly irreducible, when there exist polynomial matrices $U(\lambda)$, $V(\lambda)$, for which

$$U(\lambda)a(\lambda) + V(\lambda)c(\lambda) = I_m\,.$$

8. The above stated irreducibility criteria will be formulated alternatively.

Theorem 1.41. *A necessary and sufficient condition for the pair $(a(\lambda), b(\lambda))$ to be irreducible, is the existence of a pair $(\alpha_l(\lambda), \beta_l(\lambda))$, such that the matrix*

$$Q_l(\lambda) = \begin{bmatrix} a(\lambda) & b(\lambda) \\ \beta_l(\lambda) & \alpha_l(\lambda) \end{bmatrix}$$

becomes unimodular.

For the pair $[a(\lambda), c(\lambda)]$ to be irreducible, it is necessary and sufficient that there exists a pair $[\alpha_r(\lambda), \beta_r(\lambda)]$, such that the matrix

$$Q_r(\lambda) = \begin{bmatrix} \alpha_r(\lambda) & c(\lambda) \\ \beta_r(\lambda) & a(\lambda) \end{bmatrix}$$

becomes unimodular. ∎

9.

Lemma 1.42. *Necessary and sufficient for the irreducibility of the pair $(a(\lambda), b(\lambda))$, with the $n \times n$ and $n \times m$ polynomial matrices $a(\lambda)$ and $b(\lambda)$, is the condition*

$$\operatorname{rank} R_h(\lambda_i) = \operatorname{rank}\left[\, a(\lambda_i)\; b(\lambda_i) \,\right] = n\,, \qquad (i = 1, \ldots, q)\,, \tag{1.72}$$

where the λ_i are the different eigenvalues of the matrix $a(\lambda)$.

Proof. Sufficiency: For $\lambda = \tilde{\lambda} \neq \lambda_i$, $(i = 1, \ldots, q)$, we have $\operatorname{rank} a(\tilde{\lambda}) = n$. Therefore, together with (1.72) the relation $\operatorname{rank} R_h(\lambda) = n$ is true for all finite λ. This means, however, the pair $(a(\lambda), b(\lambda))$ is irreducible.
The *necessity* of Condition (1.72) is obvious. ∎

10.

Lemma 1.43. *Let the pair $(a(\lambda), b(\lambda))$ be given with the $n \times n$ and $n \times m$ polynomial matrices $a(\lambda)$, $b(\lambda)$. Then for the pair $(a(\lambda), b(\lambda))$ to be irreducible, it is necessary that the matrix $a(\lambda)$ has not more than m invariant polynomials different from 1.*

Proof. Assume the number of invariant polynomials different from 1 of the matrix $a(\lambda)$ be $\kappa > m$. Then it follows from (1.57), that there exists an eigenvalue λ_0 of the matrix $a(\lambda)$ with $\operatorname{rank} a(\lambda_0) = n - \kappa$. Applying Inequality (1.48), we gain $\operatorname{rank}\left[\,a(\lambda_0)\ b(\lambda_0)\,\right] \leq n - \kappa + m < n$ that means, the matrix $\left[\,a(\lambda)\ b(\lambda)\,\right]$ is not alatent and, consequently, the pair $(a(\lambda), b(\lambda))$ is not irreducible. ∎

Remark 1.44. Obviously, we could formulate adequate statements as in Lemmata 1.42 and 1.43 for vertical pairs too.

1.13 Polynomial Matrices of First Degree (Pencils)

1. For $q = 1$, $n = m$ the polynomial matrix (1.10) takes the form

$$A(\lambda) = A\lambda + B \tag{1.73}$$

with constant $n \times n$ matrices A, B. This special structure is also called a *pencil.* The pencil $A(\lambda)$ is non-singular if

$$\det(A\lambda + B) \not\equiv 0\,.$$

According to the general definition, the non-singular matrix (1.73) is called *regular* for $\det A \neq 0$ and *anomalous* for $\det A = 0$. Regular pencils arise in connection with state space representations, while anomalous pencils are configured to descriptor systems [109, 34, 182]. All introduced concepts and statements that were developed for polynomial matrices of general structure are also valid for pencils (1.73). At the same time, these matrices possess a number of important additional properties that will be investigated in this section. In what follows, we only consider non-singular pencils.

2. In accordance with the general definition, the two matrices of equal dimension

$$A(\lambda) = A\lambda + B, \quad A_1(\lambda) = A_1\lambda + B_1 \tag{1.74}$$

are called left(right)-equivalent, if there exists a unimodular matrix $p(\lambda)$ ($q(\lambda)$), such that

$$A(\lambda) = p(\lambda)A_1(\lambda), \qquad (A(\lambda) = A_1(\lambda)q(\lambda))\,.$$

The matrices (1.74) are equivalent, if they satisfy an equation

$$A(\lambda) = p(\lambda)A_1(\lambda)q(\lambda)$$

with unimodular matrices $p(\lambda)$, $q(\lambda)$. As follows from the above disclosures, the matrices (1.74) are exactly left(right)-equivalent, if their Hermitian canonical forms coincide. For the equivalence of the matrices (1.74), it is necessary and sufficient that their Smith-canonical forms coincide.

3. The matrices (1.74) are named *strictly equivalent*, if there exist constant matrices P, Q with

$$A(\lambda) = PA_1(\lambda)Q\,. \tag{1.75}$$

If in (1.74) the conditions $\det A \neq 0$, $\det A_1 \neq 0$ are valid, *i.e.* the matrices are regular, then the matrices $A(\lambda)$, $B(\lambda)$ are only in that case equivalent, when they are strictly equivalent. If $\det A = 0$ or $\det A_1 = 0$, *i.e.* the matrices (1.74) are anomalous, then the conditions for equivalence and strict equivalence do not coincide.

4. In order to formulate a criterion for the strict equivalence of anomalous matrices (1.74), following [51], we consider the $n \times n$ *Jordan block*

$$J_n(a) \triangleq \begin{bmatrix} a & 1 & 0 & \dots & 0 & 0 \\ 0 & a & 1 & \dots & 0 & 0 \\ 0 & 0 & a & \ddots & 0 & 0 \\ \vdots & \vdots & \vdots & \ddots & \ddots & \vdots \\ 0 & 0 & 0 & \dots & a & 1 \\ 0 & 0 & 0 & \dots & 0 & a \end{bmatrix}, \tag{1.76}$$

where a is a constant.

Theorem 1.45 ([51]). *Let*

$$\det A(\lambda) = \det(A\lambda + B) \neq 0$$

be given with $\det A = 0$ *and*

$$0 < \operatorname{ord} A(\lambda) = \deg\det A(\lambda) = \eta < n\,. \tag{1.77}$$

Furthermore, let

$$(\lambda - \lambda_1)^{\eta_1}, \dots, (\lambda - \lambda_q)^{\eta_q}, \qquad \eta_1 + \dots + \eta_q = \eta \tag{1.78}$$

be the entity of elementary divisors of $A(\lambda)$ *in the field* $\mathbb{C}$. *In what follows, the elementary divisors (1.78) will be called* finite elementary divisors. *Then the matrix* $A(\lambda)$ *is strictly equivalent to the matrix*

$$\tilde{A}(\lambda) = \operatorname{diag}\{\lambda I_\eta + A_\eta, I_{n-\eta} + \lambda A_\nu\} \tag{1.79}$$

with

$$\begin{aligned} A_\eta &= \operatorname{diag}\{J_{\eta_1}(\lambda_1), \dots, J_{\eta_q}(\lambda_q)\}\,, \\ A_\nu &= \operatorname{diag}\{J_{p_1}(0), \dots, J_{p_\ell}(0)\}\,, \end{aligned} \tag{1.80}$$

where $p_1, \dots, p_\ell$ *are positive integers with* $p_1 + \dots + p_\ell = n - \eta$. *The matrix* $A_\nu \in \mathbb{C}_{n-\eta,n-\eta}$ *is nilpotent, that means, there exists an integer* κ *with* $A_\nu^\kappa = O_{n-\eta,n-\eta}$. ∎

Remark 1.46. The above defined numbers $p_1, \dots, p_\ell$ are determined by the infinite elementary divisors of the matrix $A(\lambda)$, [51]. Thereby, the matrices (1.74) are strictly equivalent, if their finite and infinite elementary divisors coincide.

Remark 1.47. Matrix (1.79) can be represented as

$$\tilde{A}(\lambda) = U\lambda + V\,, \tag{1.81}$$

where

$$U = \operatorname{diag}\{I_\eta, A_\nu\}\,, \quad V = \operatorname{diag}\{A_\eta, I_{n-\eta}\}\,. \tag{1.82}$$

As is seen from (1.76) and (1.80)–(1.82) for $\eta < n$, we always obtain $\det U = 0$ and the matrix $\tilde{A}(\lambda)$ is generally spoken not row reduced.

5. As any non-singular matrix, also an anomalous matrix (1.73) can be brought into row reduced form by left equivalence transformations. Hereby, we obtain for matrices of first degree some further results.

Theorem 1.48. *Let Relation (1.77) be true for the non-singular anomalous matrix (1.73). Then there exists a unimodular matrix $P(\lambda)$, such that*

$$P(\lambda)(A\lambda + B) = \tilde{A}(\lambda) = \tilde{A}\lambda + \tilde{B} \tag{1.83}$$

is true with constant matrices

$$\tilde{A} = \begin{array}{c} n \\ \left[\begin{array}{c} \tilde{A}_1 \\ O_{n-\eta,n} \end{array}\right] \end{array} \begin{array}{l} \eta \\ n-\eta \end{array}\,, \qquad \tilde{B} = \begin{array}{c} n \\ \left[\begin{array}{c} \tilde{B}_1 \\ \tilde{B}_2 \end{array}\right] \end{array} \begin{array}{l} \eta \\ n-\eta \end{array}\,. \tag{1.84}$$

Moreover

$$\det \begin{bmatrix} \tilde{A}_1 \\ \tilde{B}_2 \end{bmatrix} \neq 0 \tag{1.85}$$

is true together with

$$\deg P(\lambda) \leq n - \eta\,. \tag{1.86}$$

Proof. We apply the row transformation algorithm of Theorem 1.9 to the matrix $A\lambda + B$. Then after a finite number of steps, we reach at a row reduced matrix $\tilde{A}(\lambda)$. Due to the fact, that the degree of the transformed matrix does not increase, we conclude $\deg \tilde{A}(\lambda) \leq 1$. The case $\deg \tilde{A}(\lambda) = 0$ is excluded, otherwise the matrix $A(\lambda)$ would be unimodular in contradiction to (1.77). Therefore, only $\deg \tilde{A}(\lambda) = 1$ is possible. Moreover, we prove

$$\tilde{A}(\lambda) = \tilde{A}\lambda + \tilde{B} = \operatorname{diag}\{\lambda^{\alpha_1}, \dots, \lambda^{\alpha_n}\}\left(\tilde{A}_0 + \tilde{A}_1\lambda^{-1}\right) \tag{1.87}$$

with $\det \tilde{A}_0 \neq 0$, where each of the numbers α_i, $(i = 1, \dots, n)$ is either 0 or 1. Due to

$$\alpha_1 + \ldots + \alpha_n = \eta \,,$$

among the numbers $\alpha_1, \ldots, \alpha_n$ are exactly η ones with the value one, and the other $n - \eta$ numbers are zero. Without loss of generality, we assume the succession

$$\alpha_1 = \alpha_2 = \ldots = \alpha_\eta = 1, \quad \alpha_{\eta+1} = \alpha_{\eta+2} = \ldots = \alpha_n = 0 \,.$$

Then the matrix $\tilde{A}$ in (1.83) takes the shape (1.84). Furthermore, if the matrix $\tilde{A}(\lambda)$ is represented in the form (1.87), then with respect to (1.79) and (1.84), we get

$$\tilde{A}_0 = \begin{bmatrix} \tilde{A}_1 \\ \tilde{B}_2 \end{bmatrix} .$$

Since the matrix $\tilde{A}(\lambda)$ is row reduced, Relation (1.85) arises.

It remains to show Relation (1.86). As follows from (1.36), each step decreases the degree of one of the rows of the transformed matrices at least by one. Hence each row of the matrix $A(\lambda)$ cannot be transformed more than once. Therefore, the number of transformation steps is at most $n - \eta$. Since however, in every step the transformation matrix $P(\lambda)$ is either constant or with degree one, Relation (1.86) holds. ∎

Corollary 1.49. *In the row-reduced form (1.83), $n - \eta$ rows of the matrix $\tilde{A}(\lambda)$ are constant. Moreover, the rank of the matrix built from these rows is equal to $n - \eta$, i.e., these rows are linearly independent.* ∎

Example 1.50. Consider the anomalous matrix

$$A(\lambda) = A\lambda + B = \begin{bmatrix} 1 & 1 & 2 \\ 1 & 1 & 2 \\ 1 & 1 & 3 \end{bmatrix} \lambda + \begin{bmatrix} 2 & 1 & 3 \\ 3 & 2 & 5 \\ 3 & 2 & 6 \end{bmatrix} = \begin{bmatrix} \lambda + 2 & \lambda + 1 & 2\lambda + 3 \\ \lambda + 3 & \lambda + 2 & 2\lambda + 5 \\ \lambda + 3 & \lambda + 2 & 3\lambda + 6 \end{bmatrix}$$

appering in [51], that is represented in the form

$$A(\lambda) = \operatorname{diag}\{\lambda, \lambda, \lambda\} \left(A_0 + A_1 \lambda^{-1} \right)$$

with

$$A_0 = A, \quad A_1 = B \,.$$

In the first transformation step (1.30), we obtain

$$\nu_1 + \nu_2 + \nu_3 = 0$$

$$2\nu_1 + 2\nu_2 + 3\nu_3 = 0 \,.$$

Now, we can choose $\nu_1 = 1$, $\nu_2 = -1$, $\nu_3 = 0$, and the matrices (1.32) and (1.35) take the form

$$P_1(\lambda) = D_1 = \begin{bmatrix} 1 & -1 & 0 \\ 0 & 1 & 0 \\ 0 & 0 & 1 \end{bmatrix}$$

hence

$$A_1(\lambda) = P_1(\lambda)A(\lambda) = \operatorname{diag}\{1, \lambda, \lambda\}\left(\begin{bmatrix} -1 & -1 & -2 \\ 1 & 1 & 2 \\ 1 & 1 & 3 \end{bmatrix} + \begin{bmatrix} 0 & 0 & 0 \\ 3 & 2 & 5 \\ 3 & 2 & 6 \end{bmatrix}\lambda^{-1}\right).$$

By appropriate manipulations, these matrices are transformed into

$$P_2(\lambda) = \begin{bmatrix} 1 & 0 & 0 \\ \lambda & 1 & 0 \\ 0 & 0 & 1 \end{bmatrix}, \quad D_2 = \begin{bmatrix} 1 & 0 & 0 \\ 1 & 1 & 0 \\ 0 & 0 & 1 \end{bmatrix}.$$

Finally, we receive over the product

$$A_2(\lambda) = P_2(\lambda)A_1(\lambda) = P_2(\lambda)P_1(\lambda)A(\lambda)$$

the row-reduced matrix

$$A_2(\lambda) = \begin{bmatrix} -1 & -1 & -2 \\ 3 & 2 & 5 \\ \lambda + 3 & \lambda + 2 & 3\lambda + 6 \end{bmatrix}.$$

□

6. Let B be a constant $n \times n$ matrix. We assign to this matrix a matrix B_λ of degree one by

$$B_\lambda = \lambda I_n - B\,,$$

which is called the *characteristic matrix* of B. For polynomial matrices of this form, all above introduced concepts and statements for polynomial matrices of general form remain valid. Hereby, the characteristic polynomial of the matrix B_λ

$$\det B_\lambda = \det(\lambda I_n - B) \stackrel{\triangle}{=} d_B(\lambda)$$

usually is named the *characteristic polynomial* of the matrix B. In the same way, we deal with the terminology of minimal polynomials, invariant polynomials, elementary divisor *etc.* Obviously,

$$\operatorname{ord} B_\lambda = \deg\det B_\lambda = n\,.$$

As a consequence from Relation (1.75) for $A_1 = A = I_n$ we formulate:

Theorem 1.51 ([51]). *For two characteristic matrices $B_\lambda = \lambda I_n - B$ and $B_{1\lambda} = \lambda I_n - B_1$ to be equivalent, it is necessary and sufficient, that the matrices B and B_1 are similar, i.e. the relation*

$$B_1 = LBL^{-1}$$

is true with a certain non-singular constant matrix L. ■

Remark 1.52. Theorem 1.51 implies the following property. If the matrix B (the matrix $\lambda I_n - B$) has the entirety of elementary divisors

$$(\lambda - \lambda_1)^{\nu_1} \cdots (\lambda - \lambda_q)^{\nu_q}, \quad \nu_1 + \ldots + \nu_q = n,$$

then the matrix B is similar to the matrix J of the form

$$J = \operatorname{diag}\{J_{\nu_1}(\lambda_1), \ldots, J_{\nu_q}(\lambda_q)\}. \tag{1.88}$$

The matrix J is said to be the *Jordan (canonical) form* or shortly, *Jordan matrix* of the corresponding matrix B. For any $n \times n$ matrix B, the Jordan matrix is uniquely determined, except the succession of the diagonal blocks.

7. Let the horizontal pair of constant matrices (A, B) with A, $n \times n$, and B, $n \times m$ be given. The pair (A, B) is called *controllable*, if the polynomial pair $(\lambda I_n - A, B)$ is irreducible. This means, that the pair (A, B) is controllable if and only if the matrix

$$R_c(\lambda) = \begin{bmatrix} \lambda I_n - A & B \end{bmatrix}$$

is alatent. It is known, see for instance [72, 69], that the pair (A, B) is controllable, if and only if

$$\operatorname{rank} Q_c(A, B) = n,$$

where the matrix $Q_c(A, B)$ is determined by

$$Q_c(A, B) = \begin{bmatrix} B & AB & \ldots & A^{n-1}B \end{bmatrix}. \tag{1.89}$$

The matrix $Q_c(A, B)$ is named *controllability matrix* of the pair (A, B).

Some statements regarding the controllability of pairs are listed now:

a) If the pair (A, B) is controllable, and the $n \times n$ matrix R is non-singular, then also the pair (A_1, B_1) with $A_1 = RAR^{-1}$, $B_1 = RB$ is controllable. Indeed, from (1.89) we obtain

$$Q_c(A_1, B_1) = \begin{bmatrix} RB & RAB & \ldots & RA^{n-1}B \end{bmatrix} == RQ_c(A, B),$$

from which follows $\operatorname{rank} Q_c(A_1, B_1) = \operatorname{rank} Q_c(A, B) = n$, because R is non-singular.

b) **Theorem 1.53.** *Let the pair (A, B) with the $n \times n$ matrix A and the $n \times m$ matrix B be given, and moreover, an $n \times n$ matrix L, which is commutative with A, i.e. $AL = LA$. Then the following statements are true:*
   1. *If the pair (A, B) is not controllable, then the pair (A, LB) is also not controllable.*
   2. *If the pair (A, B) is controllable and the matrix L is non-singular, then the pair (A, LB) is controllable.*
   3. *If the matrix L is singular, then the pair (A, LB) is not controllable.*

Proof. The controllability matrix of the pair (A, LB) has the shape

$$\begin{aligned} Q_c(A, LB) &= \left[LB \; ALB \; \ldots \; A^{n-1}LB \right] \\ &= L \left[B \; AB \; \ldots \; A^{n-1}B \right] = LQ_c(A, B) \end{aligned} \tag{1.90}$$

where $Q_c(A, B)$ is the controllability matrix (1.89). If the pair (A, B) is not controllable, then we have $\operatorname{rank} Q_c(A, B) < n$, and therefore, $\operatorname{rank} Q_c(A, LB) < n$. Thus the 1st statement is proved. If the pair (A, B) is controllable and the matrix L is non-singular, then we have $\operatorname{rank} Q_c(A, B) = n$, $\operatorname{rank} L = n$ and from (1.90) it follows $\operatorname{rank} Q_c(A, LB) = n$. Hence the 2nd statement is shown. Finally, if the matrix L is singular, then $\operatorname{rank} L < n$ and $\operatorname{rank} Q_c(A, LB) < n$ are true, which proves 3. ∎

c) Controllable pairs are structural stable - this is stated in the next theorem.

Theorem 1.54. *Let the pair (A, B) be controllable, and (A_1, B_1) be an arbitrary pair of the same dimension. Then there exists a positive number ϵ_0, such that the pair $(A + \epsilon A_1, B + \epsilon B_1)$ is controllable for all $|\epsilon| < \epsilon_0$.*

Proof. Using (1.89) we obtain

$$Q_c(A + \epsilon A_1, B + \epsilon B_1) = Q_c(A, B) + \epsilon Q_1 + \ldots + \epsilon^n Q_n \,, \tag{1.91}$$

where the Q_i, $(i = 1, \ldots, n)$ are constant matrices, that do not depend on ϵ. Since the pair (A, B) is controllable, the matrix $Q_c(A, B)$ contains a non-zero minor of n-th order. Then due to Lemma 1.39 for sufficiently small $|\epsilon|$, the corresponding minor of the matrix (1.91) also remains different from zero. ∎

Remark 1.55. Non-controllable pairs do not possess the property of structural stability. If the pair (A, B) is not controllable, then there exists a pair (A_1, B_1) of equal dimension, such that the pair $(A + \epsilon A_1, B + \epsilon B_1)$ for arbitrary small $|\epsilon| > 0$ becomes controllable.

8. The vertical pair $[A, C]$ built from the constant $m \times m$ matrix A and $n \times m$ matrix C is called *observable*, if the vertical pair of polynomial matrices $[\lambda I_m - A, \; C]$ is irreducible. Obviously, the pair $[A, C]$ is observable, if and only if the horizontal pair (A', C') is controllable, where the prime means the transposition operation. Due to this reason, observable pairs possess all the properties that have been derived above for controllable pairs. Especially, observable pairs are structural stable.

1.14 Cyclic Matrices

1. The constant $n \times n$ matrix A is said to be *cyclic*, if the assigned characteristic matrix $A_\lambda = \lambda I_n - A$ is simple in the sense of the definition in Section 1.11, see [69, 78, 191].

Cyclic matrices are provided with the important property of structural stability, as is substantiated by the next theorem.

Theorem 1.56. *Let the cyclic $n \times n$ matrix A, and an arbitrary $n \times n$ matrix B be given. Then there exists a positive number $\epsilon_0 > 0$, such that for $|\epsilon| < \epsilon_0$ all matrices $A + \epsilon B$ become cyclic.*

Proof. Let

$$\det(\lambda I_n - A) = d_A(\lambda), \quad \deg d_A(\lambda) = n\,.$$

Then we obtain

$$\det(\lambda I_n - A - \epsilon B) = d_A(\lambda) + \epsilon d_1(\lambda, \epsilon)$$

with $\deg d_1(\lambda, \epsilon) < n$ for all ϵ. Therefore, by virtue of Theorem 1.37, there exists an ϵ_0, such that for $|\epsilon| < \epsilon_0$ the matrix $\lambda I_n - A - \epsilon B$ remains simple, *i.e.* the matrix $A + \epsilon B$ is cyclic. ■

2. Square constant matrices that are not cyclic, will be called in future *composed*. Composed matrices are not equipped with the property of structural stability in the above defined sense. For any composed matrix A, we can find a matrix B, such that the sum $A+\epsilon B$ becomes cyclic, as small even $|\epsilon| > 0$ is chosen. Moreover, the sum $A+\epsilon B$ will become composed only in some special cases. This fact is illustrated by a 2×2 matrix in the next example.

Example 1.57. As follows from Theorem 1.51, any composed 2×2 matrix A is similar to the matrix

$$B = \begin{bmatrix} a & 0 \\ 0 & a \end{bmatrix} = aI_2\,, \tag{1.92}$$

where $a =$ const., so we have

$$A = LBL^{-1} = B\,.$$

Therefore, the set of all composed matrices in $\mathbb{C}_{22}$ is determined by Formula (1.92) for any a. Assume now the matrix $Q = A + F$ to be composed. Then

$$Q = \begin{bmatrix} q & 0 \\ 0 & q \end{bmatrix}$$

is true, and hence

$$F = B - Q = \begin{bmatrix} a - q & 0 \\ 0 & a - q \end{bmatrix}$$

becomes an composed matrix. When the 2×2 matrix A is composed, then the sum $A + F$ still becomes onerous, if and only if the matrix F is composed too. □

3. The property of structural stability of cyclic matrices allows a probability-theoretic interpretation. For instance, the following statement is true:

Let $A \in \mathbb{F}_{nn}$ be a composed matrix and $B \in \mathbb{F}_{nn}$ any random matrix with independent entries that are equally distributed in a certain interval $\alpha \leq b_{ik} \leq \beta$. Then the sum $A+B$ with probability 1 becomes a cyclic matrix.

4. The property of structural stability has great practical importance. Indeed, let for instance the differential equation of a certain linear process be given in the form

$$\frac{\mathrm{d}x}{\mathrm{d}t} = Ax + Bu \,, \qquad A = A_0 + \Delta A \,,$$

where x is the state vector, and A_0, ΔA are constant matrices, where A_0 is cyclic. The matrix ΔA manifests the unavoidable errors during the set up and calculation of the matrix A. From Theorem 1.56 we conclude that the matrix A remains cyclic, if the deviation ΔA satisfies the conditions of Theorem 1.56. If however, the matrix A is composed, then this property can be lost due to the imprecision characterised by the matrix ΔA, as tiny this ever has been with respect to the norm.

5. Assume

$$d(\lambda) = \lambda^n + d_1\lambda^{n-1} + \ldots + d_n \tag{1.93}$$

to be a monic polynomial. Then the $n \times n$ matrix A_F of the form

$$A_F = \begin{bmatrix} 0 & 1 & 0 & \ldots & 0 & 0 \\ 0 & 0 & 1 & \ldots & 0 & 0 \\ \vdots & \vdots & \ddots & \ddots & \vdots & \vdots \\ 0 & 0 & 0 & \ldots & 0 & 1 \\ -d_n & -d_{n-1} & -d_{n-2} & \ldots & -d_2 & -d_1 \end{bmatrix} \tag{1.94}$$

is called its *accompanying (horizontal) Frobenius matrix* with respect to the polynomial $d(\lambda)$. Moreover, we consider the *vertical accompanying Frobenius matrix*

$$\bar{A}_F = \begin{bmatrix} 0 & 0 & \ldots & 0 & -d_n \\ 1 & 0 & \ldots & 0 & -d_{n-1} \\ \vdots & \ddots & \ddots & \vdots & \vdots \\ 0 & 0 & \ddots & 0 & -d_2 \\ 0 & 0 & \ldots & 1 & -d_1 \end{bmatrix} . \tag{1.95}$$

The properties of the matrices (1.94) and (1.95) are analogue, so that we could restrict ourself to the investigation of (1.94). The characteristic matrix of A_F has the form

$$\lambda I_n - A_F = \begin{bmatrix} \lambda & -1 & 0 & \dots & 0 & 0 \\ 0 & \lambda & -1 & \dots & 0 & 0 \\ \vdots & \vdots & \ddots & \ddots & \vdots & \vdots \\ 0 & 0 & 0 & \dots & \lambda & -1 \\ d_n & d_{n-1} & d_{n-2} & \dots & d_2 & \lambda + d_1 \end{bmatrix} . \tag{1.96}$$

Appending Matrix (1.96) with the column $b = \begin{bmatrix} 0 \dots 0 \; 1 \end{bmatrix}'$, we receive the extended matrix

$$\begin{bmatrix} \lambda & -1 & 0 & \dots & 0 & 0 & 0 \\ 0 & \lambda & -1 & \dots & 0 & 0 & 0 \\ \vdots & \vdots & \ddots & \ddots & \vdots & \vdots & \vdots \\ 0 & 0 & 0 & \dots & \lambda & -1 & 0 \\ d_n & d_{n-1} & d_{n-2} & \dots & d_2 & \lambda + d_1 & 1 \end{bmatrix} .$$

This matrix is alatent, because it has a minor of n-th order that is equal to $(-1)^{n-1}$. Strength to Theorem 1.35, Matrix (1.96) is simple, and therefore, the matrix A_F is cyclic.

6. By direct calculation we recognise

$$\det(\lambda I_n - A_F) = \lambda^n + d_1 \lambda^{n-1} + \dots + d_n = d(\lambda) .$$

According to the properties of simple matrices, we conclude that the whole of invariant polynomials corresponding to the matrix A_F is presented by

$$a_1(\lambda) = a_2(\lambda) = \dots = a_{n-1}(\lambda) = 1, \quad a_n(\lambda) = f(\lambda) .$$

Let A be any cyclic $n \times n$ matrix. Then the accompanying matrix $A_\lambda = \lambda I_n - A$ is simple. Therefore, by applying equivalence transformations, A_λ might be brought into the form

$$\lambda I_n - A = p(\lambda) \operatorname{diag}\{1, 1, \dots , d(\lambda)\} q(\lambda) ,$$

where the matrices $p(\lambda)$, $q(\lambda)$ are unimodular, and $d(\lambda)$ is the characteristic polynomial of the matrix A. From the last equation, we conclude that the set of invariant polynomials of the cyclic matrix A coincides with the set of invariant polynomials of the accompanying Frobenius matrix of its characteristic polynomial $d(\lambda)$. Hereby, the matrices $\lambda I_n - A$ and $\lambda I_n - A_F$ are equivalent, hence the matrices A and A_F are similar, *i.e.*

$$A = L A_F L^{-1}$$

is true with a certain non-singular matrix L. It can be shown that in case of a real matrix A, also the matrix L could be chosen real.

7. As just defined in (1.76), let

$$J_n(a) = \begin{bmatrix} a & 1 & 0 & \dots & 0 & 0 \\ 0 & a & 1 & \dots & 0 & 0 \\ 0 & 0 & a & \ddots & 0 & 0 \\ \vdots & \vdots & \vdots & \ddots & \ddots & \vdots \\ 0 & 0 & 0 & \dots & a & 1 \\ 0 & 0 & 0 & \dots & 0 & a \end{bmatrix}$$

be a Jordan block. The matrix $J_n(a)$ turns out to be cyclic, because the matrix

$$\begin{bmatrix} \lambda - a & -1 & 0 & \dots & 0 & 0 & 0 \\ 0 & \lambda - a & -1 & \dots & 0 & 0 & 0 \\ 0 & 0 & \lambda - a & \ddots & 0 & 0 & 0 \\ \vdots & \vdots & \vdots & \ddots & \ddots & \vdots & \vdots \\ 0 & 0 & 0 & \dots & \lambda - a & -1 & 0 \\ 0 & 0 & 0 & \dots & 0 & \lambda - a & -1 \end{bmatrix}$$

is alatent.

Let us represent the polynomial (1.93) in the form

$$d(\lambda) = (\lambda - \lambda_1)^{\mu_1} \cdots (\lambda - \lambda_q)^{\mu_q} ,$$

where all numbers λ_i are different. Consider the matrix

$$J = \operatorname{diag}\{J_{\mu_1}(\lambda_1), \dots, J_{\mu_q}(\lambda_q)\} \tag{1.97}$$

and its accompanying characteristic matrix

$$\lambda I_n - J = \operatorname{diag}\{\lambda I_{\mu_1} - J_{\mu_1}(\lambda_1), \dots, \lambda I_{\mu_q} - J_{\mu_q}(\lambda_q)\} , \tag{1.98}$$

where the corresponding diagonal blocks take the shape

$$\lambda I_{\mu_i} - J_{\mu_i}(\lambda_i) = \begin{bmatrix} \lambda - \lambda_i & -1 & 0 & \dots & 0 & 0 \\ 0 & \lambda - \lambda_i & -1 & \dots & 0 & 0 \\ 0 & 0 & \lambda - \lambda_i & \ddots & 0 & 0 \\ \vdots & \vdots & \vdots & \ddots & \ddots & \vdots \\ 0 & 0 & 0 & \dots & \lambda - \lambda_i & -1 \\ 0 & 0 & 0 & \dots & 0 & \lambda - \lambda_i \end{bmatrix} . \tag{1.99}$$

Obviously, we have

$$\det[\lambda I_{\mu_i} - J_{\mu_i}(\lambda_i)] = (\lambda - \lambda_i)^{\mu_i}$$

so that from (1.98), we obtain

$$\det(\lambda I_n - J) = (\lambda - \lambda_1)^{\mu_1} \cdots (\lambda - \lambda_q)^{\mu_q} = d(\lambda)\,.$$

At the same time, using (1.98) and (1.99), we find

$$\operatorname{rank}(\lambda_i I_n - J) = n - 1, \quad (i = 1, \ldots, q)$$

that means, Matrix (1.98) is cyclic. Therefore, Matrix (1.97) is similar to the accompanying Frobenius matrix of the polynomial (1.93), thus

$$J = L A_F L^{-1}\,,$$

where L in general is a complex non-singular matrix.

1.15 Simple Realisations and Their Structural Stability

1. The triple of matrices $a(\lambda)$, $b(\lambda)$, $c(\lambda)$ of dimensions $p \times p$, $p \times m$, $n \times p$, according to [69] and others, is called a *polynomial matrix description* (PMD)

$$\tau(\lambda) = (a(\lambda), b(\lambda), c(\lambda))\,. \tag{1.100}$$

The integers n, p, m are the dimension of the PMD. In dependence on the membership of the entries of the matrices $a(\lambda)$, $b(\lambda)$, $c(\lambda)$ to the sets $\mathbb{F}[\lambda]$, $\mathbb{R}[\lambda]$, $\mathbb{C}[\lambda]$, the sets of all PMDs with dimension n, p, m are denoted by $\mathbb{F}_{npm}[\lambda]$, $\mathbb{R}_{npm}[\lambda]$, $\mathbb{C}_{npm}[\lambda]$, respectively.

A PMD (1.100) is called *minimal*, if the pairs $(a(\lambda), b(\lambda))$, $[a(\lambda), c(\lambda)]$ are irreducible.

2. A PMD of the form

$$\tau(\lambda) = (\lambda I_p - A, B, C)\,, \tag{1.101}$$

where A, B, C are constant matrices, is said to be an *elementary*. Every elementary PMD (1.101) is characterised by a triple of constant matrices A, $(p \times p)$; B, $(p \times m)$; C, $(n \times p)$. The triple (A, B, C) is called a realisation of the linear process in state space, or shortly *realisation*. The numbers n, p, m are named the *dimension* of the elementary realisation. The set of all realisations with given dimension is denoted by $\mathbb{F}_{npm}$, $\mathbb{R}_{npm}$, $\mathbb{C}_{npm}$, respectively.

Suppose the $p \times p$ matrix Q to be non-singular. Then the realisations (A, B, C) and (QAQ^{-1}, QB, CQ^{-1}) are called *similar*.

3. The realisation (A, B, C) is called *minimal*, if the pair (A, B) is controllable and the pair $[A, C]$ is observable, *i.e.* the elementary PMD (1.101) is minimal. A minimal realisation with a cyclic matrix A is called a *simple realisation*. The set of all minimal realisations of a given dimension will be symbolised by $\bar{\mathbb{F}}_{npm}$, $\bar{\mathbb{R}}_{npm}$, $\bar{\mathbb{C}}_{npm}$ respectively, and the set of all simple realisations by $\mathbb{F}^s_{npm}$, $\mathbb{R}^s_{npm}$ $\mathbb{C}^s_{npm}$. For a simple realisation $(A, B, C) \in \mathbb{R}^s_{npm}$

always exists a similar realisation $(Q_J A Q_J^{-1}, Q_J B, C Q_J^{-1}) \in \mathbb{C}^s_{npm}$, where the matrix $Q_J A Q_J^{-1}$ is of Jordan canonical form. Such a simple realisation is called a *Jordan realisation.* Moreover, for this realisation, there exists a similar realisation $(Q_F A Q_F^{-1}, Q_F B, C Q_F^{-1}) \in \mathbb{R}^s_{npm}$, where the matrix $Q_F A Q_F^{-1}$ is a Frobenius matrix of the form (1.94). Such a simple realisation is called a *Frobenius realisation.*

4. Simple realisations possess the important property of structural stability, as the next theorem states.

Theorem 1.58. *Let the realisation (A, B, C) of dimension n, p, m be simple, and (A_1, B_1, C_1) be an arbitrary realisation of the same dimension. Then there exists an $\epsilon_0 > 0$, such that the realisation $(A + \epsilon A_1, B + \epsilon B_1, C + \epsilon C_1)$ for all $|\epsilon| < \epsilon_0$ remains simple.*

Proof. Since the pair (A, B) is controllable and the pair $[A, C]$ is observable, there exists, owing to Theorem 1.54, an $\epsilon_1 > 0$, such that the pair $(A + \epsilon A_1, B + \epsilon B_1)$ becomes controllable and the pair $[A + \epsilon A_1, C + \epsilon C_1]$ observable for all $|\epsilon| < \epsilon_1$. Furthermore, due to Theorem 1.56, there exists an $\epsilon_2 > 0$, such that the matrix $A + \epsilon A_1$ becomes cyclic for all $|\epsilon| < \epsilon_2$. Consequently, for $|\epsilon| < \min(\epsilon_1, \epsilon_2) = \epsilon_0$ all realisations $(A + \epsilon A_1, B + \epsilon B_1, C + \epsilon C_1)$ are simple. ∎

Remark 1.59. Realisations that are not simple, are not provided by the property of structural stability. For instance, from the above considerations we come to the following conclusion:
Let the realisation (A, B, C) be not simple, and (A_1, B_1, C_1) be a random realisation of equal dimension, where the entries of the matrices A_1, B_1, C_1 are in the whole statistically independent and equally distributed in a certain interval $[\alpha, \beta]$. Then the realisation $(A + A_1, B + B_1, C + C_1)$ will be simple with probability 1.

5. Theorem 1.58 has fundamental importance for developing methods on base of a mathematical description of linear time-invariant multivariable systems. The dynamics of such systems are described in continuous time by state-space equations of the form

$$y = Cx, \quad \frac{\mathrm{d}x}{\mathrm{d}t} = Ax + Bu\,, \tag{1.102}$$

corresponding to the realisation (A, B, C). In practical investigations, we always will meet $A = A_0 + \Delta A$, $B = B_0 + \Delta B$, $C = C_0 + \Delta C$, where (A_0, B_0, C_0) is the nominal realisation and the realisation $(\Delta A, \Delta B, \Delta C)$ characterises inaccuracies due to finite word length *etc.* Now, if the nominal realisation is simple, then at least for sufficiently small deviations $(\Delta A, \Delta B, \Delta C)$, the simplicity is preserved. Analogue considerations are possible for the description of the dynamics of discrete-time systems, where

$$y_k = Cx_k, \quad x_{k+1} = Ax_k + Bu_k \tag{1.103}$$

is used.

If however, the nominal realisation (A, B, C) is not simple, then the structural properties will, roughly spoken, not be preserved even for tiny deviations.

6. In principle in many cases, we can find suitable bounds of disturbances for which a simple realisation remains simple. For instance, let the matrices A_1, B_1, C_1 depend continuously on a scalar parameter α, such that $A_1 = A_1(\alpha)$, $B_1 = B_1(\alpha)$, $C_1 = C_1(\alpha)$ with $A_1(0) = O_{pp}$, $B_1(0) = O_{pm}$, $C_1(0) = O_{np}$ is valid. Now, if the parameter α increases from zero to positive values, then the realisation $(A+A_1(\alpha), B+B_1(\alpha), C+C_1(\alpha)$ for $0 \leq \alpha < \alpha_0$ remains simple, where α_0 is the smallest positive number, for which at least one of the following conditions takes place:

a) The pair $(A + A_1(\alpha_0), B + B_1(\alpha_0))$ is not controllable.
b) The pair $[A + A_1(\alpha_0), C + C_1(\alpha_0)]$ is not observable.
c) The matrix $A + A_1(\alpha_0)$ is not cyclic.

2

Fractional Rational Matrices

2.1 Rational Fractions

1. A *fractional rational (rat.) function*, or shortly *rational fraction* means the relation of two polynomials

$$\ell(\lambda) = \frac{\tilde{m}(\lambda)}{\tilde{d}(\lambda)}, \tag{2.1}$$

where $\tilde{d}(\lambda) \neq 0$. In dependence on the coefficient sets for the numerator and denominator polynomials in (2.1), the corresponding set of rational fractions is designated by $\mathbb{F}(\lambda)$, $\mathbb{C}(\lambda)$ or $\mathbb{R}(\lambda)$, respectively. Hereby, rational fractions in $\mathbb{R}(\lambda)$ are named real. Over the set of rational functions, various algebraic operations can be explained.

2. Two rational fractions

$$\ell_1(\lambda) = \frac{m_1(\lambda)}{d_1(\lambda)}, \qquad \ell_2(\lambda) = \frac{m_2(\lambda)}{d_2(\lambda)} \tag{2.2}$$

are considered as equal, and we write $\ell_1(\lambda) = \ell_2(\lambda)$ for that, when

$$m_1(\lambda)d_2(\lambda) - m_2(\lambda)d_1(\lambda) = 0\,. \tag{2.3}$$

Let in particular

$$m_2(\lambda) = a(\lambda)m_1(\lambda)\,, \quad d_2(\lambda) = a(\lambda)d_1(\lambda)$$

with a polynomial $a(\lambda)$, then (2.3) is fulfilled, and the fractions

$$\ell_1(\lambda) = \frac{m_1(\lambda)}{d_1(\lambda)}, \qquad \ell_2(\lambda) = \frac{a(\lambda)m_1(\lambda)}{a(\lambda)d_1(\lambda)}$$

are equal in the sense of the above definition. Immediately, it follows that the rational fraction (2.1) does not change if the numerator and denominator are cancelled by the same factor. Any polynomial $f(\lambda)$ can be represented in the form

$$f(\lambda) = \frac{f(\lambda)}{1}\,.$$

Therefore, for polynomial rings the relations $\mathbb{F}[\lambda] \subset \mathbb{F}(\lambda)$, $\mathbb{C}[\lambda] \subset \mathbb{C}(\lambda)$, $\mathbb{R}[\lambda] \subset \mathbb{R}(\lambda)$ are true.

3. Let the fraction (2.1) be given, and $g(\lambda)$ is the GCD of the numerator $\tilde{m}(\lambda)$ and the denominator $\tilde{d}(\lambda)$, such that

$$\tilde{m}(\lambda) = g(\lambda)m_1(\lambda), \quad \tilde{d}(\lambda) = g(\lambda)d_1(\lambda)$$

with coprime $m_1(\lambda)$, $d_1(\lambda)$. Then we have

$$\ell(\lambda) = \frac{m_1(\lambda)}{d_1(\lambda)}\,. \tag{2.4}$$

Notation (2.4) is called an *irreducible form* of the rational fraction. Furthermore, assume

$$d_1(\lambda) = d_0\lambda^n + d_1\lambda^{n-1} + \ldots + d_n\,, \quad d_0 \neq 0\,.$$

Then, the numerator and denominator of (2.4) can be divided by d_0, yielding

$$\ell(\lambda) = \frac{m(\lambda)}{d(\lambda)}\,.$$

Herein the numerator and denominator are coprime polynomials, and besides the polynomial $d(\lambda)$ is monic. This representation of a rational fraction will be called its *standard form.* The standard form of a rational fraction is unique.

4. The sum of rational fractions (2.2) is defined by the formula

$$\ell_1(\lambda) + \ell_2(\lambda) = \frac{m_1(\lambda)}{d_1(\lambda)} + \frac{m_2(\lambda)}{d_2(\lambda)} = \frac{m_1(\lambda)d_2(\lambda) + m_2(\lambda)d_1(\lambda)}{d_1(\lambda)d_2(\lambda)}\,.$$

5. The product of rational fractions (2.2) is explained by the relation

$$\ell_1(\lambda)\ell_2(\lambda) = \frac{m_1(\lambda)m_2(\lambda)}{d_1(\lambda)d_2(\lambda)}\,.$$

6. In algebra, it is proved that the sets of rational fractions $\mathbb{C}(\lambda)$, $\mathbb{R}(\lambda)$ with the above explained rules for addition and multiplication form fields. The zero element of those fields proves to be the fraction $0/1$, the unit element is the rational fraction $1/1$. If we have in (2.1) $\tilde{m}(\lambda) \neq 0$, then the inverse element $\ell^{-1}(\lambda)$ is determined by the formula

$$\ell^{-1}(\lambda) = \frac{\tilde{d}(\lambda)}{\tilde{m}(\lambda)}\,.$$

7. The integer $\operatorname{ind} \ell$, for which the finite limit

$$\lim_{\lambda \to \infty} \ell(\lambda)\lambda^{\operatorname{ind} \ell} = \ell_0 \neq 0$$

exists, is called the *index* of the rational fraction (2.1). In case of $\operatorname{ind} \ell = 0$ ($\operatorname{ind} \ell > 0$), the fraction is called proper (strictly proper). In case of $\operatorname{ind} \ell \geq 0$, the fraction is said to be *at least proper.* In case of $\operatorname{ind} \ell < 0$ the fraction $\ell(\lambda)$ is named improper. If the rational fraction $\ell(\lambda)$ is represented in the form (2.1) and we introduce $\deg \tilde{m}(\lambda) = \alpha$, $\deg \tilde{d}(\lambda) = \beta$, then the fraction is proper, strictly proper or at least proper, if the corresponding relation $\alpha = \beta$, $\alpha < \beta$ or $\alpha \leq \beta$ is true. The zero rational fraction is defined as strictly proper.

8. Any fraction (2.1) can be written in the shape

$$\ell(\lambda) = \frac{r(\lambda)}{d(\lambda)} + q(\lambda) \tag{2.5}$$

with polynomials $r(\lambda)$, $q(\lambda)$, where $\deg r(\lambda) < \deg d(\lambda)$, such that the first summand at the right side of (2.5) is strictly proper. The representation (2.5) is unique. Practically, the polynomials $r(\lambda)$ and $q(\lambda)$ could be found in the following way. Using (1.6), we uniquely receive

$$m(\lambda) = d(\lambda)q(\lambda) + r(\lambda)$$

with $\deg r(\lambda) < \deg d(\lambda)$. Inserting the last relation into (2.1), we get (2.5).

9. The sum, the difference and the product of strictly proper fractions are also strictly proper. The totality of strictly proper fractions builds a commutative ring without unit element.

10. Let the strictly proper rational fraction

$$\ell(\lambda) = \frac{m(\lambda)}{d_1(\lambda)d_2(\lambda)}$$

be given, where the polynomials $d_1(\lambda)$ and $d_2(\lambda)$ are coprime. Then we can find a separation

$$\ell(\lambda) = \frac{m_1(\lambda)}{d_1(\lambda)} + \frac{m_2(\lambda)}{d_2(\lambda)}, \tag{2.6}$$

where both fractions on the right side are strictly proper. Hereby, the polynomial $m_1(\lambda)$ and $m_2(\lambda)$ are determined uniquely.

11. A separation of the form (2.6) can be generalised as follows. Let the strictly proper fraction $\ell(\lambda)$ possess the shape

$$\ell(\lambda) = \frac{m(\lambda)}{d_1(\lambda)d_2(\lambda)\cdots d_n(\lambda)},$$

where all polynomials in the denominator are two and two coprime, then there exists a unique representation of the form

$$\ell(\lambda) = \frac{m_1(\lambda)}{d_1(\lambda)} + \frac{m_1(\lambda)}{d_1(\lambda)} + \ldots + \frac{m_n(\lambda)}{d_n(\lambda)}, \tag{2.7}$$

where all fractions on the right side are strictly proper. In particular, let the strictly proper irreducible fraction

$$\ell(\lambda) = \frac{m(\lambda)}{d(\lambda)}$$

with

$$d(\lambda) = (\lambda - \lambda_1)^{\mu_1} \cdots (\lambda - \lambda_q)^{\mu_q}$$

be given, where all λ_i are different. Introduce $(\lambda - \lambda_i)^{\mu_i} = d_i(\lambda)$ and apply (2.7), then we obtain a representation of the form

$$\ell(\lambda) = \sum_{i=1}^{q} \frac{m_i(\lambda)}{(\lambda - \lambda_i)^{\mu_i}}, \qquad \deg m_i(\lambda) < \mu_i . \tag{2.8}$$

The representation (2.8) is unique.

12. Furthermore, we can show that the fractions of the form

$$\ell_i(\lambda) = \frac{m_i(\lambda)}{(\lambda - \lambda_i)^{\mu_i}}, \qquad \deg m_i(\lambda) < \mu_i$$

can be uniquely presented in the form

$$\ell_i(\lambda) = \frac{m_{i1}}{(\lambda - \lambda_i)^{\mu_i}} + \frac{m_{i2}}{(\lambda - \lambda_i)^{\mu_i - 1}} + \ldots + \frac{m_{i\mu_i}}{\lambda - \lambda_i},$$

where the m_{ij} are certain constants. Inserting this relation into (2.8), we get

$$\ell(\lambda) = \sum_{i=1}^{q} \left[\frac{m_{i1}}{(\lambda - \lambda_i)^{\mu_i}} + \frac{m_{i2}}{(\lambda - \lambda_i)^{\mu_i - 1}} + \ldots + \frac{m_{i\mu_i}}{\lambda - \lambda_i} \right], \tag{2.9}$$

which is named a *partial fraction expansion* . A representation of the form (2.9) is unique.

13. For calculating the coefficients m_{ik} of the partial fraction expansion (2.9) the formula

$$m_{ik} = \frac{1}{(k-1)!} \frac{\partial^{k-1}}{\partial\lambda^{k-1}} \frac{m(\lambda)(\lambda-\lambda_i)^{\mu_i}}{d(\lambda)}\bigg|_{\lambda=\lambda_i} \tag{2.10}$$

can be used. The coefficients (2.10) are closely connected to the expansion of the function

$$\tilde{\ell}_i(\lambda) = \frac{m(\lambda)(\lambda-\lambda_i)^{\mu_i}}{d(\lambda)}$$

into a Taylor series in powers of $(\lambda - \lambda_i)$, that exists because the function $\tilde{\ell}_i(\lambda)$ is analytical in the point $\lambda = \lambda_i$. Assume for instance

$$\tilde{\ell}_i(\lambda) = \ell_{i1} + \ell_{i2}(\lambda-\lambda_i) + \ldots + \ell_{i\mu_i}(\lambda-\lambda_i)^{\mu_i} + \ldots$$

with

$$\ell_{ik} = \frac{1}{(k-1)!} \frac{\mathrm{d}^{k-1}}{\mathrm{d}\lambda^{k-1}} \tilde{\ell}_i(\lambda)\bigg|_{\lambda=\lambda_i} .$$

Comparing this expression with (2.10) yields

$$m_{ik} = \ell_{ik} .$$

14. Let

$$\ell(\lambda) = \frac{m(\lambda)}{d_1(\lambda)d_2(\lambda)} \tag{2.11}$$

be any rational fraction, where the polynomials $d_1(\lambda)$ and $d_2(\lambda)$ are coprime. Moreover, we assume

$$\ell(\lambda) = \frac{\tilde{m}(\lambda)}{d_1(\lambda)d_2(\lambda)} + q(\lambda)$$

for a representation of $\ell(\lambda)$ in the form (2.5), where $\tilde{m}(\lambda)$, $q(\lambda)$ are polynomials with $\deg \tilde{m}(\lambda) < \deg d_1(\lambda) + \deg d_2(\lambda)$. Since $d_1(\lambda)$, $d_2(\lambda)$ are coprime, there exists a unique decomposition

$$\frac{\tilde{m}(\lambda)}{d_1(\lambda)d_2(\lambda)} = \frac{m_1(\lambda)}{d_1(\lambda)} + \frac{m_2(\lambda)}{d_2(\lambda)} ,$$

where the fractions on the right side are strictly proper. Altogether, for (2.11) we get the unique representation

$$\frac{m(\lambda)}{d_1(\lambda)d_2(\lambda)} = \frac{m_1(\lambda)}{d_1(\lambda)} + \frac{m_2(\lambda)}{d_2(\lambda)} + q(\lambda) , \tag{2.12}$$

where $\deg m_1(\lambda) < \deg d_1(\lambda)$ and $\deg m_2(\lambda) < \deg d_2(\lambda)$ are valid.

Let $g(\lambda)$ be any polynomial. Then (2.12) can be written in the shape

$$\ell(\lambda) = \left[\frac{m_1(\lambda)}{d_1(\lambda)} + g(\lambda)\right] + \left[\frac{m_2(\lambda)}{d_2(\lambda)} + q(\lambda) - g(\lambda)\right], \tag{2.13}$$

which is equivalent to

$$\ell(\lambda) = \ell_1(\lambda) + \ell_2(\lambda) = \frac{n_1(\lambda)}{d_1(\lambda)} + \frac{n_2(\lambda)}{d_2(\lambda)}, \tag{2.14}$$

where

$$\begin{aligned} n_1(\lambda) &= m_1(\lambda) + g(\lambda)d_1(\lambda), \\ n_2(\lambda) &= m_2(\lambda) + [q(\lambda) - g(\lambda)]\, d_2(\lambda). \end{aligned} \tag{2.15}$$

The representation of a rational fraction $\ell(\lambda)$ in the form (2.14) is called a *separation with respect to the polynomials* $d_1(\lambda)$ *and* $d_2(\lambda)$.

From (2.14), (2.15) it follows that a separation always is possible, but not uniquely determined. It is shown now that Formulae (2.14), (2.15) include all possible separations.

Indeed, if (2.14) holds, then the division of the fractions on the right side of (2.14) by the denominator yields

$$\ell(\lambda) = \frac{k_1(\lambda)}{d_1(\lambda)} + q_1(\lambda) + \frac{k_2(\lambda)}{d_2(\lambda)} + q_2(\lambda)$$

with $\deg k_1(\lambda) < \deg d_1(\lambda)$, $\deg k_2(\lambda) < \deg d_2(\lambda)$. Comparing the last equation with (2.12) and bear in mind the uniqueness of the representation (2.12), we get $k_1(\lambda) = m_1(\lambda)$, $k_2(\lambda) = m_2(\lambda)$ and $q_1(\lambda) + q_2(\lambda) = q(\lambda)$. While assigning $q_1(\lambda) = g(\lambda)$, $q_2(\lambda) = q(\lambda) - g(\lambda)$, we realise that the representation (2.14) takes the form (2.12).

Selecting in (2.13) $g(\lambda) = 0$, we obtain the separation

$$\ell_1(\lambda) = \frac{m_1(\lambda)}{d_1(\lambda)}, \quad \ell_2(\lambda) = \frac{m_2(\lambda)}{d_2(\lambda)} + q(\lambda), \tag{2.16}$$

where the fraction $\ell_1(\lambda)$ is strictly proper. Analogously, we find for $g(\lambda) = q(\lambda)$ the separation

$$\ell_1(\lambda) = \frac{m_1(\lambda)}{d_1(\lambda)} + q(\lambda), \quad \ell_2(\lambda) = \frac{m_2(\lambda)}{d_2(\lambda)}, \tag{2.17}$$

where the fraction $\ell_2(\lambda)$ is strictly proper. Separation (2.16) is called *minimal with respect to* $d_1(\lambda)$, and Separation (2.17) *minimal with respect to* $d_2(\lambda)$. From the above exposition, it follows that the minimal separations are uniquely determined. An important special case arises for strictly proper fractions (2.11). Then we receive in (2.12) $q(\lambda) = 0$, and the minimal separations (2.16) and (2.17) coincide.

2.2 Rational Matrices

1. A $n \times m$ matrix $L(\lambda)$ of the form

$$L(\lambda) = \begin{bmatrix} \ell_{11}(\lambda) & \dots & \ell_{1m}(\lambda) \\ \vdots & \vdots & \vdots \\ \ell_{n1}(\lambda) & \dots & \ell_{nm}(\lambda) \end{bmatrix} \tag{2.18}$$

is called a *broken rational*, or shortly *rational matrix*, if all its entries are broken rational functions of the form (2.1). If $\ell_{ik} \in \mathbb{F}(\lambda)$ (or $\mathbb{C}(\lambda), \mathbb{R}(\lambda)$), then the corresponding set of matrices (2.18) is denoted by $\mathbb{F}_{nm}(\lambda)$, (or $\mathbb{C}_{nm}(\lambda), \mathbb{R}_{nm}(\lambda)$), respectively. In the following considerations we optionally assume matrices in $\mathbb{R}_{nm}[\lambda]$ and $\mathbb{R}_{nm}(\lambda)$, that practically arise in all technological applications. But it is mentioned that most of the results derived below also hold for matrices in $\mathbb{F}_{nm}(\lambda)$, $\mathbb{C}_{nm}(\lambda)$. Rational matrices in $\mathbb{R}_{nm}(\lambda)$ are named real.

By writing all elements of Matrix (2.18) on the principal denominator, the matrix can be denoted in the form

$$L(\lambda) = \frac{\tilde{N}(\lambda)}{\tilde{d}(\lambda)}, \tag{2.19}$$

where the matrix $\tilde{N}(\lambda)$ is a polynomial matrix, and $\tilde{d}(\lambda)$ is a scalar polynomial. In description (2.19) the matrix $\tilde{N}(\lambda)$ is called the *numerator* and the polynomial $\tilde{d}(\lambda)$ the *denominator* of the matrix $L(\lambda)$. Without loss of generality, we will assume that the polynomial $\tilde{d}(\lambda)$ in (2.19) is monic and has the linear factorisation

$$\tilde{d}(\lambda) = (\lambda - \lambda_1)^{\mu_1} \cdots (\lambda - \lambda_q)^{\mu_q} . \tag{2.20}$$

The fraction (2.19) is called *irreducible*, if with respect to the factorisation (2.20)

$$\tilde{N}(\lambda_i) \neq O_{nm} , \quad (i = 1, \dots, q) .$$

However, if for at least one $1 \leq i \leq q$

$$\tilde{N}(\lambda_i) = O_{nm}$$

becomes true, then the fraction (2.19) is named *reducible*. The last equation is fulfilled, if every element of the matrix $L(\lambda)$ is divisible by $\lambda - \lambda_i$. After performing all possible cancellations, we always arrive at a representation of the form

$$L(\lambda) = \frac{N(\lambda)}{d(\lambda)}, \tag{2.21}$$

where the fraction on the right side is irreducible, and the polynomial $d(\lambda)$ is monic. This representation of a rational matrix (2.18) is named its *standard form*. The standard form of a rational matrix is uniquely determined. In future, we will always assume rational matrices in standard form if nothing else is denied.

Example 2.1. Let the rational matrix

$$L(\lambda) = \begin{bmatrix} \dfrac{5\lambda+3}{\lambda-2} & \dfrac{1}{\lambda-3} \\ \dfrac{2\lambda+1}{(\lambda-2)(\lambda-3)} & \lambda+2 \end{bmatrix} \tag{2.22}$$

be given that can be represented in the form (2.19)

$$L(\lambda) = \frac{\begin{bmatrix} (5\lambda+3)(\lambda-3) & \lambda-2 \\ 2\lambda+1 & (\lambda^2-4)(\lambda-3) \end{bmatrix}}{(\lambda-2)(\lambda-3)}, \tag{2.23}$$

where

$$\begin{aligned} N(\lambda) &= \begin{bmatrix} (5\lambda+3)(\lambda-3) & \lambda-2 \\ 2\lambda+1 & (\lambda^2-4)(\lambda-3) \end{bmatrix}, \\ d(\lambda) &= (\lambda-2)(\lambda-3). \end{aligned} \tag{2.24}$$

According to

$$N(2) = \begin{bmatrix} -13 & 0 \\ 5 & 0 \end{bmatrix}, \quad N(3) = \begin{bmatrix} 0 & 1 \\ 7 & 0 \end{bmatrix},$$

the fraction (2.23) is irreducible. Since the polynomial $d(\lambda)$ in (2.24) is monic, the expression (2.23) estabishes as the standard form of the rational matrix $L(\lambda)$. □

If the denominator $d(\lambda)$ in (2.21) has the shape (2.20), then the numbers $\lambda_1, \dots, \lambda_q$ are called the *poles* of the matrix $L(\lambda)$, and the numbers $\mu_1, \dots, \mu_q$ are their *multiplicities.*

2.3 McMillan Canonical Form

1. Let the rational $n \times m$ matrix $L(\lambda)$ be given in the standard form (2.21). The matrix $N(\lambda)$ is written in Smith canonical form (1.41):

$$N(\lambda) = p(\lambda) \begin{bmatrix} \operatorname{diag}\{a_1(\lambda), \dots, a_\rho(\lambda)\} & O_{\rho,m-\rho} \\ O_{n-\rho,\rho} & O_{n-\rho,m-\rho} \end{bmatrix} q(\lambda). \tag{2.25}$$

Then we produce from (2.21)

$$L(\lambda) = p(\lambda) M_L(\lambda) q(\lambda), \tag{2.26}$$

where

$$M_L(\lambda) = \begin{bmatrix} M_\rho(\lambda) & O_{\rho,m-\rho} \\ O_{n-\rho,\rho} & O_{n-\rho,m-\rho} \end{bmatrix} \tag{2.27}$$

and

$$M_\rho(\lambda) = \operatorname{diag}\left\{\frac{a_1(\lambda)}{d(\lambda)}, \ldots, \frac{a_\rho(\lambda)}{d(\lambda)}\right\}. \tag{2.28}$$

Executing all possible cancellations in (2.28), we arrive at

$$M_\rho(\lambda) = \operatorname{diag}\left\{\frac{\alpha_1(\lambda)}{\psi_1(\lambda)}, \ldots, \frac{\alpha_\rho(\lambda)}{\psi_\rho(\lambda)}\right\}, \tag{2.29}$$

where the $\alpha_i(\lambda)$, $\psi_i(\lambda)$, $(i = 1, \ldots, \rho)$ are coprime monic polynomials, such that $\alpha_{i+1}(\lambda)$ is divisible by $\alpha_i(\lambda)$, and $\psi_i(\lambda)$ is divisible by $\psi_{i+1}(\lambda)$.

Matrix (2.27), where $M_\rho(\lambda)$ is represented in the form (2.29), is designated as the *McMillan (canonical) form* of the rational matrix $L(\lambda)$. The McMillan form of an arbitrary rational matrix is uniquely determined.

2. The polynomial

$$\psi_L(\lambda) = \psi_1(\lambda) \cdots \psi_\rho(\lambda) \tag{2.30}$$

is said to be the *McMillan denominator* of the matrix $L(\lambda)$, and the polynomial

$$\alpha_L(\lambda) = \alpha_1(\lambda) \cdots \alpha_\rho(\lambda) \tag{2.31}$$

its *McMillan numerator*. The non-negative number

$$\operatorname{Mdeg} L(\lambda) \stackrel{\triangle}{=} \deg \psi_L(\lambda) \tag{2.32}$$

is called the *McMillan degree* of the matrix $L(\lambda)$, or shortly its degree.

3.

Lemma 2.2. *For a rational matrix $L(\lambda)$ in standard form (2.21) the fraction*

$$\Delta(\lambda) = \frac{\psi_L(\lambda)}{d(\lambda)} \tag{2.33}$$

turns out to be a polynomial.

Proof. It is shown that under the actual assumptions the polynomials $a_1(\lambda)$ and $d(\lambda)$ are coprime, such that the fraction $a_1(\lambda)/d(\lambda)$ is irreducible. Indeed, let us assume the contrary such that

$$\frac{a_1(\lambda)}{d(\lambda)} = \frac{b_1(\lambda)}{\psi_1(\lambda)},$$

where $\deg \psi_1(\lambda) < \deg d(\lambda)$. Since the polynomial $\psi_1(\lambda)$ is divisible by the polynomials $\psi_2(\lambda), \ldots, \psi_\rho(\lambda)$, we obtain from (2.29)

$$M_\rho(\lambda) = \operatorname{diag}\left\{\frac{b_1(\lambda)}{\psi_1(\lambda)}, \ldots, \frac{b_\rho(\lambda)}{\psi_1(\lambda)}\right\},$$

where $b_1(\lambda), \ldots, b_\rho(\lambda)$ are polynomials. Inserting this relation and (2.27) into (2.26), we arrive at the representation

$$L(\lambda) = \frac{N_1(\lambda)}{\psi_1(\lambda)},$$

where $N_1(\lambda)$ is a polynomial matrix, and $\deg \psi_1(\lambda) < \deg d(\lambda)$. But this inequality contradicts our assumption on the irreducibility of the standard form (2.21). This conflict proves the correctness of $\psi_1(\lambda) = d(\lambda)$, and from (2.30) arises (2.33). ■

From Lemma 2.2, for a denominator $d(\lambda)$ of the form (2.20), we deduce the relation

$$\psi_L(\lambda) = (\lambda - \lambda_1)^{\nu_1} \cdots (\lambda - \lambda_q)^{\nu_q} = d(\lambda)\psi_2(\lambda) \cdots \psi_\rho(\lambda), \tag{2.34}$$

where $\nu_i \geq \mu_i$, $(i = 1, \ldots, q)$. The number ν_i is called the *McMillan multiplicity* of the pole λ_i.

From (2.34) and (2.32) arise

$$\operatorname{Mdeg} L(\lambda) = \nu_1 + \ldots + \nu_q .$$

4.

Lemma 2.3. *For any matrix $L(\lambda)$, assuming (2.26), (2.27), we obtain*

$$\deg d(\lambda) \leq \operatorname{Mdeg} L(\lambda) \leq \rho \deg d(\lambda) .$$

Proof. The left side of the claimed inequality establishes itself as a consequence of Lemma 2.2. The right side is seen immediately from (2.28), because under the assumption that all fractions $a_i(\lambda)/d(\lambda)$ are irreducible, we obtain

$$\psi_L(\lambda) = [d(\lambda)]^\rho .$$

■

5.

Lemma 2.4. *Let $L(\lambda)$ in (2.21) be an $n \times n$ matrix with* $\operatorname{rank} N(\lambda) = n$. *Then*

$$\det L(\lambda) = \kappa \frac{\alpha_L(\lambda)}{\psi_L(\lambda)}, \quad \kappa = \text{const.} \neq 0 . \tag{2.35}$$

Proof. For $n = m$ and $\operatorname{rank} N(\lambda) = n$, from (2.26)–(2.29) it follows

$$L(\lambda) = p(\lambda) \operatorname{diag}\left\{ \frac{\alpha_1(\lambda)}{d(\lambda)}, \frac{\alpha_2(\lambda)}{\psi_2(\lambda)}, \ldots, \frac{\alpha_n(\lambda)}{\psi_n(\lambda)} \right\} q(\lambda) .$$

Calculating the determinant on the right side of this equation according to (2.30) and (2.31) yields Formula (2.35) with $\kappa = \det p(\lambda) \det q(\lambda)$. ■

2.4 Matrix Fraction Description (MFD)

1. Let the rational $n \times m$ matrix $L(\lambda)$ be given in the standard form (2.21). We suppose the existence of a non-singular $n \times n$ polynomial matrix $a_l(\lambda)$ with

$$a_l(\lambda)L(\lambda) = \frac{a_l(\lambda)N(\lambda)}{d(\lambda)} \triangleq b_l(\lambda)$$

and an $n \times m$ polynomial matrix $b_l(\lambda)$. In this case, we call the polynomial matrix $a_l(\lambda)$ a *left reducing polynomial* of the matrix $L(\lambda)$. Considering the last equation, we gain the representation

$$L(\lambda) = a_l^{-1}(\lambda)b_l(\lambda)\,, \tag{2.36}$$

which is called an *LMFD* (left matrix fraction description) of the matrix $L(\lambda)$. Analogously, if there exists a non-singular $m \times m$ matrix $a_r(\lambda)$ with

$$L(\lambda)a_r(\lambda) = \frac{N(\lambda)a_r(\lambda)}{d(\lambda)} = b_r(\lambda)$$

and a polynomial $n \times m$ matrix $b_r(\lambda)$, we call the representation

$$L(\lambda) = b_r(\lambda)a_r^{-1}(\lambda) \tag{2.37}$$

a *right MFD* (RMFD) of the matrix $L(\lambda)$, [69, 68], and the matrix $a_r(\lambda)$ is named its *right reducing polynomial.*

2. The polynomials $a_l(\lambda)$ and $b_l(\lambda)$ in the LMFD (2.36) are called *left denominator* and *right numerator*, and the polynomials $a_r(\lambda)$, $b_r(\lambda)$ of the RMFD (2.37) its *right denominator* and *left numerator*, respectively. Obviously, the set of left reducing polynomials of the matrix $L(\lambda)$ coincides with the set of its left denominators, and the same is true for the set of right reducing polynomials and the set of right denominators.

Example 2.5. Let the matrices

$$L(\lambda) = \frac{\begin{bmatrix} 2\lambda & \lambda^2 + \lambda - 2 \\ \lambda^2 - 7\lambda + 18 & -\lambda^2 + 7\lambda - 2 \end{bmatrix}}{(\lambda - 2)(\lambda - 3)}$$

and

$$a_l(\lambda) = \begin{bmatrix} \lambda - 4 & 1 \\ \lambda - 6 & \lambda \end{bmatrix}$$

be given. Then by direct calculation, we obtain

$$a_l(\lambda)L(\lambda) = \begin{bmatrix} 3 & \lambda + 1 \\ \lambda & 2 \end{bmatrix} = b_l(\lambda)\,,$$

such that $L(\lambda) = a_l^{-1}(\lambda)b_l(\lambda)$. □

3. For any matrix $L(\lambda)$ (2.21), there always exist LMFDs and RMFDs. Indeed, take

$$a_l(\lambda) = d(\lambda)I_n, \quad b_l(\lambda) = N(\lambda),$$

then the rational matrix (2.21) can be written in form of an LMFD (2.36), where

$$\det a_l(\lambda) = [d(\lambda)]^n,$$

and therefore

$$\deg \det a_l(\lambda) = \operatorname{ord} a_l(\lambda) = n \deg d(\lambda).$$

In the same way, we see that

$$a_r(\lambda) = d(\lambda)I_m, \quad b_r(\lambda) = N(\lambda)$$

is an RMFD (2.37) of Matrix (2.21), where $\operatorname{ord} a_r(\lambda) = m \deg d(\lambda)$.

However, as will be shown in future examples, in most cases we are interested in LMFDs or RMFDs with lowest possible $\operatorname{ord} a_l(\lambda)$ or $\operatorname{ord} a_r(\lambda)$.

4. In connection with the above demand, the problem arise to construct an LMFD or RMFD, where $\det a_l(\lambda)$ or $\det a_r(\lambda)$ have the minimal possible degrees. Those MFDs are called *irreducible*. In what follows, we speak about irreducible left MFDs (ILMFDs) and irreducible right MFDs (IRMFDs). The following statements are well known [69].

Statement 2.1 An LMFD (2.36) is an ILMFD, if and only if the pair $(a_l(\lambda), b_l(\lambda))$ is irreducible, *i.e.* the matrix $\begin{bmatrix} a_l(\lambda) & b_l(\lambda) \end{bmatrix}$ is alatent.

Statement 2.2 An RMFD (2.37) is an IRMFD, if and only if the pair $[a_r(\lambda), b_r(\lambda)]$ is irreducible, *i.e.* the matrix $\begin{bmatrix} a_r(\lambda) \\ b_r(\lambda) \end{bmatrix}$ is alatent.

Statement 2.3 If the $n \times m$ matrix $A(\lambda)$ possesses the two LMFDs

$$L(\lambda) = a_{l1}^{-1}(\lambda)b_{l1}(\lambda) = a_{l2}^{-1}(\lambda)b_{l2}(\lambda)$$

and the pair $(a_{l1}(\lambda), b_{l1}(\lambda))$ is irreducible, then there exists a non-singular $n \times n$ polynomial matrix $g(\lambda)$ with

$$a_{l2}(\lambda) = g(\lambda)a_{l1}(\lambda), \quad b_{l2}(\lambda) = g(\lambda)b_{l1}(\lambda).$$

Furthermore, if the pair $(a_{l2}(\lambda), b_{l2}(\lambda))$ is also irreducible, then the matrix $g(\lambda)$ is unimodular.

Remark 2.6. A corresponding statement is true for right MFDs.

5. The theoretical equipment for constructing ILMFDs and IRMFDs is founded on using the canonical form of McMillan. Indeed, from (2.27) and (2.29), we get

$$M_L(\lambda) = \tilde{a}_l^{-1}(\lambda) b(\lambda) = b(\lambda) \tilde{a}_r^{-1}(\lambda) \tag{2.38}$$

with

$$\begin{aligned}
\tilde{a}_l(\lambda) &= \operatorname{diag}\{d(\lambda), \psi_2(\lambda), \ldots, \psi_\rho(\lambda), 1, \ldots, 1\}, \\
\tilde{a}_r(\lambda) &= \operatorname{diag}\{d(\lambda), \psi_2(\lambda), \ldots, \psi_\rho(\lambda), 1, \ldots, 1\}, \\
b(\lambda) &= \begin{bmatrix} \operatorname{diag}\{\alpha_1(\lambda), , \ldots, \alpha_\rho(\lambda)\} & O_{\rho, m-\rho} \\ O_{n-\rho,\rho} & O_{n-\rho,m-\rho} \end{bmatrix}.
\end{aligned} \tag{2.39}$$

Inserting (2.38) and (2.39) in (2.26), we obtain an LMFD (2.36) and an RMFD (2.37) with

$$\begin{aligned}
a_l(\lambda) &= \tilde{a}_l(\lambda) p^{-1}(\lambda), \quad b_l(\lambda) = b(\lambda) q(\lambda), \\
a_r(\lambda) &= q^{-1}(\lambda) \tilde{a}_r(\lambda), \quad b_r(\lambda) = p(\lambda) b(\lambda).
\end{aligned} \tag{2.40}$$

In [69] is stated that the pairs $(a_l(\lambda), b_l(\lambda))$ and $[a_r(\lambda), b_r(\lambda)]$ are irreducible, *i.e.* by using (2.40), Relations (2.36) and (2.37) generate ILMFDs and IRMFDs of the matrix $L(\lambda)$.

6. If Relations (2.36) and (2.37) define ILMFDs and IRMFDs of the matrix $L(\lambda)$, then it follows from (2.40) and Statement 2.3 that the matrices $a_l(\lambda)$ and $a_r(\lambda)$ possess equal invariant polynomials different from one. Herein,

$$\det a_l(\lambda) \approx \det a_r(\lambda) \approx d(\lambda)\psi_2(\lambda) \cdots \psi_\rho(\lambda) = \psi_L(\lambda),$$

where $\psi_L(\lambda)$ is the McMillan denominator of the matrix $L(\lambda)$. Besides, the last relation together with (2.32) yields

$$\operatorname{ord} a_l(\lambda) = \operatorname{ord} a_r(\lambda) = \operatorname{Mdeg} L(\lambda). \tag{2.41}$$

Moreover, we recognise from (2.40) that the matrices $b_l(\lambda)$ and $b_r(\lambda)$ in the ILMFD (2.36) and the IRMFD (2.37) are equivalent.

7.

Lemma 2.7. *Let $\tilde{a}_l(\lambda)$ ($\tilde{a}_r(\lambda)$) be a left (right) reducing polynomial for the matrix $L(\lambda)$ with $\operatorname{ord} \tilde{a}_l(\lambda) = \kappa$ ($\operatorname{ord} \tilde{a}_r(\lambda) = \kappa$). Then*

$$\operatorname{Mdeg} L(\lambda) \leq \kappa.$$

Proof. Let us have the ILMFD (2.36). Then due to Statement 2.3, we have

$$\tilde{a}_l(\lambda) = g(\lambda) a_l(\lambda), \tag{2.42}$$

where the matrix $g(\lambda)$ is non-singular, from which directly follows the claim. ∎

8. A number of auxiliary statements about general properties of MFDs should be given now.

Lemma 2.8. *Let an LMFD*

$$L(\lambda) = a_{l1}^{-1}(\lambda)b_{l1}(\lambda)$$

be given. Then there exists an RMFD

$$L(\lambda) = b_{r1}(\lambda)a_{r1}^{-1}(\lambda)$$

with $\det a_{l1}(\lambda) \approx \det a_{r1}(\lambda)$. *The reverse statement is also true.*

Proof. Let the ILMFD and IRMFD

$$L(\lambda) = a_l^{-1}(\lambda)b_l(\lambda) = b_r(\lambda)a_r^{-1}(\lambda)$$

be given. Then with (2.42), we have

$$a_{l1}(\lambda) = g_l(\lambda)a_l(\lambda)\,,$$

where the matrix $g_l(\lambda)$ is non-singular. Let $\det g(\lambda) = h(\lambda)$ and choose the $m \times m$ matrix $g_r(\lambda)$ with $\det g_r(\lambda) \approx h(\lambda)$. Then using

$$a_{r1}(\lambda) = a_r(\lambda)g_r(\lambda), \quad b_{r1}(\lambda) = b_r(\lambda)g_r(\lambda)\,,$$

we obtain an RMFD of the desired form. ∎

9.

Lemma 2.9. *Let the PMD of the dimension* n, p, m

$$\tau(\lambda) = (a(\lambda), b(\lambda), c(\lambda))$$

be given, where the pair $(a(\lambda), b(\lambda))$ *is irreducible. Then, if we have an ILMFD*

$$c(\lambda)a^{-1}(\lambda) = a_1^{-1}(\lambda)c_1(\lambda)\,, \tag{2.43}$$

the pair $(a_1(\lambda), c_1(\lambda)b(\lambda))$ *becomes irreducible.*

On the other side, if the pair $[a(\lambda), c(\lambda)]$ *is irreducible, and we have an IRMFD*

$$a^{-1}(\lambda)b(\lambda) = b_1(\lambda)a_2^{-1}(\lambda)\,, \tag{2.44}$$

then the pair $[a_2(\lambda), c(\lambda)b_1(\lambda)]$ *becomes irreducible.*

Proof. Since the pair $(a(\lambda), b(\lambda))$ is irreducible, owing to (1.71), there exist polynomial matrices $X(\lambda)$, $Y(\lambda)$ with

$$a(\lambda)X(\lambda) + b(\lambda)Y(\lambda) = I_p\,.$$

In analogy, the irreducibility of the pair $(a_1(\lambda), c_1(\lambda))$ implies the existence of polynomial matrices $U(\lambda)$ and $V(\lambda)$ with

$$a_1(\lambda)U(\lambda) + c_1(\lambda)V(\lambda) = I_n \,. \tag{2.45}$$

Using the last two equations, we find

$$\begin{aligned} a_1(\lambda)U(\lambda) + c_1(\lambda)V(\lambda) &= a_1(\lambda)U(\lambda) + c_1(\lambda)I_pV(\lambda) \\ &= a_1(\lambda)U(\lambda) + c_1(\lambda)\left[a(\lambda)X(\lambda) + b(\lambda)Y(\lambda)\right]V(\lambda) = I_n \end{aligned}$$

which, due to (2.43), may be written in the form

$$a_1(\lambda)\left[U(\lambda) + c(\lambda)X(\lambda)V(\lambda)\right] + c_1(\lambda)b(\lambda)\left[Y(\lambda)V(\lambda)\right] = I_n \,.$$

From this equation by virtue of (1.71), it is evident that the pair $(a_1(\lambda), c_1(\lambda)b(\lambda))$ is irreducible.
In the same manner, it can be shown that the pair $[a_2(\lambda), c(\lambda)b_1(\lambda)]$ is irreducible. ∎

Remark 2.10. The reader finds in [69] an equivalent statement to Lemma 2.9 in modified form.

10.

Lemma 2.11. *Let the pair $(a_1(\lambda)a_2(\lambda), b(\lambda))$ be irreducible. Then also the pair $(a_1(\lambda), b(\lambda))$ is irreducible. Analogously, we have: If the pair $[a_1(\lambda)a_2(\lambda), c(\lambda)]$ is irreducible, then the pair $[a_2(\lambda), c(\lambda)]$ is also irreducible.*

Proof. Produce

$$L(\lambda) = a_2^{-1}(\lambda)a_1^{-1}(\lambda)b(\lambda) = [a_1(\lambda)a_2(\lambda)]^{-1}b(\lambda) \,.$$

Due to our supposition, the right side of this equation is an ILMFD. Therefore, regarding (2.41), we get

$$\operatorname{Mdeg} L(\lambda) = \operatorname{ord}[a_1(\lambda)a_2(\lambda)] = \operatorname{ord} a_1(\lambda) + \operatorname{ord} a_2(\lambda) \,. \tag{2.46}$$

Suppose the pair $(a_1(\lambda), b(\lambda))$ to be reducible. Then there would exist an ILMFD

$$a_3^{-1}(\lambda)b_1(\lambda) = a_1^{-1}(\lambda)b(\lambda) \,,$$

where $\operatorname{ord} a_3(\lambda) < \operatorname{ord} a_1(\lambda)$, and we obtain

$$L(\lambda) = a_2^{-1}(\lambda)a_3^{-1}(\lambda)b_1(\lambda) = [a_3(\lambda)a_2(\lambda)]^{-1}b_1(\lambda) \,.$$

From this equation it follows that $a_3(\lambda)a_2(\lambda)$ is a left reducing polynomial for $L(\lambda)$. Therefore, Lemma 2.7 implies

$$\operatorname{Mdeg} L(\lambda) \le \operatorname{ord}[a_3(\lambda)a_2(\lambda)] < \operatorname{ord} a_1(\lambda) + \operatorname{ord} a_2(\lambda) \,.$$

This relation contradicts (2.46), that's why the pair $(a_1(\lambda), b(\lambda))$ has to be irreducible. The second part of the Lemma is shown analogously. ∎

11. The subsequent Lemmata state further properties of the denominator and the McMillan degree.

Lemma 2.12. *Let a matrix of the form*

$$L(\lambda) = c(\lambda)a^{-1}(\lambda)b(\lambda) \tag{2.47}$$

be given with polynomial matrices $a(\lambda)$, $b(\lambda)$, $c(\lambda)$, *where the pairs* $(a(\lambda), b(\lambda))$ *and* $[a(\lambda), c(\lambda)]$ *are irreducible. Then*

$$\psi_L(\lambda) \approx \det a(\lambda)$$

is true, and thus

$$\operatorname{Mdeg} L(\lambda) = \operatorname{ord} a(\lambda)\,. \tag{2.48}$$

Proof. Build the ILMFD

$$c(\lambda)a^{-1}(\lambda) = a_1^{-1}(\lambda)c_1(\lambda)\,. \tag{2.49}$$

Since by supposition the left side of (2.49) is an IRMFD, we have

$$\det a(\lambda) \approx \det a_1(\lambda)\,. \tag{2.50}$$

Using (2.49), we obtain from (2.47)

$$L(\lambda) = a_1^{-1}(\lambda)[c_1(\lambda)b(\lambda)]\,.$$

Due to Lemma 2.9, the right side of this equation is an ILMFD and because of (2.50), we get

$$\psi_L(\lambda) \approx \det a_1(\lambda) \approx \det a(\lambda)\,.$$

Relation (2.48) now follows directly from (2.32). ∎

Lemma 2.13. *Let*

$$L(\lambda) = L_1(\lambda)L_2(\lambda) \tag{2.51}$$

be given with rational matrices $L_1(\lambda)$, $L_2(\lambda)$, $L(\lambda)$, *and* $\psi_{L_1}(\lambda)$, $\psi_{L_2}(\lambda)$, $\psi_L(\lambda)$ *should be their accompanying McMillan denominators. Then the expression*

$$\chi(\lambda) = \frac{\psi_{L_1}(\lambda)\psi_{L_2}(\lambda)}{\psi_L(\lambda)}$$

realises as a polynomial.

Proof. Let the ILMFD

$$L(\lambda) = a^{-1}(\lambda)b(\lambda) \tag{2.52}$$

and in addition

$$L_i(\lambda) = a_i^{-1}(\lambda)b_i(\lambda)\,, \quad (i = 1, 2) \tag{2.53}$$

be given. Then

$$\psi_L(\lambda) \approx \det a(\lambda), \quad \psi_{L_i}(\lambda) \approx \det a_i(\lambda), \quad (i = 1, 2). \tag{2.54}$$

Equation (2.51) with (2.53) implies

$$L(\lambda) = a_1^{-1}(\lambda) b_1(\lambda) a_2^{-1}(\lambda) b_2(\lambda). \tag{2.55}$$

Owing to Lemma 2.8, there exists an LMFD

$$a_3^{-1}(\lambda) b_3(\lambda) = b_1(\lambda) a_2^{-1}(\lambda), \tag{2.56}$$

where

$$\det a_3(\lambda) \approx \det a_2(\lambda) \approx \psi_{L_2}(\lambda).$$

Using (2.55) and (2.56), we find

$$L(\lambda) = a_1^{-1}(\lambda) a_3^{-1}(\lambda) b_3(\lambda) b_2(\lambda) = a_4^{-1}(\lambda) b_4(\lambda), \tag{2.57}$$

where

$$a_4(\lambda) = a_3(\lambda) a_1(\lambda), \quad b_4(\lambda) = b_3(\lambda) b_2(\lambda).$$

Per construction, we get

$$\det a_4(\lambda) \approx \psi_{L_1}(\lambda) \psi_{L_2}(\lambda). \tag{2.58}$$

Relations (2.52) and (2.57) define LMFDs of the matrix $L(\lambda)$, where (2.52) is an ILMFD. Therefore, the relation

$$a_4(\lambda) = g(\lambda) a(\lambda)$$

holds with an $n \times n$ polynomial matrix $g(\lambda)$. From the last equation arises that the object

$$\frac{\det a_4(\lambda)}{\det a(\lambda)} = \det g(\lambda)$$

is a polynomial. Finally, this equation together with (2.54) and (2.58) yields the claim of the Lemma. ∎

Remark 2.14. From Lemma 2.13 under supposition (2.51), we get

$$\operatorname{Mdeg}[L_1(\lambda) L_2(\lambda)] \leq \operatorname{Mdeg} L_1(\lambda) + \operatorname{Mdeg} L_2(\lambda).$$

In the following investigations, we will call the matrices $L_1(\lambda)$ and $L_2(\lambda)$ *independent*, when the equality sign takes place in the last relation.

Lemma 2.15. *Let* $L(\lambda) \in \mathbb{F}_{nm}(\lambda)$, $G(\lambda) \in \mathbb{F}_{nm}[\lambda]$ *and*

$$L_1(\lambda) = L(\lambda) + G(\lambda)$$

be given. Then, we have

$$\psi_{L_1}(\lambda) = \psi_L(\lambda) \tag{2.59}$$

and therefore,

$$\operatorname{Mdeg} L_1(\lambda) = \operatorname{Mdeg} L(\lambda). \tag{2.60}$$

Proof. Start with the ILMFD (2.52). Then the matrix

$$R_h(\lambda) = \begin{bmatrix} a(\lambda) & b(\lambda) \end{bmatrix} \tag{2.61}$$

becomes alatent. By using (2.52), we build the LMFD

$$L_1(\lambda) = a^{-1}(\lambda)\,[b(\lambda) + a(\lambda)G(\lambda)] \tag{2.62}$$

for the matrix $L_1(\lambda)$, to which the horizontal matrix

$$R_{1h}(\lambda) = \begin{bmatrix} a(\lambda) & b(\lambda) + a(\lambda)G(\lambda) \end{bmatrix}$$

is configured. The identity

$$R_{1h}(\lambda) = R_h(\lambda) \begin{bmatrix} I_n & G(\lambda) \\ O_{mn} & I_m \end{bmatrix}$$

is easily proved. The first factor on the right side is the alatent matrix $R_h(\lambda)$ and the second factor is a unimodular matrix. Therefore, the product is also alatent and consequently, (2.62) is an ILMFD, which implies

$$\operatorname{Mdeg} L_1(\lambda) = \operatorname{ord} a(\lambda) = \operatorname{Mdeg} L(\lambda)$$

and Equation (2.60) follows. ∎

Lemma 2.16. *For the matrix $L(\lambda) \in \mathbb{F}_{nm}(\lambda)$, let an ILMFD (2.52) be given, and the matrix $L_1(\lambda)$ is determined by*

$$L_1(\lambda) = L(\lambda)D(\lambda)\,,$$

where the non-singular matrix $D(\lambda) \in \mathbb{F}_{mm}[\lambda]$ should be free of eigenvalues that coincide with eigenvalues of the matrix $a(\lambda)$ in (2.52). Then the relation

$$L_1(\lambda) = a^{-1}(\lambda)[b(\lambda)D(\lambda)] \tag{2.63}$$

defines an ILMFD of the matrix $L_1(\lambda)$, and Equations (2.59), (2.60) are fulfilled.

Proof. Let an ILMFD (2.52) and the set $\lambda_1, \ldots, \lambda_q$ of eigenvalues of $a(\lambda)$ be given. Since Matrix (2.61) is alatent, we gain

$$\operatorname{rank} R_h(\lambda_i) = \operatorname{rank} \begin{bmatrix} a(\lambda_i) & b(\lambda_i) \end{bmatrix} = n, \quad (i = 1, \ldots, q)\,.$$

Consider the LMFD (2.63) and the accompanying matrix

$$R_{1h}(\lambda) = \begin{bmatrix} a(\lambda) & b(\lambda)D(\lambda) \end{bmatrix}.$$

The latent numbers of the matrix $R_{1h}(\lambda)$ belong to the set of numbers $\lambda_1, \ldots, \lambda_q$. But for any $1 \le i \le q$, we have

$$R_{1h}(\lambda_i) = R_h(\lambda_i)F(\lambda_i),$$

where the matrix $F(\lambda)$ has the form

$$F(\lambda) = \begin{bmatrix} I_n & O_{nm} \\ O_{mn} & D(\lambda) \end{bmatrix}.$$

Under the supposed conditions, $\operatorname{rank} F(\lambda_i) = n + m$ is valid, that means, the matrix $F(\lambda_i)$ is non-singular, which implies

$$\operatorname{rank} R_{1h}(\lambda_i) = n, \quad (i = 1, \dots, q).$$

Therefore, the matrix $R_{1h}(\lambda)$ satisfies Condition (1.72), and Lemma 1.42 guarantees that Relation (2.63) delivers an ILMFD of the matrix $L_1(\lambda)$. From this fact we conclude the validity of (2.59), (2.60). ∎

12.

Lemma 2.17. *Let the irreducible rational matrix*

$$L(\lambda) = \frac{N(\lambda)}{d_1(\lambda)d_2(\lambda)} \tag{2.64}$$

be given, where $N(\lambda)$ is an $n \times m$ polynomial matrix, and $d_1(\lambda)$, $d_2(\lambda)$ are coprime scalar polynomials. Moreover, let the ILMFDs

$$\tilde{L}_1(\lambda) = \frac{N(\lambda)}{d_1(\lambda)} = a_1^{-1}(\lambda)b_1(\lambda), \quad \tilde{L}_2(\lambda) = \frac{b_1(\lambda)}{d_2(\lambda)} = a_2^{-1}(\lambda)b_2(\lambda)$$

exist. Then the expression

$$L(\lambda) = [a_2(\lambda)a_1(\lambda)]^{-1}b_2(\lambda)$$

turns out to be an ILMFD of Matrix (2.64).

Proof. The proof immediately follows from Formulae (2.25)–(2.29), because the polynomials $d_1(\lambda)$ and $d_2(\lambda)$ are coprime. ∎

13.

Lemma 2.18. *Let irreducible representations of the form (2.21)*

$$L_i(\lambda) = \frac{N_i(\lambda)}{d_i(\lambda)}, \quad (i = 1, 2) \tag{2.65}$$

with $n \times m$ polynomial matrices $N_i(\lambda)$ be given, where the polynomials $d_1(\lambda)$ and $d_2(\lambda)$ are coprime. Then we have

$$\operatorname{Mdeg}[L_1(\lambda) + L_2(\lambda)] = \operatorname{Mdeg} L_1(\lambda) + \operatorname{Mdeg} L_2(\lambda). \tag{2.66}$$

Proof. Proceed from the ILMFDs

$$L_i(\lambda) = \tilde{a}_i^{-1}(\lambda)\tilde{b}_i(\lambda), \quad (i = 1, 2)\,. \tag{2.67}$$

Then under the actual assumptions, the matrices $\tilde{a}_1(\lambda)$ and $\tilde{a}_2(\lambda)$ have no common eigenvalues, and they satisfy

$$\operatorname{Mdeg} L_i(\lambda) = \operatorname{ord} \tilde{a}_i(\lambda), \quad (i = 1, 2)\,. \tag{2.68}$$

Using (2.65), we arrive at

$$L(\lambda) = L_1(\lambda) + L_2(\lambda) = \frac{N_1(\lambda)d_2(\lambda) + N_2(\lambda)d_1(\lambda)}{d_1(\lambda)d_2(\lambda)}\,,$$

where the fraction on the right side is irreducible. Consider the matrix

$$\tilde{L}_1(\lambda) = L(\lambda)d_2(\lambda) = \frac{N_1(\lambda)}{d_1(\lambda)}\,d_2(\lambda) + N_2(\lambda)\,.$$

Applying (2.67), we obtain

$$\tilde{L}_1(\lambda) = \tilde{a}_1^{-1}(\lambda)\left[\tilde{b}_1(\lambda)d_2(\lambda) + \tilde{a}_1(\lambda)N_2(\lambda)\right]\,.$$

From Lemmata 2.15–2.17, it follows that the right side of the last equation is an ILMFD, because the polynomials $d_1(\lambda)$ and $d_2(\lambda)$ are coprime. Now introduce the notation

$$\tilde{L}_2(\lambda) = \tilde{b}_1(\lambda) + \tilde{a}_1(\lambda)\frac{N_2(\lambda)}{d_2(\lambda)} = \tilde{b}_1(\lambda) + \tilde{a}_1(\lambda)\tilde{a}_2^{-1}(\lambda)\tilde{b}_2(\lambda) \tag{2.69}$$

and investigate the ILMFD

$$\tilde{a}_1(\lambda)\tilde{a}_2^{-1}(\lambda) = a_1^{-1}(\lambda)a_2(\lambda)\,. \tag{2.70}$$

The left side of this equation is an IRMFD, because the matrices $\tilde{a}_1(\lambda)$ and $\tilde{a}_2(\lambda)$ have no common eigenvalues. Therefore,

$$\operatorname{ord} \tilde{a}_2(\lambda) = \operatorname{ord} a_1(\lambda)\,, \tag{2.71}$$

and from Lemmata 2.9 and 2.15 we gather that the right side of the equation

$$\tilde{L}_2(\lambda) = a_1^{-1}(\lambda)\left[a_1(\lambda)\tilde{b}_1(\lambda) + a_2(\lambda)\tilde{b}_2(\lambda)\right] = a_1^{-1}(\lambda)b_2(\lambda)$$

is an ILMFD. This relation together with (2.69) implies

$$L(\lambda) = \left[a_1(\lambda)\tilde{a}_1(\lambda)\right]^{-1} b_2(\lambda)\,.$$

Hereby, Lemma 2.17 yields that the right side of the last equation is an ILMFD, from which by means of (2.68) and (2.71), we conclude (2.66). ∎

Corollary 2.19. *If we write with the help of (2.70)*

$$L(\lambda) = \tilde{a}_1^{-1}(\lambda)a_1^{-1}(\lambda)\left[a_1(\lambda)\tilde{b}_1(\lambda) + a_2(\lambda)\tilde{b}_2(\lambda)\right]\,,$$

then the right side is an ILMFD. ∎

2.5 Double-sided MFD (DMFD)

1. Assume in (2.64) $d_1(\lambda)$ and $d_2(\lambda)$ to be monic and coprime polynomials, *i.e.*

$$L(\lambda) = \frac{N(\lambda)}{d_1(\lambda)d_2(\lambda)}$$

is valid. Then applying (2.26)–(2.29) yields

$$L(\lambda) =$$

$$p(\lambda)\begin{bmatrix} \operatorname{diag}\left\{\frac{\alpha_1(\lambda)}{d_1(\lambda)d_2(\lambda)}, \frac{\alpha_2(\lambda)}{\varphi_2(\lambda)\xi_2(\lambda)}, \ldots, \frac{\alpha_\rho(\lambda)}{\varphi_\rho(\lambda)\xi_\rho(\lambda)}\right\} & O_{\rho,m-\rho} \\ O_{n-\rho,\rho} & O_{n-\rho,m-\rho} \end{bmatrix} q(\lambda),$$

where all fractions are irreducible, and all polynomials $\varphi_2(\lambda), \ldots, \varphi_\rho(\lambda)$ are divisors of the polynomial $d_1(\lambda)$, and all polynomials $\xi_2(\lambda), \ldots, \xi_\rho(\lambda)$ are divisor of the polynomial $d_2(\lambda)$. Furthermore, every $\varphi_i(\lambda)$ is divisible by $\varphi_{i+1}(\lambda)$, and $\xi_i(\lambda)$ by $\xi_{i+1}(\lambda)$.

2. Consider now the polynomial matrices

$$\begin{aligned} \tilde{a}_l(\lambda) &= \operatorname{diag}\{d_1(\lambda), \varphi_2(\lambda), \ \ldots \ , \varphi_\rho(\lambda), 1, \ldots, 1\}p^{-1}(\lambda), \\ \tilde{b}(\lambda) &= \begin{bmatrix} \operatorname{diag}\{\alpha_1(\lambda), \ldots, \alpha_\rho(\lambda)\} & O_{\rho,m-\rho} \\ O_{n-\rho,\rho} & O_{n-\rho,m-\rho} \end{bmatrix}, \\ \tilde{a}_r(\lambda) &= q^{-1}(\lambda)\operatorname{diag}\{d_2(\lambda), \xi_2(\lambda), \ldots, \xi_\rho(\lambda), 1, \ldots, 1\} \end{aligned} \tag{2.72}$$

with the dimensions $n \times n$, $n \times m$, $m \times m$, respectively. So we can write

$$L(\lambda) = \tilde{a}_l^{-1}(\lambda)\tilde{b}(\lambda)\tilde{a}_r^{-1}(\lambda). \tag{2.73}$$

A representation of the form (2.73) is called *double-sided or bilateral MFD* (DMFD).

3.

Lemma 2.20. *The pairs $(\tilde{a}_l(\lambda), \tilde{b}(\lambda))$ and $[\tilde{a}_r(\lambda), \tilde{b}(\lambda)]$ defined by Relations (2.72) are irreducible.*

Proof. Build the LMFD and RMFD

$$\frac{N(\lambda)\tilde{a}_r(\lambda)}{d_1(\lambda)d_2(\lambda)} = \tilde{a}_l^{-1}(\lambda)\tilde{b}(\lambda), \qquad \frac{\tilde{a}_l(\lambda)N(\lambda)}{d_1(\lambda)d_2(\lambda)} = \tilde{b}(\lambda)\tilde{a}_r^{-1}(\lambda).$$

With the help of (2.72), we immediately recognise that the right sides are ILMFD resp. IRMFD. Therefore, the pairs $(\tilde{a}_l(\lambda), \tilde{b}(\lambda))$, $[\tilde{a}_r(\lambda), \tilde{b}(\lambda)]$ are irreducible. ∎

Suppose (2.72), then under the conditions of Lemma 2.20, it follows that in the representation (2.73), the quantities $\operatorname{ord} \tilde{a}_l(\lambda)$ and $\operatorname{ord} \tilde{a}_r(\lambda)$ take their minimal values. A representation like (2.73) is named *irreducible DMFD* (IDMFD). The set of all IDMFD of the matrix $L(\lambda)$ according to given polynomials $d_1(\lambda)$, $d_2(\lambda)$ has the form

$$L(\lambda) = a_l^{-1}(\lambda) b(\lambda) a_r^{-1}(\lambda)$$

with

$$a_l(\lambda) = p(\lambda)\tilde{a}_l(\lambda), \quad b(\lambda) = p(\lambda)\tilde{b}(\lambda)q(\lambda), \quad a_r(\lambda) = \tilde{a}_r(\lambda)q(\lambda),$$

where $p(\lambda)$, $q(\lambda)$ are unimodular matrices of appropriate type.

Example 2.21. Consider the rational matrix

$$L(\lambda) = \frac{\begin{bmatrix} 5\lambda^2 - 6\lambda - 12 & -2\lambda^2 + 3\lambda + 4 \\ -2\lambda^2 - 2\lambda + 18 & \lambda^2 - 7 \end{bmatrix}}{(\lambda^2 + \lambda + 2)(\lambda - 3)}.$$

Assume $d_1(\lambda) = \lambda^2 + \lambda + 2$, $d_2(\lambda) = \lambda - 3$, then we can write

$$L(\lambda) = a_l^{-1}(\lambda) b(\lambda) a_r^{-1}(\lambda)$$

with

$$a_l(\lambda) = \begin{bmatrix} \lambda + 1 & \lambda \\ 2 & \lambda + 2 \end{bmatrix}, \quad a_r(\lambda) = \begin{bmatrix} \lambda - 1 & 1 \\ 2\lambda & 3 \end{bmatrix}, \quad b(\lambda) = \begin{bmatrix} \lambda - 2 & 0 \\ 2 & 1 \end{bmatrix}.$$

The obtained DMFD is irreducible because of $\det a_l(\lambda) = d_1(\lambda)$, $\det a_r(\lambda) = d_2(\lambda)$, and the quantities $\operatorname{ord} a_l(\lambda)$ and $\operatorname{ord} a_r(\lambda)$ take their minimal possible values. □

2.6 Index of Rational Matrices

1. As in the scalar case, we understand by the *index* of a rational $n \times m$ matrix $L(\lambda)$ that integer $\operatorname{ind} L$ for which the finite limit

$$\lim_{\lambda \to \infty} L(\lambda)\lambda^{\operatorname{ind} L} = L_0 \neq O_{nm} \tag{2.74}$$

exists. For $\operatorname{ind} L = 0$, $\operatorname{ind} L > 0$ and $\operatorname{ind} L \geq 0$ the matrix $L(\lambda)$ is called *proper, strictly proper and at least proper*, respectively. For rational matrices of the form (2.21), we have

$$\operatorname{ind} L = \deg d(\lambda) - \deg N(\lambda).$$

2. In a number of cases we also can receive the value of $\operatorname{ind} L$ from the LMFD or RMFD.

Lemma 2.22. *Suppose the matrix $L(\lambda)$ in the standard form*

$$L(\lambda) = \frac{N(\lambda)}{d(\lambda)} \tag{2.75}$$

and the relations

$$L(\lambda) = a_l^{-1}(\lambda) b_l(\lambda) = b_r(\lambda) a_r^{-1}(\lambda) \tag{2.76}$$

should define LMFD resp. RMFD of the matrix $L(\lambda)$. Then $\operatorname{ind} L$ *satisfies the inequalities*

$$\begin{aligned} \operatorname{ind} L = \deg d(\lambda) - \deg N(\lambda) &\le \deg a_l(\lambda) - \deg b_l(\lambda) \\ &\le \deg a_r(\lambda) - \deg b_r(\lambda)\,. \end{aligned} \tag{2.77}$$

Proof. From (2.75) and (2.76) we arrive at

$$d(\lambda) b_l(\lambda) = a_l(\lambda) N(\lambda)\,,$$

which results in

$$\deg[d(\lambda) b_l(\lambda)] = \deg[a_l(\lambda) N(\lambda)]\,. \tag{2.78}$$

According to

$$d(\lambda) b_l(\lambda) = [d(\lambda) I_n] b_l(\lambda)$$

and due to the regularity of the matrix $d(\lambda) I_n$, we get through (1.12)

$$\deg[d(\lambda) b_l(\lambda)] = \deg d(\lambda) + \deg b_l(\lambda)\,. \tag{2.79}$$

Moreover, using (1.11) we realise

$$\deg[a_l(\lambda) N(\lambda)] \le \deg a_l(\lambda) + \deg N(\lambda)\,. \tag{2.80}$$

Comparing (2.78)–(2.80), we obtain

$$\deg d(\lambda) + \deg b_l(\lambda) \le \deg a_l(\lambda) + \deg N(\lambda)\,,$$

which is equivalent to the first inequality in (2.77). The second inequality can be shown analogously. ∎

Corollary 2.23. *If the matrix $L(\lambda)$ is proper, i.e.* $\operatorname{ind} L = 0$, *then for any MFD (2.76) from (2.77) it follows*

$$\deg b_l(\lambda) \le \deg a_l(\lambda), \quad \deg b_r(\lambda) \le \deg a_r(\lambda).$$

If the matrix $L(\lambda)$ is even strictly proper, i.e. $\operatorname{ind} L < 0$ *is true, then we have*

$$\deg b_l(\lambda) < \deg a_l(\lambda), \quad \deg b_r(\lambda) < \deg a_r(\lambda).$$

3. A complete information about the index of $L(\lambda)$ is received in that case, where in the LMFD (2.36) [RMFD (2.37)] the matrix $a_l(\lambda)$ is row reduced [$a_r(\lambda)$ is column reduced].

Theorem 2.24. *Consider the LMFD*

$$L(\lambda) = a_l^{-1}(\lambda) b_l(\lambda) \tag{2.81}$$

with $a_l(\lambda)$, $b_l(\lambda)$ of the dimensions $n \times n$, $n \times m$, where $a_l(\lambda)$ is row reduced. Let α_i be the degree of the i-th row of $a_l(\lambda)$, and β_i the degree of the i-th row of $b_l(\lambda)$, and denote

$$\delta_i = \alpha_i - \beta_i, \quad (i = 1, \ldots, n)$$

and

$$\delta_L = \min_{1 \le i \le n} [\delta_i] .$$

Then the index of the matrix $L(\lambda)$ is determined by

$$\operatorname{ind} L = \delta_L . \tag{2.82}$$

Proof. Using (1.22), we can write

$$a_l(\lambda) = \operatorname{diag}\{\lambda^{\alpha_1}, \ldots, \lambda^{\alpha_n}\} \left(\tilde{A}_0 + \tilde{A}_1 \lambda^{-1} + \tilde{A}_2 \lambda^{-2} + \ldots \right) , \tag{2.83}$$

where the $\tilde{A}_i$, $(i = 0, 1, \ldots)$ are constant matrices with $\det \tilde{A}_0 \neq 0$. Extracting from the rows of $b_l(\lambda)$ the corresponding factors, we obtain

$$b_l(\lambda) = \operatorname{diag}\{\lambda^{\alpha_1}, \ldots, \lambda^{\alpha_n}\} \left(\tilde{B}_0 \lambda^{-\delta_L} + \tilde{B}_1 \lambda^{-\delta_L - 1} + \tilde{B}_2 \lambda^{-\delta_L - 2} + \ldots \right) ,$$

where the $\tilde{B}_i$, $(i = 0, 1, \ldots)$ are constant matrices, and $\tilde{B}_0 \neq O_{nm}$. Inserting this and (2.83) into (2.81), we find

$$L(\lambda) \lambda^{\delta_L} = \left(\tilde{A}_0 + \tilde{A}_1 \lambda^{-1} + \ldots \right)^{-1} \left(\tilde{B}_0 + \tilde{B}_1 \lambda^{-1} + \ldots \right) .$$

Now, due to $\det \tilde{A}_0 \neq 0$, it follows

$$\lim_{\lambda \to \infty} L(\lambda) \lambda^{\delta_L} = \tilde{A}_0^{-1} \tilde{B}_0 \neq O_{nm} , \tag{2.84}$$

and by (2.74) we recognise the statement (2.82) to be true. ∎

Corollary 2.25. *([69], [68]) If in the LMFD (2.81) the matrix $a_l(\lambda)$ is row reduced, then the matrix $L(\lambda)$ is proper, strictly proper or at least proper, if and only if we have $\delta_L = 0$, $\delta_L > 0$ or $\delta_L \geq 0$, respectively.* ∎

In the same way the corresponding statement for right MFD can be seen.

Theorem 2.26. *Consider the RMFD*

$$L(\lambda) = b_r(\lambda)a_r^{-1}(\lambda)$$

with $a_r(\lambda)$, $b_r(\lambda)$ of the dimensions $m \times m$, $n \times m$, where $a_r(\lambda)$ is column reduced. Let $\tilde{\alpha}_i$ be the degree of the i-th column of $a_r(\lambda)$ and $\tilde{\beta}_i$ the degree of the i-th column of $b_r(\lambda)$, and denote

$$\tilde{\delta}_i = \tilde{\alpha}_i - \tilde{\beta}_i, \quad (i = 1, \dots, m)$$

and

$$\tilde{\delta}_L = \min_{1 \le i \le m} [\tilde{\delta}_i] .$$

Then the index of the matrix $L(\lambda)$ is determined by

$$\operatorname{ind} L = \tilde{\delta}_L .$$

∎

Example 2.27. Consider the matrices

$$a_l(\lambda) = \begin{bmatrix} 2\lambda^2+1 & \lambda+2 \\ 1 & \lambda+1 \end{bmatrix}, \quad b_l(\lambda) = \begin{bmatrix} 2 & 3\lambda^2+1 \\ 5 & 7 \end{bmatrix}.$$

In this case the matrix $a_l(\lambda)$ is row reduced, where $\alpha_1 = 2$, $\alpha_1 = 1$ and $\beta_1 = 2$, $\beta_1 = 0$. Consequently, we get $\delta_1 = 0$, $\delta_2 = 1$, thus $\delta_L = 0$. Therefore, the matrix $a_l^{-1}(\lambda)b_l(\lambda)$ becomes proper. Hereby, we obtain

$$\tilde{A}_0 = \begin{bmatrix} 2 & 0 \\ 0 & 1 \end{bmatrix}, \quad \tilde{B}_0 = \begin{bmatrix} 0 & 3 \\ 0 & 0 \end{bmatrix},$$

and from (2.84) it follows

$$\lim_{\lambda \to \infty} a_l^{-1}(\lambda)b_l(\lambda) = \tilde{A}_0^{-1}\tilde{B}_0 = \begin{bmatrix} 0 & 1.5 \\ 0 & 0 \end{bmatrix}.$$

□

2.7 Strictly Proper Rational Matrices

1. According to the above definitions, Matrix (2.21) is strictly proper if $\operatorname{ind} L = \deg d(\lambda) - \deg N(\lambda) > 0$. Strictly proper rational matrices possess many properties that are analogue to the properties of scalar strictly proper rational fractions, which have been considered in Section 2.1. In particular, the sum, the difference and the product of strictly proper rational matrices are strictly proper too.

2. For any strictly proper rational $n \times m$ matrix $L(\lambda)$, there exists an indefinite set of elementary PMDs

$$\tau(\lambda) = (\lambda I_p - A, B, C) \tag{2.85}$$

i.e. realisations (A, B, C), such that

$$L(\lambda) = C(\lambda I_p - A)^{-1} B\,. \tag{2.86}$$

The right side of (2.86) is called a *standard representation* of the matrix, or simply its representation. The number p, configured in (2.86), is called its *dimension*. A representation, where the dimension p takes its minimal possible value, is called *minimal*.

A standard representation (2.86) is minimal, if and only if its elementary PMD is minimal, that means, if the pair (A, B) is controllable and the pair $[A, C]$ is observable.

The matrix $L(\lambda)$ (2.86) is called the *transfer function* (transfer matrix) of the elementary PMD (2.85), resp. of the realisation (A, B, C). The elementary PMD (2.85) and the PMD

$$\tau_1(\lambda) = (\lambda I_q - A_1, B_1, C_1) \tag{2.87}$$

are called *equivalent*, if their transfer matrices coincide.

3. Now a number of statements on the properties of strictly proper rational matrices is formulated, which will be used later.

Statement 2.4 (see [69, 68]) The minimal PMD (2.85) and (2.87) are equivalent, if and only if $p = q$. In this case, there exists a non-singular $p \times p$ matrix R with

$$A_1 = RAR^{-1}, \quad B_1 = RB, \quad C_1 = CR^{-1}, \tag{2.88}$$

i.e., the corresponding realisations are similar. ∎

Statement 2.5 Let the representation (2.86) be minimal and possess the ILMFD

$$C(\lambda I_p - A)^{-1} = a_l^{-1}(\lambda) b_l(\lambda)\,. \tag{2.89}$$

Then, as follows from Lemma 2.9, the pair $(a_l(\lambda), b_l(\lambda)B)$ is irreducible. In analogy, if we have an IRMFD

$$(\lambda I_p - A)^{-1} B = b_r(\lambda) a_r^{-1}(\lambda)\,, \tag{2.90}$$

then the pair $[a_r(\lambda), C b_r(\lambda)]$ is irreducible. ∎

Statement 2.6 If the representation (2.86) is minimal, then the matrices $a_l(\lambda)$ in the ILMFD (2.89) and $a_r(\lambda)$ in the IRMFD (2.90) possess the same invariant polynomials different from 1 as the matrix $\lambda I_p - A$. Hereby, we have

$$\psi_L(\lambda) = \det(\lambda I_p - A) \approx \det a_l(\lambda) \approx \det a_r(\lambda)\,. \tag{2.91}$$

∎

Particularly, it follows from (2.91) that $\operatorname{Mdeg} L(\lambda)$ of a strictly proper rational matrix $L(\lambda)$ is equal to the dimension of its minimal standard representation.

4.

Lemma 2.28. *Assume $n = m$ in the standard representation (2.86) and $\det L(\lambda) \not\equiv 0$. Then $p \geq n$ holds, and*

$$\det L(\lambda) = \frac{k(\lambda)}{\det(\lambda I_p - A)} \tag{2.92}$$

is valid, where $k(\lambda)$ is a scalar polynomial with

$$\deg k(\lambda) \leq p - n \operatorname{ind} L \,. \tag{2.93}$$

The case $p < n$ results in $\det L(\lambda) \equiv 0$.

Proof. In accordance with Lemma 2.8, there exists an LMFD

$$C(\lambda I_p - A)^{-1} = a_1^{-1}(\lambda) b_1(\lambda)\,,$$

where $\det a_1(\lambda) \approx \det(\lambda I_p - A)$, that's why

$$L(\lambda) = a_1^{-1}(\lambda)[b_1(\lambda)B]\,.$$

Calculating the determinants of both sides yields (2.92) with $k(\lambda) = \det[b_1(\lambda)B]$.

To prove (2.93), we write $L(\lambda)$ in the form (2.21) obtaining $\operatorname{ind} L = \deg d(\lambda) - \deg N(\lambda) > 0$. Now calculating the determinant of the right side of (2.21), we gain

$$\det L(\lambda) = \frac{\det N(\lambda)}{[d(\lambda)]^n}\,.$$

Let $\deg d(\lambda) = q$. Then $\deg N(\lambda) = q - \operatorname{ind} L$ holds, where $\deg \det N(\lambda) \leq n(q - \operatorname{ind} L)$ and $\deg[d(\lambda)]^n = nq$. From this we directly generate (2.93). For $p < n$, on account of the Binet-Cauchy formula, we come to $\det[C \operatorname{adj}(\lambda I_p - A)B] \equiv 0$. ∎

Corollary 2.29. *Consider the strictly proper $n \times n$ matrix $L(\lambda)$ and its McMillan denominator and numerator $\psi_L(\lambda)$ and $\alpha_L(\lambda)$, respectively. Then the following relation is true:*

$$\deg \alpha_L(\lambda) \leq \deg \psi_L(\lambda) - n \operatorname{ind} L\,.$$

Proof. Let (2.86) be a minimal standard representation of the matrix $L(\lambda)$. Then due to Lemma 2.4, we have

$$\det L(\lambda) = k\,\frac{\alpha_L(\lambda)}{\psi_L(\lambda)}\,, \qquad k = \text{const.}$$

and the claim immediately results from (2.93). ∎

5. Let the strictly proper rational matrix $L(\lambda)$ of the form (2.21) with

$$L(\lambda) = \frac{N(\lambda)}{d_1(\lambda)d_2(\lambda)}$$

be given, where the polynomials $d_1(\lambda)$ and $d_2(\lambda)$ are coprime. Then there exists a separation

$$L(\lambda) = \frac{N_1(\lambda)}{d_1(\lambda)} + \frac{N_2(\lambda)}{d_2(\lambda)}, \tag{2.94}$$

where $N_1(\lambda)$ and $N_2(\lambda)$ are polynomial matrices and both fractions in (2.94) are strictly proper.

The matrices $N_1(\lambda)$ and $N_2(\lambda)$ in (2.94) are uniquely determined.

In practice, the separation (2.94) can be produced by performing the separation (2.6) for every element of the matrix $L(\lambda)$.

Example 2.30. Let

$$L(\lambda) = \frac{\begin{bmatrix} \lambda+2 & \lambda \\ \lambda+3 & \lambda^2+1 \end{bmatrix}}{(\lambda-2)^2(\lambda-1)}$$

be given. By choosing $d_1(\lambda) = (\lambda-2)^2$, $d_2(\lambda) = \lambda - 1$, a separation (2.94) is found with

$$N_1(\lambda) = \begin{bmatrix} -3\lambda+10 & -\lambda+4 \\ -4\lambda+13 & -\lambda+7 \end{bmatrix}, \quad N_2(\lambda) = \begin{bmatrix} 3 & 1 \\ 4 & 2 \end{bmatrix}.$$

□

6. The separation (2.94) is extendable to a more general case. Let the strictly proper rational matrix have the form

$$L(\lambda) = \frac{N(\lambda)}{d_1(\lambda)d_2(\lambda)\cdots d_\kappa(\lambda)},$$

where all polynomials in the denominator are two-by-two coprime. Then there exists a unique representation of the form

$$L(\lambda) = \frac{N_1(\lambda)}{d_1(\lambda)} + \ldots + \frac{N_\kappa(\lambda)}{d_\kappa(\lambda)}, \tag{2.95}$$

where all fractions on the right side are strictly proper. Particularly consider (2.21), and the polynomial $d(\lambda)$ should have the form (2.20). Then under the assumption

$$d_i(\lambda) = (\lambda - \lambda_i)^{\mu_i}, \quad (i = 1, \ldots, q)$$

from (2.95), we obtain the unique representation

$$L(\lambda) = \sum_{i=1}^{q} \frac{N_i(\lambda)}{(\lambda - \lambda_i)^{\mu_i}}, \tag{2.96}$$

where $\deg N_i(\lambda) < \mu_i$.

7. By further transformations the fraction

$$L_i(\lambda) = \frac{N_i(\lambda)}{(\lambda - \lambda_i)^{\mu_i}}$$

could be written as

$$L_i(\lambda) = \frac{N_{i1}}{(\lambda - \lambda_i)^{\mu_i}} + \frac{N_{i2}}{(\lambda - \lambda_i)^{\mu_i - 1}} + \ldots + \frac{N_{i\mu_i}}{\lambda - \lambda_i}, \tag{2.97}$$

where the N_{ik}, $(k = 1, \ldots, \mu_i)$ are constant matrices. Inserting (2.97) into (2.96), we arrive at the representation

$$L(\lambda) = \sum_{i=1}^{q} \left[\frac{N_{i1}}{(\lambda - \lambda_i)^{\mu_i}} + \frac{N_{i2}}{(\lambda - \lambda_i)^{\mu_i - 1}} + \ldots + \frac{N_{i\mu_i}}{\lambda - \lambda_i} \right], \tag{2.98}$$

which is called *partial fraction expansion* of the matrix $L(\lambda)$.

8. For calculating the matrices N_{ik} in (2.97), we rely upon the analogous formula to (2.10)

$$N_{ik} = \frac{1}{(k-1)!} \left[\frac{\partial^{k-1}}{\partial \lambda^{k-1}} \frac{N(\lambda)(\lambda - \lambda_i)^{\mu_i}}{d(\lambda)} \right]_{|\lambda = \lambda_i}. \tag{2.99}$$

In practice, the coefficients in (2.99) will be determined by partial fraction expansion of the scalar entries of $L(\lambda)$.

Example 2.31. Assuming the conditions of Example 2.30, we get

$$L(\lambda) = \frac{N_{11}}{(\lambda - 2)^2} + \frac{N_{12}}{\lambda - 2} + \frac{N_{21}}{\lambda - 1},$$

where

$$N_{11} = \begin{bmatrix} 4 & 2 \\ 5 & 5 \end{bmatrix}, \quad N_{12} = \begin{bmatrix} -3 & -1 \\ -4 & -1 \end{bmatrix}, \quad N_{21} = \begin{bmatrix} 3 & 1 \\ 4 & 2 \end{bmatrix}. \qquad \square$$

9. The partial fraction expansion (2.98) can be used in some cases for solving the question on reducibility of certain rational matrices.

Indeed, it is easily shown that for the irreducibility of the strictly proper matrix (2.21), it is necessary and sufficient that in the expansion (2.98)

$$N_{i1} \neq O_{nm}, \quad (i = 1, \ldots, q) \tag{2.100}$$

must be true.

2.8 Separation of Rational Matrices

1. Let the $n \times m$ matrix $L(\lambda)$ in (2.21) be not strictly proper, that means $\operatorname{ind} L \leq 0$. Then for every element of the matrix $L(\lambda)$, the representation (2.5) can be generated, yielding

$$L(\lambda) = \frac{R(\lambda)}{d(\lambda)} + G(\lambda) = L_0(\lambda) + G(\lambda)\,, \tag{2.101}$$

where the fraction in the middle part is strictly proper, and $G(\lambda)$ is a polynomial matrix. The representation (2.101) is unique. Practically, the dissection (2.101) is done in such a way that the dissection (2.5) is applied on each element of $L(\lambda)$.

Furthermore, the strictly proper matrix $L_0(\lambda)$ on the right side of (2.101) is called the *broken part* of the matrix $L(\lambda)$, and the matrix $G(\lambda)$ its *polynomial part* .

Example 2.32. For Matrix (2.22), we obtain

$$L_0(\lambda) = \frac{\begin{bmatrix} 13(\lambda-3) & \lambda-2 \\ 2\lambda+1 & 0 \end{bmatrix}}{(\lambda-2)(\lambda-3)}\,, \qquad G(\lambda) = \begin{bmatrix} 5 & 0 \\ 0 & \lambda+2 \end{bmatrix}.$$

□

2. Let us have in (2.101)

$$L_0(\lambda) = \frac{N_0(\lambda)}{d_1(\lambda)d_2(\lambda)}\,, \tag{2.102}$$

where the polynomials $d_1(\lambda)$ and $d_2(\lambda)$ are coprime, and $\deg N_0(\lambda) < \deg d_1(\lambda) + \deg d_2(\lambda)$. Then as was shown above, there exists the unique separation

$$\frac{N_0(\lambda)}{d_1(\lambda)d_2(\lambda)} = \frac{N_1(\lambda)}{d_1(\lambda)} + \frac{N_2(\lambda)}{d_2(\lambda)}\,,$$

where the fractions on the right side are strictly proper. Inserting this separation into (2.101), we find a unique representation of the form

$$L(\lambda) = \frac{N_1(\lambda)}{d_1(\lambda)} + \frac{N_2(\lambda)}{d_2(\lambda)} + G(\lambda)\,. \tag{2.103}$$

Example 2.33. For Matrix (2.22), we generate the separation (2.103) of the shape

$$A(\lambda) = \frac{\begin{bmatrix} 13 & 0 \\ -5 & 0 \end{bmatrix}}{\lambda-2} + \frac{\begin{bmatrix} 0 & 1 \\ 7 & 0 \end{bmatrix}}{\lambda-3} + \begin{bmatrix} 5 & 0 \\ 0 & \lambda+2 \end{bmatrix}.$$

□

3. From (2.103) we learn that Matrix (2.101) can be presented in the form

$$L(\lambda) = \frac{Q_1(\lambda)}{d_1(\lambda)} + \frac{Q_2(\lambda)}{d_2(\lambda)}, \tag{2.104}$$

where

$$Q_1(\lambda) = N_1(\lambda) + d_1(\lambda)F(\lambda), \quad Q_2(\lambda) = N_2(\lambda) + d_2(\lambda)\left[G(\lambda) - F(\lambda)\right], \tag{2.105}$$

where the polynomial matrix $F(\lambda)$ is arbitrary.

The representation of the rational matrix $L(\lambda)$ from (2.102) in the form (2.104), (2.105) is called its *separation* with respect to the polynomials $d_1(\lambda)$ and $d_2(\lambda)$. It is seen from (2.105) that for coprime polynomials $d_1(\lambda)$ and $d_2(\lambda)$, the separation (2.104) is always possible, but not uniquely determined. Nevertheless, the following theorem holds.

Theorem 2.34. *The totality of pairs $Q_1(\lambda)$, $Q_2(\lambda)$ satisfying the separation (2.104), is given by Formula (2.105).*

Proof. By $\mathcal{P}$ we denote the set of all polynomial pairs $Q_1(\lambda)$, $Q_2(\lambda)$ satisfying Relation (2.104), and by $\mathcal{P}_s$ the set of all polynomial pairs produced by (2.105) when we insert there any polynomial matrices $F(\lambda)$. Since for any Pair (2.105), Relation (2.104) holds, $\mathcal{P}_s \subset \mathcal{P}$ is true. On the other side, let the matrices $\tilde{Q}_1(\lambda)$, $\tilde{Q}_2(\lambda)$ fulfill Relation (2.104). Then we obtain

$$\frac{\tilde{Q}_i(\lambda)}{d_i(\lambda)} = \frac{R_i(\lambda)}{d_i(\lambda)} + G_i(\lambda), \quad (i = 1, 2),$$

where the fractions on the right sides are strictly proper, and $G_1(\lambda)$, $G_2(\lambda)$ are polynomial matrices. Therefore,

$$L(\lambda) = \frac{R_1(\lambda)}{d_1(\lambda)} + \frac{R_2(\lambda)}{d_2(\lambda)} + G_1(\lambda) + G_2(\lambda).$$

Comparing this with (2.103), then due to the uniqueness of the expansion (2.103), we get

$$R_1(\lambda) = N_1(\lambda), \quad R_2(\lambda) = N_2(\lambda), \quad G(\lambda) = G_1(\lambda) + G_2(\lambda).$$

Denoting $G_1(\lambda) = F(\lambda)$, $G_2(\lambda) = G(\lambda) - F(\lambda)$ we find $\tilde{Q}_1(\lambda)$ and $\tilde{Q}_2(\lambda)$ satisfying Relation (2.105), *i.e.* $\mathcal{P} \subset \mathcal{P}_s$ is true. Consequently, the sets $\mathcal{P}$ and $\mathcal{P}_s$ contain each other. ∎

Example 2.35. According to (2.104) and (2.105), we find for Matrix (2.22) the set of all separations with respect to the polynomials $d_1(\lambda) = \lambda - 2$, $d_2(\lambda) = \lambda - 3$. Using the results of Example 2.33, we obtain

$$Q_1(\lambda) = \begin{bmatrix} 13 + (\lambda - 2)f_{11}(\lambda) & (\lambda - 2)f_{12}(\lambda) \\ 5 + (\lambda - 2)f_{21}(\lambda) & (\lambda - 2)f_{22}(\lambda) \end{bmatrix},$$

$$Q_2(\lambda) = \begin{bmatrix} (\lambda - 3)[5 - f_{11}(\lambda)] & 1 - (\lambda - 3)f_{12}(\lambda) \\ 7 + (\lambda - 3)f_{21}(\lambda) & (\lambda - 3)[(\lambda + 2) - f_{22}(\lambda)] \end{bmatrix},$$

where the $f_{ik}(\lambda)$, $(i, k = 1, 2)$ are some polynomials. □

4. Setting in (2.105) $F(\lambda) = O_{nm}$, we arrive at the special solution of the form

$$Q_1(\lambda) = N_1(\lambda), \quad Q_2(\lambda) = N_2(\lambda) + d_2(\lambda)G(\lambda). \tag{2.106}$$

Otherwise, taking $F(\lambda) = G(\lambda)$ results in

$$Q_1(\lambda) = N_1(\lambda) + d_1(\lambda)G(\lambda), \quad Q_2(\lambda) = N_2(\lambda). \tag{2.107}$$

For the solution (2.106), the first summand in the separation (2.104) becomes a strictly proper rational matrix, and for the solution (2.107) the second one does. The particular separations defined by Formulae (2.106) and (2.107) are called *minimal* with respect to $d_1(\lambda)$ resp. $d_2(\lambda)$. Due to their construction, the minimal separations are uniquely determined.

Example 2.36. The separation of Matrix (2.22), which is minimal with respect to $d_1(\lambda) = \lambda - 2$, is given by the matrices

$$Q_1(\lambda) = \begin{bmatrix} 13 & 0 \\ -5 & 0 \end{bmatrix}, \quad Q_2(\lambda) = \begin{bmatrix} 5(\lambda - 3) & 1 \\ 7 & (\lambda - 3)(\lambda + 2) \end{bmatrix}.$$

With respect to $d_2(\lambda) = \lambda - 3$ the separation by the matrices

$$Q_1(\lambda) = \begin{bmatrix} 5\lambda + 3 & 0 \\ -5 & \lambda^2 - 4 \end{bmatrix}, \quad Q_2(\lambda) = \begin{bmatrix} 0 & 1 \\ 7 & 0 \end{bmatrix}$$

is minimal. These minimal separations are unique per construction. □

5. If in particular the original rational matrix (2.101) is strictly proper, then $G(\lambda) = O_{nm}$ becomes true, and the minimal separations (2.106) and (2.107) coincide.

Example 2.37. For the strictly proper matrix in Example 2.30, we obtain a unique minimal separation with $Q_1(\lambda) = N_1(\lambda)$, $Q_2(\lambda) = N_2(\lambda)$, where the matrices $N_1(\lambda)$ and $N_2(\lambda)$ were already determined in Example 2.30. □

2.9 Inverses of Square Polynomial Matrices

1. Assume the $n \times n$ polynomial matrix $L(\lambda)$ to be non-singular, and $\operatorname{adj} L(\lambda)$ be its adjoint matrix, that is determined by Equation (1.8). Then the matrix

$$L^{-1}(\lambda) = \frac{\operatorname{adj} L(\lambda)}{\det L(\lambda)} \tag{2.108}$$

is said to be the *inverse* of the matrix $L(\lambda)$. Equation (1.9) implies

$$L(\lambda)L^{-1}(\lambda) = L^{-1}(\lambda)L(\lambda) = I_n\,. \tag{2.109}$$

2. The matrix $L(\lambda)$ could be written with the help of (1.40), (1.49) in the form

$$L(\lambda) = p^{-1}(\lambda)\begin{bmatrix} h_1(\lambda) & 0 & \dots & 0 \\ 0 & h_1(\lambda)h_2(\lambda) & \dots & 0 \\ \vdots & \vdots & \ddots & \vdots \\ 0 & 0 & \dots & h_1(\lambda)h_2(\lambda)\cdots h_n(\lambda) \end{bmatrix} q^{-1}(\lambda)\,,$$

where $p(\lambda)$ and $q(\lambda)$ are unimodular matrices. How the inverse matrix $L^{-1}(\lambda)$ can be calculated? For that purpose, the general Formula (2.108) is used. Denoting

$$H(\lambda) = \operatorname{diag}\{h_1(\lambda),\, h_1(\lambda)h_2(\lambda),\, \dots\,,\, h_1(\lambda)h_2(\lambda)\cdots h_n(\lambda)\}$$

we can write

$$L^{-1}(\lambda) = q(\lambda)H^{-1}(\lambda)p(\lambda)\,. \tag{2.110}$$

Now, we have to calculate the matrix $H^{-1}(\lambda)$. Obviously, the characteristic polynomial of $H(\lambda)$ amounts to

$$d_H(\lambda) = \det H(\lambda) = h_1^n(\lambda)h_2^{n-1}(\lambda)\cdots h_n(\lambda) \approx \det L(\lambda) = d_L(\lambda)\,. \tag{2.111}$$

Direct calculating the matrix of adjuncts $\operatorname{adj} H(\lambda)$ results in

$$\begin{aligned} \operatorname{adj} H(\lambda) = \operatorname{diag}\big\{&h_1^{n-1}(\lambda)h_2^{n-1}(\lambda)\cdots h_n(\lambda),\, h_1^{n-1}(\lambda)h_2^{n-2}(\lambda)\cdots h_n(\lambda),\, \dots \\ &\dots\,, h_1^{n-1}(\lambda)h_2^{n-2}(\lambda)\cdots h_{n-1}(\lambda)\big\}\,, \end{aligned} \tag{2.112}$$

from which we gain

$$H^{-1}(\lambda) = \frac{\operatorname{adj} H(\lambda)}{d_H(\lambda)}\,. \tag{2.113}$$

Herein, the numerator and denominator are constrained by Relations (2.112), (2.111).

In general, the rational matrix on the right side of (2.113) is reducible, and that's why we will write

$$H^{-1}(\lambda) = \frac{\widetilde{\operatorname{adj} H}(\lambda)}{d_{L\,\min}(\lambda)} \tag{2.114}$$

with

$$\widetilde{\operatorname{adj} H}(\lambda) = \operatorname{diag}\{h_2(\lambda)\cdots h_n(\lambda),\, h_3(\lambda)\cdots h_n(\lambda),\, \ldots,\, h_n(\lambda),\, 1\} \tag{2.115}$$

$$d_{L\,\min}(\lambda) = h_1(\lambda)h_2(\lambda)\cdots h_n(\lambda) = a_n(\lambda)\,, \tag{2.116}$$

where $a_n(\lambda)$ is the last invariant polynomial. Altogether, we receive by using (2.110)

$$L^{-1}(\lambda) = \frac{\widetilde{\operatorname{adj} L}(\lambda)}{d_{L\,\min}(\lambda)}\,, \tag{2.117}$$

where

$$\widetilde{\operatorname{adj} L}(\lambda) = q(\lambda)\widetilde{\operatorname{adj} H}(\lambda)\cdot p(\lambda)\,. \tag{2.118}$$

Matrix (2.118) is called the *monic adjoint* matrix, and the polynomial $d_{L\,\min}(\lambda)$ the *minimal polynomial* of the matrix $L(\lambda)$. The rational matrix on the right side of (2.117) will be named *monic inverse* of the polynomial matrix $L(\lambda)$.

3. Opposing (2.111) to (2.116) makes clear that among the roots of the minimal polynomial $d_{L\min}(\lambda)$ are all eigenvalues of the matrix $L(\lambda)$, however, possibly with lower multiplicity. It is remarkable that the fraction (2.117) is irreducible. The reason for that lies in the fact that the matrix $\widetilde{\operatorname{adj} H}(\lambda)$ for no value of λ becomes zero. The same can be said about Matrix (2.118), because the matrices $q(\lambda)$ and $p(\lambda)$ are unimodular.

4. Comparing (2.108) with (2.117), we find out that the fraction (2.108) is irreducible, if and only if

$$h_1(\lambda) = h_2(\lambda) = \ldots = h_{n-1}(\lambda) = 1, \quad h_n(\lambda) = a_n(\lambda) = d_{L\min}(\lambda) \approx \det L(\lambda)$$

holds, *i.e.* if the characteristic polynomial of the matrix $L(\lambda)$ is equivalent to its minimal polynomial. If the last conditions are fulfilled, then the matrix $L(\lambda)$ can be presented in the form

$$L(\lambda) = p^{-1}(\lambda)\operatorname{diag}\{1,\ \ldots\ ,1,\, a_n(\lambda)\}q^{-1}(\lambda)$$

that means, it is simple in the sense of Section 1.11, and the following theorem has been proved.

Theorem 2.38. *The inverse matrix (2.108) is irreducible, if and only if the matrix $L(\lambda)$ is simple.* ∎

5. From (2.117) we take the important equation

$$\widetilde{\operatorname{adj} L}(\lambda)L(\lambda) = L(\lambda)\widetilde{\operatorname{adj} L}(\lambda) = d_{L\,\min}(\lambda)I_n\,. \tag{2.119}$$

2.10 Transfer Matrices of Polynomial Pairs

1. The pairs $(a_l(\lambda), b_l(\lambda))$, $[a_r(\lambda), b_r(\lambda)]$ are called *non-singular*, if $\det a_l(\lambda) \not\equiv 0$ resp. $\det a_r(\lambda) \not\equiv 0$. For a non-singular pair $(a_l(\lambda), b_l(\lambda))$, the rational matrix

$$w_l(\lambda) = a_l^{-1}(\lambda) b_l(\lambda) \tag{2.120}$$

can be explained, and for the non-singular pair $[a_r(\lambda), b_r(\lambda)]$, we build the rational matrix

$$w_r(\lambda) = b_r(\lambda) a_r^{-1}(\lambda) . \tag{2.121}$$

Matrix (2.120) or (2.121) is called the *transfer matrix (transfer function)* of the corresponding pair. Applying the general Formula (2.108), we obtain

$$w_l(\lambda) = \frac{\operatorname{adj} a_l(\lambda)\, b_l(\lambda)}{d_{a_l}(\lambda)}, \quad w_r(\lambda) = \frac{b_r(\lambda) \operatorname{adj} a_r(\lambda)}{d_{a_r}(\lambda)} \tag{2.122}$$

with the notation $d_{a_l}(\lambda) = \det a_l(\lambda)$, $d_{a_r}(\lambda) = \det a_r(\lambda)$.

2.

Definition 2.39. *The transfer matrices $w_l(\lambda)$ and $w_r(\lambda)$ are called* irreducible, *if the rational matrices on the right side of (2.122) are irreducible.*

Now, we collect some facts on the reducibility of transfer matrices.

Lemma 2.40. *If the matrices $a_l(\lambda)$, $a_r(\lambda)$ are not simple, then the transfer matrices (2.120), (2.121) are reducible.*

Proof. If the matrices $a_l(\lambda)$, $a_r(\lambda)$ are not simple, then owing to (2.116), we conclude that the matrices $a_l^{-1}(\lambda)$, $a_r^{-1}(\lambda)$ are reducible, and therefore, also the fractions

$$w_l(\lambda) = \frac{\widetilde{\operatorname{adj} a_l}(\lambda)\, b_l(\lambda)}{d_{a_l \min}(\lambda)}, \quad w_r(\lambda) = \frac{b_r(\lambda) \widetilde{\operatorname{adj} a_r}(\lambda)}{d_{a_r \min}(\lambda)} . \tag{2.123}$$

But this means, fractions (2.120), (2.121) are reducible. ∎

The matrices (2.123) are said to be the *monic transfer matrices.*

3.

Lemma 2.41. *If the pairs $(a_l(\lambda), b_l(\lambda))$, $[a_r(\lambda), b_r(\lambda)]$ are reducible, i.e. the matrices*

$$R_h(\lambda) = \begin{bmatrix} a_l(\lambda) & b_l(\lambda) \end{bmatrix}, \quad R_v(\lambda) = \begin{bmatrix} a_r(\lambda) \\ b_r(\lambda) \end{bmatrix} \tag{2.124}$$

are latent, then the fractions (2.120), (2.121) are reducible.

Proof. If the pair $(a_l(\lambda), b_l(\lambda))$ is reducible, then by virtue of the results in Section 1.12, we obtain

$$a_l(\lambda) = g(\lambda)a_{l1}(\lambda), \quad b_l(\lambda) = g(\lambda)b_{l1}(\lambda) \tag{2.125}$$

with $\operatorname{ord} g(\lambda) > 0$ and polynomial matrices $a_{l1}(\lambda)$, $b_{l1}(\lambda)$, where due to

$$\det a_l(\lambda) = \det g(\lambda) \det a_{l1}(\lambda)\,,$$

the relation

$$\deg \det a_{l1}(\lambda) < \deg \det a_l(\lambda)$$

holds. From (2.125), we gain

$$w_l(\lambda) = a_{l1}^{-1}(\lambda)b_{l1}(\lambda) = \frac{\operatorname{adj} a_{l1}(\lambda)\; b_{l1}(\lambda)}{\det a_{l1}(\lambda)}\,.$$

The denominator of this rational matrix possesses a lower degree than that of (2.122), what implies that the fraction $w_l(\lambda)$ in (2.122) is reducible. For the vertical pair $[a_r(\lambda), b_r(\lambda)]$, we carry out the proof in the same way. ∎

4. Let the matrices $a_l(\lambda)$ and $a_r(\lambda)$ be not simple. Then using (2.117), we receive the monic transfer matrix (2.123).

Theorem 2.42. *If the pairs $(a_l(\lambda), b_l(\lambda))$, $[a_r(\lambda), b_r(\lambda)]$ are irreducible, then the monic transfer matrices (2.123) are irreducible.*

Proof. Let the pair $(a_l(\lambda), b_l(\lambda))$ be irreducible. Then the matrix $R_h(\lambda)$ in (2.124) is alatent. Therefore, an arbitrary fixed $\lambda = \tilde{\lambda}$ yields

$$\operatorname{rank} R_h(\tilde{\lambda}) = \operatorname{rank}\left[\, a_l(\tilde{\lambda})\; b_l(\tilde{\lambda}) \,\right] = n. \tag{2.126}$$

Multiplying the matrix $R_h(\lambda)$ from left by the monic adjoint matrix $\widetilde{\operatorname{adj} a_l}(\lambda)$, with benefit from (2.119), we find

$$\widetilde{\operatorname{adj} a_l}(\lambda)R_h(\lambda) = \left[\, d_{a_l \min}(\lambda)I_n \;\; \widetilde{\operatorname{adj} a_l}(\lambda)b_l(\lambda) \,\right]. \tag{2.127}$$

Now, let $\lambda = \lambda_0$ be any root of the polynomial $d_{a_l \min}(\lambda)$, then due to $d_{a_l \min}(\lambda_0) = 0$ in (2.127)

$$\widetilde{\operatorname{adj} a_l}(\lambda_0)R_h(\lambda_0) = \left[\, O_{nn} \;\; \widetilde{\operatorname{adj} a_l}(\lambda_0)b_l(\lambda_0) \,\right]$$

is preserved. If we assume that the matrix $w_l(\lambda)$ in (2.123) is reducible, then for a certain root $\tilde{\lambda}_0$, we obtain

$$\widetilde{\operatorname{adj} a_l}(\tilde{\lambda}_0)b_l(\tilde{\lambda}_0) = O_{nm}$$

and therefore

$$\widetilde{\operatorname{adj} a_l}(\tilde\lambda_0) R_h(\tilde\lambda_0) = O_{n,n+m} \,. \tag{2.128}$$

But from Relations (2.115), (2.118), we know rank $[\,\widetilde{\operatorname{adj} a_l}(\tilde\lambda_0)\,] \ge 1$. Moreover, from (2.126) we get rank $R_h(\tilde\lambda_0) = n$, and owing to the Sylvester inequality (1.44), we conclude

$$\operatorname{rank}\left[\widetilde{\operatorname{adj} a_l}(\tilde\lambda_0) R_h(\tilde\lambda_0)\right] \ge 1 \,.$$

Consequently, Equation (2.128) cannot be fulfilled and therefore, the fraction $w_l(\lambda)$ in (2.123) is irreducible. The proof for the irreducibility of the matrix $w_r(\lambda)$ in (2.123) runs analogously. ∎

Remark 2.43. The reverse statement of the just proven Theorem 2.42 is in general not true, as the next example illustrates.

Example 2.44. Consider the pair $(a_l(\lambda), b_l(\lambda))$ with

$$a_l(\lambda) = \begin{bmatrix} \lambda & 0 \\ 0 & \lambda \end{bmatrix}, \quad b_l(\lambda) = \begin{bmatrix} 1 \\ 1 \end{bmatrix}. \tag{2.129}$$

In this case, we have

$$a_l^{-1}(\lambda) = \frac{\begin{bmatrix} 1 & 0 \\ 0 & 1 \end{bmatrix}}{\lambda} \,,$$

which means

$$\widetilde{\operatorname{adj} a_l}(\lambda) = \begin{bmatrix} 1 & 0 \\ 0 & 1 \end{bmatrix}, \quad d_{a_l \min}(\lambda) = \lambda \,.$$

So we arrive at

$$w_l(\lambda) = \frac{\widetilde{\operatorname{adj} a_l}(\lambda) b_l(\lambda)}{d_{a_l \min}(\lambda)} = \frac{\begin{bmatrix} 1 \\ 1 \end{bmatrix}}{\lambda} \tag{2.130}$$

and the fraction on the right side is irreducible. Nevertheless, the pair (2.129) is not irreducible, because the matrix

$$R_h(\lambda) = \begin{bmatrix} \lambda & 0 & 1 \\ 0 & \lambda & 1 \end{bmatrix}$$

for $\lambda = 0$ has only rank 1. On the other side, we immediately recognise that the pair

$$a_{l1}(\lambda) = \begin{bmatrix} \lambda & 0 \\ -1 & 1 \end{bmatrix}, \quad b_{l1}(\lambda) = \begin{bmatrix} 1 \\ 0 \end{bmatrix}$$

is an ILMFD of the transfer matrix (2.130), because the matrix

$$R_{h1}(\lambda) = \begin{bmatrix} \lambda & 0 & 1 \\ -1 & 1 & 0 \end{bmatrix}$$

possesses rank 2 for all λ. □

5.

Theorem 2.45. *For the transfer matrices (2.122) to be irreducible, it is necessary and sufficient that the pairs* $(a_l(\lambda), b_l(\lambda))$, $[a_r(\lambda), b_r(\lambda)]$ *are irreducible and the matrices* $a_l(\lambda)$, $a_r(\lambda)$ *are simple.*

Proof. The necessity follows from the above considerations.

That the condition is also sufficient, we see by assuming $a_l(\lambda)$ to be simple. Then $d_{a_l\min}(\lambda) \approx d_{a_l}(\lambda)$, $\operatorname{adj} a_l(\lambda) \approx \widetilde{\operatorname{adj} a_l}(\lambda)$ hold, and Theorem 2.42 yields that the fraction

$$w_l(\lambda) = \frac{\operatorname{adj} a_l(\lambda)\; b_l(\lambda)}{d_{a_l}(\lambda)}$$

is irreducible. For the second fraction in (2.122), the statement is proven analogously. ∎

2.11 Transfer Matrices of PMDs

1. A PMD of the dimension n, p, m

$$\tau(\lambda) = (a(\lambda), b(\lambda), c(\lambda)) \tag{2.131}$$

is called *regular*, if the matrix $a(\lambda)$ is non-singular. All descriptor systems of interest belong to the set of regular PMDs. A regular PMD (2.131) is related to a rational transfer matrix

$$w_\tau(\lambda) = c(\lambda) a^{-1}(\lambda) b(\lambda) \tag{2.132}$$

that is named the *transfer function (-matrix) of the PMD* (2.131). Using (2.108), the transfer matrix can be presented in the form

$$w_\tau(\lambda) = \frac{c(\lambda) \operatorname{adj} a(\lambda)\; b(\lambda)}{\det a(\lambda)}\,. \tag{2.133}$$

When in the general case $a(\lambda)$ is not simple, then by virtue of (2.117), we obtain

$$w_\tau(\lambda) = \frac{c(\lambda) \widetilde{\operatorname{adj} a(\lambda)}\; b(\lambda)}{d_{a\min}(\lambda)}\,. \tag{2.134}$$

The rational matrix on the right side of (2.134) is called the *monic transfer matrix of the PMD* (2.131).

2.

Theorem 2.46. *For a minimal PMD (2.131), the monic transfer matrix (2.134) is irreducible.*

Proof. Construct the ILMFD

$$c(\lambda)a^{-1}(\lambda) = a_1^{-1}(\lambda)c_1(\lambda). \tag{2.135}$$

Since the left side of (2.135) is an IRMFD,

$$\det a(\lambda) \approx \det a_1(\lambda). \tag{2.136}$$

Furthermore, also the minimal polynomials of the matrices $a(\lambda)$ and $a_1(\lambda)$ coincide, because they possess the same sequences of invariant polynomials different from one. Therefore, we have

$$d_{a\,\min}(\lambda) = d_{a_1\,\min}(\lambda). \tag{2.137}$$

Utilising (2.132) and (2.135), we can write

$$w_\tau(\lambda) = a_1^{-1}(\lambda)[c_1(\lambda)b(\lambda)]. \tag{2.138}$$

The right side of (2.138) is an ILMFD, what follows from the minimality of the PMD (2.131) and Lemma 2.9. But then, employing Lemma 2.8 yields the fraction

$$w_\tau(\lambda) = \frac{\widetilde{\operatorname{adj} a_1(\lambda)}\, c_1(\lambda)b(\lambda)}{d_{a_1\,\min}(\lambda)}$$

to be irreducible, and this implies, owing to (2.137), the irreducibility of the right side of (2.134). ∎

3.

Theorem 2.47. *For the right side of Relation (2.133) to be irreducible, it is necessary and sufficient that the PMD (2.131) is minimal and the matrix $a(\lambda)$ is simple.*

Proof. The necessity results from Lemmata 2.40 and 2.41.
Sufficiency: If the matrix $a(\lambda)$ is simple, then

$$\det a(\lambda) \approx d_{a\,\min}(\lambda),$$

and the irreducibility of the right side of (2.133) follows from Theorem 2.46. ∎

4. Let in addition to the PMD (2.131) be given a regular PMD of dimension n, q, m

$$\tilde{\tau}(\lambda) = (\tilde{a}(\lambda), \tilde{b}(\lambda), \tilde{c}(\lambda)) . \tag{2.139}$$

The PMD (2.131) and (2.139) are called *equivalent*, if their transfer functions coincide, that means

$$c(\lambda)a^{-1}(\lambda)b(\lambda) = \tilde{c}(\lambda)\tilde{a}^{-1}(\lambda)\tilde{b}(\lambda) . \tag{2.140}$$

Lemma 2.48. *Assume the PMDs (2.131) and (2.139) be equivalent and the PMD (2.131) be minimal. Then the expression*

$$\Delta(\lambda) = \frac{\det \tilde{a}(\lambda)}{\det a(\lambda)} \tag{2.141}$$

turns out to be a polynomial.

Proof. Lemma 2.8 implies the existence of the LMFD

$$\tilde{c}(\lambda)\tilde{a}^{-1}(\lambda) = a_2^{-1}(\lambda)c_2(\lambda) ,$$

where

$$\det \tilde{a}(\lambda) \approx \det a_2(\lambda) . \tag{2.142}$$

Utilising (2.140), from this we gain the LMFD of the matrix $w_\tau(\lambda)$

$$w_\tau(\lambda) = a_2^{-1}(\lambda)[c_2(\lambda)\tilde{b}(\lambda)] . \tag{2.143}$$

On the other side, the minimality of the PMD (2.131) allows to conclude that the right side of (2.138) is an ILMFD of the matrix $w_\tau(\lambda)$. Comparing (2.138) with (2.143), we obtain $a_2(\lambda) = g(\lambda)a(\lambda)$, where $g(\lambda)$ is a polynomial matrix. Therefore, the expression

$$\Delta_1(\lambda) = \frac{\det a_2(\lambda)}{\det a_1(\lambda)} = \det g(\lambda)$$

proves to be a polynomial. Taking into account (2.136) and (2.142), we realise that the right side of Equation (2.141) becomes a polynomial. Hereby, $\Delta(\lambda) \approx \Delta_1(\lambda)$ holds. ∎

Corollary 2.49. *If the PMDs (2.131) and (2.139) are equivalent and minimal, then*

$$\det a(\lambda) \approx \det \tilde{a}(\lambda) .$$

Proof. Lemma 2.48 offers under the given suppositions that

$$\frac{\det a(\lambda)}{\det \tilde{a}(\lambda)}, \quad \frac{\det \tilde{a}(\lambda)}{\det a(\lambda)}$$

are polynomials, this proves the claim. ∎

5.

Lemma 2.50. *Consider a regular PMD (2.131) and its corresponding transfer matrix (2.132). Moreover, let the ILMFD and IRMFD*

$$w_\tau(\lambda) = p_l^{-1}(\lambda)q_l(\lambda) = q_r(\lambda)p_r^{-1}(\lambda) \tag{2.144}$$

exist. Then the expressions

$$\Delta_l(\lambda) = \frac{\det a(\lambda)}{\det p_l(\lambda)}, \quad \Delta_r(\lambda) = \frac{\det a(\lambda)}{\det p_r(\lambda)} \tag{2.145}$$

turn out to be polynomials. Besides, the sets of poles of each of the matrices

$$w_1(\lambda) = p_l(\lambda)c(\lambda)a^{-1}(\lambda), \quad w_2(\lambda) = a^{-1}(\lambda)b(\lambda)p_r(\lambda)$$

are contained in the set of roots of the polynomial $\Delta_l(\lambda) \approx \Delta_r(\lambda)$.

Proof. Consider the PMDs

$$\begin{aligned} \tau_1(\lambda) &= (p_l(\lambda), q_l(\lambda), I_n)\,, \\ \tau_2(\lambda) &= (p_r(\lambda), I_m, q_r(\lambda))\,. \end{aligned} \tag{2.146}$$

Per construction, the PMDs (2.131) and (2.146) are equivalent, where the PMD (2.146) is minimal. Therefore, due to Lemma 2.48, the functions (2.145) are polynomials. Now we build the LMFD

$$c(\lambda)a^{-1}(\lambda) = a_3^{-1}(\lambda)c_3(\lambda)\,, \tag{2.147}$$

where

$$\det a_3(\lambda) \approx \det a(\lambda)\,. \tag{2.148}$$

As above, we have an LMFD of the transfer matrix $w_\tau(\lambda)$

$$w_\tau(\lambda) = a_3^{-1}(\lambda)[c_3(\lambda)b(\lambda)]\,. \tag{2.149}$$

This relation together with (2.144) determines two LMFDs of the transfer matrix $w_\tau(\lambda)$, where (2.144) is an ILMFD. Therefore,

$$a_3(\lambda) = g_l(\lambda)p_l(\lambda) \tag{2.150}$$

holds with a non-singular $n \times n$ polynomial matrix $g_l(\lambda)$. Inversion of both sides of the last equation leads to

$$a_3^{-1}(\lambda) = p_l^{-1}(\lambda)g_l^{-1}(\lambda)\,. \tag{2.151}$$

Moreover, from (2.150) through (2.148), we receive

$$\det g_l(\lambda) = \frac{\det a_3(\lambda)}{\det p_l(\lambda)} \approx \frac{\det a(\lambda)}{\det p_l(\lambda)} = \Delta_l(\lambda)\,. \tag{2.152}$$

From (2.147) and (2.151), we earn

$$p_l(\lambda)c(\lambda)a^{-1}(\lambda) = g_l^{-1}(\lambda)c_3(\lambda) = \frac{\operatorname{adj} g_l(\lambda)\; c_3(\lambda)}{\det g_l(\lambda)}\,,$$

and with the aid of (2.152), this yields the proof for a left MFD. The relation for a right MFD is proven analogously. ∎

2.12 Subordination of Rational Matrices

1. Let us have the rational $n \times m$ matrix $w(\lambda)$ and the ILMFD

$$w(\lambda) = p_l^{-1}(\lambda) q_l(\lambda) . \tag{2.153}$$

Furthermore, let the rational $n \times s$ matrix $w_1(\lambda)$ be given.

Definition 2.51. *The matrix $w_1(\lambda)$ is said to be* subordinated from left *to the matrix $w(\lambda)$, and we write*

$$w_1(\lambda) \underset{l}{\prec} w(\lambda) \tag{2.154}$$

for that, when the polynomial $p_l(\lambda)$ is a left-cancelling polynomial for $w_1(\lambda)$, i.e. the product

$$q_{l1}(\lambda) \overset{\triangle}{=} p_l(\lambda) w_1(\lambda)$$

is a polynomial.

2. In analogy, if the $n \times m$ matrix $w(\lambda)$ has an IRMFD

$$w(\lambda) = q_r(\lambda) p_r^{-1}(\lambda)$$

and the $s \times m$ matrix $w_2(\lambda)$ is of such a kind, that the product

$$q_{r1}(\lambda) \overset{\triangle}{=} w_2(\lambda) p_r(\lambda)$$

turns out as a polynomial matrix, then the matrix $w_2(\lambda)$ is said to be *subordinated from right* to the matrix $w(\lambda)$, and we denote this fact by

$$w_2(\lambda) \underset{r}{\prec} w(\lambda) . \tag{2.155}$$

3.

Lemma 2.52. *Let the right side of (2.153) define an ILMFD of the matrix $w(\lambda)$, and Condition (2.154) should be fulfilled. Let $\psi_w(\lambda)$ and $\psi_{w_1}(\lambda)$ be the McMillan denominators of the matrices $w(\lambda)$ resp. $w_1(\lambda)$. Then the fraction*

$$\Delta(\lambda) = \frac{\psi_w(\lambda)}{\psi_{w_1}(\lambda)}$$

proves to be a polynomial.

Proof. Take the ILMFD of the matrix $w_1(\lambda)$:

$$w_1(\lambda) = p_{l1}^{-1}(\lambda) q_{l1}(\lambda) .$$

Since the polynomial $p_l(\lambda)$ is left cancelling for $w_1(\lambda)$, a factorisation $p_l(\lambda) = g_1(\lambda) p_{l1}(\lambda)$ with an $n \times n$ polynomial matrix $g_1(\lambda)$ is possible. Besides, we obtain

$$\det g_1(\lambda) = \frac{\det p_l(\lambda)}{\det p_{l1}(\lambda)} \approx \frac{\psi_w(\lambda)}{\psi_{w_1}(\lambda)} = \Delta(\lambda) .$$

∎

Remark 2.53. A corresponding statement holds, when (2.155) is true.

4.

Lemma 2.54. *Assume (2.154) be valid, and* $Q(\lambda)$, $Q_1(\lambda)$ *be any polynomial matrices of appropriate dimension. Then*

$$w_1(\lambda) + Q_1(\lambda) \underset{l}{\prec} w(\lambda) + Q(\lambda) .$$

Proof. Start with the ILMFD (2.153). Then the expression

$$w(\lambda) + Q(\lambda) = p_l^{-1}(\lambda) \left[q_l(\lambda) + p_l Q_1(\lambda) \right] ,$$

due to Lemma 2.15, is also an ILMFD. Hence owing to (2.154), the product $p_l(\lambda)[w_1(\lambda) + Q_1(\lambda)]$ turns out as a polynomial, that's what the lemma claims. ∎

Remark 2.55. An analogous statement is true for subordination from right. Therefore, when the matrix $w_1(\lambda)$ is subordinated to the matrix $w(\lambda)$, then the broken part of $w_1(\lambda)$ is subordinated to the broken part of $w(\lambda)$. The reverse is also true.

5.

Theorem 2.56. *Consider the strictly proper* $n \times m$ *matrix* $w(\lambda)$, *and its minimal realisation*

$$w(\lambda) = C(\lambda I_p - A)^{-1} B . \qquad (2.156)$$

Then for holding the relation

$$w_1(\lambda) \underset{l}{\prec} w(\lambda) , \qquad (2.157)$$

where the rational $n \times s$ *matrix* $w_1(\lambda)$ *is strictly proper, it is necessary and sufficient, that there exists a constant* $p \times s$ *matrix* B_1, *which guarantees*

$$w_1(\lambda) = C(\lambda I_p - A)^{-1} B_1 .$$

Proof. Sufficiency: Build the ILMFD

$$C(\lambda I_p - A)^{-1} = a_1^{-1}(\lambda) b_1(\lambda) .$$

Then owing to Lemma 2.9, the expression

$$w(\lambda) = a_1^{-1}(\lambda)[b_1(\lambda) B] \qquad (2.158)$$

defines an ILMFD of the matrix $w(\lambda)$. Hence the product

$$a_1(\lambda) w_1(\lambda) = b_1(\lambda) B_1$$

proves to be a polynomial matrix, that's what (2.157) declares.
Necessity: Build the matrix $\tilde{w}(\lambda)$ of the form

$$\tilde{w}(\lambda) = \overset{\quad s \qquad\quad m}{\left[\, w_1(\lambda) \;\; w(\lambda) \,\right]} n\,. \tag{2.159}$$

We will show

$$\text{Mdeg}\,\tilde{w}(\lambda) = \text{Mdeg}\,w(\lambda) = p\,. \tag{2.160}$$

The equality $\text{Mdeg}\,w(\lambda) = p$ immediately follows because the realisation (2.156) is minimal. It remains to show $\text{Mdeg}\,\tilde{w}(\lambda) = \text{Mdeg}\,w(\lambda)$. For this purpose, we multiply (2.159) from left by the matrix $a_1(\lambda)$. Then taking into account (2.157) and (2.158), we realise that

$$a_1(\lambda)\tilde{w}(\lambda) = \left[\, a_1(\lambda)w_1(\lambda) \; a_1(\lambda)w(\lambda) \,\right]$$

is a polynomial matrix. Using $\text{ord}\,a_1(\lambda) = p$, Lemma 2.7 yields

$$\text{Mdeg}\,\tilde{w}(\lambda) \le p\,.$$

Now, we will prove that the inequality cannot happen. Indeed, assume $\text{Mdeg}\,\tilde{w}(\lambda) = \kappa < p$, then there exists a polynomial $\tilde{a}(\lambda)$ with $\text{ord}\,\tilde{a}(\lambda) = \deg\det\tilde{a}(\lambda) = \kappa$, and

$$\tilde{a}(\lambda)\tilde{w}(\lambda) = \left[\, \tilde{a}(\lambda)w_1(\lambda) \; \tilde{a}(\lambda)w(\lambda) \,\right]$$

becomes a polynomial matrix. When this happens, also $\tilde{a}(\lambda)w(\lambda)$ becomes a polynomial matrix, and regarding to Lemma 2.7, we have $\text{Mdeg}\,w(\lambda) \le \kappa < p$. But this is impossible, due to our supposition $\text{Mdeg}\,w(\lambda) = p$, and the correctness of (2.160) is proven.

Since the matrix $\tilde{w}(\lambda)$ is strictly proper and $\text{Mdeg}\,\tilde{w}(\lambda) = p$ holds, there exists a minimal realisation

$$\tilde{w}(\lambda) = \tilde{C}(\lambda I_p - \tilde{A})^{-1}\tilde{B} \tag{2.161}$$

with constant $p \times p$, $n \times p$ and $p \times (s+m)$ matrices $\tilde{A}$, $\tilde{C}$, $\tilde{B}$. Bring the matrix $\tilde{B}$ into the form

$$\tilde{B} = \overset{\;s \;\; m}{\left[\, \tilde{B}_1 \; \tilde{B}_2 \,\right]} p\,.$$

Then from (2.161), we gain

$$\tilde{w}(\lambda) = \left[\, \tilde{C}(\lambda I_p - \tilde{A})^{-1}\tilde{B}_1 \; \tilde{C}(\lambda I_p - \tilde{A})^{-1}\tilde{B}_2 \,\right].$$

When we relate this and (2.159), we find

$$w_1(\lambda) = \tilde{C}(\lambda I_p - \tilde{A})^{-1}\tilde{B}_1\,, \tag{2.162}$$

$$w(\lambda) = \tilde{C}(\lambda I_p - \tilde{A})^{-1}\tilde{B}_2\,. \tag{2.163}$$

Expressions (2.156) and (2.163) define realisations of the matrix $w(\lambda)$ of the same dimension p. However, since realisation (2.156) is minimal, also realisation (2.163) has to be minimal. According to (2.88), we can find a non-singular matrix R with

$$\tilde{A} = RAR^{-1}, \quad \tilde{B}_2 = RB, \quad \tilde{C} = CR^{-1}.$$

From this and (2.162), it follows

$$w_1(\lambda) = C(\lambda I_p - A)^{-1}B_1, \quad B_1 = R^{-1}\tilde{B}_1,$$

and the theorem is proven. ■

Remark 2.57. A corresponding theorem can be proven for subordination from right.

Theorem 2.58. *Let (2.156) and the rational $q \times m$ matrix $w_1(\lambda)$ be given. Then for holding the relation*

$$w_1(\lambda) \underset{r}{\prec} w(\lambda),$$

it is necessary and sufficient, that there exists a constant $q \times p$ matrix C_1 with

$$w_1(\lambda) = C_1(\lambda I_p - A)^{-1}B.$$

■

6.

Theorem 2.59. *Consider the rational matrices*

$$F(\lambda), \quad G(\lambda), \quad H(\lambda) = F(\lambda)G(\lambda) \tag{2.164}$$

and the ILMFD

$$F(\lambda) = a_1^{-1}(\lambda)b_1(\lambda), \quad G(\lambda) = a_2^{-1}(\lambda)b_2(\lambda). \tag{2.165}$$

Furthermore, let us have the ILMFD

$$b_1(\lambda)a_2^{-1}(\lambda) = a_3^{-1}(\lambda)b_3(\lambda). \tag{2.166}$$

Then the relation

$$F(\lambda) \underset{l}{\prec} H(\lambda) \tag{2.167}$$

is true, if and only if the matrix

$$R_h(\lambda) = \begin{bmatrix} a_3(\lambda)a_1(\lambda) & b_3(\lambda)b_2(\lambda) \end{bmatrix} \tag{2.168}$$

is alatent, i.e. the pair $(a_3(\lambda)a_1(\lambda), b_3(\lambda)b_2(\lambda))$ is irreducible.

Proof. Sufficiency: Start with the ILMFD

$$H(\lambda) = a^{-1}(\lambda)b(\lambda). \tag{2.169}$$

Then from (2.165) and (2.166), we obtain

$$H(\lambda) = a_1^{-1}(\lambda)b_1(\lambda)a_2^{-1}(\lambda)b_2(\lambda) = a_1^{-1}(\lambda)a_3^{-1}(\lambda)b_3(\lambda)b_2(\lambda)$$
$$= [a_3(\lambda)a_1(\lambda)]^{-1}b_3(\lambda)b_2(\lambda)\,. \tag{2.170}$$

Let Matrix (2.168) be alatent. Then the right side of (2.170) is an ILMFD and with the aid of (2.169), we get $a(\lambda) = g(\lambda)a_3(\lambda)a_1(\lambda)$, where $g(\lambda)$ is a unimodular matrix. Besides

$$a(\lambda)F(\lambda) = g(\lambda)a_3(\lambda)b_1(\lambda)$$

is a polynomial matrix, and hence (2.167) is true.
Necessity: Assume (2.167), then we have

$$a(\lambda) = h(\lambda)a_1(\lambda)\,, \tag{2.171}$$

where $h(\lambda)$ is a non-singular polynomial matrix. This relation leads us to

$$H(\lambda) = a^{-1}(\lambda)b(\lambda) = a_1^{-1}(\lambda)h^{-1}(\lambda)b(\lambda)\,. \tag{2.172}$$

Comparing the expressions for $H(\lambda)$ in (2.170) and (2.172), we find

$$a_3^{-1}(\lambda)b_3(\lambda)b_2(\lambda) = h^{-1}(\lambda)b(\lambda)\,. \tag{2.173}$$

But the matrix $\begin{bmatrix} a_3(\lambda) & b_3(\lambda)b_2(\lambda) \end{bmatrix}$ due to Lemma 2.9 is alatent, and the matrix $\begin{bmatrix} h(\lambda) & b(\lambda) \end{bmatrix}$ with respect to (2.171) and owing to Lemma 2.11 is alatent. Therefore, the left as well as the right side of (2.173) present ILMFDs of the same rational matrix. Then from Statement 2.3 on page 64 arise

$$h(\lambda) = \varphi(\lambda)a_3(\lambda), \quad b(\lambda) = \varphi(\lambda)b_3(\lambda)b_2(\lambda)\,, \tag{2.174}$$

where the matrix $\varphi(\lambda)$ is unimodular. Applying (2.172) and (2.174), we arrive at the ILMFD

$$H(\lambda) = [\varphi(\lambda)a_3(\lambda)a_1(\lambda)]^{-1}b(\lambda)\,.$$

This expression and (2.170) define two LMFDs of the same matrix $H(\lambda)$. Since the matrix $\varphi(\lambda)$ is unimodular, we have

$$\operatorname{ord}[\varphi(\lambda)a_3(\lambda)a_1(\lambda)] = \operatorname{ord}[a_3(\lambda)a_1(\lambda)]\,,$$

and the right side of (2.170) is an ILMFD too. Therefore, Matrix (2.168) is alatent. ∎

A corresponding statement holds for subordination from right.

Theorem 2.60. *Consider the rational matrices (2.164) and the IRMFDs*

$$F(\lambda) = \tilde b_1(\lambda)\tilde a_1^{-1}(\lambda), \quad G(\lambda) = \tilde b_2(\lambda)\tilde a_2^{-1}(\lambda)\,.$$

Moreover, let the IRMFD

$$\tilde a_1^{-1}(\lambda)\tilde b_2(\lambda) = \tilde b_3(\lambda)\tilde a_3^{-1}(\lambda)$$

be given. Then the relation

$$G(\lambda) \underset{r}{\prec} H(\lambda)$$

is true, if and only if the pair $[\tilde a_2(\lambda)\tilde a_3(\lambda), \tilde b_1(\lambda)\tilde b_3(\lambda)]$ *is irreducible.* ∎

7. The following theorem states an important special case, where the conditions of Theorems 2.59 and 2.60 are fulfilled.

Theorem 2.61. *If for the rational $n \times p$ and $p \times m$ matrices $F(\lambda)$ and $G(\lambda)$ the relation*

$$\operatorname{Mdeg}[F(\lambda)G(\lambda)] = \operatorname{Mdeg} F(\lambda) + \operatorname{Mdeg} G(\lambda) \tag{2.175}$$

holds, i.e. the matrices $F(\lambda)$ and $G(\lambda)$ are independent, then the relations

$$F(\lambda) \underset{l}{\prec} F(\lambda)G(\lambda), \quad G(\lambda) \underset{r}{\prec} F(\lambda)G(\lambda) \tag{2.176}$$

take place.

Proof. Let us have the ILMFD (2.165), then $\operatorname{Mdeg} F(\lambda) = \operatorname{ord} a_1(\lambda)$ and $\operatorname{Mdeg} G(\lambda) = \operatorname{ord} a_2(\lambda)$. Besides, the pair $[a_2(\lambda), b_1(\lambda)]$ is irreducible, that can be seen by assuming the contrary. In case of $\operatorname{ord} a_3(\lambda) < \operatorname{ord} a_2(\lambda)$ in (2.166), we would obtain from (2.170)

$$\operatorname{Mdeg} H(\lambda) \le \operatorname{ord} a_1(\lambda) + \operatorname{ord} a_3(\lambda) < \operatorname{ord} a_1(\lambda) + \operatorname{ord} a_2(\lambda)$$

which contradicts (2.175). The irreducibility of the pairs $[a_2(\lambda), b_1(\lambda)]$ and (2.175) implies that the right part of (2.170) is an ILMFD. Owing to Theorem 2.59, the first relation in (2.176) is shown. The second relation in (2.137) is seen analogously. ∎

8.

Remark 2.62. Under the conditions of Theorem 2.59 using (2.171) and (2.174), we obtain

$$a(\lambda)F(\lambda) = \varphi(\lambda)a_3(\lambda)b_1(\lambda)\,,$$

that means, the factor $\varphi(\lambda)a_3(\lambda)$ is a left divisor of the polynomial matrix $a(\lambda)F(\lambda)$. Analogously, we conclude from the conditions of Theorem 2.60, when the IRMFD $H(\lambda) = \tilde{b}(\lambda)\tilde{a}^{-1}(\lambda)$ is present, that

$$G(\lambda)\tilde{a}(\lambda) = \tilde{b}_2(\lambda)\tilde{a}_3(\lambda)\psi(\lambda)$$

takes place with a unimodular matrix $\psi(\lambda)$. We learn from this equation that under the conditions of Theorem 2.60, the polynomial matrix $\tilde{a}_3(\lambda)\psi(\lambda)$ is a right divisor of the polynomial matrix $G(\lambda)\tilde{a}(\lambda)$.

2.13 Dominance of Rational Matrices

1. Consider the rational block matrix

$$w(\lambda) = \begin{bmatrix} w_{11}(\lambda) & \dots & w_{1m}(\lambda) \\ \vdots & \vdots & \vdots \\ w_{n1}(\lambda) & \dots & w_{nm}(\lambda) \end{bmatrix}, \tag{2.177}$$

where the $w_{ik}(\lambda)$ are rational matrices of appropriate dimensions. Let $\psi(\lambda)$ and $\psi_{ik}(\lambda)$, $(i = 1, \dots, n;\ k = 1, \dots, m)$ be the McMillan denominators of the matrix $w(\lambda)$ resp. of its blocks $w_{ik}(\lambda)$. Hereinafter, we abbreviate McMillan denominator by MMD.

Lemma 2.63. *All expressions*

$$d_{ik}(\lambda) = \frac{\psi(\lambda)}{\psi_{ik}(\lambda)} \tag{2.178}$$

turn out to be polynomials.

Proof. At first assume only a block row

$$w(\lambda) \stackrel{\triangle}{=} w^z(\lambda) = \begin{bmatrix} w_1(\lambda) \dots w_m(\lambda) \end{bmatrix}, \tag{2.179}$$

and we should have an ILMFD

$$w^z(\lambda) = a^{-1}(\lambda)b(\lambda)$$

for it. Then per construction

$$\det a(\lambda) \approx \psi^z(\lambda),$$

where $\psi^z(\lambda)$ is the MMD of the row (2.179). Besides, the polynomial $a(\lambda)$ is canceling from left for all matrices $w_i(\lambda)$, $(i = 1, \dots, m)$, that means

$$w_i(\lambda) \underset{l}{\prec} w^z(\lambda), \quad (i = 1, \dots, m).$$

Therefore, the relations

$$d_i^z(\lambda) \stackrel{\triangle}{=} \frac{\psi^z(\lambda)}{\psi_i(\lambda)}, \quad (i = 1, \dots, m)$$

where the $\psi_i(\lambda)$ are the MMDs of the matrices $w_i(\lambda)$, owing to Lemma 2.52, become polynomials. In the same way can be seen that for a block column

$$w(\lambda) \stackrel{\triangle}{=} w^s(\lambda) = \begin{bmatrix} \tilde{w}_1(\lambda) \\ \vdots \\ \tilde{w}_n(\lambda) \end{bmatrix} \tag{2.180}$$

the expressions

$$d_k^s(\lambda) = \frac{\tilde{\psi}^s(\lambda)}{\tilde{\psi}_k(\lambda)}, \quad (k = 1, \dots, n)$$

become polynomials, where $\tilde{\psi}^s(\lambda)$, $\tilde{\psi}_k(\lambda)$ are the MMDs of the column (2.180) and of its elements.

Denote by $w_i^z(\lambda)$, $(i = 1, \dots, n)$, $\tilde{w}_k^s(\lambda)$, $(k = 1, \dots, m)$ all rows and columns of Matrix (2.177), and by $\psi_i^z(\lambda)$, $(i = 1, \dots, n)$, $\tilde{\psi}_k^s(\lambda)$, $(k = 1, \dots, m)$ their MMDs. With respect to the above shown, the relations

$$\frac{\psi(\lambda)}{\psi_i^z(\lambda)}, \quad (i = 1, \dots, n); \qquad \frac{\psi(\lambda)}{\tilde{\psi}_k^s(\lambda)}, \quad (k = 1, \dots, m)$$

are polynomials. Therefore, all relations

$$d_{ik}(\lambda) = \frac{\psi(\lambda)}{\psi_{ik}(\lambda)} = \frac{\psi(\lambda)}{\psi_i^z(\lambda)} \frac{\psi_i^z(\lambda)}{\psi_{ik}(\lambda)}$$

become polynomials. ∎

Corollary 2.64. *For any Matrix (2.177), the inequalities*

$$\operatorname{Mdeg} w(\lambda) \geq \operatorname{Mdeg} w_{ik}(\lambda), \quad (i = 1, \dots, n; \ k = 1, \dots, m)$$

are true. ∎

2. The element $w_{ik}(\lambda)$ in Matrix (2.177) is said to be *dominant*, if the equality

$$\psi(\lambda) = \psi_{ik}(\lambda)$$

takes place.

Lemma 2.65. *The element $w_{ik}(\lambda)$ is dominant in Matrix (2.177), if and only if*

$$\operatorname{Mdeg} w(\lambda) = \operatorname{Mdeg} w_{ik}(\lambda) .$$

Proof. The necessity of this condition is obvious. That it is also sufficient follows from the fact that expression (2.178) is a polynomial, and from the equations

$$\operatorname{Mdeg} w(\lambda) = \deg \psi(\lambda), \quad \operatorname{Mdeg} w_{ik}(\lambda) = \deg \psi_{ik}(\lambda) .$$ ∎

3.

Theorem 2.66. *A necessary and sufficient condition for the matrix $w_2(\lambda)$ to be dominant in the block row*

$$w(\lambda) = \begin{bmatrix} w_1(\lambda) & w_2(\lambda) \end{bmatrix}$$

is that it meets the relation

$$w_1(\lambda) \underset{l}{\prec} w_2(\lambda) .$$

A necessary and sufficient condition for the matrix $w_2(\lambda)$ *to be dominant in the block column*

$$w(\lambda) = \begin{bmatrix} w_1(\lambda) \\ w_2(\lambda) \end{bmatrix}$$

is that it meets the relation

$$w_1(\lambda) \underset{r}{\prec} w_2(\lambda) .$$

Proof. The proof immediately arises from the proof of Theorem 2.56. ∎

4.

Theorem 2.67. *Consider the strictly proper rational block matrix* $G(\lambda)$ *of the shape*

$$G(\lambda) = \overset{\quad\;\; \ell \qquad\;\; m}{\begin{bmatrix} K(\lambda) & L(\lambda) \\ M(\lambda) & N(\lambda) \end{bmatrix}} \begin{matrix} i \\ n \end{matrix} \tag{2.181}$$

and the minimal realisation

$$N(\lambda) = C(\lambda I_p - A)^{-1} B . \tag{2.182}$$

Then the matrix $N(\lambda)$ *is dominant in* $G(\lambda)$*, i.e.*

$$\operatorname{Mdeg} N(\lambda) = \operatorname{Mdeg} G(\lambda) = p , \tag{2.183}$$

if and only if there exist constant $i \times p$ *and* $p \times \ell$ *matrices* C_1 *and* B_1 *with*

$$\begin{aligned} K(\lambda) &= C_1(\lambda I_p - A)^{-1} B_1 , \\ L(\lambda) &= C_1(\lambda I_p - A)^{-1} B , \\ M(\lambda) &= C(\lambda I_p - A)^{-1} B_1 . \end{aligned} \tag{2.184}$$

Proof. Necessity: Let (2.183) be valid. Since the matrix $G(\lambda)$ is strictly proper, there exists a minimal realisation

$$G(\lambda) = \tilde{C}(\lambda I_p - \tilde{A})^{-1} \tilde{B} \tag{2.185}$$

with constant matrices $\tilde{A}$, $\tilde{B}$ and $\tilde{C}$ of the dimensions $p \times p$, $p \times (\ell + m)$ and $(i + n) \times m$, respectively. Assume

$$\tilde{C} = \overset{p}{\begin{bmatrix} \tilde{C}_1 \\ \tilde{C}_2 \end{bmatrix}} \begin{matrix} i \\ n \end{matrix} \qquad \tilde{B} = \overset{\;\; \ell \quad m}{\begin{bmatrix} \tilde{B}_1 & \tilde{B}_2 \end{bmatrix}} \; p$$

and substitute this expression in (2.185), then we obtain

$$G(\lambda) = \begin{bmatrix} \tilde{C}_1(\lambda I_p - \tilde{A})^{-1}\tilde{B}_1 & \tilde{C}_1(\lambda I_p - \tilde{A})^{-1}\tilde{B}_2 \\ \tilde{C}_2(\lambda I_p - \tilde{A})^{-1}\tilde{B}_1 & \tilde{C}_2(\lambda I_p - \tilde{A})^{-1}\tilde{B}_2 \end{bmatrix}. \tag{2.186}$$

Relating (2.181) to (2.186), we find

$$N(\lambda) = \tilde{C}_2(\lambda I_p - \tilde{A})^{-1}\tilde{B}_2\,. \tag{2.187}$$

Both Equations (2.182) and (2.187) are realisations of $N(\lambda)$ and possess the same dimension. Since (2.182) is a minimal realisation, so (2.187) has to be minimal too. Therefore, Relations (2.88) can be used that will lead us to

$$\tilde{A} = RAR^{-1}, \quad \tilde{B}_2 = RB, \quad \tilde{C}_2 = CR^{-1}$$

with a non-singular matrix R. Inserting this into (2.186), Relation (2.184) is achieved with

$$C_1 = \tilde{C}_1 R\,, \quad B_1 = \tilde{R}^{-1}\tilde{B}_1\,,$$

that proves the necessity of the conditions of the theorem.
Sufficiency: Suppose Conditions (2.182), (2.184) to be true. Then,

$$G(\lambda) = \begin{bmatrix} C_1 \\ C \end{bmatrix} (\lambda I_p - A)^{-1} \begin{bmatrix} B_1 & B \end{bmatrix}$$

holds. Consider the matrix

$$G_2(\lambda) = \begin{bmatrix} M(\lambda) & N(\lambda) \end{bmatrix} = C(\lambda I_p - A)^{-1} \begin{bmatrix} B_1 & B \end{bmatrix}. \tag{2.188}$$

The realisation on the right side of (2.188) is minimal, because of $\operatorname{Mdeg} N(\lambda) = p$. Therefore, the pair $(\lambda I_p - A, \begin{bmatrix} B_1 & B \end{bmatrix})$ is irreducible, and $\operatorname{Mdeg} G_2(\lambda) = p$.

Utilising (2.181) and (2.188), the matrix $G(\lambda)$ can be written in the form

$$G(\lambda) = \begin{bmatrix} G_1(\lambda) \\ G_2(\lambda) \end{bmatrix}$$

with

$$G_1(\lambda) = C_1(\lambda I_p - A)^{-1} \begin{bmatrix} B_1 & B \end{bmatrix},$$

where the pair $[\lambda I_p - A, C_1]$ is, roughly said, non-irreducible. Therefore, according to Theorem 2.58, we obtain

$$G_1(\lambda) \underset{r}{\prec} G_2(\lambda)\,,$$

and with account of Theorem 2.66

$$\operatorname{Mdeg} G(\lambda) = \operatorname{Mdeg} G_2(\lambda) = p\,.$$

■

Corollary 2.68. *Under Conditions (2.181)–(2.184), the relations*

$$K(\lambda) \underset{l}{\prec} L(\lambda), \quad K(\lambda) \underset{r}{\prec} M(\lambda) \tag{2.189}$$

are true.

Proof. Assume the ILMFD

$$C_1(\lambda I_p - A)^{-1} = \alpha_l^{-1}(\lambda)\beta_l(\lambda)\,.$$

Owing to Lemma 2.9, the right side of

$$C_1(\lambda I_p - A)^{-1}B_2 = \alpha_l^{-1}(\lambda)[\beta_l(\lambda)B_2]$$

is an ILMFD of the matrix $L(\lambda)$. Therefore, the product

$$\alpha_l(\lambda)K(\lambda) = \beta_l(\lambda)B_1$$

becomes a polynomial matrix. Herewith the first relation in (2.189) is shown. The second part can be proven analogously. ∎

3

Normal Rational Matrices

3.1 Normal Rational Matrices

1. Consider the rational $n \times m$ matrix $A(\lambda)$ in the standard form (2.21)

$$A(\lambda) = \frac{N(\lambda)}{d(\lambda)}, \quad \deg d(\lambda) = p \tag{3.1}$$

and, furthermore, let be given certain ILMFD and IRMFD

$$A(\lambda) = a_l^{-1}(\lambda)b_l(\lambda) = b_r(\lambda)a_r^{-1}(\lambda)\,, \tag{3.2}$$

where, due to the irreducibility of the MFD, we have

$$\operatorname{ord} a_l(\lambda) = \operatorname{ord} a_r(\lambda) = \operatorname{Mdeg} A(\lambda)\,,$$

and $\operatorname{Mdeg} A(\lambda)$ is the degree of the McMillan denominator of the matrix $A(\lambda)$. At first, Relation (2.34) implies $\operatorname{Mdeg} A(\lambda) \geq p$. In the following disclosure, matrices will play an important role for which

$$\operatorname{Mdeg} A(\lambda) = p\,. \tag{3.3}$$

Since $\det a_l(\lambda) \approx \det a_r(\lambda)$ is valid and both polynomials are divisible by $d(\lambda)$, Relation (3.3) is equivalent to

$$d(\lambda) = \psi_A(\lambda)\,, \tag{3.4}$$

where $\psi_A(\lambda)$ is the McMillan denominator of $A(\lambda)$. Further on, rational matrices satisfying (3.3), (3.4) will be called *normal matrices* .

2. For a normal matrix (3.1), it is possible to build IMFDs (3.2), such that

$$\det a_l(\lambda) \approx d(\lambda), \quad \det a_r(\lambda) \approx d(\lambda)\,.$$

If both ILMFD and IRMFD satisfy such conditions, the pair is called a *complete MFD*. Thus, normal rational matrices are rational matrices that possess a complete MFD.

It is emphasised that a complete MFD is always irreducible. Indeed, from (2.34) is seen that for any matrix $A(\lambda)$ in form (3.1) it always follows $\deg \psi_A(\lambda) \geq \deg d(\lambda)$. Therefore, if we have any matrix $A(\lambda)$ satisfying (3.3), then the polynomials $\det a_l(\lambda)$ and $\det a_r(\lambda)$ possess the minimal possible degree, and hence the complete MFD is irreducible.

3. A general characterisation of the set of normal rational matrices yields the next theorem.

Theorem 3.1. *Let in (3.1) be* $\min(n,m) \geq 2$. *Then for the fact that the irreducible rational matrix (3.1) becomes normal, it is necessary and sufficient, that every minor of second order of the polynomial matrix $N(\lambda)$ is divisible without remainder by the denominator $d(\lambda)$. Moreover, if Relations (3.2) define a complete MFD, then the matrices $a_l(\lambda)$ and $a_r(\lambda)$ are simple.*

Proof. Necessity: To consider a concrete case, assume a left MFD. Let

$$A(\lambda) = \frac{N(\lambda)}{d(\lambda)} = a_l^{-1}(\lambda) b_l(\lambda)$$

be part of a complete MFD. Then the matrix $a_l(\lambda)$ is simple, because from (2.34) it follows that Equations (3.3), (3.4) can be fulfilled only for $\psi_2(\lambda) = \psi_3(\lambda) = \ldots = \psi_\rho(\lambda) = 1$. Therefore, (2.39) delivers the representation

$$a_l(\lambda) = \mu(\lambda) \operatorname{diag}\{1, \ldots, 1, d(\lambda)\} \nu(\lambda) \tag{3.5}$$

with unimodular matrices $\mu(\lambda)$, $\nu(\lambda)$. We take from (3.5), that the matrix $a_l(\lambda)$ is simple, and furthermore from (3.5), we obtain

$$a_l^{-1}(\lambda) = \frac{\nu^{-1}(\lambda) \operatorname{diag}\{d(\lambda), \ldots, d(\lambda), 1\} \mu^{-1}(\lambda)}{d(\lambda)} \stackrel{\triangle}{=} \frac{Q(\lambda)}{d(\lambda)}\,. \tag{3.6}$$

All minors of second order of the matrix $\operatorname{diag}\{d(\lambda), \ldots, d(\lambda), 1\}$ are divisible by $d(\lambda)$. Thus, by the Binet-Cauchy theorem this property passes to the numerator of the fraction on the right side of (3.6). But, due to the Binet-Cauchy theorem, the matrix $N(\lambda) = Q(\lambda) b_l(\lambda)$ possesses the shown property. Hence the necessity of the condition of the theorem is proven.

Sufficiency: Assume that all minors of second order of the matrix $N(\lambda)$ are divisible by $d(\lambda)$. Then we learn from (1.40)–(1.42), that this matrix can be presented in the form

$$N(\lambda) = p(\lambda) S_N(\lambda) q(\lambda)\,, \tag{3.7}$$

where $p(\lambda)$, $q(\lambda)$ are unimodular, and the matrix $S_N(\lambda)$ has the appropriate Smith canonical form. Thus, from (1.49) we receive

$$S_N(\lambda) = \begin{bmatrix} \begin{matrix} g_1(\lambda) & 0 & \dots & 0 \\ 0 & g_1(\lambda)g_2(\lambda)d(\lambda) & \dots & 0 \\ \vdots & \vdots & \ddots & \vdots \\ 0 & 0 & \dots & g_1(\lambda)\cdots g_\rho(\lambda)d(\lambda) \end{matrix} & O_{\rho,m-\rho} \\ O_{n-\rho,\rho} & O_{n-\rho,m-\rho} \end{bmatrix}, \qquad (3.8)$$

where the polynomial $g_1(\lambda)$ and the denominator $d(\lambda)$ are coprime, because in the contrary the fraction (3.1) would be reducible. According to (3.7) and (3.8), the matrix $A(\lambda)$ of (3.1) can be written in the shape

$$A(\lambda) = p(\lambda) \begin{bmatrix} \begin{matrix} \frac{g_1(\lambda)}{d(\lambda)} & 0 & \dots & 0 \\ 0 & g_1(\lambda)g_2(\lambda) & \dots & 0 \\ \vdots & \vdots & \ddots & \vdots \\ 0 & 0 & \dots & g_1(\lambda)\cdots g_\rho(\lambda) \end{matrix} & O_{\rho,m-\rho} \\ O_{n-\rho,\rho} & O_{n-\rho,m-\rho} \end{bmatrix} q(\lambda),$$

where the fraction $\dfrac{g_1(\lambda)}{d(\lambda)}$ is irreducible. Therefore, choosing

$$a_l(\lambda) = \operatorname{diag}\{d(\lambda), 1, \dots, 1\} p^{-1}(\lambda)$$

$$b_l(\lambda) = \begin{bmatrix} \begin{matrix} g_1(\lambda) & 0 & \dots & 0 \\ 0 & g_1(\lambda)g_2(\lambda) & \dots & 0 \\ \vdots & \vdots & \ddots & \vdots \\ 0 & 0 & \dots & g_1(\lambda)\cdots g_\rho(\lambda) \end{matrix} & O_{\rho,m-\rho} \\ O_{n-\rho,\rho} & O_{n-\rho,m-\rho} \end{bmatrix} q(\lambda),$$

we obtain the LMFD

$$A(\lambda) = a_l^{-1}(\lambda) b_l(\lambda),$$

which is complete, because $\det a_l(\lambda) \approx d(\lambda)$ is true. ∎

Corollary 3.2. *It follows from (3.5), (3.6) that for a simple $n \times n$ matrix $a(\lambda)$, the rational matrix $a^{-1}(\lambda)$ is normal, and* vice versa. ∎

Corollary 3.3. *From Equations (3.7), (3.8) we learn that for $k \geq 2$, all minors of k-th order of the numerator of a normal matrix $N(\lambda)$ are divisible by $d^{k-1}(\lambda)$.* ∎

Remark 3.4. Irreducible rational matrix rows or columns are always normal. Let for instance the column

$$A(\lambda) = \frac{1}{d(\lambda)} \begin{bmatrix} a_1(\lambda) \\ \vdots \\ a_n(\lambda) \end{bmatrix} \qquad (3.9)$$

with polynomials $a_i(\lambda)$, $(i = 1, \dots, n)$ be given. Then by applying left elementary operations, $A(\lambda)$ can be brought into the form

$$A(\lambda) = \frac{1}{d(\lambda)}\, c(\lambda) \begin{bmatrix} \ell(\lambda) \\ \vdots \\ 0 \end{bmatrix},$$

where $c(\lambda)$ is a unimodular $n \times n$ matrix, and $\ell(\lambda)$ is the GCD of the polynomials $a_1(\lambda), \dots, a_n(\lambda)$. The polynomials $\ell(\lambda)$ and $d(\lambda)$ are coprime, because in other case the rational matrix (3.9) could be cancelled. Choose

$$a_l(\lambda) = \operatorname{diag}\{d(\lambda), 1, \dots, 1\} c^{-1}(\lambda), \quad b_l(\lambda) = \begin{bmatrix} \ell(\lambda) \\ \vdots \\ 0 \end{bmatrix},$$

then obviously we have

$$A(\lambda) = a_l^{-1}(\lambda) b_l(\lambda),$$

and this LMFD is complete, because of $\det a_l(\lambda) \approx d(\lambda)$.

4. A general criterion for calculating the normality of the rational matrix (3.1) directly from its elements yields the following theorem.

Theorem 3.5. *Let the fraction (3.1) be irreducible, and furthermore*

$$d(\lambda) = (\lambda - \lambda_1)^{\mu_1} \cdots (\lambda - \lambda_q)^{\mu_q}, \quad \mu_1 + \ldots + \mu_q = p. \tag{3.10}$$

Then a necessary and sufficient condition for the matrix $A(\lambda)$ to be normal is the fact that each of its minors of second order possess poles in the points $\lambda = \lambda_i$ $(i = 1, \dots, q)$ with multiplicity not higher than μ_i.

Proof. Necessity: Let $N(\lambda) = \left[n_{ij}(\lambda) \right]$ and

$$A\begin{pmatrix} i & j \\ k & \ell \end{pmatrix} = \det \begin{bmatrix} \dfrac{n_{ik}(\lambda)}{d(\lambda)} & \dfrac{n_{i\ell}(\lambda)}{d(\lambda)} \\ \dfrac{n_{jk}(\lambda)}{d(\lambda)} & \dfrac{n_{j\ell}(\lambda)}{d(\lambda)} \end{bmatrix} \tag{3.11}$$

be a minor of the matrix $A(\lambda)$ that is generated by the elements of the rows with numbers i, j and columns with numbers k, ℓ. Obviously

$$A\begin{pmatrix} i & j \\ k & \ell \end{pmatrix} = \frac{n_{ik}(\lambda) n_{j\ell}(\lambda) - n_{jk}(\lambda) n_{i\ell}(\lambda)}{d^2(\lambda)} \tag{3.12}$$

is true. If the matrix $A(\lambda)$ is normal, then, due to Theorem 3.1, the numerator of the last fraction is divisible by $d(\lambda)$. Thus we have

$$A\begin{pmatrix} i & j \\ k & \ell \end{pmatrix} = \frac{a^{ij}_{k\ell}(\lambda)}{d(\lambda)}, \tag{3.13}$$

where $a^{ij}_{k\ell}(\lambda)$ is a certain polynomial. It is seen from (3.13) and (3.10) that the minor (3.11) possess in $\lambda = \lambda_i$ poles of order μ_i or lower.

Sufficiency: Conversely, if for every minor (3.11) the representation (3.13) is correct, then the numerator of each fraction (3.12) is divisible by $d(\lambda)$, that means, every minor of second order of the matrix $N(\lambda)$ is divisible by $d(\lambda)$, or in other words, the matrix $A(\lambda)$ is normal. ∎

5.

Theorem 3.6. *If the matrix $A(\lambda)$ (3.1) is normal, and (3.10) is assumed, then*

$$\operatorname{rank} N(\lambda_i) = 1, \quad (i = 1, \ldots, q). \tag{3.14}$$

Thereby, if the polynomial (3.10) has only single roots, i.e. $q = p$, $\mu_1 = \mu_2 = \ldots = \mu_p = 1$, then Condition (3.14) is not only necessary but also sufficient for the normality the matrix $A(\lambda)$.

Proof. Equation (3.6) implies $\operatorname{rank} Q(\lambda_i) = 1$, $(i = 1, \ldots, q)$. Therefore, the matrix

$$N(\lambda_i) = Q(\lambda_i) b_l(\lambda_i)$$

is either the zero matrix or it has rank 1. The first possibility is excluded, otherwise the fraction (3.1) would have been reducible. Hence we get (3.14). If all roots λ_i are simple, and (3.14) holds, then every minor of second order of the matrix $N(\lambda)$ is divisible by $(\lambda - \lambda_i)$, $(i = 1, \ldots, q)$. Since in the present case $d(\lambda) = (\lambda - \lambda_1)(\lambda - \lambda_2) \cdots (\lambda - \lambda_p)$ is true, so every minor of second order of $N(\lambda)$ is divisible by $d(\lambda)$, it means, that the matrix $A(\lambda)$ is normal. ∎

6. We learn from Theorems 3.1–3.6 that the elements of a normal matrix $A(\lambda)$ are constrained by a number of strict equations that consist between them, which ensure that all minors of second order are divisible by the denominator. Even small deviations, sometimes only in one element, cause that these equations are violated, and the matrix $A(\lambda)$ is no longer normal, with the consequence that the order of the McMillan denominator of this matrix grows abruptly. As a whole, this leads to incorrect solutions during the construction of the IMFD and the corresponding realisations in state space. The above said gives evidence of the *structural instability* of normal matrices, and from that we conclude immediately the instability of the numeric operations with such matrices. On the other side, it is shown below, that in practical problems, the frequency domain models for real objects are described essentially by normal transfer matrices. Therefore, the methods for practical solution of control problems have to be supplied by additional tools, which help to overcome the mentioned structural and numeric instabilities to reach correct results.

Example 3.7. Consider the rational matrix

$$A(\lambda) = \frac{N(\lambda)}{d(\lambda)}$$

with

$$N(\lambda) = \begin{bmatrix} \lambda - 1 & 1 \\ \epsilon & \lambda - 2 \end{bmatrix}, \quad d(\lambda) = (\lambda - 1)(\lambda - 2),$$

where ϵ is a constant. Due to $\det N(\lambda) = (\lambda - 1)(\lambda - 2) - \epsilon$, the matrix $A(\lambda)$ proves to be normal if and only if $\epsilon = 0$. It is easily checked that

$$\begin{bmatrix} \lambda - 1 & 1 \\ \epsilon & \lambda - 2 \end{bmatrix} = \begin{bmatrix} 1 & 0 \\ \lambda - 2 & -1 \end{bmatrix} \begin{bmatrix} 1 & 0 \\ 0 & \lambda^2 - 3\lambda + 2 - \epsilon \end{bmatrix} \begin{bmatrix} \lambda - 1 & 1 \\ 1 & 0 \end{bmatrix},$$

where the first and last matrix on the right side are unimodular. Thus for $\epsilon \neq 0$, the matrix $A(\lambda)$ has the McMillan canonical form

$$M_A(\lambda) = \begin{bmatrix} \dfrac{1}{(\lambda - 1)(\lambda - 2)} & 0 \\ 0 & 1 - \dfrac{\epsilon}{(\lambda - 1)(\lambda - 2)} \end{bmatrix}.$$

In the present case we have $\psi_1(\lambda) = (\lambda-1)(\lambda-2) = d(\lambda)$, $\psi_2(\lambda) = (\lambda-1)(\lambda-2) = d(\lambda)$. Hence the McMillan denominator is $\psi_A(\lambda) = (\lambda - 1)^2(\lambda - 2)^2$ and $\operatorname{Mdeg} A(\lambda) = 4$. However, if $\epsilon = 0$ is true, then we get

$$M_A(\lambda) = \begin{bmatrix} \dfrac{1}{(\lambda - 1)(\lambda - 2)} & 0 \\ 0 & 1 \end{bmatrix}.$$

In this case, we obtain $\psi_1(\lambda) = (\lambda - 1)(\lambda - 2) = d(\lambda)$, $\psi_2(\lambda) = 1$ and the McMillan denominator $\psi_A(\lambda) = (\lambda-1)(\lambda-2)$ which yields $\operatorname{Mdeg} A(\lambda) = 2$. □

3.2 Algebraic Properties of Normal Matrices

1. In this section we give some general algebraic properties of normal matrices that will be used further.

Theorem 3.8. *Let two normal matrices*

$$A_1(\lambda) = \frac{N_1(\lambda)}{d_1(\lambda)}, \quad A_2(\lambda) = \frac{N_2(\lambda)}{d_2(\lambda)} \tag{3.15}$$

of dimensions $n \times \ell$ resp. $\ell \times m$ be given. Then, if the fraction

$$A(\lambda) = A_1(\lambda)A_2(\lambda) = \frac{N_1(\lambda)N_2(\lambda)}{d_1(\lambda)d_2(\lambda)} \tag{3.16}$$

is irreducible, the matrix $A(\lambda)$ becomes normal.

Proof. If $n = 1$ or $m = 1$ is true, then the statement follows from the remark after Theorem 3.1. Now, let $\min(n, m) \geq 2$ and assume $N(\lambda) = N_1(\lambda)N_2(\lambda)$. Due to the theorem of Binet-Cauchy, every minor of second order of the matrix $N(\lambda)$ is a bilinear form of the minors of second order of the matrices $N_1(\lambda)$ and $N_2(\lambda)$, and consequently divisible by the product $d_1(\lambda)d_2(\lambda)$. Therefore, the fraction (3.16) is normal, because it is also irreducible. ∎

2.

Theorem 3.9. *Let the matrices (3.15) have the same dimension, and the polynomials $d_1(\lambda)$ and $d_2(\lambda)$ be coprime. Then the matrix*

$$A(\lambda) = A_1(\lambda) + A_2(\lambda) \tag{3.17}$$

is normal.

Proof. From (3.15) and (3.17), we generate

$$A(\lambda) = \frac{d_2(\lambda)N_1(\lambda) + d_1(\lambda)N_2(\lambda)}{d_1(\lambda)d_2(\lambda)}\,. \tag{3.18}$$

The fraction (3.18) is irreducible, because the sum (3.17) has its poles at the zeros of $d_1(\lambda)$ and $d_2(\lambda)$ with the same multiplicity.

Denote

$$A_1(\lambda) = \left[\frac{\alpha_{ik}(\lambda)}{d_1(\lambda)}\right], \quad A_2(\lambda) = \left[\frac{\beta_{ik}(\lambda)}{d_2(\lambda)}\right].$$

Then the minor (3.11) for the matrix $A(\lambda)$ has the shape

$$A\begin{pmatrix} i & j \\ k & \ell \end{pmatrix} = \det \begin{bmatrix} \frac{\alpha_{ik}(\lambda)}{d_1(\lambda)} + \frac{\beta_{ik}(\lambda)}{d_2(\lambda)} & \frac{\alpha_{i\ell}(\lambda)}{d_1(\lambda)} + \frac{\beta_{i\ell}(\lambda)}{d_2(\lambda)} \\ \frac{\alpha_{jk}(\lambda)}{d_1(\lambda)} + \frac{\beta_{jk}(\lambda)}{d_2(\lambda)} & \frac{\alpha_{j\ell}(\lambda)}{d_1(\lambda)} + \frac{\beta_{j\ell}(\lambda)}{d_2(\lambda)} \end{bmatrix}. \tag{3.19}$$

Applying the summation theorem for determinants, and using the normality of $A_1(\lambda)$, $A_2(\lambda)$ after cancellation, we obtain the expression

$$A\begin{pmatrix} i & j \\ k & \ell \end{pmatrix} = \frac{b_{k\ell}^{ij}(\lambda)}{d_1(\lambda)d_2(\lambda)}$$

with certain polynomials $b_{k\ell}^{ij}(\lambda)$. It follows from this expression that the poles of the minor (3.19) can be found under the roots of the denominators of Matrix (3.18), and they possess no higher multiplicity. Since this rational matrix is irreducible, Theorem 3.5 yields that the matrix $A(\lambda)$ is normal. ∎

Corollary 3.10. *If $A(\lambda)$ is a normal $n \times m$ rational matrix, and $G(\lambda)$ is an $n \times m$ polynomial matrix, then the rational matrix*

$$A_1(\lambda) = A(\lambda) + G(\lambda)$$

is normal. ∎

3. For normal matrices the reverse to Theorem 2.42 is true.

Theorem 3.11. *Let the polynomial $n \times n$ matrix $a_l(\lambda)$ be simple and $b_l(\lambda)$ be any $n \times m$ polynomial matrix. If under this condition, the fraction*

$$A(\lambda) = a_l^{-1}(\lambda) b_l(\lambda) = \frac{\operatorname{adj} a_l(\lambda)\ b_l(\lambda)}{\det a_l(\lambda)} \tag{3.20}$$

is irreducible, then the pair $(a_l(\lambda), b_l(\lambda))$ is irreducible and the matrix $A(\lambda)$ is normal.

Proof. It is sufficient to consider the case $\min\{n, m\} \geq 2$. Since the matrix $a_l(\lambda)$ is simple, with the help of Corollary 3.2, it follows that the matrix $a_l^{-1}(\lambda)$ is normal and all minors of second order of the matrix $\operatorname{adj} a_l(\lambda)$ are divisible by $\det a_l(\lambda)$. Due to the theorem of Binet-Cauchy, this property transfers to the numerator on the right side of (3.20), that's why the matrix $A(\lambda)$ is normal. By using (3.6), we also obtain

$$A(\lambda) = \frac{Q(\lambda) b_l(\lambda)}{d(\lambda)} .$$

Here per construction, we have $\det a_l(\lambda) \approx d(\lambda)$. Comparing this equation with (3.20), we realise that the middle part of (3.20) proves to be a complete LMFD, and consequently the pair $(a_l(\lambda), b_l(\lambda))$ is irreducible. ∎

Analogously, the following statement for right MFD can be shown:

Corollary 3.12. *If the polynomial $m \times m$ matrix $a_r(\lambda)$ is simple, $\det a_r(\lambda) \approx d(\lambda)$, $b_r(\lambda)$ is any polynomial $n \times m$ matrix, and the fraction*

$$b_r(\lambda) a_r^{-1}(\lambda) = \frac{R(\lambda)}{d(\lambda)} = A(\lambda)$$

is irreducible, then the pair $[a_r(\lambda), b_r(\lambda)]$ is irreducible and the left side defines a complete RMFD of the matrix $A(\lambda)$.

4. Let us investigate some general properties of the MFD of the product of normal matrices. Consider some normal matrices (3.15), where their product (3.16) should exist and be irreducible. Moreover, let us have the complete LMFD

$$A_1(\lambda) = a_1^{-1}(\lambda) b_1(\lambda), \quad A_2(\lambda) = a_2^{-1}(\lambda) b_2(\lambda) ,$$

where the matrices $a_1(\lambda)$, $a_2(\lambda)$ are simple, and $\det a_1(\lambda) \approx d_1(\lambda)$, $\det a_2(\lambda) \approx d_2(\lambda)$ are valid. Applying these representations, we can write

$$A(\lambda) = a_1^{-1}(\lambda) b_1(\lambda) a_2^{-1}(\lambda) b_2(\lambda) . \tag{3.21}$$

Notice, that the fraction $L(\lambda) = b_1(\lambda) a_2^{-1}(\lambda)$ owing to the irreducibility of $A(\lambda)$ is also irreducible. Hence as a result of Corollary 3.12, the fraction $L(\lambda)$ is normal, and there exists the complete LMFD

$$b_1(\lambda)a_2^{-1}(\lambda) = a_3^{-1}(\lambda)b_3(\lambda)\,,$$

where $\det a_3(\lambda) \approx d_2(\lambda)$. From this and (3.21), we get

$$A(\lambda) = a_l^{-1}(\lambda)b_l(\lambda)$$
$$\text{with}\quad a_l(\lambda) = a_3(\lambda)a_1(\lambda), \quad b_l(\lambda) = b_3(\lambda)b_2(\lambda)\,.$$

Per construction, $\det a_l(\lambda) \approx d_1(\lambda)d_2(\lambda)$ is valid, that's why the last relations define a complete LMFD, the matrix $a_l(\lambda)$ is simple and the pair $(a_3(\lambda)a_1(\lambda),\, b_3(\lambda)b_2(\lambda))$ is irreducible. Hereby, we still obtain

$$\operatorname{Mdeg}[A_1(\lambda)A_2(\lambda)] = \operatorname{Mdeg} A_1(\lambda) + \operatorname{Mdeg} A_2(\lambda)\,.$$

Hence the following theorem has been proven:

Theorem 3.13. *If the matrices (3.15) are normal and the product (3.16) is irreducible, then the matrices $A_1(\lambda)$ and $A_2(\lambda)$ are independent in the sense of Section 2.4.* ■

5. From Theorems 3.13 and 2.61 we conclude the following statement, which is formulated in the terminology of subordination of matrices in the sense of Section 2.12.

Theorem 3.14. *Let us have the normal matrices (3.15), and their product (3.16) should be irreducible. Then*

$$A_1(\lambda) \underset{l}{\prec} A_1(\lambda)A_2(\lambda), \quad A_2(\lambda) \underset{r}{\prec} A_1(\lambda)A_2(\lambda)\,.$$

■

6.

Theorem 3.15. *Let the separation*

$$A(\lambda) = \frac{N(\lambda)}{d_1(\lambda)d_2(\lambda)} = \frac{N_1(\lambda)}{d_1(\lambda)} + \frac{N_2(\lambda)}{d_2(\lambda)} \tag{3.22}$$

exist, where the matrix $A(\lambda)$ is normal and the polynomials $d_1(\lambda)$, $d_2(\lambda)$ are coprime. Then each of the fractions on the right side of (3.22) is normal.

Proof. At first we notice that the fractions on the right side of (3.22) are irreducible, otherwise the fraction $A(\lambda)$ would be reducible. The fraction

$$A_1(\lambda) = \frac{N(\lambda)}{d_1(\lambda)}$$

is also normal, because it is irreducible and the minors of second order of the numerator are divisible by the denominator $d_1(\lambda)$. Therefore,

$$A_1(\lambda) = a_1^{-1}(\lambda)b_1(\lambda)\,, \tag{3.23}$$

where the matrix $a_1(\lambda)$ is simple and $\det a_1(\lambda) \approx d_1(\lambda)$. Multiplying both sides of Equation (3.22) from left by $a_1(\lambda)$ and considering (3.23), we get

$$\frac{b_1(\lambda)}{d_2(\lambda)} = a_1(\lambda)\frac{N_1(\lambda)}{d_1(\lambda)} + a_1(\lambda)\frac{N_2(\lambda)}{d_2(\lambda)}\,,$$

this means

$$\frac{b_1(\lambda)}{d_2(\lambda)} - a_1(\lambda)\frac{N_2(\lambda)}{d_2(\lambda)} = a_1(\lambda)\frac{N_1(\lambda)}{d_1(\lambda)}\,.$$

The left side of the last equation is analytical at the zeros of the polynomials $d_1(\lambda)$, and the right side at the zeros of $d_2(\lambda)$. Consequently

$$\frac{a_1(\lambda)N_1(\lambda)}{d_1(\lambda)} = L(\lambda)$$

has to be a polynomial matrix $L(\lambda)$ and

$$\frac{N_1(\lambda)}{d_1(\lambda)} = a_1^{-1}(\lambda)L(\lambda)\,.$$

The fraction on the right side is irreducible, otherwise the fraction (3.22) has been irreducible. The matrix $a_1(\lambda)$ is simple, and therefore the last fraction owing to Theorem 3.11 is normal. In analogy it may be shown that the matrix $N_2(\lambda)/d_2(\lambda)$ is normal. ∎

3.3 Normal Matrices and Simple Realisations

1. At the first sight, normal rational matrices seem to be quite artificial constructions, because their elements are bounded by a number of crisp equations. However, in this section we will demonstrate that even normal matrices for the most of real problems will give the correct description of multidimensional LTI objects in the frequency domain.

2. We will use the terminology and the notation of Section 1.15, and consider an arbitrary realisation (A, B, C) of dimension n, p, m. Doing so, the realisation (A, B, C) is called minimal, if the pair (A, B) is controllable and the pair $[A, C]$ is observable. A minimal realisation is called *simple* if the matrix A is cyclic. As shown in Section 1.15, the property of simplicity of the realisation (A, B, C) is structural stable, and it is conserved at least for sufficiently small deviations in the matrices A, B, C. Realisations, that are not simple, are not supplied with the property of structural stability. Practically, this means that correct models of real linear objects in state space amounts to simple realisations.

3. As explained in chapter 2, every realisation (A, B, C) is assigned to a strictly proper rational $n \times m$ matrix $w(\lambda)$ by the relation

$$w(\lambda) = C(\lambda I_p - A)^{-1}B \tag{3.24}$$

equivalently expressed by

$$w(\lambda) = \frac{C \operatorname{adj}(\lambda I_p - A)B}{d_A(\lambda)}, \tag{3.25}$$

where $\operatorname{adj}(\lambda I_p - A)$ is the adjoint matrix and $d_A(\lambda) = \det(\lambda I_p - A)$. As is taken from (3.24), every realisation (A, B, C) is uniquely related to a transfer matrix. Conversely, every strictly proper rational $n \times m$ matrix

$$w(\lambda) = \frac{N(\lambda)}{d(\lambda)} \tag{3.26}$$

is configured to an infinite set of realisations (A, B, C) of dimensions n, q, m with $q \geq \operatorname{Mdeg} w(\lambda)$, where $\operatorname{Mdeg} w(\lambda)$ means the McMillan-degree of the matrix $w(\lambda)$. Realisations, where the number q takes its minimal value, as before will be called minimal. The realisation (A, B, C) is minimal, if and only if the pair (A, B) is controllable and the pair $[A, C]$ is observable.

In general, minimal realisations of arbitrary matrices $w(\lambda)$ are not simple, and therefore, they do not possess the property of structural stability. In this case small deviations in the coefficients of the linear objects (1.102), (1.103) lead to essential changes in their transfer functions. In this connection, the question arises, for which class of matrices the corresponding minimal realisations will be simple. The answer to this question lies in the following statement.

4.

Theorem 3.16. *The transfer matrix (3.25) of the realisation (A, B, C) is irreducible, if and only if the realisation (A, B, C) is simple.*

Proof. As follows from Theorem 2.45, for the irreducibility of the fractions (3.25), it is necessary and sufficient that the elementary PMD

$$\tau(\lambda) = (\lambda I_p - A, B, C) \tag{3.27}$$

is minimal and the matrix $\lambda I_p - A$ is simple, which is equivalent to the demand for simplicity of the realisation (A, B, C). ∎

Theorem 3.17. *If the realisation (A, B, C) of dimension (n, p, m) is simple, then the corresponding transfer matrix (3.24) is normal.*

Proof. Assume that the realisation (A, B, C) is simple. Then the elementary PMD (3.27) is also simple, and the fraction on the right side of (3.25) is irreducible. Hereby, due to the simplicity of the matrix $\lambda I_p - A$, the rational matrix

$$(\lambda I_p - A)^{-1} = \frac{\operatorname{adj}(\lambda I_p - A)}{\det(\lambda I_p - A)}$$

becomes normal, what means, it is irreducible and all minors of 2nd order of the matrix $\operatorname{adj}(\lambda I_p - A)$ are divisible by $\det(\lambda I_p - A)$. But then for $\min\{m, n\} \geq 2$ owing to the theorem of Binet-Cauchy, also the minors of 2nd order of the matrix

$$Q(\lambda) = C \operatorname{adj}(\lambda I_p - A) B$$

possess this property, and this means that Matrix (3.24) is normal. ∎

Theorem 3.18. *For a strictly proper rational matrix to possess a simple realisation, it is necessary and sufficient, that this matrix is normal.*

Proof. Necessity: When the rational matrix (3.26) allows a simple realisation (A, B, C), then it is normal by virtue of Theorem 3.17.
Sufficiency: Let the irreducible matrix (3.26) be normal and $\deg d(\lambda) = p$. Then there exists a complete LMFD

$$w(\lambda) = a^{-1}(\lambda) b(\lambda)$$

for it with $\operatorname{ord} a(\lambda) = p$, and consequently $\operatorname{Mdeg} w(\lambda) = p$. From this we conclude, that Matrix (3.26) allows a minimal realisation (A, B, C) of dimension (n, p, m). We now assume that the matrix A is not cyclic. Then the fraction

$$\frac{C \operatorname{adj}(\lambda I_p - A) B}{\det(\lambda I_p - A)}$$

would be reducible. Hereby, Matrix (3.26) would permit the representation

$$w(\lambda) = \frac{N_1(\lambda)}{d_1(\lambda)},$$

where $\deg d_1(\lambda) < \deg d(\lambda)$. But this contradicts the supposition on the irreducibility of Matrix (3.26). Therefore, the matrix A must be cyclic and the matrix $\lambda I_p - A$ simple, hence the minimal realisation has to be simple. ∎

3.4 Structural Stable Representation of Normal Matrices

1. The notation of normal matrices in the form (3.1) is structural unstable, because it looses for arbitrary small errors in its coefficients the property that its minors of 2nd order of the numerator are divisible by the denominator. In that case, the quantity $\operatorname{Mdeg} A(\lambda)$ will abruptly increase. Especially, if Matrix

(3.1) is strictly proper, then the dimensions of the matrices in its minimal realisation in state space will also abruptly increase, *i.e.* the dynamical properties with respect to the original system will change drastically. In this section, a structural stable representation (S-representation) of normal rational matrices will be introduced, . Regarding normality, the S-representation is invariant related to parameter deviations in the transfer matrix, originated for instance by modeling or rounding errors.

2.

Theorem 3.19 ([144, 145]). *The irreducible rational $n \times m$ matrix*

$$A(\lambda) = \frac{N(\lambda)}{d(\lambda)} \tag{3.28}$$

is normal, if and only if its numerator permits the representation

$$N(\lambda) = P(\lambda)Q'(\lambda) + d(\lambda)G(\lambda) \tag{3.29}$$

with an $n \times 1$ polynomial column $P(\lambda)$, an $m \times 1$ polynomial column $Q(\lambda)$, and an $n \times m$ polynomial matrix $G(\lambda)$.

Proof. Sufficiency: Let us have

$$P(\lambda) = \begin{bmatrix} p_1(\lambda) \\ \vdots \\ p_n(\lambda) \end{bmatrix}, \quad Q(\lambda) = \begin{bmatrix} q_1(\lambda) \\ \vdots \\ q_m(\lambda) \end{bmatrix}, \quad G(\lambda) = \begin{bmatrix} g_{11}(\lambda) & \dots & g_{1m}(\lambda) \\ \vdots & \dots & \vdots \\ g_{n1}(\lambda) & \dots & g_{nm}(\lambda) \end{bmatrix}$$

with scalar polynomials $p_i(\lambda)$, $q_i(\lambda)$, $g_{ik}(\lambda)$. The minor (3.11) of Matrix (3.29) possesses the form

$$\begin{aligned} N\begin{pmatrix} i & j \\ k & \ell \end{pmatrix} &= \det \begin{bmatrix} p_i(\lambda)q_k(\lambda) + d(\lambda)g_{ik}(\lambda) & p_i(\lambda)q_\ell(\lambda) + d(\lambda)g_{i\ell}(\lambda) \\ p_j(\lambda)q_k(\lambda) + d(\lambda)g_{jk}(\lambda) & p_j(\lambda)q_\ell(\lambda) + d(\lambda)g_{j\ell}(\lambda) \end{bmatrix} \\ &= d(\lambda)n_{k\ell}^{ij}(\lambda) \end{aligned}$$

where $n_{k\ell}^{ij}(\lambda)$ is a polynomial. Therefore, an arbitrary minor is divisible by $d(\lambda)$, and thus Matrix (3.28) is normal.
Necessity: Let Matrix (3.28) be normal. Then all minors of second order of its numerator $N(\lambda)$ are divisible by the denominator $d(\lambda)$. Applying (3.7) and (3.8), we find out that the matrix $N(\lambda)$ allows the representation

$$N(\lambda) = p(\lambda) \begin{bmatrix} g_1(\lambda) & 0 & \dots & 0 & \\ 0 & g_1(\lambda)g_2(\lambda)d(\lambda) & \dots & 0 & \\ \vdots & \vdots & \ddots & \vdots & O_{\rho,m-\rho} \\ 0 & 0 & \dots & g_1(\lambda)\cdots g_\rho(\lambda)d(\lambda) & \\ & & O_{n-\rho,\rho} & & O_{n-\rho,m-\rho} \end{bmatrix} q(\lambda) \tag{3.30}$$

where the $g_i(\lambda)$, $(i = 1, \dots, \rho)$ are monic polynomials and $p(\lambda)$, $q(\lambda)$ are unimodular matrices. Relation (3.30) can be arranged in the form

$$N(\lambda) = N_1(\lambda) + d(\lambda)N_2(\lambda)\,, \tag{3.31}$$

where

$$N_1(\lambda) = g_1(\lambda)p(\lambda)\begin{bmatrix} 1 & O_{1,m-1} \\ O_{n-1,1} & O_{n-1,m-1} \end{bmatrix} q(\lambda)\,, \tag{3.32}$$

$$N_2(\lambda) = g_1(\lambda)g_2(\lambda)p(\lambda)\left[\begin{array}{cccccc|c} 0 & 0 & 0 & \dots & & 0 & \\ 0 & 1 & 0 & \dots & & 0 & \\ 0 & 0 & g_3(\lambda) & \dots & & 0 & O_{\rho,m-\rho} \\ \vdots & \vdots & \vdots & \ddots & & \vdots & \\ 0 & 0 & 0 & \dots & g_3(\lambda)\cdots g_\rho(\lambda) & & \\ & & & O_{n-\rho,\rho} & & & O_{n-\rho,m-\rho} \end{array}\right] q(\lambda)\,. \tag{3.33}$$

Obviously, we have

$$N_1(\lambda) = g_1(\lambda)P_1(\lambda)Q_1(\lambda)\,, \tag{3.34}$$

where $P_1(\lambda)$ is the first column of the matrix $p(\lambda)$ and $Q_1(\lambda)$ is the first row of $q(\lambda)$. Inserting (3.32)–(3.34) into (3.31), we arrive at the representation (3.29), where for instance

$$P(\lambda) = g_1(\lambda)P_1(\lambda), \quad Q(\lambda) = Q_1'(\lambda), \quad G(\lambda) = N_2(\lambda)$$

can be used. ■

3. Inserting (3.29) into (3.28) yields

$$A(\lambda) = \frac{P(\lambda)Q'(\lambda)}{d(\lambda)} + G(\lambda)\,. \tag{3.35}$$

The representation of a normal rational matrix in the form (3.35) is called its *structural stable representation* or *S-representation*. Notice, that the S-representation of a normal matrix is structural stable (invariant) according to variations of the vectors $P(\lambda)$, $Q(\lambda)$, the matrix $G(\lambda)$ and of the polynomial $d(\lambda)$, because the essential structural specialities of Matrix (3.35) still hold.

4. Assume

$$P(\lambda) = d(\lambda)L_1(\lambda) + \tilde{P}_1(\lambda)\,,$$

$$Q(\lambda) = d(\lambda)L_2(\lambda) + \tilde{Q}_1(\lambda)\,,$$

where

$$\deg \tilde{P}_1(\lambda) < \deg d(\lambda), \quad \deg \tilde{Q}_1(\lambda) < \deg d(\lambda)\,. \tag{3.36}$$

Then from (3.23), we obtain

$$N(\lambda) = \tilde{P}_1(\lambda)\tilde{Q}_1'(\lambda) + d(\lambda)G_1(\lambda)$$

with

$$G_1(\lambda) = L_1(\lambda)\tilde{Q}_1'(\lambda) + \tilde{P}_1(\lambda)L_2'(\lambda) + d(\lambda)L_1(\lambda)L_2'(\lambda) + G(\lambda).$$

Altogether from (3.35), we get

$$A(\lambda) = \frac{\tilde{P}_1(\lambda)\tilde{Q}_1'(\lambda)}{d(\lambda)} + G_1(\lambda), \tag{3.37}$$

where the vectors $\tilde{P}_1(\lambda)$ and $\tilde{Q}_1(\lambda)$ satisfy Relation (3.36). Representation (3.37) is named as *minimal S-representation* of a normal matrix.

A minimal S-representation (3.34) also turns out to be structural stable, if the parameter variations do not violate Condition (3.36).

5. The proof of Theorem 3.19 has constructive character, so it yields a practical method for calculating the vectors $P(\lambda)$, $Q(\lambda)$ and the matrix $G(\lambda)$, which appear in the S-representations (3.35) and (3.37). However, at first the matrix $N(\lambda)$ has to be given the shape (3.30), which is normally connected with extensive calculations. In constructing the S-representation of a normal matrix, the next statement allows essential simplification in most practical cases.

Theorem 3.20. *Let the numerator $N(\lambda)$ of a normal matrix (3.28) be given in the form*

$$N(\lambda) = g(\lambda)\phi(\lambda)\begin{bmatrix} \alpha(\lambda) & 1 \\ L(\lambda) & \beta(\lambda) \end{bmatrix}\psi(\lambda), \tag{3.38}$$

where $g(\lambda)$ is a scalar polynomial, which is equal to the GCD of the elements of $N(\lambda)$. The polynomial matrices $\phi(\lambda)$, $\psi(\lambda)$ are unimodular and $L(\lambda)$ is an $(n-1)\times(m-1)$ polynomial matrix. Furthermore, let

$$\alpha(\lambda) = \begin{bmatrix} \alpha_m(\lambda) & \ldots & \alpha_2(\lambda) \end{bmatrix},$$

$$\beta'(\lambda) = \begin{bmatrix} \beta_2(\lambda) & \ldots & \beta_n(\lambda) \end{bmatrix}$$

be row vectors. Then

$$L(\lambda) = \beta(\lambda)\alpha(\lambda) + d(\lambda)G_2(\lambda), \tag{3.39}$$

where $G_2(\lambda)$ is an $(n-1)\times(m-1)$ polynomial matrix. Hereby, we have

$$N(\lambda) = g(\lambda)P(\lambda)Q'(\lambda) + g(\lambda)d(\lambda)G(\lambda), \tag{3.40}$$

where

$$P(\lambda) = \phi(\lambda)\begin{bmatrix} 1 \\ \beta(\lambda) \end{bmatrix}, \quad Q'(\lambda) = \begin{bmatrix} \alpha(\lambda) & 1 \end{bmatrix}\psi(\lambda), \tag{3.41}$$

and moreover

$$G(\lambda) = \phi(\lambda) \begin{bmatrix} O_{1,m-1} & 0 \\ G_2(\lambda) & O_{n-1,1} \end{bmatrix} \psi(\lambda) . \tag{3.42}$$

Doing so, Matrix (3.28) takes the S-representation

$$A(\lambda) = \frac{g(\lambda)P(\lambda)Q'(\lambda)}{d(\lambda)} + g(\lambda)G(\lambda) . \tag{3.43}$$

Proof. Introduce the unimodular matrices

$$N_\beta(\lambda) = \begin{bmatrix} 1 & O_{1,n-1} \\ -\beta(\lambda) & I_{n-1} \end{bmatrix}, \quad N_\alpha(\lambda) = \begin{bmatrix} O_{m-1,1} & I_{m-1} \\ 1 & -\alpha(\lambda) \end{bmatrix} . \tag{3.44}$$

By direct calculation, we obtain

$$N_\beta^{-1}(\lambda) = \begin{bmatrix} 1 & O_{1,n-1} \\ \beta(\lambda) & I_{n-1} \end{bmatrix}, \quad N_\alpha^{-1}(\lambda) = \begin{bmatrix} \alpha(\lambda) & 1 \\ I_{m-1} & O_{m-1,1} \end{bmatrix} . \tag{3.45}$$

Easily,

$$N_\beta(\lambda) \begin{bmatrix} \alpha(\lambda) & 1 \\ L(\lambda) & \beta(\lambda) \end{bmatrix} N_\alpha(\lambda) = \begin{bmatrix} 1 & O_{1,m-1} \\ O_{n-1,1} & \tilde{R}(\lambda) \end{bmatrix} \triangleq B(\lambda) \tag{3.46}$$

is established, where

$$\tilde{R}(\lambda) = L(\lambda) - \beta(\lambda)\alpha(\lambda) . \tag{3.47}$$

The polynomials $g(\lambda)$, $d(\lambda)$ are coprime, otherwise the fraction (3.28) would be reducible. Thereby, all minors of second order of the matrix

$$N_g(\lambda) = \phi(\lambda) \begin{bmatrix} \alpha(\lambda) & 1 \\ L(\lambda) & \beta(\lambda) \end{bmatrix} \psi(\lambda) \tag{3.48}$$

are divisible by $d(\lambda)$. With regard to

$$N_g(\lambda) = \phi(\lambda)N_\beta^{-1}(\lambda)B(\lambda)N_\alpha^{-1}(\lambda)\psi(\lambda)$$

and the observation that the matrices $\phi(\lambda)N_\beta^{-1}(\lambda)$ and $N_\alpha^{-1}(\lambda)\psi(\lambda)$ are unimodular, we realise that the matrices $N_g(\lambda)$ and $B(\lambda)$ are equivalent. Since all minors of second order of the matrix $B(\lambda)$ are divisible by $d(\lambda)$, we get immediately that all elements of the matrix $\tilde{R}(\lambda)$ are also divisible by $d(\lambda)$, which runs into the equality

$$\tilde{R}(\lambda) = L(\lambda) - \beta(\lambda)\alpha(\lambda) = d(\lambda)G_2(\lambda)$$

that is equivalent to (3.39). Inserting the relation

$$L(\lambda) = \beta(\lambda)\alpha(\lambda) + d(\lambda)G_2(\lambda)$$

into (3.38), we get

$$N(\lambda) = g(\lambda)\phi(\lambda)\begin{bmatrix} \alpha(\lambda) & 1 \\ \beta(\lambda)\alpha(\lambda) + d(\lambda)G_2(\lambda) & \beta(\lambda) \end{bmatrix}\psi(\lambda)$$

$$= g(\lambda)\phi(\lambda)\begin{bmatrix} \alpha(\lambda) & 1 \\ \beta(\lambda)\alpha(\lambda) & \beta(\lambda) \end{bmatrix}\psi(\lambda) + g(\lambda)d(\lambda)\phi(\lambda)\begin{bmatrix} O_{1,m-1} & 0 \\ G_2(\lambda) & O_{n-1,1} \end{bmatrix}\psi(\lambda). \tag{3.49}$$

Using

$$\begin{bmatrix} \alpha(\lambda) & 1 \\ \beta(\lambda)\alpha(\lambda) & \beta(\lambda) \end{bmatrix} = \begin{bmatrix} 1 \\ \beta(\lambda) \end{bmatrix}\begin{bmatrix} \alpha(\lambda) & 1 \end{bmatrix},$$

we generate from (3.49) Formulae (3.40)–(3.42). Relation (3.43) is held by substituting (3.49) into (3.28). ■

Remark 3.21. Let

$$\phi(\lambda) = \begin{bmatrix} \phi_1(\lambda) & \dots & \phi_n(\lambda) \end{bmatrix},$$

$$\psi'(\lambda) = \begin{bmatrix} \psi'_1(\lambda) & \dots & \psi'_m(\lambda) \end{bmatrix},$$

where $\phi_i(\lambda)$, $(i = 1, \dots, n)$, $\psi_i(\lambda)$, $(i = 1, \dots, m)$ are columns or rows, respectively. Then from (3.41), it follows

$$P(\lambda) = \phi_1(\lambda) + \phi_2(\lambda)\beta_2(\lambda) + \dots + \phi_n(\lambda)\beta_n(\lambda),$$

$$Q(\lambda) = \psi'_1(\lambda)\alpha_m(\lambda) + \psi'_2(\lambda)\alpha_{m-1}(\lambda) + \dots + \psi'_m(\lambda).$$

Remark 3.22. Equation (3.42) delivers

$$\operatorname{rank} G(\lambda) \le \min\{n-1, m-1\}.$$

6.

Example 3.23. Generate the S-representation of a normal matrix (3.28) with

$$N(\lambda) = \begin{bmatrix} -\lambda+1 & 2 & 1 \\ 0 & (\lambda+1)(\lambda-2) & \lambda-2 \end{bmatrix}, \quad d(\lambda) = (\lambda-1)(\lambda-2).$$

In the present case, the matrix $A(\lambda)$ possesses only the two single poles $\lambda_1 = 1$ and $\lambda_2 = 2$. Hereby, we have

$$N(\lambda_1) = \begin{bmatrix} 0 & 2 & 1 \\ 0 & -2 & -1 \end{bmatrix}, \quad \operatorname{rank} N(\lambda_1) = 1,$$

$$N(\lambda_2) = \begin{bmatrix} -1 & 2 & 1 \\ 0 & 0 & 0 \end{bmatrix}, \quad \operatorname{rank} N(\lambda_2) = 1,$$

that's why the matrix $A(\lambda)$, owing to Theorem 3.6 is normal. For construction of the S-representation, Theorem 3.20 is used. In the present case, we have

$$g(\lambda) = 1, \quad \phi(\lambda) = I_2\,, \quad \psi(\lambda) = I_3\,,$$
$$\beta(\lambda) = \lambda - 2, \quad \alpha(\lambda) = \begin{bmatrix} -\lambda + 1 & 2 \end{bmatrix}, \quad L(\lambda) = \begin{bmatrix} 0 & (\lambda+1)(\lambda-2) \end{bmatrix}.$$

Applying (3.39), we produce

$$L(\lambda) - \beta(\lambda)\alpha(\lambda) = \begin{bmatrix} d(\lambda) & d(\lambda) \end{bmatrix} - d(\lambda)\begin{bmatrix} 1 & 1 \end{bmatrix},$$

and therefore

$$G_2(\lambda) = \begin{bmatrix} 1 & 1 \end{bmatrix}.$$

On the basis of (3.41), we find

$$P(\lambda) = \begin{bmatrix} 1 \\ \lambda - 2 \end{bmatrix}, \quad Q'(\lambda) = \begin{bmatrix} -\lambda + 1 & 2 & 1 \end{bmatrix}.$$

Moreover, due to (3.42), we get

$$G(\lambda) = \begin{bmatrix} 0 & 0 & 0 \\ 1 & 1 & 0 \end{bmatrix}.$$

With these results, we obtain

$$A(\lambda) = \frac{\begin{bmatrix} 1 \\ \lambda - 2 \end{bmatrix} \begin{bmatrix} 1 - \lambda & 2 & 1 \end{bmatrix}}{(\lambda - 1)(\lambda - 2)} + \begin{bmatrix} 0 & 0 & 0 \\ 1 & 1 & 0 \end{bmatrix}.$$

Regarding $\deg P(\lambda) = \deg Q(\lambda) = 1$, the generated S-representation is minimal. □

3.5 Inverses of Characteristic Matrices of Jordan and Frobenius Matrices

1. In this section S-representations for matrices of the shape $(\lambda I_p - A)^{-1} = A_\lambda^{-1}$ will be constructed, where $A = J_p(a)$ is a Jordan block (1.76) or $A = A_F$ is a Frobenius matrix (1.94). In the first case, the matrix A_λ is called the *characteristic Jordan matrix* , and in the second case the *characteristic Frobenius matrix.*

2. Consider the upper Jordan block with the eigenvalue a

$$J_p(a) = \begin{bmatrix} a & 1 & \dots & 0 & 0 \\ 0 & a & \dots & 0 & 0 \\ \vdots & \vdots & \ddots & \vdots & \vdots \\ 0 & 0 & \dots & a & 1 \\ 0 & 0 & \dots & 0 & a \end{bmatrix} \tag{3.50}$$

and the corresponding characteristic matrix

$$J_p(\lambda, a) = \lambda I_p - J_p(a)\,.$$

Now, a direct calculation of the adjoint matrix yields

$$\operatorname{adj} J_p(\lambda, a) = \begin{bmatrix} (\lambda-a)^{p-1} & (\lambda-a)^{p-2} & \dots & \lambda-a & 1 \\ 0 & (\lambda-a)^{p-1} & \dots & (\lambda-a)^2 & \lambda-a \\ \vdots & \vdots & \ddots & \vdots & \vdots \\ 0 & 0 & \dots & (\lambda-a)^{p-1} & (\lambda-a)^{p-2} \\ 0 & 0 & \dots & 0 & (\lambda-a)^{p-1} \end{bmatrix}. \tag{3.51}$$

Matrix (3.51) has the shape (3.38) with

$$g(\lambda) = 1, \qquad \phi(\lambda) = \psi(\lambda) = I_p\,,$$
$$\alpha(\lambda) = \begin{bmatrix} (\lambda-a)^{p-1} & \dots & \lambda-a \end{bmatrix},$$
$$\beta'(\lambda) = \begin{bmatrix} (\lambda-a) & \dots & (\lambda-a)^{p-1} \end{bmatrix}.$$

Consistent with Theorem 3.20, we get

$$\operatorname{adj} J_p(\lambda, a) = P(\lambda)Q'(\lambda) + d(\lambda)G(\lambda)\,, \tag{3.52}$$

where

$$P'(\lambda) = \begin{bmatrix} 1 & \beta'(\lambda) \end{bmatrix}, \quad Q'(\lambda) = \begin{bmatrix} \alpha(\lambda) & 1 \end{bmatrix}$$

and

$$d(\lambda) = \det J_p(\lambda, a) = (\lambda-a)^p\,.$$

For determining the polynomial matrix $G(\lambda)$, take care of

$$P(\lambda)Q'(\lambda) = \begin{bmatrix} 1 \\ \lambda-a \\ \vdots \\ (\lambda-a)^{p-1} \end{bmatrix} \begin{bmatrix} (\lambda-a)^{p-1} & \dots & (\lambda-a) & 1 \end{bmatrix}$$

$$= \begin{bmatrix} (\lambda-a)^{p-1} & (\lambda-a)^{p-2} & \dots & 1 \\ (\lambda-a)^{p} & (\lambda-a)^{p-1} & \dots & \lambda-a \\ \vdots & \vdots & \ddots & \vdots \\ (\lambda-a)^{2p-2} & (\lambda-a)^{2p-3} & \dots & (\lambda-a)^{p-1} \end{bmatrix}.$$

As a result, we obtain

$$\operatorname{adj} J_p(\lambda, a) - P(\lambda)Q'(\lambda) = \begin{bmatrix} 0 & 0 & \dots & 0 & 0 \\ -(\lambda-a)^p & 0 & \dots & 0 & 0 \\ -(\lambda-a)^{p+1} & -(\lambda-a)^p & \dots & 0 & 0 \\ \vdots & \vdots & \ddots & \vdots & \vdots \\ -(\lambda-a)^{2p-2} & -(\lambda-a)^{2p-3} & \dots & -(\lambda-a)^p & 0 \end{bmatrix}$$

$$= -(\lambda-a)^p \begin{bmatrix} 0 & 0 & \dots & 0 & 0 \\ 1 & 0 & \dots & 0 & 0 \\ \lambda-a & 1 & \dots & 0 & 0 \\ \vdots & \vdots & \ddots & \vdots & \vdots \\ (\lambda-a)^{p-2} & (\lambda-a)^{p-3} & \dots & 1 & 0 \end{bmatrix}.$$

Substituting this into (3.52) yields

$$G(\lambda) = -\begin{bmatrix} 0 & 0 & \dots & 0 & 0 \\ 1 & 0 & \dots & 0 & 0 \\ \lambda-a & 1 & \dots & 0 & 0 \\ \vdots & \vdots & \ddots & \vdots & \vdots \\ (\lambda-a)^{p-2} & (\lambda-a)^{p-3} & \dots & 1 & 0 \end{bmatrix},$$

and applying (3.43) delivers the S-representation

$$J_p^{-1}(\lambda, a) = \frac{P(\lambda)Q'(\lambda)}{(\lambda-a)^p} + G(\lambda).$$

Since $\deg P(\lambda) < p$ and $\deg Q(\lambda) < p$, the produced S-representation is minimal.

3. Now consider the problem to find the S-representation for

$$A_{\lambda F}^{-1} = (\lambda I_p - A_F)^{-1},$$

where

$$A_F = \begin{bmatrix} 0 & 1 & 0 & \dots & 0 \\ 0 & 0 & 1 & \dots & 0 \\ \vdots & \vdots & \vdots & \ddots & \vdots \\ 0 & 0 & 0 & \dots & 1 \\ -d_p & -d_{p-1} & -d_{p-2} & \dots & -d_1 \end{bmatrix} \tag{3.53}$$

is the lower Frobenius normal form of dimension $p \times p$. Its characteristic matrix has obviously the shape

$$A_{\lambda F} = \begin{bmatrix} \lambda & -1 & 0 & \dots & 0 \\ 0 & \lambda & -1 & \dots & 0 \\ \vdots & \vdots & \ddots & \ddots & \vdots \\ 0 & 0 & 0 & \ddots & -1 \\ d_p & d_{p-1} & d_{p-2} & \dots & \lambda + d_1 \end{bmatrix}. \tag{3.54}$$

The adjoint matrix for (3.54) is calculated by

$$\operatorname{adj}(\lambda I_p - A_F) = \begin{bmatrix} d_1(\lambda) \; d_2(\lambda) & \dots & d_{p-1}(\lambda) & 1 \\ & & & \lambda \\ & L_F(\lambda) & & \vdots \\ & & & \lambda^{p-1} \end{bmatrix}. \tag{3.55}$$

Here and in what follows, we denote

$$\begin{aligned} d(\lambda) &= \lambda^p + d_1\lambda^{p-1} + \ldots + d_{p-1}\lambda + d_p\,, \\ d_1(\lambda) &= \lambda^{p-1} + d_1\lambda^{p-2} + \ldots + d_{p-2}\lambda + d_{n-1}\,, \\ d_2(\lambda) &= \lambda^{p-2} + d_1\lambda^{p-3} + \ldots + d_{p-2}\,, \\ \vdots \quad & \quad \vdots \\ d_{p-1}(\lambda) &= \lambda + d_1\,, \\ d_p(\lambda) &= 1\,, \end{aligned} \tag{3.56}$$

and $L_F(\lambda)$ is a certain $(p-1) \times (m-1)$ polynomial matrix. Relation (3.55) with Theorem 3.20 implies

$$\operatorname{adj}(\lambda I_p - A_F) = P_F(\lambda)Q'_F(\lambda) + d(\lambda)G_F(\lambda)\,, \tag{3.57}$$

where

$$\begin{aligned} P'_F(\lambda) &= \begin{bmatrix} 1 & \lambda & \dots & \lambda^{p-1} \end{bmatrix}, \\ Q'_F(\lambda) &= \begin{bmatrix} d_1(\lambda) & \dots & d_{p-1}(\lambda) & 1 \end{bmatrix}. \end{aligned} \tag{3.58}$$

It remains to calculate the matrix $G_F(\lambda)$. Denote

$$\begin{aligned} &\operatorname{adj}(\lambda I_p - A_F) = [a_{ik}(\lambda)], \quad P_F(\lambda)Q'_F(\lambda) = [b_{ik}(\lambda)], \\ &G_F(\lambda) = [g_{ik}(\lambda)], \quad (i,k = 1,\ldots,p)\,, \end{aligned} \tag{3.59}$$

then from (3.57), we obtain

$$b_{ik}(\lambda) = -d(\lambda)g_{ik}(\lambda) + a_{ik}(\lambda)\,.$$

Per construction, $\deg d(\lambda) = p$, $\deg a_{ik}(\lambda) \leq p-1$. Bringing this face to face with (1.6), we recognise that $-g_{ik}(\lambda)$ is the integral part and $a_{ik}(\lambda)$ is the rest, when dividing the polynomial $b_{ik}(\lambda)$ by $d(\lambda)$. Utilising (3.58), we arrive at

$$b_{ik}(\lambda) = \lambda^{i-1} d_k(\lambda)\,. \tag{3.60}$$

Due to $\deg d_k(\lambda) = p-k$, we obtain $\deg b_{ik}(\lambda) = p-k+i-1$. Thus, for $k \geq i$, we get $g_{ik}(\lambda) = 0$. Substituting $i = k+\ell$, $(\ell = 1,\dots,p-k)$ and taking into account (3.60) and (3.56), we receive

$$b_{k+\ell,k}(\lambda) = \lambda^{\ell-1} d(\lambda) + d_{k\ell}(\lambda)\,,$$

where $\deg d_{k\ell}(\lambda) < p$. From this, we read

$$g_{ik}(\lambda) = -\lambda^{\ell-1}\,, \qquad (i = k+\ell;\, \ell = 1,\dots,p-k)\,.$$

Altogether, this leads to

$$G_F(\lambda) = -\begin{bmatrix} 0 & 0 & 0 & \dots & 0 & 0 \\ -1 & 0 & 0 & \dots & 0 & 0 \\ -\lambda & -1 & 0 & \dots & 0 & 0 \\ \vdots & \vdots & \vdots & \vdots & \vdots & \vdots \\ -\lambda^{p-2} & -\lambda^{p-3} & -\lambda^{p-4} & -\dots & -1 & 0 \end{bmatrix} \tag{3.61}$$

and the wanted S-representation

$$(\lambda I_p - A_F)^{-1} = \frac{P_F(\lambda) Q'_F(\lambda)}{d(\lambda)} + G_F(\lambda)\,. \tag{3.62}$$

Per construction, $\deg P_F(\lambda) < p$, $\deg Q_F(\lambda) < p$ is valid, so the produced S-representation (3.62) is minimal.

3.6 Construction of Simple Jordan Realisations

1. Suppose the strictly proper normal rational matrix

$$A(\lambda) = \frac{N(\lambda)}{d(\lambda)}\,, \qquad \deg d(\lambda) = p \tag{3.63}$$

and let (A_0, B_0, C_0) be one of its simple realisations. Then any simple realisation of the matrix $A(\lambda)$ has the form $(QA_0Q^{-1}, QB_0, C_0Q^{-1})$ with a certain non-singular matrix Q. Keeping in mind that all simple matrices of the same dimension with the same characteristic polynomial are similar, the matrix Q can be selected in such a way that the equation $QA_0Q^{-1} = A_1$ for a cyclic matrix A_1 fulfills a prescribed form. Especially, we can achieve $A_1 = J$, where J is a Jordan matrix (1.97), and every distinct root of the polynomials $d(\lambda)$ is configured to exactly one Jordan block. The corresponding simple realisation (J, B_J, C_J) is named a *Jordan realisation*. But, if we choose $A_1 = A_F$, where

A_F is a Frobenius matrix (3.53), then the corresponding simple realisation (A_F, B_F, C_F) is called a *Frobenius realisation.* These two simple realisations are said to be *canonical.*

In this section the question is considered, how to produce a Jordan realisation from a given normal rational matrix in S-representation.

2. Suppose the normal strictly proper rational matrix (3.63) in S-representation

$$A(\lambda) = \frac{P(\lambda)Q'(\lambda)}{d(\lambda)} + G(\lambda)\,. \tag{3.64}$$

Then the following theorem gives the answer to the question, how to construct a simple Jordan realisation.

Theorem 3.24. *Suppose the normal $n \times m$ matrix $A(\lambda)$ in S-representation (3.64) with*

$$d(\lambda) = (\lambda - \lambda_1)^{\mu_1} \cdots (\lambda - \lambda_q)^{\mu_q}, \quad \mu_1 + \ldots + \mu_q = p\,. \tag{3.65}$$

Then a simple Jordan realisation of the matrix $A(\lambda)$ is attained by the following steps:

1) For each j, $(j = 1, \ldots, q)$ calculate the vectors

$$\begin{aligned} P_{jk} &= \frac{1}{(k-1)!} \left. \frac{\mathrm{d}^{k-1} P(\lambda)}{\mathrm{d}\lambda^{k-1}} \right|_{\lambda=\lambda_j}, \\ & \qquad\qquad\qquad\qquad\qquad\qquad (k = 1, \ldots, \mu_j) \\ Q_{jk} &= \frac{1}{(k-1)!} \left. \left[\frac{\mathrm{d}^{k-1}}{\mathrm{d}\lambda^{k-1}} \frac{Q(\lambda)(\lambda - \lambda_j)^{\mu_j}}{d(\lambda)} \right] \right|_{\lambda=\lambda_j}. \end{aligned} \tag{3.66}$$

2) For each j, $(j = 1, \ldots, q)$ build the matrices

$$\begin{aligned} \tilde{P}_j &= \begin{bmatrix} P_{j1} \; P_{j2} \ldots P_{j\mu_j} \end{bmatrix} \qquad (n \times \mu_j)\,, \\ \tilde{Q}_j &= \begin{bmatrix} Q'_{j\mu_j} \\ Q'_{j,\mu_j-1} \\ \vdots \\ Q'_{j1} \end{bmatrix} \qquad (\mu_j \times m)\,. \end{aligned} \tag{3.67}$$

3) Put together the matrices

$$\begin{aligned} P_J &= \begin{bmatrix} \tilde{P}_1 \; \tilde{P}_2 \ldots \tilde{P}_q \end{bmatrix} \qquad (n \times p)\,, \\ Q_J &= \begin{bmatrix} \tilde{Q}_1 \\ \tilde{Q}_2 \\ \vdots \\ \tilde{Q}_q \end{bmatrix} \qquad (p \times m)\,. \end{aligned} \tag{3.68}$$

4) Build the simple Jordan matrix J (1.88) according to the polynomial (3.65). Then

$$A(\lambda) = P_J(\lambda I_p - J)^{-1} Q_J\,, \tag{3.69}$$

and the realisation (J, Q_J, P_J) is a simple Jordan realisation.

The proof of Theorem 3.24 is prepared by two Lemmata.

Lemma 3.25. *Let $J_\mu(a)$ be an upper Jordan block (3.50), and $J_\mu(\lambda, a)$ be its corresponding characteristic matrix. Introduce for fixed μ the $\mu \times \mu$ matrices $H_{\mu i}$, $(i = 0, \ldots, \mu - 1)$ of the following shape:*

$$H_{\mu 0} = I_\mu\,, \quad H_{\mu 1} = \begin{bmatrix} 0 & 1 & 0 & \ldots & 0 \\ 0 & 0 & 1 & \ldots & 0 \\ \vdots & \vdots & \ddots & \ddots & \vdots \\ 0 & 0 & 0 & \ddots & 1 \\ 0 & 0 & 0 & \ldots & 0 \end{bmatrix}, \quad \ldots, \quad H_{\mu,\mu-1} = \begin{bmatrix} O_{1,\mu-1} & 1 \\ O_{\mu-1,\mu-1} & O_{\mu-1,1} \end{bmatrix}. \tag{3.70}$$

Then,

$$\begin{aligned} &\operatorname{adj}\left[\lambda I_\mu - J_\mu(a)\right] \\ &\quad = (\lambda - a)^{\mu-1} H_{\mu 0} + (\lambda - a)^{\mu-2} H_{\mu 1} + \ldots + (\lambda - a) H_{\mu,\mu-2} + H_{\mu,\mu-1}\,. \end{aligned} \tag{3.71}$$

Proof. The proof deduces immediately from (3.51) and (3.70). ∎

Lemma 3.26. *Assume the constant $n \times \mu$ and $\mu \times m$ matrices U and V with*

$$U = \begin{bmatrix} u_1 & \ldots & u_\mu \end{bmatrix}, \quad V = \begin{bmatrix} v'_\mu \\ \vdots \\ v'_1 \end{bmatrix},$$

where u_i, v'_i, $(i = 1, \ldots, \mu)$ are columns or rows, respectively. Then the equation

$$U \operatorname{adj}\left[\lambda I_\mu - J_\mu(a)\right] V = L_1 + (\lambda - a) L_2 + \ldots + (\lambda - a)^{\mu-1} L_\mu \tag{3.72}$$

is true, where

$$\begin{aligned} L_1 &= u_1 v'_1\,, \\ L_2 &= u_1 v'_2 + u_2 v'_1\,, \\ &\vdots \\ L_\mu &= u_1 v'_\mu + u_2 v'_{\mu-1} + \ldots + u_\mu v'_1 = UV\,. \end{aligned} \tag{3.73}$$

Proof. The proof follows directly by inserting (3.71) and (3.70) into the left side of (3.72). ∎

Remark 3.27. Concluding in reverse direction, it comes out that, under assumption (3.73), the right side of Relation (3.72) is equal to the left one.

Proof (of Theorem 3.24). From (3.64), we obtain

$$A(\lambda) = \frac{N(\lambda)}{d(\lambda)}, \tag{3.74}$$

where

$$N(\lambda) = P(\lambda)Q'(\lambda) + d(\lambda)G(\lambda). \tag{3.75}$$

Since Matrix (3.74) is strictly proper, it can be developed into partial fractions (2.98). Applying (3.65) and (2.96)–(2.97), this expansion can be expressed in the form

$$A(\lambda) = \sum_{j=1}^{q} A_j(\lambda), \tag{3.76}$$

where

$$A_j(\lambda) = \frac{A_{j1}}{(\lambda-\lambda_j)^{\mu_j}} + \frac{A_{j2}}{(\lambda-\lambda_j)^{\mu_j-1}} + \ldots + \frac{A_{j,\mu_j}}{(\lambda-\lambda_j)}, \quad (j = 1,\ldots,q). \tag{3.77}$$

The constant matrices A_{jk}, $(k = 1,\ldots,\mu_j)$ appearing in (3.77) are determined by the Taylor expansion at the point $\lambda = \lambda_j$:

$$\frac{N(\lambda)(\lambda-\lambda_j)^{\mu_j}}{d(\lambda)} = A_{j1} + (\lambda-\lambda_j)A_{j2} + \ldots + (\lambda-\lambda_j)^{\mu_j-1}A_{j,\mu_j} + (\lambda-\lambda_j)^{\mu_j}R_j(\lambda), \tag{3.78}$$

where $R_j(\lambda)$ is a rational matrix that is analytical in the point $\lambda = \lambda_j$. Utilising (3.74), (3.75), we can write

$$\frac{N(\lambda)(\lambda-\lambda_j)^{\mu_j}}{d(\lambda)} = P(\lambda)Q_j'(\lambda) + (\lambda-\lambda_j)^{\mu_j}G(\lambda), \tag{3.79}$$

where

$$Q_j(\lambda) = \frac{Q(\lambda)}{d_j(\lambda)}, \quad d_j(\lambda) = \frac{d(\lambda)}{(\lambda-\lambda_j)^{\mu_j}}.$$

Conformable with (3.78), for the determination of the matrices A_{jk}, $(k = 1,\ldots,\mu_j)$, we have to find the first μ_j terms of the separation on the right side of (3.79) in the Taylor series. Obviously, the matrices A_{jk}, $(k = 1,\ldots,\mu_j)$ do not depend on the matrix $G(\lambda)$.

Near the point $\lambda = \lambda_j$, suppose the developments

$$P(\lambda) = P_{j1} + (\lambda-\lambda_j)P_{j2} + \ldots + (\lambda-\lambda_j)^{\mu_j-1}P_{j,\mu_j} + \ldots,$$

$$Q_j(\lambda) = Q_{j1} + (\lambda-\lambda_j)Q_{j2} + \ldots + (\lambda-\lambda_j)^{\mu_j-1}Q_{j,\mu_j} + \ldots,$$

where the vectors P_{jk} and Q_{jk} are determined by (3.66). Then we get

$$P(\lambda)Q_j'(\lambda) = P_{j1}Q_{j1}' + (\lambda - \lambda_j)(P_{j1}Q_{j2}' + P_{j2}Q_{j1}') + \\ + (\lambda - \lambda_j)^2(P_{j1}Q_{j3}' + P_{j2}Q_{j2}' + P_{j3}Q_{j1}') + \dots \,.$$

Comparing this with (3.78) delivers

$$\begin{aligned} A_{j1} &= P_{j1}Q_{j1}' \,, \\ A_{j2} &= P_{j1}Q_{j2}' + P_{j2}Q_{j1}' \,, \\ \vdots \quad &\vdots \qquad \vdots \\ A_{j,\mu_j} &= P_{j1}Q_{j,\mu_j}' + P_{j2}Q_{j,\mu_j-1}' + \dots + P_{j,\mu_j}Q_{j1}' \,. \end{aligned}$$

Substituting this into (3.77) leads to

$$\begin{aligned} A_j(\lambda) &= (\lambda - \lambda_j)^{-\mu_j}\left[A_{j1} + (\lambda - \lambda_j)A_{j2} + \dots + (\lambda - \lambda_j)^{\mu_j - 1}A_{j,\mu_j}\right] \\ &= (\lambda - \lambda_j)^{-\mu_j}\Big[P_{j1}Q_{j1}' + (\lambda - \lambda_j)(P_{j1}Q_{j2}' + P_{j2}Q_{j1}') + \dots \\ &\qquad \dots + (\lambda - \lambda_j)^{\mu_j - 1}(P_{j1}Q_{j,\mu_j}' + \dots + P_{j,\mu_j}Q_{j1}')\Big] \,. \end{aligned}$$

Taking into account Remark 3.27, we obtain from the last expression

$$\begin{aligned} A_j(\lambda) &= (\lambda - \lambda_j)^{-\mu_j}\tilde{P}_j \operatorname{adj}\left[\lambda I_{\mu_j} - J_{\mu_j}(\lambda_j)\right]\tilde{Q}_j \\ &= \tilde{P}_j\left[\lambda I_{\mu_j} - J_{\mu_j}(\lambda_j)\right]^{-1}\tilde{Q}_j \,, \end{aligned}$$

where the matrices $\tilde{P}_j$, $\tilde{Q}_j$ are committed by (3.67). From the last equations and (3.76), it follows

$$A(\lambda) = \sum_{j=1}^{q} \tilde{P}_j\left[\lambda I_{\mu_j} - J_{\mu_j}(\lambda_j)\right]^{-1}\tilde{Q}_j = P_J(\lambda I_p - J)^{-1}Q_J \,,$$

where P_J, Q_J are the matrices (3.68) and

$$\lambda I_p - J = \operatorname{diag}\left[\lambda I_{\mu_1} - J_{\mu_1}(\lambda_1), \lambda I_{\mu_2} - J_{\mu_2}(\lambda_2), \dots, \lambda I_{\mu_q} - J_{\mu_q}(\lambda_q)\right],$$

which is equivalent to Formula (3.69). Since in the present case the $p \times p$ matrix J possesses the minimal possible dimension, the realisation (J, B_J, C_J) is minimal. Therefore, the pair (J, B_J) is controllable, and the pair $[J, C_J]$ is observable. Finally, per construction, the matrix J is cyclic. Hence (J, B_J, C_J) is a simple Jordan realisation of the matrix $A(\lambda)$. ∎

3.

Example 3.28. Find the the Jordan realisation of the strictly proper normal matrix

$$A(\lambda) = \frac{\begin{bmatrix} (\lambda-1)^2 & 1 \\ 0 & \lambda-2 \end{bmatrix}}{(\lambda-1)^2(\lambda-2)} . \tag{3.80}$$

Using the notation in Section 3.4

$$g(\lambda) = 1, \quad \phi(\lambda) = \psi(\lambda) = I_2 , \quad \beta(\lambda) = \lambda - 2, \quad \alpha(\lambda) = (\lambda-1)^2 ,$$

$$d(\lambda) = (\lambda-1)^2(\lambda-2), \quad L(\lambda) = 0, \quad \lambda_1 = 1, \; \lambda_2 = 2$$

is performed, and applying (3.41) yields

$$P(\lambda) = \begin{bmatrix} 1 \\ \lambda - 2 \end{bmatrix}, \quad Q'(\lambda) = \begin{bmatrix} (\lambda-1)^2 & 1 \end{bmatrix} .$$

For constructing a simple Jordan realisation, we have to find the vectors (3.66). For the root $\lambda_1 = 1$, we introduce the notation

$$P_1(\lambda) = P(\lambda) = \begin{bmatrix} 1 \\ \lambda - 2 \end{bmatrix}, \quad Q'_1(\lambda) = \frac{Q'(\lambda)}{\lambda - 2} = \begin{bmatrix} \frac{(\lambda-1)^2}{\lambda-2} & \frac{1}{\lambda-2} \end{bmatrix} .$$

Using (3.66), we obtain

$$P_{11} = P(\lambda)\,|_{\lambda=\lambda_1} = \begin{bmatrix} 1 \\ -1 \end{bmatrix}, \quad P_{12} = \left. \frac{\mathrm{d}P_1(\lambda)}{\mathrm{d}\lambda} \right|_{\lambda=\lambda_1} = \begin{bmatrix} 0 \\ 1 \end{bmatrix},$$

and furthermore

$$Q'_{11} = Q'_1(\lambda)\,|_{\lambda=\lambda_1} = \begin{bmatrix} 0 & -1 \end{bmatrix}, \quad Q'_{12} = \left. \frac{\mathrm{d}Q'_1(\lambda)}{\mathrm{d}\lambda} \right|_{\lambda=\lambda_1} = \begin{bmatrix} 0 & -1 \end{bmatrix} .$$

Then (3.67) ensures

$$\tilde{P}_1 = \begin{bmatrix} 1 & 0 \\ -1 & 1 \end{bmatrix}, \quad \tilde{Q}_1 = \begin{bmatrix} 0 & -1 \\ 0 & -1 \end{bmatrix} .$$

For the single root $\lambda_2 = 2$, we denote

$$P_2(\lambda) = P(\lambda) = \begin{bmatrix} 1 \\ \lambda - 2 \end{bmatrix}, \quad Q'_2(\lambda) = \frac{Q'(\lambda)}{(\lambda-1)^2} = \begin{bmatrix} 1 & \frac{1}{(\lambda-1)^2} \end{bmatrix}$$

and for $\lambda = 2$, we get

$$P_{21} = P(\lambda)\,|_{\lambda=\lambda_2} = \begin{bmatrix} 1 \\ 0 \end{bmatrix}, \quad Q'_{21} = Q'_2(\lambda)\,|_{\lambda=\lambda_2} = \begin{bmatrix} 1 & 1 \end{bmatrix} .$$

Applying (3.68) yields

$$P_J = \begin{bmatrix} 1 & 0 & 1 \\ -1 & 1 & 0 \end{bmatrix}, \quad Q_J = \begin{bmatrix} 0 & -1 \\ 0 & -1 \\ 1 & 1 \end{bmatrix} .$$

Thus the simple Jordan realisation of Matrix (3.80) possesses the shape (J, B_J, C_J) with

$$J = \begin{bmatrix} 1 & 1 & 0 \\ 0 & 1 & 0 \\ 0 & 0 & 2 \end{bmatrix} .$$

□

3.7 Construction of Simple Frobenius Realisations

1. For constructing a simple Jordan-realisation, the roots of the polynomials $d(\lambda)$ have to be calculated, and this task can be connected with honest numerical problems. From this point of view, it is much easier to produce the realisation (A_F, B_F, C_F), where the matrix A_F has the Frobenius normal form (3.53), that turns out to be the accompanying matrix for the polynomial $d(\lambda)$ in (3.56). The assigned characteristic matrix to A_F has the shape (3.54), and the S-representation of the matrix $(\lambda I_p - A_F)^{-1}$ is determined by Relations (3.62). For a given realisation (A_F, B_F, C_F), the transfer matrix $A(\lambda)$ has the shape (3.63) with

$$N(\lambda) = C_F \operatorname{adj}(\lambda I_p - A_F) B_F \,.$$

Taking advantage from (3.57), the last equation can be represented in the form

$$N(\lambda) = P(\lambda)Q'(\lambda) + d(\lambda)\tilde{G}(\lambda) \tag{3.81}$$

with

$$P(\lambda) = C_F \begin{bmatrix} 1 \\ \lambda \\ \vdots \\ \lambda^{n-1} \end{bmatrix}, \qquad \begin{aligned} Q'(\lambda) &= \begin{bmatrix} d_1(\lambda) & \dots & d_{n-1}(\lambda) & 1 \end{bmatrix} B_F \,, \\ G(\lambda) &= C_F G_F(\lambda) B_F \,. \end{aligned} \tag{3.82}$$

Inserting (3.81) and (3.82) into (3.63), we get the wanted S-representation. Per construction, Relation (3.36) is fulfilled, *i.e.* the obtained S-representation is minimal. Therefore, Formulae (3.81), (3.82) forth the possibility of direct transfer from the Frobenius realisation to the corresponding minimal S-representation of its transfer matrix.

2.

Example 3.29. Assume the Frobenius realisation with

$$A_F = \begin{bmatrix} 0 & 1 & 0 \\ 0 & 0 & 1 \\ -2 & -1 & -1 \end{bmatrix}, \quad B_F = \begin{bmatrix} 1 & 0 \\ 2 & 3 \\ 0 & 1 \end{bmatrix}, \quad C_F = \begin{bmatrix} 1 & 2 & 0 \\ 3 & 1 & -1 \end{bmatrix}. \tag{3.83}$$

Here A_F is the accompanying matrix for the polynomial

$$d(\lambda) = \lambda^3 + \lambda^2 + \lambda + 2 \,,$$

so we get the coefficients $d_1 = 1$, $d_2 = 1$, $d_3 = 2$. In this case, the polynomials (3.56) have the shape

$$d_1(\lambda) = \lambda^2 + \lambda + 1, \quad d_2(\lambda) = \lambda + 1 \,.$$

Hence recall (3.58) for the considered example, we receive

$$P'_F(\lambda) = \begin{bmatrix} 1 \ \lambda \ \lambda^2 \end{bmatrix}, \quad Q'_F(\lambda) = \begin{bmatrix} \lambda^2+\lambda+1 & \lambda+1 & 1 \end{bmatrix}.$$

Then using (3.82), we obtain

$$\begin{aligned} P(\lambda) &= \begin{bmatrix} 1 & 2 & 0 \\ 3 & 1 & -1 \end{bmatrix} \begin{bmatrix} 1 \\ \lambda \\ \lambda^2 \end{bmatrix} = \begin{bmatrix} 2\lambda+1 \\ -\lambda^2+\lambda+3 \end{bmatrix}, \\ Q'(\lambda) &= \begin{bmatrix} \lambda^2+\lambda+1 & \lambda+1 & 1 \end{bmatrix} \begin{bmatrix} 1 & 0 \\ 2 & 3 \\ 0 & 1 \end{bmatrix} = \begin{bmatrix} \lambda^2+3\lambda+3 & 3\lambda+4 \end{bmatrix}. \end{aligned} \tag{3.84}$$

Moreover, a direct calculation with the help of (3.61) yields

$$G(\lambda) = \begin{bmatrix} 1 & 2 & 0 \\ 3 & 1 & -1 \end{bmatrix} \begin{bmatrix} 0 & 0 & 0 \\ -1 & 0 & 0 \\ -\lambda & -1 & 0 \end{bmatrix} \begin{bmatrix} 1 & 0 \\ 2 & 3 \\ 0 & 1 \end{bmatrix} = \begin{bmatrix} -2 & 0 \\ \lambda+1 & 3 \end{bmatrix},$$

which together with (3.84) gives the result

$$\begin{aligned} A(\lambda) &= \frac{P(\lambda)Q'(\lambda)}{d(\lambda)} + G(\lambda) \\ &= \frac{\begin{bmatrix} 2\lambda+1 \\ -\lambda^2+\lambda+3 \end{bmatrix} \begin{bmatrix} \lambda^2+3\lambda+3 & 3\lambda+4 \end{bmatrix}}{\lambda^3+\lambda^2+\lambda+2} + \begin{bmatrix} -2 & 0 \\ \lambda+1 & 3 \end{bmatrix}. \end{aligned}$$

□

3. It is remarkable that there exists a rather easy way from a minimal S-representation (3.64) to the matrix $A(\lambda)$ of the simple Frobenius realisation.

Theorem 3.30. *Let for the strictly proper normal $n \times m$ matrix $A(\lambda)$ be given the minimal S-representation (3.64), where*

$$d(\lambda) = \lambda^s + d_1\lambda^{s-1} + \ldots + d_s. \tag{3.85}$$

Since the S-representation (3.64) is minimal, it follows

$$\begin{aligned} P(\lambda) &= N_1 + N_2\lambda + \ldots N_s\lambda^{s-1}, \\ Q(\lambda) &= M_1 + M_2\lambda + \ldots M_s\lambda^{s-1}, \end{aligned} \tag{3.86}$$

where the N_i, M_i, $(i = 1, \ldots, s)$ are constant vectors of dimensions $n \times 1$ and $m \times 1$, respectively. Introduce the columns $B_1, \ldots, B_s$ recursively by

$$\begin{aligned} B_1 &= M_s, \\ B_2 &= M_{s-1} - d_1B_1, \\ B_3 &= M_{s-2} - d_1B_2 - d_2B_1, \\ &\vdots \qquad \vdots \\ B_s &= M_1 - d_1B_{s-1} - d_2B_{s-2} - \ldots - d_{s-1}B_1. \end{aligned} \tag{3.87}$$

With account to (3.86) and (3.87), build the matrices

$$C_F = \begin{bmatrix} N_1 \dots N_s \end{bmatrix}, \quad B_F = \begin{bmatrix} B'_1 \\ \vdots \\ B'_s \end{bmatrix}. \tag{3.88}$$

Then the matrix

$$\tilde{A}(\lambda) = C_F(\lambda I_s - A_F)^{-1} B_F, \tag{3.89}$$

where A_F is the accompanying Frobenius matrix for the polynomial (3.85), defines the minimal standard realisation of the matrix $A(\lambda)$, that means, the realisation (A_F, B_F, C_F) is the simple Frobenius realisation of the matrix $A(\lambda)$.

Proof. Using (3.62), we obtain from (3.89)

$$\tilde{A}(\lambda) = \frac{C_F P_F(\lambda) Q'_F(\lambda) B_F}{d(\lambda)} + C_F G_F(\lambda) B_F. \tag{3.90}$$

From (3.88) and (3.58), we get

$$C_F P_F(\lambda) = N_1 + N_2\lambda + \dots + N_s\lambda^{s-1} = P(\lambda), \tag{3.91}$$

and also

$$Q'_F(\lambda) B_F = d_1(\lambda) B'_1 + d_2(\lambda) B'_2 + \dots + d_{s-1}(\lambda) B'_{s-1} + B'_s,$$

where $d_1(\lambda), \dots, d_{s-1}(\lambda)$ are the polynomials (3.56). Substituting (3.56) into the last equation, we find

$$\begin{aligned} Q'_F(\lambda) B_F &= (\lambda^{s-1} + d_1\lambda^{s-2} + \dots + d_{s-1}) B'_1 + \dots + (\lambda + d_1) B'_{s-1} + B'_s \\ &= \lambda^{s-1} B'_1 + \lambda^{s-2}(d_1 B'_1 + B'_2) + \dots + (d_{s-1} B'_1 + d_{s-2} B'_2 + \dots + B'_s) \end{aligned} \tag{3.92}$$

such that from (3.87), it follows

$$\begin{aligned} M_s &= B_1, \\ M_{s-1} &= B_2 + d_1 B_1, \\ M_{s-2} &= B_3 + d_1 B_2 + d_2 B_1, \\ &\vdots \\ M_1 &= B_s + d_1 B_{s-1} + d_2 B_{s-2} + \dots + d_{s-1} B_1, \end{aligned} \tag{3.93}$$

and from (3.92) with (3.86), we find

$$Q'_F(\lambda) B_F = \lambda^{s-1} M'_s + \dots + \lambda M'_2 + M'_1 = Q'(\lambda). \tag{3.94}$$

Finally, by virtue of this and (3.91), we produce from (3.90)

$$\tilde{A}(\lambda) = \frac{P(\lambda)Q'(\lambda)}{d(\lambda)} + C_F G_F(\lambda) B_F \,. \tag{3.95}$$

Comparing this expressions with (3.64) and paying attention to the fact that the matrices $A(\lambda)$ and $\tilde{A}(\lambda)$ are strictly proper, and the matrices $G(\lambda)$ and $C_F G_F(\lambda) B_F$ are polynomial matrices, we obtain

$$A(\lambda) = \tilde{A}(\lambda)\,.$$

■

Example 3.31. Under the conditions of Example 3.29, we obtain

$$P(\lambda) = \begin{bmatrix} 1 \\ 3 \end{bmatrix} + \begin{bmatrix} 2 \\ 1 \end{bmatrix} \lambda + \begin{bmatrix} 0 \\ -1 \end{bmatrix} \lambda^2 ,$$

so that with regard to (3.86), we configure

$$N_1 = \begin{bmatrix} 1 \\ 3 \end{bmatrix}, \quad N_2 = \begin{bmatrix} 2 \\ 1 \end{bmatrix}, \quad N_3 = \begin{bmatrix} 0 \\ -1 \end{bmatrix},$$

that agrees with (3.84). Moreover, (3.84) yields

$$Q(\lambda) = \begin{bmatrix} 3 \\ 4 \end{bmatrix} + \begin{bmatrix} 3 \\ 3 \end{bmatrix} \lambda + \begin{bmatrix} 1 \\ 0 \end{bmatrix} \lambda^2$$

and thus

$$M_1 = \begin{bmatrix} 3 \\ 4 \end{bmatrix}, \quad M_2 = \begin{bmatrix} 3 \\ 3 \end{bmatrix}, \quad M_3 = \begin{bmatrix} 1 \\ 0 \end{bmatrix}.$$

Applying (3.93), we obtain

$$\begin{aligned}
B_1' &= M_3' = \begin{bmatrix} 1 & 0 \end{bmatrix}, \\
B_2' &= M_2' - d_1 B_1' = \begin{bmatrix} 3 & 3 \end{bmatrix} - \begin{bmatrix} 1 & 0 \end{bmatrix} = \begin{bmatrix} 2 & 3 \end{bmatrix}, \\
B_3' &= M_1' - d_1 B_2' - d_2 B_1' = \begin{bmatrix} 3 & 4 \end{bmatrix} - \begin{bmatrix} 2 & 3 \end{bmatrix} - \begin{bmatrix} 1 & 0 \end{bmatrix} = \begin{bmatrix} 0 & 1 \end{bmatrix},
\end{aligned}$$

that with respect to (3.88) can be written as

$$B_F = \begin{bmatrix} B_1' \\ B_2' \\ B_3' \end{bmatrix} = \begin{bmatrix} 1 & 0 \\ 2 & 3 \\ 0 & 1 \end{bmatrix}.$$

This result is again consistent with (3.83). □

It is referred to the fact that analogue formulae to (3.87), (3.93) for realisations with vertical Frobenius matrices were dedicated in different way in [165].

Remark 3.32. It is important that Formulae (3.87), (3.93) only depend on the coefficients of the characteristic polynomial (3.85), but not on its roots, that's why the practical handling of these formulae is less critical.

3.8 Construction of S-representations from Simple Realisations. General Case

1. If the simple realisation (A, B, C) of a normal rational $n \times m$ transfer matrix $A(\lambda)$ is known, then the corresponding S-representation of $A(\lambda)$ can be build on basis of the general considerations in Section 3.4. Indeed, let the simple realisation (A, B, C) be given, so

$$A(\lambda) = \frac{C\,\mathrm{adj}(\lambda I_p - A)B}{\det(\lambda I_p - A)} = \frac{N(\lambda)}{d_A(\lambda)}\,. \tag{3.96}$$

is valid, and by equivalence transformations, the representation

$$\mathrm{adj}(\lambda I_p - A) = \phi(\lambda)\begin{bmatrix} \alpha(\lambda) & 1 \\ L(\lambda) & \beta(\lambda) \end{bmatrix}\psi(\lambda)$$

can be generated with unimodular matrices $\phi(\lambda)$, $\psi(\lambda)$. Then for constructing the S-representation, Theorem 3.20 is applicable. Using Theorem 3.20, the last equation yields

$$\mathrm{adj}(\lambda I_p - A) = P_a(\lambda)Q_a'(\lambda) + d_A(\lambda)G_a(\lambda)$$

that leads to

$$C\,\mathrm{adj}(\lambda I_p - A)B = P(\lambda)Q'(\lambda) + d_A(\lambda)G(\lambda)\,,$$

where

$$P(\lambda) = CP_a(\lambda), \quad Q(\lambda) = B'Q_a(\lambda), \quad G(\lambda) = CG_a(\lambda)B\,.$$

The last relations proves to be an S-representation of the matrix (3.96)

$$A(\lambda) = \frac{P(\lambda)Q'(\lambda)}{d_A(\lambda)} + G(\lambda)$$

that could easily transformed into a minimal S-representation.

2. For calculating the adjoint matrix $\mathrm{adj}(\lambda I_p - A)$, we can benefit from some general relations in [51]. Assume

$$d_A(\lambda) = \lambda^p - q_1\lambda^{p-1} - \ldots - q_p\,.$$

Then the adjoint matrix $\mathrm{adj}(\lambda I_p - A)$ is determined by the formula

$$\mathrm{adj}(\lambda I_p - A) = \lambda^{p-1}I_p + \lambda^{p-2}F_1 + \ldots + F_{p-1}\,, \tag{3.97}$$

where

$$F_1 = A - q_1 I_p\,, \quad F_2 = A^2 - q_1 A - q_2 I_p\,, \ \ldots$$

or generally

$$F_k = A^k - q_1 A^{k-1} - \ldots - q_k I_p \,.$$

The matrices $F_1, \ldots, F_{p-1}$ can be calculated successively by the recursion

$$F_k = AF_{k-1} - q_k I_p \,, \quad (k = 1, 2, \ldots, p-1;\; F_0 = I_p) \,.$$

After this, the solution can be checked by the equation

$$AF_{p-1} - q_p I_p = O_{pp} \,.$$

3.

Example 3.33. Suppose the simple realisation (A, B, C) with

$$A = \begin{bmatrix} 0 & -1 \\ 0 & 1 \end{bmatrix}, \quad B = \begin{bmatrix} 0 & 1 \\ 1 & -1 \end{bmatrix}, \quad C = \begin{bmatrix} 1 & 0 \\ -1 & 1 \end{bmatrix}. \tag{3.98}$$

In the present case, we have

$$d_A(\lambda) = \lambda^2 - \lambda$$

so we read $p = 2$ and $q_1 = 1$, $q_2 = 0$. Using (3.97), we find

$$\operatorname{adj}(\lambda I_2 - A) = \lambda I_2 + F_1$$

with

$$F_1 = A - I_2 = \begin{bmatrix} -1 & -1 \\ 0 & 0 \end{bmatrix}.$$

Formula (3.97) delivers

$$\operatorname{adj}(\lambda I_2 - A) = -\begin{bmatrix} -\lambda + 1 & 1 \\ 0 & -\lambda \end{bmatrix},$$

which is easily produced by direct calculation. Applying Theorem 3.20, we get

$$-\operatorname{adj}(\lambda I_2 - A) = P(\lambda)Q'(\lambda) + d_A G(\lambda) \,,$$

where

$$P(\lambda) = \begin{bmatrix} 1 \\ -\lambda \end{bmatrix}, \quad Q(\lambda) = \begin{bmatrix} -\lambda + 1 \\ 1 \end{bmatrix}, \quad G(\lambda) = \begin{bmatrix} 0 & 0 \\ -1 & 0 \end{bmatrix}.$$

Therefore, the S-representation of the matrix $A(\lambda)$ for the realisation (3.98) takes the shape

$$A(\lambda) = \frac{\begin{bmatrix} -1 \\ \lambda + 1 \end{bmatrix} \begin{bmatrix} 1 & -\lambda \end{bmatrix}}{\lambda^2 - \lambda} + \begin{bmatrix} 0 & 0 \\ 0 & 1 \end{bmatrix}. \tag{3.99}$$

The obtained S-representation is minimal. □

3.9 Construction of Complete MFDs for Normal Matrices

1. Let the normal matrix in standard form (2.21)

$$A(\lambda) = \frac{N(\lambda)}{d(\lambda)} \tag{3.100}$$

be given with $\deg d(\lambda) = p$. Then in accordance with Section 3.1, Matrix (3.100) allows the irreducible complete MFD

$$A(\lambda) = a_l^{-1}(\lambda) b_l(\lambda) = b_r(\lambda) a_r^{-1}(\lambda) \tag{3.101}$$

for which

$$\operatorname{ord} a_l(\lambda) = \operatorname{ord} a_r(\lambda) = \operatorname{Mdeg} L(\lambda) = p\,.$$

In principle, for building a complete MFD (3.101), the general methods from Section 2.4 can be applied. However, with respect to numerical effort and numeric stability, essentially more effective methods can be developed when we profit from the special structure of normal matrices while constructing complete MFDs.

2.

Theorem 3.34. *Let the numerator of Matrix (3.100) be brought into the form (3.38)*

$$N(\lambda) = g(\lambda)\phi(\lambda) \begin{bmatrix} \alpha(\lambda) & 1 \\ L(\lambda) & \beta(\lambda) \end{bmatrix} \psi(\lambda)\,. \tag{3.102}$$

Then the pair of matrices

$$a_l(\lambda) = \begin{bmatrix} d(\lambda) & O_{1,n-1} \\ -\beta(\lambda) & I_{n-1} \end{bmatrix} \phi^{-1}(\lambda)\,, \quad b_l(\lambda) = a_l(\lambda) A(\lambda) \tag{3.103}$$

proves to be a complete LMFD, and the pair of matrices

$$a_r(\lambda) = \psi^{-1}(\lambda) \begin{bmatrix} O_{m-1,1} & I_{m-1} \\ d(\lambda) & -\alpha(\lambda) \end{bmatrix}, \quad b_r(\lambda) = A(\lambda) a_r(\lambda)$$

is a complete RMFD of Matrix (3.100).

Proof. Applying Relations (3.44)–(3.49), Matrix (3.102) is represented in the form

$$N(\lambda) = g(\lambda)\phi(\lambda) \begin{bmatrix} 1 & O_{1,n-1} \\ \beta(\lambda) & I_{n-1} \end{bmatrix} \begin{bmatrix} 1 & O_{1,m-1} \\ O_{n-1,1} & d(\lambda) G_2(\lambda) \end{bmatrix} \begin{bmatrix} \alpha(\lambda) & 1 \\ I_{m-1} & O_{m-1,1} \end{bmatrix} \psi(\lambda)$$

and with respect to (3.100), we get

$$A(\lambda) = g(\lambda)\phi(\lambda) \begin{bmatrix} 1 & O_{1,n-1} \\ \beta(\lambda) & I_{n-1} \end{bmatrix} \begin{bmatrix} \dfrac{1}{d(\lambda)} & O_{1,m-1} \\ O_{n-1,1} & G_2(\lambda) \end{bmatrix} \begin{bmatrix} \alpha(\lambda) & 1 \\ I_{m-1} & O_{m-1,1} \end{bmatrix} \psi(\lambda).$$

Multiplying this from left with the matrix $a_l(\lambda)$ in (3.103), and considering

$$\begin{bmatrix} d(\lambda) & O_{1,n-1} \\ -\beta(\lambda) & I_{n-1} \end{bmatrix} \begin{bmatrix} 1 & O_{1,n-1} \\ \beta(\lambda) & I_{n-1} \end{bmatrix} = \begin{bmatrix} d(\lambda) & O_{1,n-1} \\ O_{n-1,1} & I_{n-1} \end{bmatrix},$$

we find out that the product

$$\begin{aligned} a_l(\lambda)A(\lambda) &= g(\lambda) \begin{bmatrix} 1 & O_{1,m-1} \\ O_{n-1,1} & G_2(\lambda) \end{bmatrix} \begin{bmatrix} \alpha(\lambda) & 1 \\ I_{m1} & O_{m-1,1} \end{bmatrix} \psi(\lambda) \\ &= g(\lambda) \begin{bmatrix} \alpha(\lambda) & 1 \\ G_2(\lambda) & O_{m-1,1} \end{bmatrix} \psi(\lambda) = b_l(\lambda) \end{aligned} \tag{3.104}$$

proves to be a polynomial matrix. Per construction, we have $\det a_l(\lambda) \approx d(\lambda)$ and $\operatorname{ord} a_l(\lambda) = \deg d(\lambda)$, that's why the LMFD is complete. For a right MFD the proof runs analogously. ∎

3.

Example 3.35. Let us have a normal matrix (3.100) with

$$N(\lambda) = \begin{bmatrix} \lambda & 2\lambda - 3 \\ \lambda^2 + \lambda + 1 & 2\lambda^2 - 5 \end{bmatrix}, \quad d(\lambda) = \lambda^2 - 4\lambda + 3.$$

In this case, we find

$$N(\lambda) = \begin{bmatrix} 0 & 1 \\ 1 & \lambda + 1 \end{bmatrix} \begin{bmatrix} \lambda - 2 & 1 \\ 2\lambda - 3 & \lambda \end{bmatrix} \begin{bmatrix} 0 & 1 \\ 1 & 0 \end{bmatrix},$$

so we get

$$\phi(\lambda) = \begin{bmatrix} 0 & 1 \\ 1 & \lambda + 1 \end{bmatrix}, \quad \psi(\lambda) = \begin{bmatrix} 0 & 1 \\ 1 & 0 \end{bmatrix}, \quad g(\lambda) = 1,$$

and with respect to (3.103), (3.104), we obtain immediately

$$a_l(\lambda) = \begin{bmatrix} d(\lambda) & 0 \\ -\lambda & 1 \end{bmatrix} \begin{bmatrix} -(\lambda + 1) & 1 \\ 1 & 0 \end{bmatrix} = \begin{bmatrix} -d(\lambda)(\lambda + 1) & d(\lambda) \\ \lambda^2 + \lambda + 1 & -\lambda \end{bmatrix},$$

$$b_l(\lambda) = a_l(\lambda)A(\lambda) = \begin{bmatrix} 1 & \lambda - 2 \\ 0 & -1 \end{bmatrix}.$$

In the present case, we have $\deg a_l(\lambda) = 3$. The degree of the matrix $a_l(\lambda)$ can be decreased, if we build the row-reduced form. The extended matrix according to the above pair has the shape

$$R_h(\lambda) = \begin{bmatrix} -\lambda^3 + 3\lambda^2 + \lambda - 3 & \lambda^2 - 4\lambda + 3 & 1 & \lambda - 2 \\ \lambda^2 + \lambda + 1 & -\lambda & 0 & -1 \end{bmatrix}.$$

Multiplying the matrix $R_h(\lambda)$ from left with the unimodular matrix

$$\phi(\lambda) = \begin{bmatrix} 1 & \lambda - 4 \\ 0.5\lambda & 0.5\lambda^2 - 2\lambda + 1 \end{bmatrix},$$

we arrive at

$$\phi(\lambda)R_h(\lambda) = \begin{bmatrix} -2\lambda - 7 & 3 & 1 & 2 \\ -2.5\lambda + 1 & 0.5\lambda & 0.5\lambda & \lambda - 1 \end{bmatrix}.$$

This matrix corresponds to the complete LMFD, where

$$a_l(\lambda) = \begin{bmatrix} -2\lambda - 7 & 3 \\ -2.5\lambda + 1 & 0.5\lambda \end{bmatrix}, \quad b_l(\lambda) = \begin{bmatrix} 1 & 2 \\ 0.5\lambda & \lambda - 1 \end{bmatrix}$$

and the matrix $a_l(\lambda)$ is row-reduced. Thus $\deg a_l(\lambda) = 1$, and this degree cannot be decreased. □

4. For a known S-representation of the normal matrix

$$A(\lambda) = \frac{P(\lambda)Q'(\lambda)}{d(\lambda)} + G(\lambda)\,, \tag{3.105}$$

a complete MFD can be built by the following theorem.

Theorem 3.36. *Suppose the ILMFD and IRMFD*

$$\frac{P(\lambda)}{d(\lambda)} = a_l^{-1}(\lambda)\tilde{b}_l(\lambda), \qquad \frac{Q'(\lambda)}{d(\lambda)} = \tilde{b}_r(\lambda)a_r^{-1}(\lambda)\,. \tag{3.106}$$

Then the expressions

$$\begin{aligned} A(\lambda) &= a_l^{-1}(\lambda)\left[\tilde{b}_l(\lambda)Q'(\lambda) + a_l(\lambda)G(\lambda)\right] = a_l^{-1}(\lambda)b_l(\lambda)\,, \\ A(\lambda) &= \left[P(\lambda)\tilde{b}_r(\lambda) + G(\lambda)a_r(\lambda)\right]a_r^{-1}(\lambda) = b_r(\lambda)a_r^{-1}(\lambda) \end{aligned} \tag{3.107}$$

define a complete MFD of Matrix (3.105).

Proof. Due to Remark 3.4, the matrix $P(\lambda)/d(\lambda)$ is normal. Therefore, for the ILMFD (3.106) $\det a_l(\lambda) \approx \det d(\lambda)$ is true, and the first row in (3.107) proves to be a complete LMFD of the matrix $A(\lambda)$. In analogy, we realise that a complete RMFD stands in the second row. ∎

5.

Example 3.37. For Matrix (3.99) in Example 3.33,

$$\frac{P(\lambda)}{d(\lambda)} = \frac{\begin{bmatrix} -1 \\ \lambda+1 \end{bmatrix}}{\lambda^2-\lambda}, \qquad \frac{Q'(\lambda)}{d(\lambda)} = \frac{\begin{bmatrix} 1 & -\lambda \end{bmatrix}}{\lambda^2-\lambda}$$

is performed. It is easily checked, that in this case, we can choose

$$a_l(\lambda) = \begin{bmatrix} -2\lambda & -\lambda \\ \lambda+1 & 1 \end{bmatrix}, \quad \tilde{b}_l(\lambda) = \begin{bmatrix} -1 \\ 0 \end{bmatrix},$$

$$a_r(\lambda) = \begin{bmatrix} -\lambda & \lambda \\ -\lambda & 1 \end{bmatrix}, \qquad \tilde{b}_r(\lambda) = \begin{bmatrix} 1 & 0 \end{bmatrix}$$

and according to (3.107), we build the matrices

$$b_l(\lambda) = \begin{bmatrix} -1 & 0 \\ 0 & 1 \end{bmatrix}, \quad b_r(\lambda) = \begin{bmatrix} -1 & 0 \\ 1 & 1 \end{bmatrix}.$$

□

3.10 Normalisation of Rational Matrices

1. During the construction of complete LMFD, RMFD and simple realisations for normal rational matrices, we have to take into account the structural peculiarities, and the equations that exist between their elements. Indeed, even arbitrarily small inaccuracies during the calculation of the elements of a normal matrix (3.100), most likely will lead to a situation, where the divisibility of all minors of second order by the denominator is violated, and the resulting matrix $\tilde{A}(\lambda)$ is no longer normal. After that, also the irreducible MFD, built from the matrix $\tilde{A}(\lambda)$ will not be complete, and the values of $\operatorname{ord} a_l(\lambda)$ and $\operatorname{ord} a_r(\lambda)$ in the configured IMFDs will get too large. Also the matrix $\tilde{A}$ that is assigned by the minimal realisation $(\tilde{A}, \tilde{B}, \tilde{C})$ according to the matrix $\tilde{A}(\lambda)$, would have too high dimension. As a consequence of these errors after transition to an IMFD or to corresponding minimal realisations, we would obtain linear models with totally different dynamic behavior than that of the original object, which is described by the transfer matrix (3.100).

2. Let us illustrate the above remarks by a simple example.

Example 3.38. Consider the nominal transfer matrix

$$A(\lambda) = \frac{\begin{bmatrix} \lambda-a & 0 \\ 0 & \lambda-b \end{bmatrix}}{(\lambda-a)(\lambda-b)}, \quad a \neq b \tag{3.108}$$

that proves to be normal. Assume that, due to practical calculations, the approximated matrix

$$\tilde{A}(\lambda) = \frac{\begin{bmatrix} \lambda - a + \epsilon & 0 \\ 0 & \lambda - b \end{bmatrix}}{(\lambda - a)(\lambda - b)} \tag{3.109}$$

is built, that is normal only for $\epsilon = 0$. For the nominal matrix (3.108), there exists the simple realisation (A, B, C) with

$$A = \begin{bmatrix} a & 0 \\ 0 & b \end{bmatrix}, \quad B = \begin{bmatrix} 0 & 1 \\ 1 & 0 \end{bmatrix}, \quad C = \begin{bmatrix} 0 & 1 \\ 1 & 0 \end{bmatrix}. \tag{3.110}$$

All other simple realisations of Matrix (3.108) are produced from (3.110) by similarity transformations. Realisation (3.110) corresponds to the system of differential equations of second order

$$\begin{aligned} \dot{x}_1 &= a x_1 + u_2 \\ \dot{x}_2 &= b x_2 + u_1 \\ y_1 &= x_2, \quad y_2 = x_1 \,. \end{aligned} \tag{3.111}$$

For $\epsilon \neq 0$, we find the minimal realisation $(A_\epsilon, B_\epsilon, C_\epsilon)$ for (3.109), where

$$A_\epsilon = \begin{bmatrix} a & 0 & 0 \\ 0 & a & 0 \\ 0 & 0 & b \end{bmatrix}, \quad B_\epsilon = \begin{bmatrix} \dfrac{\epsilon}{a-b} & 0 \\ 0 & 1 \\ 1 - \dfrac{\epsilon}{a-b} & 0 \end{bmatrix}, \quad C_\epsilon = \begin{bmatrix} 1 & 0 & 1 \\ 0 & 1 & 0 \end{bmatrix}. \tag{3.112}$$

All other minimal realisations of Matrix (3.109) are held from (3.112) by similarity transformations.

Realisation (3.112) is assigned to the differential equation of third order

$$\begin{aligned} \dot{x}_1 &= a x_1 + \frac{\epsilon}{a-b} u_1 \\ \dot{x}_2 &= a x_2 + u_2 \\ \dot{x}_3 &= b x_3 + \left(1 - \frac{\epsilon}{a-b}\right) u_1 \\ y_1 &= x_1 + x_3, \quad y_2 = x_2 \,. \end{aligned}$$

For $\epsilon = 0$, these equations do not turn into (3.111), and the component x_1 looses controllability. Moreover, for $\epsilon = 0$ and $a > 0$, the object is no more stabilisable, though the nominal object (3.111) was stabilisable.

In constructing the MFDs, similarly different solutions are held for $\epsilon = 0$ and $\epsilon \neq 0$. Indeed, if the numerator of the perturbed matrix (3.109) is written in Smith canonical form, then we obtain for $b - a + \epsilon \neq 0$

$$\begin{bmatrix} \lambda - a + \epsilon & 0 \\ 0 & \lambda - b \end{bmatrix}$$

$$= \begin{bmatrix} \lambda - a + \epsilon & \frac{-1}{b-a+\epsilon} \\ b - \lambda & \frac{1}{b-a+\epsilon} \end{bmatrix} \begin{bmatrix} 1 & 0 \\ 0 & (\lambda - a + \epsilon)(\lambda - b) \end{bmatrix} \begin{bmatrix} \frac{\lambda-a+\epsilon}{b-a+\epsilon} & \frac{\lambda-b}{b-a+\epsilon} \\ 1 & 1 \end{bmatrix}.$$

Thus, the McMillan canonical form of (3.109) becomes

$$\tilde{A}(\lambda) = \begin{bmatrix} \lambda - a + \epsilon & \frac{-1}{b-a+\epsilon} \\ b - \lambda & \frac{1}{b-a+\epsilon} \end{bmatrix} \begin{bmatrix} \frac{1}{(\lambda-a)(\lambda-b)} & 0 \\ 0 & \frac{\lambda-a+\epsilon}{\lambda-a} \end{bmatrix} \begin{bmatrix} \frac{\lambda-a+\epsilon}{b-a+\epsilon} & \frac{\lambda-b}{b-a+\epsilon} \\ 1 & 1 \end{bmatrix}. \tag{3.113}$$

For $\epsilon \neq 0$, the McMillan denominator $\psi_{\tilde{A}}(\lambda)$ of Matrix (3.109) results to

$$\psi_{\tilde{A}}(\lambda) = (\lambda - a)^2(\lambda - b).$$

Hence the irreducible left MFD is built with the matrices

$$a_l(\lambda) = \begin{bmatrix} (\lambda - a)(\lambda - b) & 0 \\ 0 & \lambda - a \end{bmatrix} \begin{bmatrix} \frac{1}{b-a+\epsilon} & \frac{1}{b-a+\epsilon} \\ \lambda - b & \lambda - a + \epsilon \end{bmatrix},$$

$$b_l(\lambda) = \begin{bmatrix} \frac{\lambda-a+\epsilon}{b-a+\epsilon} & \frac{\lambda-b}{b-a+\epsilon} \\ \lambda - a + \epsilon & \lambda - a + \epsilon \end{bmatrix}.$$

For $\epsilon = 0$ the situation changes. Then from (3.113), it arises

$$A(\lambda) = \begin{bmatrix} \lambda - a & -\frac{1}{b-a} \\ b - \lambda & \frac{1}{b-a} \end{bmatrix} \begin{bmatrix} \frac{1}{(\lambda-a)(\lambda-b)} & 0 \\ 0 & 1 \end{bmatrix} \begin{bmatrix} \frac{\lambda-a}{b-a} & \frac{\lambda-b}{b-a} \\ 1 & 1 \end{bmatrix}$$

and we arrive at the LMFD

$$a_l(\lambda) = \begin{bmatrix} \frac{(\lambda-a)(\lambda-b)}{b-a} & \frac{(\lambda-a)(\lambda-b)}{b-a} \\ \lambda - b & \lambda - a \end{bmatrix},$$

$$b_l(\lambda) = \begin{bmatrix} \frac{\lambda-a}{b-a} & \frac{\lambda-b}{b-a} \\ 1 & 1 \end{bmatrix},$$

and for that, according to the general theory, we get $\operatorname{ord} a_l(\lambda) = 2$. □

3. In connection with the above considerations, the following problem arises. Suppose a simple realisation (A, B, C) of dimension n, p, m. Then its assigned (ideal) transfer matrix

$$A(\lambda) = \frac{C \operatorname{adj}(\lambda I_p - A)B}{\det(\lambda I_p - A)} = \frac{N(\lambda)}{d(\lambda)} \tag{3.114}$$

is normal. However, due to inevitable inaccuracies, we could have the real transfer matrix

$$\tilde{A}(\lambda) = \frac{\tilde{N}(\lambda)}{\tilde{d}(\lambda)}, \tag{3.115}$$

which practically always deviates from a normal matrix. Even more, if the random calculation errors are independent, then Matrix (3.115) with probability 1 is not normal. Hence the transition from Matrix (3.115) to its minimal realisation leads to a realisation $(\tilde{A}, \tilde{B}, \tilde{C})$ of dimension n, q, m with $q > p$, that means, to an object of higher order with non-predictable dynamic properties.

4. Analogue difficulties arise during the solution of identification problems for linear MIMO systems in the frequency domain [108, 4, 120]. Let for instance the real object be described by the simple realisation (A, B, C). Any identification procedure in the frequency domain will only give an approximate transfer matrix (3.115). Even perfect preparation of the identification conditions cannot avoid that the coefficients of the estimated transfer matrix (3.115) will slightly deviate from the coefficients of the exact matrix (3.114). But this deviation suffices that the probability for Matrix (3.115) to become normal turns to zero. Therefore, the formal transition from Matrix (3.115) to the corresponding minimal realisation will lead to a system of higher order, *i.e.* the identification problem is incorrectly solved.

5. Situations of this kind also arise during the application of frequency domain methods for design of linear MIMO systems [196, 6, 48, 206, 95], ... The algorithm of the optimal controller normally bases on the demand that it is described by a simple realisation (A_0, B_0, C_0). The design method is usually supplied by numerical calculations, so that the transfer matrix of the optimal controller $\tilde{A}_0(\lambda)$ will practically not be normal. Therefore, the really produced realisation $(\tilde{A}_0, \tilde{B}_0, \tilde{C}_0)$ of the optimal controller will have an increased order. Due to this fact, the system with this controller may show a unintentional behavior, especially it might become (internally) unstable.

6. As a consequence of the outlined problems, the following general task is stated [144, 145].

Normalisation problem. Suppose a rational matrix (3.115), the coefficients of which deviate slightly from the coefficients of a certain normal matrix (3.114). Then, find a normal matrix

$$A_\nu(\lambda) = \frac{N_\nu(\lambda)}{d_\nu(\lambda)}$$

the coefficients of which differ only a bit from the coefficients of Matrix (3.115).

A possible approach for the solution of the normalisation problem consists in the following reflection. By equivalence transformation, the numerator of the rational matrix $\tilde{A}(\lambda)$ can be brought into the form (3.102)

$$\tilde{N}(\lambda) = \tilde{g}(\lambda)\tilde{\phi}(\lambda)\begin{bmatrix} \tilde{\alpha}(\lambda) & 1 \\ \tilde{L}(\lambda) & \tilde{\beta}(\lambda) \end{bmatrix}\tilde{\psi}(\lambda).$$

Let $\tilde{d}(\lambda)$ be the denominator of the approximated strictly proper matrix (3.115). Then owing to (3.40), from the above considerations, it follows immediately the representation

$$A_\nu(\lambda) = \frac{\hat{g}(\lambda)P_\nu(\lambda)Q'_\nu(\lambda)}{d_\nu(\lambda)} + G_\nu(\lambda),$$

where

$$d_\nu(\lambda) = \tilde{d}(\lambda), \quad P_\nu(\lambda) = \hat{\phi}(\lambda)\begin{bmatrix} 1 \\ \hat{\beta}(\lambda) \end{bmatrix}, \quad Q'_\nu(\lambda) = \begin{bmatrix} \hat{\alpha}(\lambda) & 1 \end{bmatrix}\hat{\psi}(\lambda) \tag{3.116}$$

and the matrix $G_\nu(\lambda)$ is determined in such a way that the matrix $A_\nu(\lambda)$ becomes strictly proper.

Example 3.39. Apply the normalisation procedure to Matrix (3.109) for $\epsilon \neq 0$ and $b - a + \epsilon \neq 0$. Notice that

$$\begin{bmatrix} \lambda - a + \epsilon & 0 \\ 0 & \lambda - b \end{bmatrix} = \begin{bmatrix} b - a + \epsilon & -1 \\ 0 & 1 \end{bmatrix}\begin{bmatrix} \frac{\lambda - b}{b-a+\epsilon} & 1 \\ \lambda - b & -\lambda + b \end{bmatrix}\begin{bmatrix} 1 & 1 \\ 1 & 0 \end{bmatrix}$$

is true, *i.e.* in the present case, we can choose

$$\tilde{g}(\lambda) = 1, \quad \tilde{\beta}(\lambda) = -\lambda + b, \quad \tilde{\alpha}(\lambda) = \frac{\lambda - b}{b - a + \epsilon}$$

and

$$\tilde{\phi}(\lambda) = \begin{bmatrix} b - a + \epsilon & -1 \\ 0 & 1 \end{bmatrix}, \quad \tilde{\psi}(\lambda) = \begin{bmatrix} 1 & 1 \\ 1 & 0 \end{bmatrix}.$$

Using (3.116), we finally find

$$P_\nu(\lambda) = \begin{bmatrix} b - a + \epsilon & -1 \\ 0 & 1 \end{bmatrix}\begin{bmatrix} 1 \\ -\lambda + b \end{bmatrix} = \begin{bmatrix} \lambda - a + \epsilon \\ -\lambda + b \end{bmatrix},$$

$$Q'_\nu(\lambda) = \begin{bmatrix} \frac{\lambda - b}{b-a+\epsilon} & 1 \end{bmatrix}\begin{bmatrix} 1 & 1 \\ 1 & 0 \end{bmatrix} = \begin{bmatrix} \frac{\lambda-b}{b-a+\epsilon} + 1 & \frac{\lambda-b}{b-a+\epsilon} \end{bmatrix}.$$

Selecting the denominator $d_\nu(\lambda) = (\lambda - a)(\lambda - b)$, we get

$$G_\nu(\lambda) = \frac{1}{b - a + \epsilon}\begin{bmatrix} -1 & -1 \\ 1 & 1 \end{bmatrix}.$$

□

Part II

General MIMO Control Problems

4

Assignment of Eigenvalues and Eigenstructures by Polynomial Methods

In this chapter, and later on if possible, the fundamental results are formulated for real polynomials or real rational matrices, because this case dominates in technical applications, and its handling is more comfortable.

4.1 Problem Statement

1. Suppose the horizontal pair $(a(\lambda), b(\lambda))$ with $a(\lambda) \in \mathbb{R}_{nn}[\lambda]$, $b \in \mathbb{R}_{nm}[\lambda]$. For the theory and in many applications, the following problem is important.

For a given pair $(a(\lambda), b(\lambda))$, find a pair $(\alpha(\lambda), \beta(\lambda))$ with $\alpha(\lambda) \in \mathbb{R}_{mm}[\lambda]$, $\beta(\lambda) \in \mathbb{R}_{mn}[\lambda]$ such that the set of eigenvalues of the matrix

$$Q(\lambda, \alpha, \beta) = \begin{bmatrix} a(\lambda) & -b(\lambda) \\ -\beta(\lambda) & \alpha(\lambda) \end{bmatrix} \tag{4.1}$$

takes predicted values $\lambda_1, \ldots, \lambda_q$ with the multiplicities $\mu_1, \ldots, \mu_q$. In what follows, the pair $(a(\lambda), b(\lambda))$ is called the process to control, or shortly the *process*, and the pair $(\alpha(\lambda), \beta(\lambda))$ the *controller* . Matrix (4.1) is designated as the *characteristic matrix* of the closed loop, or shortly the characteristic matrix. Denote

$$d(\lambda) = (\lambda - \lambda_1)^{\mu_1} \cdots (\lambda - \lambda_q)^{\mu_q}, \tag{4.2}$$

then the problem of eigenvalue assignment is formulated as follows.

Eigenvalue assignment. For a given process $(a(\lambda), b(\lambda))$ and prescribed polynomial $d(\lambda)$, find all controllers $(\alpha(\lambda), \beta(\lambda))$ that ensure

$$\det Q(\lambda, \alpha, \beta) = \det \begin{bmatrix} a(\lambda) & -b(\lambda) \\ -\beta(\lambda) & \alpha(\lambda) \end{bmatrix} \approx d(\lambda). \tag{4.3}$$

In what follows, the polynomial $\det Q(\lambda, \alpha, \beta)$ is designated as the *characteristic polynomial* of the closed loop. For a given process and polynomial

$d(\lambda)$, Relation (4.3) can be seen as an equation depending on the controller $(\alpha(\lambda), \beta(\lambda))$.

2. Let the just formulated task of eigenvalue assignment be solvable for a given process with a certain polynomial $d(\lambda)$. Suppose Ω_d to be the set of controllers satisfying Equation (4.3). Assume $a_1(\lambda), \ldots, a_{n+m}(\lambda)$ to be the sequence of invariant polynomials of the matrix $Q(\lambda, \alpha, \beta)$. In principle, for different controllers in the set Ω_d, these sequences will be different, because such a sequence $a_1(\lambda), \ldots, a_{n+m}(\lambda)$ only has to meet the three demands: All polynomials $a_i(\lambda)$ are monic, each polynomial $a_{i+1}(\lambda)$ is divisible by $a_i(\lambda)$, and

$$a_1(\lambda) \cdots a_{n+m}(\lambda) = d(\lambda)\,.$$

Assume particularly

$$a_1(\lambda) = a_2(\lambda) = \ldots = a_{n+m-1}(\lambda) = 1, \quad a_{n+m}(\lambda) = d(\lambda).$$

Then the matrix $Q(\lambda, \alpha, \beta)$ is simple. In connection with the above said the following task seems substantiated.

> **Structural eigenvalue assignment.** For a given process $(a(\lambda), b(\lambda))$ and scalar polynomial $d(\lambda)$, the eigenvalue assignment (4.3) should deliver the solution set Ω_d. Find the subset $\tilde{\Omega}_d \subset \Omega_d$, where the matrix $Q(\lambda, \alpha, \beta)$ possesses a prescribed sequence of invariant polynomials $a_1(\lambda), \ldots, a_{n+m}(\lambda)$.

3. In many cases, it is useful to formulate the control problem more general, when the process to control is described by a PMD

$$\tau(\lambda) = (a(\lambda), b(\lambda), c(\lambda)) \in \mathbb{R}_{npm}[\lambda]\,, \tag{4.4}$$

which then is called as a *PMD process*. Introduce the matrix $Q_\tau(\lambda, \alpha, \beta)$ of the shape

$$Q_\tau(\lambda, \alpha, \beta) = \begin{bmatrix} a(\lambda) & O_{pn} & -b(\lambda) \\ -c(\lambda) & I_n & O_{nm} \\ O_{mp} & -\beta(\lambda) & \alpha(\lambda) \end{bmatrix}, \tag{4.5}$$

which is called the *characteristic matrix* of the closed loop with PMD process. Then the eigenvalue assignment can be formulated as follows:

> **Eigenvalue assignment for a PMD process.** For a given PMD process (4.4) and polynomial $d(\lambda)$ of the form (4.2), find the set of all controllers $(\alpha(\lambda), \beta(\lambda))$ for which the relation
>
> $$\det Q_\tau(\lambda, \alpha, \beta) \approx d(\lambda) \tag{4.6}$$
>
> is fulfilled.

4. Let the task of eigenvalue assignment for a PMD process be solvable, and Ω_τ be the configured set of controllers $(\alpha(\lambda), \beta(\lambda))$. For different controllers in Ω_τ, the sequence of the invariant polynomials of Matrix (4.5) can be different. Therefore, also the next task is of interest.

> **Structural eigenvalue assignment for a PMD process.** For a given PMD process (4.4) and polynomial $d(\lambda)$, the set of solutions $(\alpha(\lambda), \beta(\lambda))$ of the eigenvalue assignment (4.6) is designated by Ω_τ. Find the subset $\tilde{\Omega}_\tau \subset \Omega_\tau$, where the matrix $Q_\tau(\lambda, \alpha, \beta)$ possesses a prescribed sequence of invariant polynomials.

In the present chapter, the general solution of the eigenvalue assignment problem is derived, where the processes are given as polynomial pairs or as PMDs. Moreover, the structure of the set of invariant polynomials is stated, which can be prescribed for this task. Although, the following results are formulated for real matrices, they could be transferred practically without changes to the complex case. In the considerations below, the eigenvalue assignment problem is also called *modal control problem*, and the determination of the structured eigenvalues is also named *structural modal control problem.*

4.2 Basic Controllers

1. In this section, the important question is investigated, how to design the controller $(\alpha(\lambda), \beta(\lambda))$ that Matrix (4.1) becomes unimodular, *i.e.*

$$a_1(\lambda) = a_2(\lambda) = \ldots = a_{n+m}(\lambda) = 1\,.$$

In the following, such controllers are called *basic controllers.*

It follows directly from Theorem 1.41 that for the existence of a basic controller for the process $(a(\lambda, b(\lambda))$, it is necessary and sufficient that the pair $(a(\lambda), b(\lambda))$ is irreducible, *i.e.* the matrix

$$R_h(\lambda) = \begin{bmatrix} a(\lambda) \; b(\lambda) \end{bmatrix}$$

is alatent. If a process meets this condition, it is called *irreducible.*

2. The next theorem presents a general expression for the set of all basic controllers for a given irreducible pair.

Theorem 4.1. *Let $(\alpha_0^*(\lambda), \beta_0^*(\lambda))$ be a certain basic controller for the process $(a(\lambda), b(\lambda))$. Then the set of all basic controllers $(\alpha_0(\lambda), \beta_0(\lambda))$ is determined by the formula*

$$\begin{aligned} \alpha_0(\lambda) &= D(\lambda)\alpha_0^*(\lambda) - M(\lambda)b(\lambda)\,, \\ \beta_0(\lambda) &= D(\lambda)\beta_0^*(\lambda) - M(\lambda)a(\lambda)\,, \end{aligned} \tag{4.7}$$

where $M(\lambda) \in \mathbb{R}_{mn}[\lambda]$ is an arbitrary, and $D(\lambda) \in \mathbb{R}_{mm}[\lambda]$ is an arbitrary, but unimodular matrix.

Proof. The set of all basic controllers is denoted by $\mathcal{R}_0$, and the set of all pairs satisfying Condition (4.7) by $\mathcal{R}_p$. At first, we will show $\mathcal{R}_0 \subset \mathcal{R}_p$.

Let $(\alpha_0^*(\lambda), \beta_0^*(\lambda))$ be a certain basic controller, and

$$Q_l(\lambda, \alpha_0^*, \beta_0^*) = \begin{bmatrix} a(\lambda) & -b(\lambda) \\ -\beta_0^*(\lambda) & \alpha_0^*(\lambda) \end{bmatrix} \tag{4.8}$$

be its configured characteristic matrix, which is unimodular. Introduce

$$Q_l^{-1}(\lambda, \alpha_0^*, \beta_0^*) \triangleq Q_r(\lambda, \alpha_0^*, \beta_0^*) = \overset{\begin{matrix} \quad n & \quad\quad m \end{matrix}}{\begin{bmatrix} \alpha_r^*(\lambda) & b_r(\lambda) \\ \beta_r^*(\lambda) & a_r(\lambda) \end{bmatrix}} \begin{matrix} n \\ m \end{matrix} \,. \tag{4.9}$$

Owing to the properties of the inverse matrix (2.109), we have the relations

$$\begin{aligned} a(\lambda)\alpha_r^*(\lambda) - b(\lambda)\beta_r^*(\lambda) &= I_n \,, \\ a(\lambda)b_r(\lambda) - b(\lambda)a_r(\lambda) &= O_{nm} \,. \end{aligned} \tag{4.10}$$

Let $(\alpha_0(\lambda), \beta_0(\lambda))$ be any other basic controller, and

$$Q_l(\lambda, \alpha_0, \beta_0) = \begin{bmatrix} a(\lambda) & -b(\lambda) \\ -\beta_0(\lambda) & \alpha_0(\lambda) \end{bmatrix} \tag{4.11}$$

be its configured characteristic matrix. Then due to (4.10),

$$Q_l(\lambda, \alpha_0, \beta_0) Q_r(\lambda, \alpha_0^*, \beta_0^*) = \begin{bmatrix} I_n & O_{nm} \\ M(\lambda) & D(\lambda) \end{bmatrix} \tag{4.12}$$

where

$$\begin{aligned} D(\lambda) &= -\beta_0(\lambda)b_r(\lambda) + \alpha_0(\lambda)a_r(\lambda) \,, \\ M(\lambda) &= -\beta_0(\lambda)\alpha_r^*(\lambda) + \alpha_0(\lambda)\beta_r^*(\lambda) \,. \end{aligned}$$

From (4.12) with regard to (4.9), we receive

$$Q_l(\lambda, \alpha_0, \beta_0) = \begin{bmatrix} I_n & O_{nm} \\ M(\lambda) & D(\lambda) \end{bmatrix} Q_l(\lambda, \alpha_0^*, \beta_0^*) \,,$$

which directly delivers Formulae (4.7). Calculating (4.12), we get

$$\det D(\lambda) = \det Q_l(\lambda, \alpha_0, \beta_0) \det Q_r(\lambda, \alpha_0^*, \beta_0^*) = \text{const.} \,,$$

i.e. the matrix D is unimodular. Therefore, every basic controller $(\alpha_0(\lambda), \beta_0(\lambda)$ permits a representation (4.7), that's why $\mathcal{R}_0 \subset \mathcal{R}_p$ is true.

On the other side, if (4.7) is valid, then from (4.11), it follows

$$Q_l(\lambda, \alpha_0, \beta_0) = \begin{bmatrix} a(\lambda) & -b(\lambda) \\ -D(\lambda)\beta_0^*(\lambda) + M(\lambda)a(\lambda) & D(\lambda)\alpha_0^*(\lambda) - M(\lambda)b(\lambda) \end{bmatrix}$$

$$= \begin{bmatrix} I_n & O_{nm} \\ M(\lambda) & D(\lambda) \end{bmatrix} \begin{bmatrix} a(\lambda) & -b(\lambda) \\ -\beta_0^*(\lambda) & \alpha_0^*(\lambda) \end{bmatrix}$$

$$= \begin{bmatrix} I_n & O_{nm} \\ M(\lambda) & D(\lambda) \end{bmatrix} Q(\lambda, \alpha_0^*, \beta_0^*) .$$

Since $D(\lambda)$ is unimodular, also this matrix has to be unimodular, *i.e.* $\mathcal{R}_0 \subset \mathcal{R}_p$ is proven, and therefore the sets $\mathcal{R}_0$ and $\mathcal{R}_p$ coincide. ∎

3. As emerges from (4.7), before constructing the set of all basic controllers, at first we have to find one sample of them. Usually, search procedures for such a controller found on the following considerations.

Lemma 4.2. *For the irreducible process* $(a(\lambda), b(\lambda))$, *there exist an* $m \times m$ *polynomial matrix* $a_r(\lambda)$ *and an* $n \times m$ *polynomial matrix* $b_r(\lambda)$, *such that the equation*

$$a(\lambda)b_r(\lambda) = b(\lambda)a_r(\lambda) \tag{4.13}$$

is fulfilled, where the pair $[a_r(\lambda), b_r(\lambda)]$ *is irreducible.*

Proof. Since the process $(a(\lambda), b(\lambda))$ is irreducible, there exists a basic controllers $(\alpha_0^*(\lambda), \beta_0^*(\lambda))$, such that the matrix

$$Q(\lambda, \alpha_0^*, \beta_0^*) = \begin{bmatrix} a(\lambda) & -b(\lambda) \\ -\beta_0^*(\lambda) & \alpha_0^*(\lambda) \end{bmatrix}$$

becomes unimodular. Thus, the inverse matrix

$$Q^{-1}(\lambda, \alpha_0^*, \beta_0^*) = \begin{bmatrix} \alpha_{0r}^*(\lambda) & b_r(\lambda) \\ \beta_{0r}^*(\lambda) & a_r(\lambda) \end{bmatrix}$$

is also unimodular. Then from (4.10), it follows Statement (4.13). Moreover, the pair $[a_r(\lambda), b_r(\lambda)]$ is irreducible thanks to Theorem 1.32. ∎

Remark 4.3. If the matrix $a(\lambda)$ is non-singular, *i.e.* $\det a(\lambda) \not\equiv 0$, then there exists the transfer matrix of the processes

$$w(\lambda) = a^{-1}(\lambda)b(\lambda) .$$

The right side of this equation proves to be an ILMFD of the matrix $w(\lambda)$. If we consider an arbitrary IRMFD

$$w(\lambda) = b_r(\lambda)a_r^{-1}(\lambda) ,$$

then Equation (4.13) holds, and the pair $[a_r(\lambda), b_r(\lambda)]$ is irreducible. Therefore, Lemma 4.2 is a generalisation of this property in case the matrix $a(\lambda)$ is singular.

In what follows, the original pair $(a(\lambda), b(\lambda))$ is called *left process model*, and any pair $[a_r(\lambda), b_r(\lambda)]$ satisfying (4.13), is named *right process model.* If in this case, the pair $[a_r(\lambda), b_r(\lambda)]$ is irreducible, then the right process model should also be designated as irreducible.

Lemma 4.4. *Let $[a_r(\lambda), b_r(\lambda)]$ be an irreducible right process model. Then any pair $(\alpha_0(\lambda), \beta_0(\lambda)$ satisfying the Diophantine equation*

$$-\beta_0(\lambda)b_r(\lambda) + \alpha_0(\lambda)a_r(\lambda) = P(\lambda) \tag{4.14}$$

with a unimodular matrix $P(\lambda)$, turns out to be a basic controller for the left process model $(a(\lambda), b(\lambda))$.

Proof. Since the pair $(a(\lambda), b(\lambda))$ is irreducible, there exists a vertical pair $[\alpha_r(\lambda), \beta_r(\lambda)]$ with

$$a(\lambda)\alpha_r(\lambda) - b(\lambda)\beta_r(\lambda) = I_n \,.$$

The pair $[\alpha_0(\lambda), \beta_0(\lambda)]$ should satisfy Condition (4.14). Then build the product

$$\begin{bmatrix} a(\lambda) & -b(\lambda) \\ -\beta_0(\lambda) & \alpha_0(\lambda) \end{bmatrix} \begin{bmatrix} \alpha_r(\lambda) & b_r(\lambda) \\ \beta_r(\lambda) & a_r(\lambda) \end{bmatrix} = \begin{bmatrix} I_n & O_{nm} \\ M(\lambda) & P(\lambda) \end{bmatrix},$$

where $M(\lambda)$ is a polynomial matrix. Since the matrix $P(\lambda)$ is unimodular, the matrix on the right side becomes unimodular. Thus, both matrices on the left side are unimodular, and $(\alpha_0(\lambda), \beta_0(\lambda))$ proves to be a basic controller. ∎

Corollary 4.5. *An arbitrary pair $(\alpha_0(\lambda), \beta_0(\lambda))$ satisfying the Diophantine equation*

$$-\beta_0(\lambda)b_r(\lambda) + \alpha_0(\lambda)a_r(\lambda) = I_m \,,$$

proves to be a basic controller.

Remark 4.6. It is easily shown that the set of all pairs $(\alpha_0(\lambda), \beta_0(\lambda))$ satisfying Equation (4.14) for all possible unimodular matrices $P(\lambda)$ generate the complete set of basic controllers.

4.3 Recursive Construction of Basic Controllers

1. As arises from Lemma 4.4, a basic controller $(\alpha_0(\lambda), \beta_0(\lambda))$ can be found as solution of the Diophantine matrix equation (4.14). In the present section, an alternative method for finding a basic controller is described that leads to a recursive solution of simpler scalar Diophantine equations, and does not need the matrices $a_r(\lambda)$, $b_r(\lambda)$, arising in (4.14).

2. For finding this approach, the polynomial equation

$$\sum_{i=1}^{n} a_i(\lambda)x_i(\lambda) = c(\lambda) \tag{4.15}$$

is considered, where the $a_i(\lambda)$, $(i = 1, \dots, n)$, $c(\lambda)$ are known polynomials, and $x_i(\lambda)$, $(i = 1, \dots, n)$ are unknown scalar polynomials. We will say, that the polynomials $a_i(\lambda)$ are *in all coprime*, if their monic GCD is equal to 1. The next lemma is a corollary from a more general statement in [79].

Lemma 4.7. *A necessary and sufficient condition for the solvability of Equation (4.15) is, that the greatest common divisor of the polynomials $a_i(\lambda)$, $(i = 1, \dots, n)$ is a divisor of the polynomial $c(\lambda)$.*

Proof. Necessity: Suppose $\gamma(\lambda)$ as a GCD of the polynomials $a_i(\lambda)$. Then

$$a_i(\lambda) = \gamma(\lambda)a_{1i}(\lambda), \quad (i = 1, \dots, n) \tag{4.16}$$

where the polynomials $a_{1i}(\lambda)$, $(i = 1, \dots, n)$ are in all coprime. Substituting (4.16) into (4.15), we obtain

$$\gamma(\lambda)\left[\sum_{i=1}^{n} a_{1i}(\lambda)x_i(\lambda)\right] = c(\lambda),$$

from which it is clear that the polynomial $c(\lambda)$ must be divisible by $\gamma(\lambda)$.

Sufficiency: The proof is done by complete induction. The statement should be valid for one $n = k > 0$, and then it is shown that it is also valid for $n = k + 1$. Consider the equation

$$\sum_{i=1}^{k+1} a_i(\lambda)x_i(\lambda) = c(\lambda). \tag{4.17}$$

Without loss of generality assume that the coefficients $a_i(\lambda)$, $(i = 1, \dots, k+1)$ are in all coprime, otherwise both sides of Equation (4.17) could be divided by the common factor.

Let $\kappa(\lambda)$ be the GCD of the polynomials $a_1(\lambda), \dots, a_k(\lambda)$. Then (4.16) is true, where the polynomials $a_{i1}(\lambda)$, $(i = 1, \dots, k)$ are in all coprime. Herein, the polynomials $\kappa(\lambda)$ and $a_{k+1}(\lambda)$ are also coprime, otherwise the coefficients of Equation (4.17) would not be in all coprime. Hence the Diophantine equation

$$\kappa(\lambda)u(\lambda) + a_{k+1}(\lambda)x_{k+1}(\lambda) = c(\lambda) \tag{4.18}$$

is solvable. Let $\tilde{u}(\lambda)$, $\tilde{x}_{k+1}(\lambda)$ be a certain solution of Equation (4.18). Investigate the equation

$$\sum_{i=1}^{k} a_{1i}(\lambda)x_i(\lambda) = \tilde{u}(\lambda) \tag{4.19}$$

which is solvable due to the induction supposition, since all coefficients $a_{1i}(\lambda)$ are in all coprime. Let $\tilde{x}_i(\lambda)$, $(i = 1, \ldots, k)$ be any solution of Equation (4.19). Then applying (4.16) and (4.18), we get

$$\sum_{i=1}^{k+1} a_i(\lambda)\tilde{x}_i(\lambda) = c(\lambda)$$

this means, the totality of polynomials $\tilde{x}_i(\lambda)$, $(i = 1, \ldots, k+1)$ presents a solution of Equation (4.15). Since the statement of the theorem holds for $k = 2$, we have proved by complete induction that it is also true for all $k \geq 2$.
■

3. The idea of the proof consists in constructing a solution of (4.15) by reducing the problem to the case of two variables. It can be used to generate successively the solutions of Diophantine equations with several unknowns, where in every step a Diophantine equation with two unknowns is solved.

Example 4.8. Find a solution of the equation

$$(\lambda - 1)(\lambda - 2)x_1(\lambda) + (\lambda - 1)(\lambda - 3)x_2(\lambda) + (\lambda - 2)(\lambda - 3)x_3(\lambda) = 1\,.$$

Here, the coefficients are in all coprime, though they are not coprime by twos. In the present case, the auxiliary equation (4.18) could be given the shape

$$(\lambda - 1)u(\lambda) + (\lambda - 2)(\lambda - 3)x_3(\lambda) = 1\,.$$

A special solution takes the form

$$\tilde{u}(\lambda) = 2 - 0.5\lambda\,, \quad \tilde{x}_3(\lambda) = 0.5\,.$$

Equation (4.19) can be represented in the form

$$(\lambda - 2)x_1(\lambda) + (\lambda - 3)x_2(\lambda) = 2 - 0.5\lambda\,.$$

As a special solution for the last equation, we find

$$\tilde{x}_1(\lambda) = 0.5\,, \quad \tilde{x}_2(\lambda) = -1\,.$$

Thus, as a special solution of the original equation, we obtain

$$\tilde{x}_1(\lambda) = 0.5\,, \quad \tilde{x}_2(\lambda) = -1\,, \quad \tilde{x}_3(\lambda) = 0.5\,.$$

□

4.

Lemma 4.9. *Suppose the* $n \times (n+1)$ *polynomial matrix*

$$A(\lambda) = \begin{bmatrix} a_{11}(\lambda) & \dots & a_{1,n+1}(\lambda) \\ \vdots & \dots & \vdots \\ a_{n1}(\lambda) & \dots & a_{n,n+1}(\lambda) \end{bmatrix} \tag{4.20}$$

with $\operatorname{rank} A(\lambda) = n$*. Furthermore, denote* $D_A(\lambda)$ *as the monic GCD of the minors of* n*-th order of the matrix* $A(\lambda)$*. Then there exist scalar polynomials* $d_1(\lambda), \dots, d_{n+1}(\lambda)$*, such that the* $(n+1) \times (n+1)$ *polynomial matrix*

$$A_1(\lambda) = \begin{bmatrix} & A(\lambda) & \\ d_1(\lambda) & \dots & d_{n+1}(\lambda) \end{bmatrix}$$

satisfies the relation

$$\det A_1(\lambda) = D_A(\lambda) . \tag{4.21}$$

Proof. Denote $B_i(\lambda)$ as that $n \times n$ polynomial matrix which is held from $A(\lambda)$ by cutting its i-th column. Then the expansion of the determinant by the last row delivers

$$\det A_1(\lambda) = \sum_{i=1}^{n+1} (-1)^{n+1+i} d_i(\lambda) \Delta_i(\lambda)$$

with $\Delta_i(\lambda) = \det B_i(\lambda)$. Thus, Relation (4.21) is equivalent to the Diophantine equation

$$\sum_{i=1}^{n+1} (-1)^{n+1+i} d_i(\lambda) \Delta_i(\lambda) = D_A(\lambda) . \tag{4.22}$$

By definition, the polynomial $D_A(\lambda)$ is the GCD of the polynomials $\Delta_i(\lambda)$. Hence we can write

$$\Delta_i(\lambda) = D_A(\lambda) \Delta_{1i}(\lambda) ,$$

where the polynomials $\Delta_{1i}(\lambda)$, $(i = 1, \dots, n+1)$ are in all coprime. By virtue of this relation, Equation (4.22) can take the form

$$\sum_{i=1}^{n+1} (-1)^{n+1+i} d_i(\lambda) \Delta_{1i}(\lambda) = 1 .$$

Since the polynomials $\Delta_{1i}(\lambda)$ are in all coprime, this equation is solvable thanks to Lemma 4.7. Multiplying both sides by $D_A(\lambda)$, we conclude that Equation (4.22) is solvable. ∎

5.

Theorem 4.10. [193] *Suppose the non-degenerated* $n \times m$ *polynomial matrix* $\tilde{A}(\lambda)$, $m > n+1$, *where*

$$\tilde{A}(\lambda) = \left[\begin{array}{cccc|ccc} a_{11}(\lambda) & \dots & a_{1n}(\lambda) & a_{1,n+1}(\lambda) & a_{1,n+2}(\lambda) & \dots & a_{1m}(\lambda) \\ \vdots & \dots & \vdots & \vdots & \vdots & \dots & \vdots \\ a_{n1}(\lambda) & \dots & a_{nn}(\lambda) & a_{n,n+1}(\lambda) & a_{n,n+2}(\lambda) & \dots & a_{nm}(\lambda) \end{array}\right]$$

$$= \left[\begin{array}{c|ccc} & a_{1,n+2}(\lambda) & \dots & a_{1m}(\lambda) \\ A(\lambda) & \vdots & \dots & \vdots \\ & a_{n,n+2}(\lambda) & \dots & a_{nm}(\lambda) \end{array}\right].$$

Assume that the submatrix $A(\lambda)$ *on the left of the line has the form (4.20). Let* $D_{\tilde{A}}(\lambda)$ *be the monic GCD of the minors of* n*-th order of the matrix* $\tilde{A}(\lambda)$, *and* $D_A(\lambda)$ *the monic GCD of the minors of* n*-th order of the matrix* $A(\lambda)$. *The polynomials* $d_1(\lambda), \dots, d_{n+1}(\lambda)$ *should be a solution of Equation (4.22). Then the monic GCD of the minors of* n*-th order of the matrix*

$$A_d(\lambda) = \left[\begin{array}{cccc|ccc} a_{11}(\lambda) & \dots & a_{1n}(\lambda) & a_{1,n+1}(\lambda) & a_{1,n+2}(\lambda) & \dots & a_{1m}(\lambda) \\ \vdots & \dots & \vdots & \vdots & \vdots & \dots & \vdots \\ a_{n1}(\lambda) & \dots & a_{nn}(\lambda) & a_{n,n+1}(\lambda) & a_{n,n+2}(\lambda) & \dots & a_{nm}(\lambda) \\ d_1(\lambda) & \dots & d_n(\lambda) & d_{n+1}(\lambda) & d_{n+2}(\lambda) & \dots & d_m(\lambda) \end{array}\right] \tag{4.23}$$

satisfies the condition

$$D_{A_d}(\lambda) = D_{\tilde{A}}(\lambda)$$

for any polynomials $d_{n+2}(\lambda), \dots, d_{n+m}(\lambda)$. ■

Corollary 4.11. *If the matrix* $\tilde{A}(\lambda)$ *is alatent, then also Matrix (4.23) is alatent.*

6. Suppose an irreducible pair $(a(\lambda), b(\lambda))$, where $a(\lambda)$ has the dimension $n \times n$ and $b(\lambda)$ dimension $n \times m$. Then by successive repeating the procedure explained in Theorem 4.10, the unimodular matrix

$$Q_l(\lambda, \alpha_0, \beta_0) = \begin{bmatrix} a_{11}(\lambda) & \dots & a_{1n}(\lambda) & -b_{11}(\lambda) & \dots & -b_{1m}(\lambda) \\ \vdots & \dots & \vdots & \vdots & \dots & \vdots \\ a_{n1}(\lambda) & \dots & a_{nn}(\lambda) & -b_{n1}(\lambda) & \dots & -b_{nm}(\lambda) \\ -\beta_{11}(\lambda) & \dots & -\beta_{1n}(\lambda) & \alpha_{11}(\lambda) & \dots & \alpha_{1m}(\lambda) \\ \vdots & \dots & \vdots & \vdots & \dots & \vdots \\ -\beta_{m1}(\lambda) & \dots & -\beta_{mn}(\lambda) & \alpha_{m1}(\lambda) & \dots & \alpha_{mm}(\lambda) \end{bmatrix}$$

is produced. The last m rows of this matrix present a certain basic controller

$$\alpha_0(\lambda) = \begin{bmatrix} \alpha_{11}(\lambda) & \dots & \alpha_{1m}(\lambda) \\ \vdots & \dots & \vdots \\ \alpha_{m1}(\lambda) & \dots & \alpha_{mm}(\lambda) \end{bmatrix}, \quad \beta_0(\lambda) = \begin{bmatrix} \beta_{11}(\lambda) & \dots & \beta_{1n}(\lambda) \\ \vdots & \dots & \vdots \\ \beta_{m1}(\lambda) & \dots & \beta_{mn}(\lambda) \end{bmatrix}.$$

For a transition to the matrix $Q_l(\lambda, \alpha_0, \beta_0)$ in every step essentially a scalar Diophantine equation must be solved with several unknowns of the type (4.15). As shown above, the solution of such equations amounts to the successive solution of simple Diophantine equations of the form (4.18).

7.

Example 4.12. Determine a basic controller for the process $(a(\lambda), b(\lambda))$ with

$$a(\lambda) = \begin{bmatrix} \lambda - 1 & \lambda + 1 \\ 0 & 1 \end{bmatrix}, \quad b(\lambda) = \begin{bmatrix} 0 & -\lambda \\ -\lambda & 0 \end{bmatrix}. \tag{4.24}$$

The pair $(a(\lambda), b(\lambda))$ establishes to be irreducible, because the matrix

$$\begin{bmatrix} \lambda - 1 & \lambda + 1 & 0 & \lambda \\ 0 & 1 & \lambda & 0 \end{bmatrix}$$

is alatent, which is easily checked. Hence the design problem for a basic controller is solvable. In the first step, we search the polynomials $d_1(\lambda)$, $d_1(\lambda)$, $d_3(\lambda)$, so that the matrix

$$A_1(\lambda) = \begin{bmatrix} \lambda - 1 & \lambda + 1 & 0 \\ 0 & 1 & \lambda \\ d_1(\lambda) & d_2(\lambda) & d_3(\lambda) \end{bmatrix}$$

becomes unimodular. Without loss in generality, we assume $\det A_1(\lambda) = 1$. Thus we arrive at the Diophantine equation

$$\lambda(\lambda + 1)d_1(\lambda) - \lambda(\lambda - 1)d_2(\lambda) + (\lambda - 1)d_3(\lambda) = 1\,.$$

A special solution of this equation is represented in the form

$$d_1(\lambda) = \frac{1}{2}, \quad d_2(\lambda) = 0, \quad d_3(\lambda) = -\left(\frac{\lambda}{2} + 1\right).$$

The not designated polynomial $d_4(\lambda)$ can be chosen arbitrarily. Take for instance $d_4(\lambda) = 0$, so the alatent matrix becomes

$$A_2(\lambda) = \begin{bmatrix} \lambda - 1 & \lambda + 1 & 0 & \lambda \\ 0 & 1 & \lambda & 0 \\ \frac{1}{2} & 0 & -\frac{\lambda}{2} - 1 & 0 \end{bmatrix}.$$

It remains the task to complete $A_2(\lambda)$ that it becomes a unimodular matrix. For that, we attempt

$$A_3(\lambda) = \begin{bmatrix} \lambda-1 & \lambda+1 & 0 & \lambda \\ 0 & 1 & \lambda & 0 \\ \frac{1}{2} & 0 & \frac{-\lambda}{2} - 1 & 0 \\ \tilde{d}_1(\lambda) & \tilde{d}_2(\lambda) & \tilde{d}_3(\lambda) & \tilde{d}_4(\lambda) \end{bmatrix}$$

with unknown polynomials $\tilde{d}_i(\lambda)$, $(i = 1, \ldots, 4)$. Assume $\det A_3(\lambda) = 1$, so we obtain the Diophantine equation

$$\lambda\left(\frac{\lambda}{2}+1\right)\tilde{d}_1(\lambda) - \frac{\lambda^2}{2}\tilde{d}_2(\lambda) + \frac{\lambda}{2}\tilde{d}_3(\lambda) + \tilde{d}_4(\lambda) = 1$$

which has the particular solution

$$\tilde{d}_1(\lambda) = \tilde{d}_2(\lambda) = \tilde{d}_3(\lambda) = 0, \quad \tilde{d}_4(\lambda) = 1\,.$$

In summary, we obtain the unimodular matrix

$$A_3(\lambda) = \begin{bmatrix} \lambda-1 & \lambda+1 & 0 & \lambda \\ 0 & 1 & \lambda & 0 \\ \frac{1}{2} & 0 & -\frac{\lambda}{2} - 1 & 0 \\ 0 & 0 & 0 & 1 \end{bmatrix},$$

where we read the basic controller $(\alpha_0^*(\lambda), \beta_0^*(\lambda))$ with

$$\alpha_0^*(\lambda) = \begin{bmatrix} -\frac{\lambda}{2} - 1 & 0 \\ 0 & 1 \end{bmatrix}, \qquad \beta_0^*(\lambda) = \begin{bmatrix} -\frac{1}{2} & 0 \\ 0 & 0 \end{bmatrix}.$$

Using this solution and Formula (4.7), we construct the set of all basic controllers for the process (4.24). □

Example 4.13. Find a basic controller for the process $(a(\lambda), b(\lambda))$ with

$$a(\lambda) = \begin{bmatrix} \lambda-1 & \lambda+1 \\ \lambda-1 & \lambda+1 \end{bmatrix}, \quad b(\lambda) = \begin{bmatrix} -1 \\ 0 \end{bmatrix}. \tag{4.25}$$

In the present case the matrix $a(\lambda)$ is singular, nevertheless, a basic controller can be found because the matrix

$$R_h(\lambda) = \begin{bmatrix} a(\lambda) & -b(\lambda) \end{bmatrix} = \begin{bmatrix} \lambda-1 & \lambda+1 & 1 \\ \lambda-1 & \lambda+1 & 0 \end{bmatrix}$$

is alatent. In accordance with the derived methods, we search for polynomials $d_1(\lambda)$, $d_2(\lambda)$, $d_3(\lambda)$ such that the condition

$$\det \begin{bmatrix} \lambda - 1 & \lambda + 1 & 1 \\ \lambda - 1 & \lambda + 1 & 0 \\ d_1(\lambda) & d_2(\lambda) & d_3(\lambda) \end{bmatrix} = 1$$

is satisfied, which is equivalent to the Diophantine equation

$$-(\lambda + 1)d_1(\lambda) + (\lambda - 1)d_2(\lambda) = 1$$

which has the particular solution $d_1(\lambda) = d_2(\lambda) = -0.5$. Thus, we can take $\alpha_0^*(\lambda) = d_3(\lambda)$, $\beta_0^*(\lambda) = \begin{bmatrix} 0.5 & 0.5 \end{bmatrix}$, where $d_3(\lambda)$ is an arbitrary polynomial. □

4.4 Dual Models and Dual Bases

1. For the further investigations, we want to modify the introduced notation. The initial irreducible process $(a(\lambda), b(\lambda))$ is written as $(a_l(\lambda), b_l(\lambda))$, and is called, as before, a left process model. Any basic controller $\alpha_0(\lambda), \beta_0(\lambda))$ is written in the form $(\alpha_{0l}(\lambda), \beta_{0l}(\lambda))$, and it is called a left basic controller. Hereby, the assigned unimodular matrix (4.8) that presents itself in the form

$$Q_l(\lambda, \alpha_{0l}, \beta_{0l}) = \begin{bmatrix} a_l(\lambda) & -b_l(\lambda) \\ -\beta_{0l}(\lambda) & \alpha_{0l}(\lambda) \end{bmatrix} \tag{4.26}$$

is named a *left basic matrix*. However, if the pair $[a_r(\lambda), b_r(\lambda)]$ is an irreducible right process model, then the vertical pair $[\alpha_{0r}(\lambda), \beta_{0r}(\lambda)]$, for which the matrix

$$Q_r(\lambda, \alpha_{0r}, \beta_{0r}) = \begin{bmatrix} \alpha_{0r}(\lambda) & b_r(\lambda) \\ \beta_{0r}(\lambda) & a_r(\lambda) \end{bmatrix} \tag{4.27}$$

becomes unimodular, is said to be a *right basic controller*, and the configured matrix (4.27) is called a *right basic matrix*. .

2. Using (4.27), we find

$$Q_r'(\lambda, \alpha_{0r}, \beta_{0r}) = \begin{bmatrix} \alpha_{0r}'(\lambda) & \beta_{0r}'(\lambda) \\ b_r'(\lambda) & a_r'(\lambda) \end{bmatrix} \approx \begin{bmatrix} \beta_{0r}'(\lambda) & \alpha_{0r}'(\lambda) \\ a_r'(\lambda) & b_r'(\lambda) \end{bmatrix} \approx \begin{bmatrix} a_r'(\lambda) & b_r'(\lambda) \\ \beta_{0r}'(\lambda) & \alpha_{0r}'(\lambda) \end{bmatrix},$$

where the symbol $\approx$ stands for the equivalence of the polynomial matrices. Now, if $[\alpha_{0r}^*(\lambda), \beta_{0r}^*(\lambda)]$ is any right basic controller, then applying Theorem 4.1 and the last relation, the set of all right basic controllers is expressed by the formula

$$\begin{aligned} \alpha_{0r}(\lambda) &= \alpha_{0r}^*(\lambda)D_r(\lambda) - b_r(\lambda)M_r(\lambda), \\ \beta_{0r}(\lambda) &= \beta_{0r}^*(\lambda)D_r(\lambda) - a_r(\lambda)M_r(\lambda), \end{aligned} \tag{4.28}$$

where the $m \times n$ polynomial matrix $M_r(\lambda)$ is arbitrary, and $D_r(\lambda)$ is any unimodular $n \times n$ polynomial matrix.

3. The basic matrices (4.26) and (4.27) are called *dual* , if the equation

$$Q_r(\lambda, \alpha_{0r}, \beta_{0r}) = Q_l^{-1}(\lambda, \alpha_{0l}, \beta_{0l}) \tag{4.29}$$

holds, or equivalently, if

$$Q_l(\lambda, \alpha_{0l}, \beta_{0l}) Q_r(\lambda, \alpha_{0r}, \beta_{0r}) = Q_r(\lambda, \alpha_{0r}, \beta_{0r}) Q_l(\lambda, \alpha_{0l}, \beta_{0l}) = I_{n+m} \,. \tag{4.30}$$

The processes $(a_l(\lambda), b_l(\lambda), [a_r(\lambda), b_r(\lambda)]$ configured by Equations (4.29) and (4.30), as well as the basic controllers $(\alpha_{0l}(\lambda), \beta_{0l}(\lambda)), [\alpha_{0r}(\lambda), \beta_{0r}(\lambda)]$ will also be named as dual.

From (4.30) and (4.26), (4.27) emerge two groups of equations, respectively for left or right dual models as well as left and right dual basic controllers

$$\begin{aligned} a_l(\lambda)\alpha_{0r}(\lambda) - b_l(\lambda)\beta_{0r}(\lambda) &= I_n \,, & a_l(\lambda)b_r(\lambda) - b_l(\lambda)a_r(\lambda) &= O_{nm} \,, \\ -\beta_{0l}(\lambda)\alpha_{0r}(\lambda) + \alpha_{0l}(\lambda)\beta_{0r}(\lambda) &= O_{nm} \,, & -\beta_{0l}(\lambda)b_r(\lambda) + \alpha_{0l}(\lambda)a_r(\lambda) &= I_m \end{aligned} \tag{4.31}$$

and

$$\begin{aligned} \alpha_{0r}(\lambda)a_l(\lambda) - b_r(\lambda)\beta_{0l}(\lambda) &= I_n \,, & -\alpha_{0r}(\lambda)b_l(\lambda) + b_r(\lambda)\alpha_{0l}(\lambda) &= O_{nm} \,, \\ \beta_{0r}(\lambda)a_l(\lambda) - a_r(\lambda)\beta_{0l}(\lambda) &= O_{nm} \,, & -\beta_{0r}(\lambda)b_l(\lambda) + a_r(\lambda)\alpha_{0l}(\lambda) &= I_m \,. \end{aligned} \tag{4.32}$$

Relations (4.31) and (4.32) are called direct and reverse *Bezout identity*, respectively [69].

Remark 4.14. The validity of the relations of anyone of the groups (4.31) or (4.32) is necessary and sufficient for the validity of Formulae (4.29), (4.30). Therefore, each of the groups of Relations (4.31) or (4.32) follows from the other one.

4. Applying the new notation, Formula (4.7) can be expressed in the form

$$\begin{aligned} \left[-\beta_{0l}(\lambda) \;\; \alpha_{0l}(\lambda) \right] &= \left[M_l(\lambda) \;\; D_l(\lambda) \right] Q_l(\lambda, \alpha_{0l}^*, \beta_{0l}^*) \\ &= \left[M_l(\lambda) \;\; D_l(\lambda) \right] \begin{bmatrix} a_l(\lambda) & -b_l(\lambda) \\ -\beta_{0l}^*(\lambda) & \alpha_{0l}^*(\lambda) \end{bmatrix}, \end{aligned}$$

which results in

$$\begin{aligned} \left[M_l(\lambda) \;\; D_l(\lambda) \right] &= \left[-\beta_{0l}(\lambda) \;\; \alpha_{0l}(\lambda) \right] Q_l^{-1}(\lambda, \alpha_{0l}^*, \beta_{0l}^*) \\ &= \left[-\beta_{0l}(\lambda) \;\; \alpha_{0l}(\lambda) \right] Q_r(\lambda, \alpha_{0l}^*, \beta_{0l}^*) \end{aligned} \tag{4.33}$$

with the dual basic matrix

$$Q_r(\lambda, \alpha_{0r}^*, \beta_{0r}^*) = \begin{bmatrix} \alpha_{0r}^*(\lambda) & b_r(\lambda) \\ \beta_{0r}^*(\lambda) & a_r(\lambda) \end{bmatrix}.$$

Alternatively, (4.33) can be written in the form

$$D_l(\lambda) = -\beta_{0l}(\lambda)b_r(\lambda) + \alpha_{0l}(\lambda)a_r(\lambda)\,,$$
$$M_l(\lambda) = -\beta_{0l}(\lambda)\alpha_{0r}^*(\lambda) + \alpha_{0l}(\lambda)\beta_{0r}^*(\lambda)\,.$$

Analogously, Formula (4.28) can be presented in the form

$$\begin{bmatrix}\alpha_{0r}(\lambda)\\ \beta_{0r}(\lambda)\end{bmatrix} = Q_r(\lambda, \alpha_{0r}^*, \beta_{0r}^*)\begin{bmatrix}D_r(\lambda)\\ -M_r(\lambda)\end{bmatrix},$$

where we derive

$$\begin{bmatrix}D_r(\lambda)\\ -M_r(\lambda)\end{bmatrix} = Q_l(\lambda, \alpha_{0l}^*, \beta_{0l}^*)\begin{bmatrix}\alpha_{0r}(\lambda)\\ \beta_{0r}(\lambda)\end{bmatrix}$$

or

$$D_r(\lambda) = a_l(\lambda)\alpha_{0r}(\lambda) - b_l(\lambda)\beta_{0r}(\lambda)\,,$$
$$M_r(\lambda) = \beta_{0l}^*(\lambda)\alpha_{0r}(\lambda) - \alpha_{0l}^*(\lambda)\beta_{0r}(\lambda)\,.$$

5.

Theorem 4.15. *Let the left and right irreducible models* $(a_l(\lambda), b_l(\lambda))$, $[a_r(\lambda), b_r(\lambda)]$ *of an object be given, which satisfy the condition*

$$a_l(\lambda)b_r(\lambda) = b_l(\lambda)a_r(\lambda) \tag{4.34}$$

and, moreover, an arbitrary left basic controller $(\alpha_{0l}(\lambda), \beta_{0l}(\lambda))$. *Then a necessary and sufficient condition for the existence of a right basic controller* $[\alpha_{0r}(\lambda), \beta_{0r}(\lambda)]$ *that is dual to the controller* $(\alpha_{0l}(\lambda), \beta_{0l}(\lambda))$, *is that the pair* $(\alpha_{0l}(\lambda), \beta_{0l}(\lambda))$ *is a solution of the Diophantine equation*

$$\alpha_{0l}(\lambda)a_r(\lambda) - \beta_{0l}(\lambda)b_r(\lambda) = I_m\,.$$

Proof. The necessity follows from the Bezout identity (4.31). To prove the sufficiency, we notice that due to the irreducibility of the pair $(a_l(\lambda), b_l(\lambda))$, there exists a pair $[\tilde{\alpha}_{0r}(\lambda), \tilde{\beta}_{0r}(\lambda)]$, that fulfills the relation

$$a_l(\lambda)\tilde{\alpha}_{0r}(\lambda) - b_l(\lambda)\tilde{\beta}_{0r}(\lambda) = I_n\,. \tag{4.35}$$

Now, build the product

$$\begin{bmatrix}a_l(\lambda) & -b_l(\lambda)\\ -\beta_{0l}(\lambda) & \alpha_{0l}(\lambda)\end{bmatrix}\begin{bmatrix}\tilde{\alpha}_{0r}(\lambda) & b_r(\lambda)\\ \tilde{\beta}_{0r}(\lambda) & a_r(\lambda)\end{bmatrix} = \begin{bmatrix}I_n & O_{nm}\\ -\beta_{0l}(\lambda)\tilde{\alpha}_{0r}(\lambda) + \alpha_{0l}(\lambda)\tilde{\beta}_{0r}(\lambda) & I_m\end{bmatrix}$$

from which we held

$$\begin{bmatrix} a_l(\lambda) & -b_l(\lambda) \\ -\beta_{0l}(\lambda) & \alpha_{0l}(\lambda) \end{bmatrix} \begin{bmatrix} \alpha_{0r}(\lambda) & b_r(\lambda) \\ \beta_{0r}(\lambda) & a_r(\lambda) \end{bmatrix} = \begin{bmatrix} I_n & O_{nm} \\ O_{mn} & I_m \end{bmatrix}, \tag{4.36}$$

where

$$\alpha_{0r}(\lambda) = \tilde{\alpha}_{0r}(\lambda) + b_r(\lambda)\left[\beta_{0l}(\lambda)\tilde{\alpha}_{0r}(\lambda) - \alpha_{0l}(\lambda)\tilde{\beta}_{0r}(\lambda)\right],$$

$$\beta_{0r}(\lambda) = \tilde{\beta}_{0r}(\lambda) + a_r(\lambda)\left[\beta_{0l}(\lambda)\tilde{\alpha}_{0r}(\lambda) - \alpha_{0l}(\lambda)\tilde{\beta}_{0r}(\lambda)\right].$$

It arises from (4.36), that the last pair is a right basic controller, which is dual to the left basic controller $(\alpha_{0l}(\lambda), \beta_{0l}(\lambda))$. ∎

Remark 4.16. In analogy, it can be shown that for a right basic controller $[\alpha_{0r}(\lambda), \beta_{0r}(\lambda)]$, there exists a dual left basic controller $(\tilde{\alpha}_{0l}(\lambda), \tilde{\beta}_{0l}(\lambda))$, if and only if the relation

$$a_l(\lambda)\alpha_{0r}(\lambda) - b_l(\lambda)\beta_{0r}(\lambda) = I_n \tag{4.37}$$

is fulfilled. Hereby, if the pair $(\tilde{\alpha}_{0l}(\lambda), \tilde{\beta}_{0l})$ satisfies the condition

$$\tilde{\alpha}_{0l}(\lambda)a_r(\lambda) - \tilde{\beta}_{0l}(\lambda)b_r(\lambda) = I_m,$$

then the formulae

$$\begin{aligned} \alpha_{0l}(\lambda) &= \tilde{\alpha}_{0l}(\lambda) - \left[\tilde{\beta}_{0l}(\lambda)\alpha_{0r}(\lambda) - \tilde{\alpha}_{0l}(\lambda)\beta_{0r}(\lambda)\right] b_l(\lambda), \\ \beta_{0l}(\lambda) &= \tilde{\beta}_{0l}(\lambda) - \left[\tilde{\beta}_{0l}(\lambda)\alpha_{0r}(\lambda) - \tilde{\alpha}_{0l}(\lambda)\beta_{0r}(\lambda)\right] a_l(\lambda) \end{aligned} \tag{4.38}$$

define a left basic controller, that is dual to the controller $[\alpha_{0r}(\lambda), \beta_{0r}(\lambda)]$.

6. The next theorem supplies a parametrisation of the set of all pairs of dual basic controllers.

Theorem 4.17. *Suppose two dual basic controllers* $(\alpha^*_{0l}(\lambda), \beta^*_{0l}(\lambda))$ *and* $[\alpha^*_{0r}(\lambda), \beta^*_{0r}(\lambda)]$. *Then the set of all pairs of dual basic controllers* $(\alpha_{0l}(\lambda), \beta_{0l}(\lambda))$, $[\alpha_{0r}(\lambda), \beta_{0r}(\lambda)]$ *is determined by the relations*

$$\begin{aligned} \alpha_{0l}(\lambda) &= \alpha^*_{0l}(\lambda) - M(\lambda)b_l(\lambda), \qquad & \beta_{0l}(\lambda) &= \beta^*_{0l}(\lambda) - M(\lambda)a_l(\lambda), \\ \alpha_{0r}(\lambda) &= \alpha^*_{0r}(\lambda) - b_r(\lambda)M(\lambda), \qquad & \beta_{0r}(\lambda) &= \beta^*_{0r}(\lambda) - a_r(\lambda)M(\lambda), \end{aligned} \tag{4.39}$$

where $M(\lambda)$ *is any polynomial matrix of appropriate dimension.*

Proof. In order to determine the set of all pairs of dual controllers, we at first notice that from (4.7) and (4.28) it follows that the relations

$$\begin{aligned} \alpha_{0l}(\lambda) &= D_l(\lambda)\alpha^*_{0l}(\lambda) - M_l(\lambda)b_l(\lambda), \qquad & \beta_{0l}(\lambda) &= D_l(\lambda)\beta^*_{0l}(\lambda) - M_l(\lambda)a_l(\lambda), \\ \alpha_{0r}(\lambda) &= \alpha^*_{0r}(\lambda)D_r(\lambda) - b_r(\lambda)M_r(\lambda), \qquad & \beta_{0r}(\lambda) &= \beta^*_{0r}(\lambda)D_r(\lambda) - a_r(\lambda)M_r(\lambda) \end{aligned} \tag{4.40}$$

hold, from which we get

$$
\begin{aligned}
Q_l(\lambda, \alpha_{0l}, \beta_{0l}) &= \begin{bmatrix} I_n & O_{nm} \\ M_l(\lambda) & D_l(\lambda) \end{bmatrix} Q_l(\lambda, \alpha_{0l}^*, \beta_{0l}^*) \\
Q_r(\lambda, \alpha_{0r}, \beta_{0r}) &= Q_r(\lambda, \alpha_{0r}^*, \beta_{0r}^*) \begin{bmatrix} D_r(\lambda) & O_{nm} \\ -M_r(\lambda) & I_m \end{bmatrix}.
\end{aligned} \tag{4.41}
$$

For the duality of the controllers $(\alpha_{0l}(\lambda), \beta_{0l}(\lambda))$ and $[\alpha_{0r}(\lambda), \beta_{0r}(\lambda)]$, it is necessary and sufficient that the matrices (4.41) satisfy Relation (4.29). But from (4.41), owing to the duality of the controllers $(\alpha_{0l}^*(\lambda), \beta_{0l}^*(\lambda))$ and $[\alpha_{0r}^*(\lambda), \beta_{0r}^*(\lambda)]$, we get

$$
Q_l(\lambda, \alpha_{0l}, \beta_{0l}) Q_r(\lambda, \alpha_{0r}, \beta_{0r}) = \begin{bmatrix} D_r(\lambda) & O_{nm} \\ M_l(\lambda) D_r(\lambda) - D_l(\lambda) M_r(\lambda) & D_l(\lambda) \end{bmatrix}
$$

and Relation (4.29) is fulfilled, if and only if

$$
D_r(\lambda) = I_n\,, \quad D_l(\lambda) = I_m\,, \quad M_l(\lambda) = M_r(\lambda) = M(\lambda)\,.
$$

∎

Corollary 4.18. *Each solution of Equation (4.35) uniquely corresponds to a right dual controller, and each solution of Equation (4.36) uniquely corresponds to a left dual controller.* ∎

Remark 4.19. Theorems 4.15 and 4.17 indicate that the pairs of left and right process models, used for building the dual basic controllers, may be chosen arbitrarily, as long as Condition (4.34) holds. If the pairs $(a_l(\lambda), b_l(\lambda))$, $[a_r(\lambda), b_r(\lambda)]$ satisfy Condition (4.34), and the $n \times n$ polynomial matrix $p(\lambda)$ and the $m \times m$ polynomial matrix $q(\lambda)$ are unimodular, then the pairs $(p(\lambda)a_l(\lambda), p(\lambda)b_l(\lambda))$, $[a_r(\lambda)q(\lambda), b_r(\lambda)q(\lambda)]$ fulfill this condition. Therefore, we can reach for instance that in (4.34), the matrix $a_l(\lambda)$ is row reduced and the matrix $a_r(\lambda)$ is column reduced.

Remark 4.20. From Theorems 4.15 and 4.17, it follows that as a first right basic controller any solution $[\alpha_{0r}(\lambda), \beta_{0r}(\lambda)]$ of the Diophantine Equation (4.37) can be used. Then the corresponding dual left basic controller is found by Formula (4.38). After that, the complete set of all pairs of dual basic controllers is constructed by Relations (4.40).

4.5 Eigenvalue Assignment for Polynomial Pairs

1. As stated in Section 4.1, the eigenvalue assignment problem for the pair $(a_l(\lambda), b_l(\lambda))$ amounts to finding the set of controllers $(\alpha_l(\lambda), \beta_l(\lambda))$ which satisfy the condition

$$
\det Q_l(\lambda, \alpha_l, \beta_l) \approx d(\lambda)\,, \tag{4.42}
$$

where $d(\lambda)$ is a prescribed monic polynomial and

$$Q_l(\lambda, \alpha_l, \beta_l) = \begin{bmatrix} a_l(\lambda) & -b_l(\lambda) \\ -\beta_l(\lambda) & \alpha_l(\lambda) \end{bmatrix}. \tag{4.43}$$

The general solution for the formulated problem in case of an irreducible process provides the following theorem.

Theorem 4.21. *Let the process $(a_l(\lambda), b_l(\lambda))$ be irreducible. Then Equation (4.42) is solvable for any polynomial $d(\lambda)$. Thereby, if $(\alpha_{0l}(\lambda), \beta_{0l}(\lambda))$ is a certain basic controller for the process $(a_l(\lambda), b_l(\lambda))$, then the set of all controllers $(\alpha_l(\lambda), \beta_l(\lambda))$ satisfying (4.42) can be represented in the form*

$$\begin{aligned} \alpha_l(\lambda) &= D_l(\lambda)\alpha_{0l}(\lambda) - M_l(\lambda)b_l(\lambda), \\ \beta_l(\lambda) &= D_l(\lambda)\beta_{0l}(\lambda) - M_l(\lambda)a_l(\lambda), \end{aligned} \tag{4.44}$$

where the $m \times n$ polynomial matrix $M_l(\lambda)$ is arbitrary, and for the $m \times m$ polynomial matrix $D_l(\lambda)$ the condition

$$\det D(\lambda) \approx d(\lambda)$$

is valid. Besides, the pair $(\alpha_l(\lambda), \beta_l(\lambda))$ is irreducible, if and only if the pair $(D_l(\lambda), M_l(\lambda))$ is irreducible.

Proof. Denote the set of solutions of Equation (4.42) by $\mathcal{N}_0$, and the set of pairs (4.44) by $\mathcal{N}_p$. Let $(\alpha_{0l}(\lambda), \beta_{0l}(\lambda))$ be a certain basic controller. Then the matrices

$$\begin{aligned} Q_l(\lambda, \alpha_{0l}, \beta_{0l}) &= \begin{bmatrix} a_l(\lambda) & -b_l(\lambda) \\ -\beta_{0l}(\lambda) & \alpha_{0l}(\lambda) \end{bmatrix}, \\ Q_l(\lambda, \alpha_{0l}, \beta_{0l})^{-1} = Q_r(\lambda, \alpha_{0r}, \beta_{0r}) &= \begin{bmatrix} \alpha_{0r}(\lambda) & b_r(\lambda) \\ \beta_{0r}(\lambda) & a_r(\lambda) \end{bmatrix} \end{aligned} \tag{4.45}$$

are unimodular, and the condition

$$Q_l(\lambda, \alpha_{0l}, \beta_{0l})Q_r(\lambda, \alpha_{0r}, \beta_{0r}) = I_{n+m}$$

holds. Let $(\alpha_l(\lambda), \beta_l(\lambda))$ be a controller satisfying Equation (4.42). Then using (4.34), (4.43), (4.45) and the Bezout identity (4.31), we get

$$\begin{aligned} Q_l(\lambda, \alpha_l, \beta_l)Q_r(\lambda, \alpha_{0r}, \beta_{0r}) &= \begin{bmatrix} a_l(\lambda) & -b_l(\lambda) \\ -\beta_l(\lambda) & \alpha_l(\lambda) \end{bmatrix} \begin{bmatrix} \alpha_{0r}(\lambda) & b_r(\lambda) \\ \beta_{0r}(\lambda) & a_r(\lambda) \end{bmatrix} \\ &= \begin{bmatrix} I_n & O_{nm} \\ M_l(\lambda) & D_l(\lambda) \end{bmatrix} \end{aligned} \tag{4.46}$$

with

$$\begin{aligned} D_l(\lambda) &= -\beta_l(\lambda) b_r(\lambda) + \alpha_l(\lambda) a_r(\lambda) \,, \\ M_l(\lambda) &= -\beta_l(\lambda) \alpha_{0r}(\lambda) + \alpha_l(\lambda) \beta_{0r}(\lambda) \,. \end{aligned} \tag{4.47}$$

Applying (4.46) and (4.47), we find

$$Q_l(\lambda, \alpha_l, \beta_l) = N_l(\lambda) Q_l(\lambda, \alpha_{0l}, \beta_{0l}) \,, \tag{4.48}$$

where

$$N_l(\lambda) = \begin{bmatrix} I_n & O_{nm} \\ M_l(\lambda) & D_l(\lambda) \end{bmatrix} , \tag{4.49}$$

where we read (4.44). Calculating the determinant on both sides of (4.48) shows that

$$\det Q_l(\lambda, \alpha_l, \beta_l) \approx \det D_l(\lambda) \approx d(\lambda) \,.$$

Thus $\mathcal{N}_0 \subset \mathcal{N}_p$ was proven. By reversing the conclusions, we deduce as in Theorem 4.1 that also $\mathcal{N}_p \subset \mathcal{N}_0$ is true. Therefore, the sets $\mathcal{N}_0$ and $\mathcal{N}_p$ coincide.

Notice that Formulae (4.44) may be written in the shape

$$\begin{aligned} \begin{bmatrix} -\beta_l(\lambda) & \alpha_l(\lambda) \end{bmatrix} &= \begin{bmatrix} M_l(\lambda) & D_l(\lambda) \end{bmatrix} Q_l(\lambda, \alpha_{0l}, \beta_{0l}) \\ &= \begin{bmatrix} M_l(\lambda) & D_l(\lambda) \end{bmatrix} \begin{bmatrix} a_l(\lambda) & -b_l(\lambda) \\ -\beta_{0l}(\lambda) & \alpha_{0l}(\lambda) \end{bmatrix} . \end{aligned}$$

Since the matrix $Q_l(\lambda, \alpha_{0l}, \beta_{0l})$ is unimodular, the matrices $\begin{bmatrix} -\beta_l(\lambda) & \alpha_l(\lambda) \end{bmatrix}$ and $\begin{bmatrix} M_l(\lambda) & D_l(\lambda) \end{bmatrix}$ are right-equivalent, and that's why the pair $(\alpha_l(\lambda), \beta_l(\lambda))$ is irreducible, if and only if the pair $(D_l(\lambda), M_l(\lambda))$ is irreducible. ∎

2.

Example 4.22. For a prescribed polynomial $d(\lambda)$, the solution set of the eigenvalue assignment problem for the process (4.24) in Example 4.12 has the form

$$\alpha_l(\lambda) = \begin{bmatrix} d_{11}(\lambda) & d_{12}(\lambda) \\ d_{21}(\lambda) & d_{22}(\lambda) \end{bmatrix} \begin{bmatrix} -(0.5\lambda + 1) & 0 \\ 0 & 1 \end{bmatrix} - \begin{bmatrix} m_{11}(\lambda) & m_{12}(\lambda) \\ m_{21}(\lambda) & m_{22}(\lambda) \end{bmatrix} \begin{bmatrix} 0 & -\lambda \\ -\lambda & 0 \end{bmatrix} ,$$

$$\beta_l(\lambda) = \begin{bmatrix} d_{11}(\lambda) & d_{12}(\lambda) \\ d_{21}(\lambda) & d_{22}(\lambda) \end{bmatrix} \begin{bmatrix} -0.5 & 0 \\ 0 & 0 \end{bmatrix} - \begin{bmatrix} m_{11}(\lambda) & m_{12}(\lambda) \\ m_{21}(\lambda) & m_{22}(\lambda) \end{bmatrix} \begin{bmatrix} \lambda - 1 & \lambda + 1 \\ 0 & 1 \end{bmatrix} .$$

Here the $m_{ik}(\lambda)$ are arbitrary polynomials and $d_{ik}(\lambda)$ are arbitrary polynomials bound by the condition

$$d_{11}(\lambda) d_{22}(\lambda) - d_{21}(\lambda) d_{12}(\lambda) \approx d(\lambda) \,.$$

□

Example 4.23. The set of solutions of Equation (4.42) for the process (4.25) in Example 4.13 has the form

$$\alpha_l(\lambda) = kd(\lambda)d_3(\lambda) + m_1(\lambda)\}\,,$$
$$\beta_l(\lambda) = 0.5kd(\lambda)\begin{bmatrix}1 & 1\end{bmatrix} - \begin{bmatrix}m_1(\lambda) & m_2(\lambda)\end{bmatrix}\begin{bmatrix}\lambda - 1 & \lambda + 1\end{bmatrix},$$

where k is a constant and $d_3(\lambda)$, $m_1(\lambda)$, $m_2(\lambda)$ are any polynomials. □

3. Now, consider the question, how the solution of Equation (4.42) looks like when the process $(a_l(\lambda), b_l(\lambda))$ is reducible. In this case, with respect to the results in Section 1.12, there exists a latent square $n \times n$ polynomial matrix $q(\lambda)$, such that

$$a_l(\lambda) = q(\lambda)a_{l1}(\lambda), \quad b_l(\lambda) = q(\lambda)b_{l1}(\lambda) \tag{4.50}$$

is true with an irreducible pair $(a_{l1}(\lambda), b_{l1}(\lambda))$. The solvability conditions for Equation (4.42) in case (4.50) states the following theorem.

Theorem 4.24. *Let (4.50) be valid and* $\det q(\lambda) = \gamma(\lambda)$. *Then a necessary and sufficient condition for the solvability of Equation (4.42) is, that the polynomial* $d(\lambda)$ *is divisible by* $\gamma(\lambda)$. *Thus, if* $(\tilde{\alpha}_{0l}(\lambda), \tilde{\beta}_{0l}(\lambda))$ *is a certain basic controller for the process* $(a_{l1}(\lambda), b_{l1}(\lambda))$, *then the set of all controllers satisfying Equation (4.42) is bound by the relations*

$$\begin{aligned}\alpha_l(\lambda) &= \tilde{D}_l(\lambda)\tilde{\alpha}_{0l}(\lambda) - \tilde{M}_l(\lambda)b_{l1}(\lambda)\,,\\ \beta_l(\lambda) &= \tilde{D}_l(\lambda)\tilde{\beta}_{0l}(\lambda) - \tilde{M}_l(\lambda)a_{l1}(\lambda)\,,\end{aligned} \tag{4.51}$$

where the $m \times n$ *polynomial matrix* $\tilde{M}_l(\lambda)$ *is arbitrary, and the* $m \times m$ *polynomial matrix* $\tilde{D}_l(\lambda)$ *satisfies the condition* $\det \tilde{D}_l(\lambda) \approx \tilde{d}(\lambda)$. *Here, the polynomial* $\tilde{d}(\lambda)$ *is determined by*

$$\tilde{d}(\lambda) = \frac{d(\lambda)}{\gamma(\lambda)}\,. \tag{4.52}$$

Proof. Let (4.50) be true. Then (4.42) can be presented in the shape

$$\det\left\{\begin{bmatrix} q(\lambda) & O_{nm} \\ O_{mn} & I_m \end{bmatrix} \tilde{Q}_l(\lambda, \alpha_l, \beta_l)\right\} \approx d(\lambda)\,, \tag{4.53}$$

where

$$\tilde{Q}_l(\lambda, \alpha_l, \beta_l) = \begin{bmatrix} a_{l1}(\lambda) & -b_{l1}(\lambda) \\ -\beta_l(\lambda) & \alpha_l(\lambda) \end{bmatrix}.$$

Calculating the determinants, we find

$$\gamma(\lambda)\det \tilde{Q}_l(\lambda, \alpha_l, \beta_l) \approx d(\lambda)\,,$$

i.e. for the solvability of Equation (4.53), it is necessary that the polynomial $d(\lambda)$ is divisible by $\gamma(\lambda)$. If this condition is ensured and (4.52) is used, then Equation (4.53) leads to

$$\det \tilde{Q}_l(\lambda, \alpha, \beta) \approx \tilde{d}(\lambda) .$$

Since the pair $(a_{l1}(\lambda), b_{l1}(\lambda))$ is irreducible, Equation (4.53) is always solvable, thanks to Theorem 4.17, and its solution has the shape (4.51). ∎

4. Let $(a(\lambda), b(\lambda))$ be an irreducible process and $(\alpha_l(\lambda), \beta_l(\lambda))$ such a controller, that $\det Q_l(\lambda, \alpha_l, \beta_l) = d(\lambda) \not\equiv 0$ becomes true. Furthermore, let $(\alpha_{0l}(\lambda), \beta_{0l}(\lambda))$ be a certain basic controller. Then owing to Theorem 4.17, there exist $m \times m$ and $m \times n$ polynomial matrices $D_l(\lambda)$ and $M_l(\lambda)$, such that

$$\begin{aligned} \alpha_l(\lambda) &= D_l(\lambda)\alpha_{0l}(\lambda) - M_l(\lambda)b_l(\lambda) , \\ \beta_l(\lambda) &= D_l(\lambda)\beta_{0l}(\lambda) - M_l(\lambda)a_l(\lambda) , \end{aligned} \tag{4.54}$$

where $\det D_l(\lambda) \approx d(\lambda)$. Relations (4.54) are called the *basic representation* of the controllers $(\alpha_l(\lambda), \beta_l(\lambda))$ with respect to the basis $(\alpha_{0l}(\lambda), \beta_{0l}(\lambda))$.

Theorem 4.25. *The basic representation (4.54) is unique in the sense, that from the validity of (4.54) and the relation*

$$\begin{aligned} \alpha_l(\lambda) &= D_{l1}(\lambda)\alpha_{0l}(\lambda) - M_{l1}(\lambda)b_l(\lambda) , \\ \beta_l(\lambda) &= D_{l1}(\lambda)\beta_{0l}(\lambda) - M_{1l}(\lambda)a_l(\lambda) , \end{aligned} \tag{4.55}$$

we can conclude $D_{l1}(\lambda) = D_l(\lambda)$, $M_{l1}(\lambda) = M_l(\lambda)$.

Proof. Suppose (4.54) and (4.55) are fulfilled at the same time. Subtracting (4.55) from (4.54), we get

$$\begin{aligned} \left[D_l(\lambda) - D_{l1}(\lambda)\right] \alpha_{0l}(\lambda) - \left[M_l(\lambda) - M_{l1}(\lambda)\right] b_l(\lambda) &= O_{mm} , \\ \left[D_l(\lambda) - D_{l1}(\lambda)\right] \beta_{0l}(\lambda) - \left[M_l(\lambda) - M_{l1}(\lambda)\right] a_l(\lambda) &= O_{mn} , \end{aligned}$$

which is equivalent to

$$\left[\, M_l(\lambda) - M_{l1}(\lambda) \;\; D_l(\lambda) - D_{l1}(\lambda) \,\right] Q_l(\lambda, \alpha_{0l}, \beta_{0l}) = O_{m,m+n} .$$

From this, it follows immediately $M_{l1}(\lambda) = M_l(\lambda)$, $D_{l1}(\lambda) = D_l(\lambda)$, because the matrix $Q_l(\lambda, \alpha_{0l}, \beta_{0l})$ is unimodular. ∎

4.6 Eigenvalue Assignment by Transfer Matrices

1. In case of $\det \alpha_l(\lambda) \not\equiv 0$, it means, the pair $(\alpha_l(\lambda), \beta_l(\lambda))$ is not singular, the transfer function of the controller

$$w_\varrho(\lambda) = \alpha_l^{-1}(\lambda)\beta_l(\lambda) \tag{4.56}$$

might be included into our considerations. Its standard form (2.21) can be written as

$$w_\varrho(\lambda) = \frac{M_\varrho(\lambda)}{d_\varrho(\lambda)} \tag{4.57}$$

for which Relation (4.56) defines a certain LMFD. Conversely, if the transfer function of the controller is given in the standard form (4.57), then various LMFD (4.56) and the corresponding characteristic matrices

$$Q_l(\lambda, \alpha_l, \beta_l) = \begin{bmatrix} a_l(\lambda) & -b_l(\lambda) \\ -\beta_l(\lambda) & \alpha_l(\lambda) \end{bmatrix} \tag{4.58}$$

can be investigated. Besides, every LMFD (4.56) is uniquely related to a characteristic polynomial $\Delta(\lambda) = \det Q_l(\lambda, \alpha_l, \beta_l)$.

In future, we will say that the transfer matrix $w_\varrho(\lambda)$ is a *solution of the eigenvalue assignment* for the process $(a_l(\lambda), b_l(\lambda))$, if it allows an LMFD (4.56) such that the corresponding pair $(\alpha_l(\lambda), \beta_l(\lambda))$ satisfies Equation (4.42).

2. The set of transfer matrices (4.57) that supply the solution of the eigenvalue assignment is generally characterised by the next theorem.

Theorem 4.26. *Let the pair $(a_l(\lambda), b_l(\lambda))$ be irreducible and $(\alpha_{0l}(\lambda), \beta_{0l}(\lambda))$ be an appropriate left basic controller. Then for the fact that the transfer matrix (4.56) is a solution of Equation (4.42), it is necessary and sufficient that it allows a representation of the form*

$$w_\varrho(\lambda) = [\alpha_{0l}(\lambda) - \phi(\lambda)b_l(\lambda)]^{-1}[\beta_{0l}(\lambda) - \phi(\lambda)a_l(\lambda)]\,, \tag{4.59}$$

where $\phi(\lambda)$ is a broken rational $m \times n$ matrix, for which exists an LMFD

$$\phi(\lambda) = D_l^{-1}(\lambda)M_l(\lambda)\,, \tag{4.60}$$

where $\det D_l(\lambda) \approx d(\lambda)$ *is true and the polynomial matrix $M_l(\lambda)$ is arbitrary.*

Proof. Sufficiency: Suppose the LMFD (4.60). Then from (4.59) we get

$$w_\varrho(\lambda) = [D_l(\lambda)\alpha_{0l}(\lambda) - M_l(\lambda)b_l(\lambda)]^{-1}[D_l(\lambda)\beta_{0l}(\lambda) - M_l(\lambda)a_l(\lambda)]\,. \tag{4.61}$$

Thus, the set of equations

$$\begin{aligned} \alpha_l(\lambda) &= D_l(\lambda)\alpha_{0l}(\lambda) - M_l(\lambda)b_l(\lambda)\,, \\ \beta_l(\lambda) &= D_l(\lambda)\beta_{0l}(\lambda) - M_l(\lambda)a_l(\lambda) \end{aligned} \tag{4.62}$$

describes a controller satisfying Relation (4.42).

Necessity: If (4.56) and $\det Q_l(\lambda, \alpha_l, \beta_l)) \approx d(\lambda)$ are true, then for the matrices $\alpha_l(\lambda)$ and $\beta_l(\lambda)$, we can find a basic representation (4.54), and under the invertability condition for the matrix $\alpha_l(\lambda)$, we obtain (4.59), so the proof is carried out. ∎

Corollary 4.27. *From (4.59) we learn that the transfer matrices $w_\varrho(\lambda)$, defined as the solution set of Equation (4.42), depend on a matrix parameter, namely the fractional rational matrix $\phi(\lambda)$.*

3. Let the transfer function of the controller be given in the form (4.59). Then under Condition (4.60), it can be represented in form of the LMFD (4.56), where the matrices $\alpha_l(\lambda)$, $\beta_l(\lambda)$ are determined by (4.62). For applications, the question of the irreducibility of the pair (4.62) is important.

Theorem 4.28. *The pair (4.62) is exactly then irreducible, when the pair $[D_l(\lambda), M_l(\lambda)]$ is irreducible, i.e. the right side of (4.59) is an ILMFD.*

Proof. The proof follows directly from Theorem 4.17. ∎

4. Let the process $(a_l(\lambda), b_l(\lambda))$ and a certain fractional rational $m \times n$ matrix $w_\varrho(\lambda)$ be given, for which the expression (4.56) defines a certain LMFD. Thus, if

$$\det Q_l(\lambda, \alpha_l, \beta_l)) = \det \begin{bmatrix} a_l(\lambda) & -b_l(\lambda) \\ -\beta_l(\lambda) & \alpha_l(\lambda) \end{bmatrix} \approx d(\lambda)\,,$$

then, owing to Theorem 4.26, the matrix $w_\varrho(\lambda)$ can be represented in the form (4.59), (4.61), where $(\alpha_{0l}(\lambda), \beta_{0l}(\lambda))$ is a certain basic controller. Under these circumstances, the notation (4.59) of the matrix $w_\varrho(\lambda)$ is called its *basic representation* with respect to the basis $(\alpha_{0l}(\lambda), \beta_{0l}(\lambda))$.

Theorem 4.29. *For a fixed basic controller $(\alpha_{0l}(\lambda), \beta_{0l}(\lambda))$, the basic representation (4.59) is unique in the sense, that the validity of (4.59) and*

$$w_\varrho(\lambda) = [\alpha_{0l}(\lambda) - \phi_1(\lambda) b_l(\lambda)]^{-1} [\beta_{0l}(\lambda) - \phi_1(\lambda) a_l(\lambda)] \tag{4.63}$$

at the same time implies the equality $\phi(\lambda) = \phi_1(\lambda)$.

Proof. Without loss of generality, we suppose that the right side of (4.60) is an ILMFD. Then owing to Theorem 4.26, the right side of (4.61) is an ILMFD of the matrix $w_\varrho(\lambda)$. In addition let us have the LMFD

$$\phi_1(\lambda) = D_1^{-1}(\lambda) M_1(\lambda)\,.$$

Then from (4.63) for the matrix $w_\varrho(\lambda)$, we obtain the LMFD

$$w_\varrho(\lambda) = [D_1(\lambda)\alpha_{0l}(\lambda) - M_1(\lambda) b_l(\lambda)]^{-1} [D_1(\lambda)\beta_{0l}(\lambda) - M_1(\lambda) a_l(\lambda)]\,. \tag{4.64}$$

This relation and (4.61) define two different LMFDs of the matrix $w_\varrho(\lambda)$. By supposition the LMFD (4.61) is irreducible, so with respect to Statement 2.3 on page 64, we come out with

$$\begin{aligned} D_1(\lambda)\alpha_{0l}(\lambda) - M_1(\lambda) b_l(\lambda) &= U(\lambda)[D_l(\lambda)\alpha_{0l}(\lambda) - M_l(\lambda) b_l(\lambda)]\,, \\ D_1(\lambda)\beta_{0l}(\lambda) - M_1(\lambda) a_l(\lambda) &= U(\lambda)\left[D_l(\lambda)\beta_{0l}(\lambda) - M_l(\lambda) a_l(\lambda)\right]\,, \end{aligned}$$

where $U(\lambda)$ is a non-singular $m \times m$ polynomial matrix. These relations can be written as

$$\left[\, M_1(\lambda) - U(\lambda)M_l(\lambda) \;\; D_1(\lambda) - U(\lambda)D_l(\lambda) \,\right] Q_l(\lambda, \alpha_{0l}, \beta_{0l}) = O_{m,m+n}\,, \tag{4.65}$$

where

$$Q_l(\lambda, \alpha_{0l}, \beta_{0l}) = \begin{bmatrix} a_l(\lambda) & -b_l(\lambda) \\ -\beta_{0l}(\lambda) & \alpha_{0l}(\lambda) \end{bmatrix}.$$

Since this matrix is designed unimodular, it follows from (4.65)

$$M_1(\lambda) = U(\lambda)M_l(\lambda), \quad D_1(\lambda) = U(\lambda)D_l(\lambda)\,.$$

Thus we derive

$$\phi(\lambda) = D_l^{-1}(\lambda)M_l(\lambda) = D_1^{-1}(\lambda)M_1(\lambda) = \phi_1(\lambda)\,,$$

which completes the proof. ■

4.7 Structural Eigenvalue Assignment for Polynomial Pairs

1. The solution of the structural eigenvalue assignment for an irreducible process $(a_l(\lambda), b_l(\lambda))$ by the controller $(\alpha_l(\lambda), \beta_l(\lambda))$ bases on the following statement.

Theorem 4.30. *Let the process $(a_l(\lambda), b_l(\lambda))$ be irreducible and the controller $(\alpha_l(\lambda), \beta_l(\lambda))$ should have the basic representation (4.54). Then the matrices*

$$Q_l(\lambda, \alpha_l, \beta_l) = \begin{bmatrix} a_l(\lambda) & -b_l(\lambda) \\ -\beta_l(\lambda) & \alpha_l(\lambda) \end{bmatrix}, \quad S(\lambda) = \begin{bmatrix} I_n & O_{nm} \\ O_{mn} & D_l(\lambda) \end{bmatrix} \tag{4.66}$$

are equivalent, and this fact does not depend on the matrix $M_l(\lambda)$.

Proof. Notice

$$\begin{bmatrix} I_n & O_{nm} \\ M_l(\lambda) & D_l(\lambda) \end{bmatrix} = \begin{bmatrix} I_n & O_{nm} \\ M_l(\lambda) & I_m \end{bmatrix} \begin{bmatrix} I_n & O_{nm} \\ O_{mn} & D_l(\lambda) \end{bmatrix},$$

then Relations (4.48), (4.49) can be written in the form

$$Q_l(\lambda, \alpha_l, \beta_l) = \begin{bmatrix} I_n & O_{nm} \\ M_l(\lambda) & I_m \end{bmatrix} S(\lambda) Q_l(\lambda, \alpha_{0l}, \beta_{0l})\,.$$

The first and the last factor on the right side are unimodular matrices and therefore, the matrices (4.66) are equivalent. ■

Theorem 4.31. *Let* $a_1(\lambda), \dots, a_{n+m}(\lambda)$ *and* $b_1(\lambda), \dots, b_m(\lambda)$ *be the sequences of invariant polynomials of the matrices* $Q_l(\lambda, \alpha_l, \beta_l)$ *and* $D_l(\lambda)$*, respectively. Then the equations*

$$a_1(\lambda) = a_2(\lambda) = \ldots = a_n(\lambda) = 1 \tag{4.67}$$

and furthermore

$$a_{n+i}(\lambda) = b_i(\lambda), \quad (i = 1, \dots, m). \tag{4.68}$$

Proof. Assume $b_1(\lambda), \dots, b_m(\lambda)$ be the sequence of invariant polynomials of $D_l(\lambda)$. Then the sequence of invariant polynomials of $S(\lambda)$ is equal to $1, \dots, 1, b_1(\lambda), \dots, b_m(\lambda)$. But the matrices $D_l(\lambda)$ and $Q_l(\lambda, \alpha_l, \beta_l)$ are equivalent, hence their sequences of invariant polynomials coincide, that means, Equations (4.67) and (4.68) are correct. ∎

Corollary 4.32. *Theorem 4.31 supplies a constructive procedure for the design of closed systems with a prescribed sequence of invariant polynomials of the characteristic matrix. Indeed, let a sequence of monic polynomials* $b_1(\lambda), \dots, b_m(\lambda)$ *with*

$$b_1(\lambda) \cdots b_m(\lambda) \approx d(\lambda)$$

be given and for all $i = 2, \dots, m$*, the polynomial* $b_i(\lambda)$ *is divisible by* $b_{i-1}(\lambda)$*. Then we take*

$$D_l(\lambda) = p(\lambda) \operatorname{diag}\{b_1(\lambda), \dots, b_m(\lambda)\} q(\lambda),$$

where $p(\lambda)$*,* $q(\lambda)$ *are unimodular matrices. After that, independently of the selection of* $M_l(\lambda)$ *in (4.54), the sequence of the last* m *invariant polynomials of the matrix* $Q_l(\lambda, \alpha_l, \beta_l)$ *coincides with the sequence* $b_1(\lambda), \dots, b_m(\lambda)$*.* ∎

Corollary 4.33. *If the process* $(a_l(\lambda), b_l(\lambda))$ *is irreducible, then there exists a set of controllers* Ω_s *for which the matrix* $Q_l(\lambda, \alpha_l, \beta_l)$ *becomes simple. This happens exactly when the matrix* $D_l(\lambda)$ *is simple, i.e. it allows the representation*

$$D_l(\lambda) = p(\lambda) \operatorname{diag}\{1, \dots, 1, d(\lambda)\} q(\lambda)$$

with unimodular matrices $p(\lambda)$*,* $q(\lambda)$*.* ∎

Corollary 4.34. *Let irreducible left and right models of the process* $(a_l(\lambda), b_l(\lambda))$ *and* $[a_r(\lambda), b_r(\lambda)]$ *be given. Then the sequence of invariant polynomials* $a_{n+1}(\lambda), \dots, a_{n+m}(\lambda)$ *of the characteristic matrix* $Q_l(\lambda, \alpha_l, \beta_l)$ *coincides with the sequence of invariant polynomials of the matrix*

$$D_l(\lambda) = -\beta_l(\lambda) b_r(\lambda) + \alpha_l(\lambda) a_r(\lambda),$$

which is a direct consequence of Theorem 4.30 and Equations (4.47). ∎

4.8 Eigenvalue and Eigenstructure Assignment for PMD Processes

1. In the present section, the generale solution for the eigenvalue assignment (4.6) for non-singular PMD processes is developed. Moreover, the set of sequences of invariant polynomials is described for which the structural eigenvalue assignment is solvable.

Theorem 4.35. *Let the PMD (4.4) be non-singular and minimal, and*

$$w_\tau(\lambda) = c(\lambda)a^{-1}(\lambda)b(\lambda) \tag{4.69}$$

should be its corresponding transfer matrix. Furthermore, let us have the ILMFD

$$w_\tau(\lambda) = a_l^{-1}(\lambda)b_l(\lambda)\,. \tag{4.70}$$

Then the eigenvalue assignment problem (4.5) is solvable for any polynomial $d(\lambda)$. Besides, the set of pairs $(\alpha_l(\lambda), \beta_l(\lambda))$ that are solutions of (4.6) coincides with the set of pairs that are determined as solutions of the eigenvalue assignment for the irreducible pair $(a_l(\lambda), b_l(\lambda))$, and these may be produced on the base of Theorem 4.17.

Preparing the proof, some auxiliary statements are given.

Lemma 4.36. *For the non-singular PMD (4.4), formula*

$$\det Q_\tau(\lambda, \alpha, \beta) = \det a(\lambda) \det\left[\alpha(\lambda) - \beta(\lambda)w_\tau(\lambda)\right] \tag{4.71}$$

holds, where the matrix $Q_\tau(\lambda, \alpha, \beta)$ is established in (4.5).

Proof. The matrix $Q_\tau(\lambda, \alpha, \beta)$ is brought into the form

$$Q_\tau(\lambda, \alpha, \beta) = \begin{bmatrix} A(\lambda) & -B(\lambda) \\ -C(\lambda) & D(\lambda) \end{bmatrix}, \tag{4.72}$$

where

$$A(\lambda) = \begin{bmatrix} a(\lambda) & O_{pn} \\ -c(\lambda) & I_n \end{bmatrix}, \qquad B(\lambda) = \begin{bmatrix} b(\lambda) \\ O_{nm} \end{bmatrix}, \tag{4.73}$$

$$C(\lambda) = \begin{bmatrix} O_{mp} & \beta(\lambda) \end{bmatrix}, \qquad D(\lambda) = \alpha(\lambda)\,. \tag{4.74}$$

Under the taken propositions, we have

$$\det A(\lambda) = \det a(\lambda) \not\equiv 0\,. \tag{4.75}$$

Therefore, the well-known formula [51]

$$\det Q_\tau(\lambda, \alpha, \beta) = \det A(\lambda) \det\left[D(\lambda) - C(\lambda)A^{-1}(\lambda)B(\lambda)\right] \tag{4.76}$$

is applicable. Observing

$$A^{-1}(\lambda) = \begin{bmatrix} a^{-1}(\lambda) & O_{pn} \\ c(\lambda)a^{-1}(\lambda) & I_n \end{bmatrix}$$

and (4.72)–(4.75), we obtain (4.71). ∎

Lemma 4.37. *Let the non-singular PMD (4.4) and its corresponding transfer matrix (4.69) be given, for which Relation (4.70) defines an ILMFD. Consider the matrix*

$$Q_l(\lambda, \alpha, \beta) = \begin{bmatrix} a_l(\lambda) & -b_l(\lambda) \\ -\beta(\lambda) & \alpha(\lambda) \end{bmatrix}, \tag{4.77}$$

where the matrices $\alpha(\lambda)$ and $\beta(\lambda)$ are defined as in (4.5). If under this condition, the PMD (4.4) is minimal, then

$$\det Q_\tau(\lambda, \alpha, \beta) \approx \det Q_l(\lambda, \alpha, \beta)\,. \tag{4.78}$$

Proof. Applying Formula (4.76) to Matrix (4.77), we find

$$\det Q_l(\lambda, \alpha, \beta) = \det a_l(\lambda) \left[\alpha(\lambda) - \beta(\lambda) w_\tau(\lambda)\right]\,. \tag{4.79}$$

Consider now the ILMFD

$$c(\lambda)a^{-1}(\lambda) = a_1^{-1}(\lambda)c_1(\lambda)\,.$$

Since the left side is an IRMFD, the relation

$$\det a_1(\lambda) \approx \det a(\lambda)$$

holds. Thus due to Lemma 2.9, the expression

$$w_\tau(\lambda) = a^{-1}(\lambda)[c_1(\lambda)b(\lambda)]$$

defines an ILMFD of the matrix $w_\tau(\lambda)$. This expression and (4.70) define at the same time ILMFDs of the matrix $w_\tau(\lambda)$, so we have

$$\det a_1(\lambda) \approx \det a(\lambda) \approx \det a_l(\lambda)\,.$$

Using this and (4.70) from (4.79), we obtain the statement (4.78). ∎

Proof of Theorem 4.35. The minimality of the PMD (4.4) and Lemma 4.37 imply that the sets of solutions of (4.6) and of the equation

$$\det Q_l(\lambda, \alpha, \beta) \approx d(\lambda)$$

coincide. ∎

2. The next theorem supplies the solution of the eigenvalue assignment for the case, when the PMD (4.4) is not minimal.

Theorem 4.38. *Let the non-singular PMD (4.4) be not minimal, and Relation (4.70) should describe an ILMFD of the transfer matrix $w_\tau(\lambda)$. Then the relation*

$$\chi(\lambda) = \frac{\det a(\lambda)}{\det a_l(\lambda)} \tag{4.80}$$

turns out to be a polynomial. Thereby, Equation (4.6) is exactly then solvable, when

$$d(\lambda) = \chi(\lambda) d_1(\lambda)\,, \tag{4.81}$$

where $d_1(\lambda)$ is any polynomial. If (4.81) is true, then the set of controllers that are solutions of Equation (4.5) coincide with the set of solutions of the equation

$$\det \begin{bmatrix} a_l(\lambda) & -b_l(\lambda) \\ -\beta(\lambda) & \alpha(\lambda) \end{bmatrix} \approx d_1(\lambda)\,. \tag{4.82}$$

This solution set can be constructed with the help of Theorem 4.17.

Proof. Owing to Lemma 2.48, Relation (4.80) is a polynomial. With the help of (4.71) and (4.80), we gain

$$\begin{aligned} \det Q_\tau(\lambda, \alpha, \beta) &= \det a(\lambda) \det[\alpha(\lambda) - \beta(\lambda) w_\tau(\lambda)] \\ &= \chi(\lambda) \det a_l(\lambda) \det[\alpha(\lambda) - \beta(\lambda) w_\tau(\lambda)]\,. \end{aligned}$$

Using (4.79), we find out that Equation (4.5) leads to

$$\chi(\lambda) \det Q_l(\lambda, \alpha, \beta) \approx d(\lambda)\,. \tag{4.83}$$

From (4.83), it is immediately seen that Equation (4.6) needs Condition (4.81) be fulfilled for its solvability. Conversely, if (4.81) is fulfilled, Equation (4.83) leads to Equation (4.82). ∎

3. The solution of the structural eigenvalue assignment for minimal PMD (4.4) supplies the following theorem.

Theorem 4.39. *Let the non-singular PMD (4.4) be minimal, and Relation (4.70) should define an ILMFD of the transfer matrix (4.69). Furthermore, let $(\alpha_0(\lambda), \beta_0(\lambda))$ be a basic controller for the pair $(a_l(\lambda), b_l(\lambda))$, and the set of pairs*

$$\begin{aligned} \alpha(\lambda) &= N(\lambda)\alpha_0(\lambda) - M(\lambda) b_l(\lambda)\,, \\ \beta(\lambda) &= N(\lambda)\beta_0(\lambda) - M(\lambda) a_l(\lambda) \end{aligned} \tag{4.84}$$

should determine the set of solutions of the eigenvalue assignment (4.6). Moreover, let $q_1(\lambda), \ldots, q_{p+n+m}(\lambda)$ be the sequence of invariant polynomials of the *polynomial matrix* $Q_\tau(\lambda, \alpha, \beta)$, *and* $\nu_1(\lambda), \ldots, \nu_m(\lambda)$ *be the sequence of invariant polynomials of the polynomial matrix* $N(\lambda)$*. Then*

$$\begin{aligned} q_1(\lambda) = q_2(\lambda) = \ldots \; &= q_{p+n}(\lambda) = 1\,, \\ q_{p+n+i}(\lambda) = \nu_i(\lambda), \qquad &(i = 1, \ldots, m)\,. \end{aligned} \tag{4.85}$$

Proof. a) It is shown that under the conditions of Theorem 4.39, the pair $(A(\lambda), B(\lambda))$ defined by Relation (4.73) is irreducible. Indeed, let $(\alpha_0(\lambda), \beta_0(\lambda))$ be a basic controller for the pair $(a_l(\lambda), b_l(\lambda))$ that is determined by the ILMFD (4.70). Then owing to Lemma 4.37, we have

$$\det \begin{bmatrix} A(\lambda) & -B(\lambda) \\ -C_0(\lambda) & D_0(\lambda) \end{bmatrix} = \text{const.} \neq 0\,,$$

where

$$C_0(\lambda) = \begin{bmatrix} O_{mp} \; \beta_0(\lambda) \end{bmatrix}, \quad D_0(\lambda) = \alpha_0(\lambda)\,, \tag{4.86}$$

and due to Theorem 1.41, the pair $(A(\lambda), B(\lambda))$ is irreducible.

b) Equation (4.6) is written in the form

$$\det \begin{bmatrix} A(\lambda) & -B(\lambda) \\ -C(\lambda) & D(\lambda) \end{bmatrix} \approx d(\lambda)\,. \tag{4.87}$$

Since the pair $A(\lambda), B(\lambda))$ is irreducible, it follows from Theorem 4.17 that for any polynomial $d(\lambda)$, Equation (4.87) is solvable and the set of solutions can be presented in the shape

$$\begin{aligned} D(\lambda) &= N_1(\lambda) D_0(\lambda) - M_1(\lambda) \begin{bmatrix} b(\lambda) \\ O_{nm} \end{bmatrix}, \\ C(\lambda) &= N_1(\lambda) C_0(\lambda) - M_1(\lambda) \begin{bmatrix} a(\lambda) & O_{pn} \\ -c(\lambda) & I_n \end{bmatrix}, \end{aligned} \tag{4.88}$$

where the $m \times (p+n)$ polynomial matrix $M_1(\lambda)$ is arbitrary, but the $m \times m$ polynomial matrix $N_1(\lambda)$ has to fulfill the single condition

$$\det N_1(\lambda) \approx d(\lambda)\,.$$

c) On the other side, due to Theorem 4.38, the set of pairs $(\alpha(\lambda), \beta(\lambda))$ satisfying Equation (4.87) coincides with the set of solutions of the equation

$$\det \begin{bmatrix} a_l(\lambda) & -b_l(\lambda) \\ -\beta(\lambda) & \alpha(\lambda) \end{bmatrix} \approx d(\lambda)$$

that has the form (4.84), where the $m \times n$ polynomial matrix $M(\lambda)$ is arbitrary, and the $m \times m$ polynomial matrix $N(\lambda)$ has to satisfy the condition

$$\det N(\lambda) \approx d(\lambda)\,.$$

Assume in (4.88)

$$M_1(\lambda) = \overset{\quad p \qquad\quad n}{\left[\, \tilde{M}_1(\lambda)\ \tilde{M}_2(\lambda) \,\right]}\, m\,.$$

Then with the help of (4.74), (4.86), Relation (4.88) can be presented in the shape

$$\alpha(\lambda) = N_1(\lambda)\alpha_0(\lambda) - \tilde{M}_1(\lambda)b(\lambda)\,,$$

$$\left[O_{mp}\ \beta(\lambda)\right] = N_1(\lambda)\left[O_{mp}\ \beta_0(\lambda)\right] - \left[\tilde{M}_1(\lambda)a(\lambda) - \tilde{M}_2(\lambda)c(\lambda)\ \tilde{M}_2(\lambda)\right].$$

In order to avoid a contradiction between these equations with (4.84), it is necessary and sufficient that the condition

$$N_1(\lambda) = N(\lambda) \tag{4.89}$$

holds, and moreover,

$$\tilde{M}_1(\lambda)b(\lambda) = M(\lambda)b_l(\lambda), \quad \tilde{M}_1(\lambda)a(\lambda) - \tilde{M}_2(\lambda)c(\lambda) = O_{mn}\,,$$
$$\tilde{M}_2(\lambda) = M(\lambda)a_l(\lambda)$$

are fulfilled. Now we directly conclude that these relations are satisfied for

$$\tilde{M}_1(\lambda) = M(\lambda)a_l(\lambda)c(\lambda)a^{-1}(\lambda), \quad \tilde{M}_2(\lambda) = M(\lambda)a_l(\lambda)\,. \tag{4.90}$$

Besides, due to Lemma 2.9, the product $a_l(\lambda)c(\lambda)a^{-1}(\lambda)$ is a polynomial matrix . Substituting the last relations and (4.89) into (4.88), we find

$$\begin{aligned} D(\lambda) &= N(\lambda)\alpha_0(\lambda) - M(\lambda)\left[a_l(\lambda)c(\lambda)a^{-1}(\lambda)\ a_l(\lambda)\right]\begin{bmatrix} b(\lambda) \\ O_{nm} \end{bmatrix}, \\ C(\lambda) &= N(\lambda)C_0(\lambda) - M(\lambda)\left[a_l(\lambda)c(\lambda)a^{-1}(\lambda)\ a_l(\lambda)\right]\begin{bmatrix} a(\lambda) & O_{pn} \\ -c(\lambda) & I_n \end{bmatrix}. \end{aligned} \tag{4.91}$$

From this and Theorem 4.8, Equations (4.85) emerge immediately. ∎

Corollary 4.40. *In order to get a simple matrix $Q_\tau(\lambda, \alpha, \beta)$ under the conditions of Theorem 4.11, it is necessary and sufficient that the matrix $N(\lambda)$ in Formula (4.91) is simple.*

4. The structure of the characteristic matrix $Q_\tau(\lambda, \alpha, \beta)$ for the case, when the non-singular PMD (4.4) is not minimal, decides the following theorem.

Theorem 4.41. *For the non-singular PMD (4.4), the factorisations*

$$a(\lambda) = d_1(\lambda)a_1(\lambda), \qquad b(\lambda) = d_1(\lambda)b_1(\lambda) \tag{4.92}$$

should be valid, where $d_1(\lambda)$, $a_1(\lambda)$ are $p \times p$ polynomial matrices, $b_1(\lambda)$ is a $p \times m$ polynomial matrix and the pair $(a_1(\lambda), b_1(\lambda))$ is irreducible. Moreover, suppose

$$a_1(\lambda) = a_2(\lambda)d_2(\lambda), \qquad c(\lambda) = c_1(\lambda)d_2(\lambda) \tag{4.93}$$

with $p \times p$ polynomial matrices $d_2(\lambda)$, $a_2(\lambda)$, the $n \times p$ polynomial matrix $c_1(\lambda)$ and the irreducible pair $[a_2(\lambda), c_1(\lambda)]$. Then the following statements are true:

a) The PMD $\tau_1(\lambda) = (a_2(\lambda), b_1(\lambda), c_1(\lambda))$ *is equivalent to the PMD (4.4) and minimal.*

b) The relation

$$\xi(\lambda) = \frac{\det a(\lambda)}{\det a_2(\lambda)} \tag{4.94}$$

turns out to be a polynomial with

$$\xi(\lambda) \approx \chi(\lambda) = \frac{\det a(\lambda)}{\det a_l(\lambda)}, \tag{4.95}$$

where $\chi(\lambda)$ *is the polynomial (4.80).*

c) The relation

$$Q_\tau(\lambda, \alpha, \beta) = G_l(\lambda) Q_{\tau_1}(\lambda, \alpha, \beta) G_r(\lambda) \tag{4.96}$$

is true with

$$G_l(\lambda) = \operatorname{diag}\{d_1(\lambda), 1, \ldots, 1\}, \quad G_r(\lambda) = \operatorname{diag}\{d_2(\lambda), 1, \ldots, 1\}, \tag{4.97}$$

and the matrix $Q_{\tau_1}(\lambda, \alpha, \beta)$ *has the shape*

$$Q_{\tau_1}(\lambda, \alpha, \beta) = \begin{bmatrix} a_2(\lambda) & O_{pn} & -b_1(\lambda) \\ -c_1(\lambda) & I_n & O_{nm} \\ O_{mp} & -\beta(\lambda) & \alpha(\lambda) \end{bmatrix}. \tag{4.98}$$

d) Formula

$$\xi(\lambda) \approx \det d_1(\lambda) \det d_2(\lambda) \tag{4.99}$$

is valid.

e) Let $\tilde{q}_1(\lambda), \ldots, \tilde{q}_{p+n+m}(\lambda)$ *be the sequence of invariant polynomials of the matrix* $Q_{\tau_1}(\lambda, \alpha, \beta)$ *and* $\nu_1(\lambda), \ldots, \nu_m(\lambda)$ *be the sequence of invariant polynomials of the matrix* $N(\lambda)$ *in the representation (4.91), where instead of* $a_l(\lambda)$, $b(\lambda)$, $c(\lambda)$ *we have to write* $a_2(\lambda)$, $b_1(\lambda)$, $c_1(\lambda)$. *Then*

$$\begin{aligned} \tilde{q}_1(\lambda) = \tilde{q}_2(\lambda) = \ldots &= \tilde{q}_{p+n}(\lambda) = 1, \\ \tilde{q}_{p+n+i}(\lambda) &= \nu_i(\lambda), \quad (i = 1, \ldots, m). \end{aligned} \tag{4.100}$$

Proof. a) Using (4.92) and (4.93), we find

$$w_\tau(\lambda) = c(\lambda) a^{-1}(\lambda) b(\lambda) = c_1(\lambda) a_2^{-1}(\lambda) b_1(\lambda) = w_{\tau_1}(\lambda),$$

where $w_{\tau_1}(\lambda)$ is the transfer function of the PMD $\tau_1(\lambda)$, this means, the PMD $\tau(\lambda)$ and $\tau_1(\lambda)$ are equivalent. It is demonstrated that the PMD $\tau_1(\lambda)$ is minimal. Since the pair $[a_2(\lambda), c_1(\lambda)]$ is irreducible per construction, it is sufficient to show that the pair $(a_2(\lambda), b_1(\lambda))$ is irreducible. Per construction, the pair

$$(a_1(\lambda), b_1(\lambda)) = (a_2(\lambda) d_2(\lambda), b_1(\lambda))$$

is irreducible. Hence due to Lemma 2.11, also the pair $(a_2(\lambda), b_1(\lambda))$ is irreducible.

b) From (4.92) and (4.93), we recognise that Relation (4.94) is a polynomial. Since the PMD $\tilde{\tau}(\lambda) = (a_l(\lambda), b_l(\lambda), I_n)$ and $\tau_1(\lambda)$ are equivalent and minimal, Corollary 2.49 implies

$$\det a_l(\lambda) \approx \det a_2(\lambda)$$

and this yields (4.95).

c) Relations (4.96)–(4.98) can be taken immediately from (4.6), (4.92), (4.93).

d) Applying Formula (4.71) to Matrix (4.98), we obtain

$$\det Q_{\tau_1}(\lambda, \alpha, \beta) = \det a_2(\lambda) \det[\alpha(\lambda) - \beta(\lambda) w_\tau(\lambda)] .$$

Therefore, from (4.71) with the help of (4.94), we receive

$$\det Q_\tau(\lambda, \alpha, \beta) = \xi(\lambda) \det Q_{\tau_1}(\lambda, \alpha, \beta) .$$

On the other side from (4.96)–(4.98), it follows

$$\det Q_\tau(\lambda, \alpha, \beta) = \det G_l(\lambda) \det G_r(\lambda) \det Q_{\tau_1}(\lambda, \alpha, \beta).$$

Bringing face to face the last two equations proves Relation (4.99).

e) Since the PMD $\tau_1(\lambda)$ is minimal and Relation (4.84) holds, Formula (4.100) follows from Theorem 4.39. ∎

Corollary 4.42. *If one of the matrices $d_1(\lambda)$ or $d_2(\lambda)$ is not simple, then the matrix $Q_\tau(\lambda, \alpha, \beta)$ cannot be made simple with the help of any controller $(\alpha(\lambda), \beta(\lambda))$.*

Proof. Let for instance the matrix $d_1(\lambda)$ be not simple, then we learned from the considerations in Section 1.11 that there exists an eigenvalue $\tilde{\lambda}$ with $\operatorname{def} G_l(\tilde{\lambda}) > 1$. Hence considering (4.96), (4.97), we get $\operatorname{def} Q_\tau(\tilde{\lambda}, \alpha, \beta) > 1$, *i.e.*, the matrix $Q_\tau(\lambda, \alpha, \beta)$ is not simple. If $d_2(\lambda)$ is not simple, we conclude analogously. ∎

Corollary 4.43. *Let the matrices $d_1(\lambda)$ and $d_2(\lambda)$ be simple and possess no eigenvalues in common. Then for the simplicity of the matrix $Q_\tau(\lambda, \alpha, \beta)$, it suffices that the matrix $N(\lambda)$ in (4.84) is simple and has no common eigenvalues with the matrix $\tilde{d}(\lambda) = d_1(\lambda) d_2(\lambda)$.*

Proof. Let $\mu_1, \ldots, \mu_q$ and $\nu_1, \ldots, \nu_s$ be the different eigenvalues the matrices $d_1(\lambda)$ and $d_2(\lambda)$, respectively. Then the matrices $G_l(\lambda)$ and $G_r(\lambda)$ in (4.97) are also simple, where the eigenvalues of $G_l(\lambda)$ are the numbers $\mu_1, \ldots, \mu_q$, but the eigenvalues of the matrix $G_r(\lambda)$ are the numbers $\nu_1, \ldots, \nu_s$. Let the matrix $N(\lambda)$ in (4.84) be simple and should possess the eigenvalues $n_1, \ldots, n_k$ that are disjunct with all the values μ_i, $(i = 1, \ldots, q)$ and ν_j, $(j = 1, \ldots, s)$. Then from Corollary 4.40, it follows that the matrix $Q_{\tau_1}(\lambda, \alpha, \beta)$ is simple, and possesses the set of eigenvalues $\{n_1, \ldots, n_k\}$. From (4.96), we recall that the set of eigenvalues of the matrix $Q_\tau(\lambda, \alpha, \beta)$ is built from the unification of the sets $\{\mu_1, \dots, \mu_q\}$, $\{\nu_1, \dots, \nu_s\}$ and $\{n_1, \dots, n_k\}$. Using (4.96), we find out that for all appropriate i

$$\operatorname{def} Q_\tau(\mu_i, \alpha, \beta) = 1, \quad \operatorname{def} Q_\tau(\nu_i, \alpha, \beta) = 1, \quad \operatorname{def} Q_\tau(n_i, \alpha, \beta) = 1\,,$$

and together with the results of Section 1.11, this yields that the matrix $Q_\tau(\lambda, \alpha, \beta)$ is simple. ■

5

Fundamentals for Control of Causal Discrete-time LTI Processes

5.1 Finite-dimensional Discrete-time LTI Processes

1. Generalised discrete linear processes can be represented by the abstract Fig. 5.1, where $\{u\}$, $\{y\}$ are input and output vector sequences

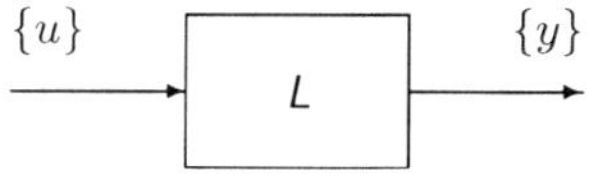

Fig. 5.1. Generalised discrete LTI process

$$\{u\} = \begin{bmatrix} \{u\}_1 \\ \vdots \\ \{u\}_m \end{bmatrix}, \qquad \{y\} = \begin{bmatrix} \{y\}_1 \\ \vdots \\ \{y\}_n \end{bmatrix}$$

and their components $\{u\}_i$, $\{y\}_i$ are scalar sequences

$$\{u\}_i = \{u_{i,0}, u_{i,1}, \dots \}, \qquad \{y\}_i = \{y_{i,0}, y_{i,1}, \dots \}.$$

The above vector sequences can also be represented in the form

$$\{u\} = \{u_0, u_1, \dots \}, \qquad \{y\} = \{y_0, y_1, \dots \}, \tag{5.1}$$

where

$$u_s = \begin{bmatrix} u_{s,1} \\ \vdots \\ u_{s,m} \end{bmatrix}, \qquad y_s = \begin{bmatrix} y_{s,1} \\ \vdots \\ y_{s,n} \end{bmatrix}. \tag{5.2}$$

Furthermore, in Fig. 5.1, the letter L symbolises a certain system of linear equations that consists between the input and output sequences. If L stands for a system with a finite number of linear difference equation with constant coefficients, then the corresponding process is called a *finite-dimensional discrete-time LTI object*. In this section exclusively such objects will be considered and they will be called shortly as LTI objects.

2. Compatible with the introduced concepts, the LTI object in Fig. 5.1 is configured to a system of scalar difference equations

$$\sum_{p=1}^{n} a_{ip}^{(0)} y_{p,k+\ell} + \ldots + \sum_{p=1}^{n} a_{ip}^{(\ell)} y_{p,k} = \sum_{r=1}^{m} b_{ir}^{(0)} u_{r,k+s} + \ldots + \sum_{r=1}^{m} b_{ir}^{(s)} u_{r,k} \tag{5.3}$$
$$(i = 1, \ldots, n;\ k = 0, 1, \ldots),$$

where the $a_{ip}^{(j)}$, $b_{ir}^{(j)}$ are constant real coefficients. Introducing into the considerations the constant matrices

$$\tilde{a}_j = \left[a_{ip}^{(j)} \right], \qquad \tilde{b}_j = \left[b_{ir}^{(j)} \right]$$

and using the notation (5.1), the system of scalar Equations (5.3) can be written in form of the vector difference equation

$$\tilde{a}_0 y_{k+\ell} + \ldots + \tilde{a}_\ell\, y_k = \tilde{b}_0 u_{k+s} + \ldots + \tilde{b}_s u_k\,, \qquad (k = 0, 1, \ldots)\,, \tag{5.4}$$

which connects the components of the input and output sequences.

Introduce the shifted vector sequences of the form

$$\{u_j\} = \{u_j, u_{j+1}, \ldots\}, \qquad \{y_j\} = \{y_j, y_{j+1}, \ldots\}\,. \tag{5.5}$$

Then the system of Equations (5.4) can be written as equation connecting their shifted sequences:

$$\tilde{a}_0\{y_\ell\} + \ldots + \tilde{a}_\ell\{y\} = \tilde{b}_0\{u_s\} + \ldots + \tilde{b}_s\{u\}\,. \tag{5.6}$$

It is easily checked that Equations (5.4) and (5.6) are equivalent. This can be done by substituting the expressions (5.5) into (5.6) and comparing the corresponding components of the sequences on the left and right side.

3. Next the right-shift (forward shift) operator q is introduced by the relations

$$q y_k = y_{k+1}\,, \quad q u_k = u_{k+1}\,. \tag{5.7}$$

Herewith, (5.4) can be written in the form

$$\tilde{a}(q) y_k = \tilde{b}(q) u_k\,, \tag{5.8}$$

where

$$\tilde{a}(q) = \tilde{a}_0 q^\ell + \ldots + \tilde{a}_\ell\,, \quad \tilde{b}(q) = \tilde{b}_0 q^s + \ldots + \tilde{b}_s \tag{5.9}$$

are polynomial matrices. However, if the operator q is defined by the relation

$$q\{y_k\} = \{y_{k+1}\}, \qquad q\{u_k\} = \{u_{k+1}\},$$

then we come to equations that depend on the sequences

$$\tilde{a}(q)\{y\} = \tilde{b}(q)\{u\}. \tag{5.10}$$

4. In what follows, Equations (5.10) or (5.8) will be called a *forward model* of the LTI process. The matrix $\tilde{a}(q)$ is named *eigenoperator*, and the matrix $\tilde{b}(q)$ *inputoperator* of the forward model.

If not mentioned otherwise, we always suppose

$$\det \tilde{a}(q) \not\equiv 0 \tag{5.11}$$

i.e., the matrix $\tilde{a}(q)$ is non-singular. If (5.11) is valid, the LTI process is said to be *non-singular.*

Moreover, we assume that in the equations of the non-singular processes (5.8), (5.9) always $\tilde{a}_0 \neq O_{nn}$ is true, and at least one of the matrices $\tilde{a}_\ell$ or $\tilde{b}_s$ is a nonzero matrix.

If under the mentioned propositions, the relation

$$\det \tilde{a}_0 \neq 0 \tag{5.12}$$

is valid, then the LTI process is called *normal.* If however, instead of (5.12)

$$\det \tilde{a}_0 = 0 \tag{5.13}$$

is true, then the LTI process is named *anomalous*, [39]. For instance, descriptor processes can be modelled by anomalous systems [34].

5. For a given input sequence $\{u\}$, Relations (5.4) can be regarded as a difference equation for the unknown output sequence $\{y\}$. If (5.13) is allowed, then in general the difference equation (5.4) cannot be written as recursion for y_k, and we have to define, what we will understand by a solution.

> **Solution of a not necessarily normal difference equation.** For a known input sequence as a solution of Equation (5.4) $\{u\}$, we understand any sequence that is defined for all $k \geq 0$, and the elements of which satisfy Relation (5.4) for all $k \geq 0$.

Suppose the non-singular matrix $\chi(q)$. Multiplying both sides of Equation (5.8) from left by $\chi(q)$, we obtain

$$\chi(q)\tilde{a}(q)y_k = \chi(q)\tilde{b}(q)u_k\,, \qquad (k = 0, 1, \dots) \tag{5.14}$$

and using (5.7), this can be written in an analogue form to (5.4). Equation (5.14) is said to be *derived* from the output Equation (5.4), and Equation (5.4) itself is called *original.*

Example 5.1. Consider the equations of the LTI processes in the form (5.3)

$$\begin{aligned} y_{1,k+2} + y_{1,k} + y_{2,k} &= u_{k+1} \\ y_{1,k+1} + 2y_{2,k+1} &= 2u_k \, . \end{aligned} \tag{5.15}$$

Denote $y_k = \begin{bmatrix} y_{1,k} & y_{2,k} \end{bmatrix}'$, then (5.15) is written in the form

$$\tilde{a}_0 y_{k+2} + \tilde{a}_1 y_{k+1} + \tilde{a}_2 y_k = \tilde{b}_0 u_{k+2} + \tilde{b}_1 u_{k+1} + \tilde{b}_2 u_k \, , \tag{5.16}$$

where

$$\tilde{a}_0 = \begin{bmatrix} 1 & 0 \\ 0 & 0 \end{bmatrix}, \quad \tilde{a}_1 = \begin{bmatrix} 0 & 0 \\ 1 & 1 \end{bmatrix}, \quad \tilde{a}_2 = \begin{bmatrix} 1 & 1 \\ 0 & 0 \end{bmatrix},$$

$$\tilde{b}_0 = \begin{bmatrix} 0 \\ 0 \end{bmatrix}, \quad \tilde{b}_1 = \begin{bmatrix} 1 \\ 0 \end{bmatrix}, \quad \tilde{b}_2 = \begin{bmatrix} 0 \\ 2 \end{bmatrix},$$

so that we obtain (5.9) with

$$\tilde{a}(q) = \begin{bmatrix} q^2 + q & 1 \\ q & 2q \end{bmatrix}, \quad \tilde{b}(q) = \begin{bmatrix} q \\ 2 \end{bmatrix}.$$

Since $\det \tilde{a}(q) = 2q^3 + 2q^2 - q \not\equiv 0$, the LTI process (5.15) is non-singular. Besides, due to $\det \tilde{a}_0 = 0$, the process is anomalous.

Assume

$$\chi(q) = \begin{bmatrix} q & 1 \\ 1 & -q \end{bmatrix},$$

then (5.14), that is derived from the original Equation (5.16), takes the form

$$\begin{bmatrix} q^3 + q^2 + q & 3q \\ q & 1 - 2q^2 \end{bmatrix} y_k = \begin{bmatrix} q^2 + 2 \\ -q \end{bmatrix} u_k$$

or, by means of (5.7), it is equivalently written as the system of equations

$$\begin{aligned} y_{1,k+3} + y_{1,k+2} + y_{1,k+1} + 3y_{2,k+1} &= u_{k+2} + 2u_k \\ y_{1,k+1} - 2y_{2,k+2} + y_{2,k} &= -u_{k+1} \, . \end{aligned} \qquad (k = 0, 1, \ldots) \tag{5.17}$$

Hereby, Equations (5.15) are called original with respect to (5.17). □

6.

Lemma 5.2. *For any matrix $\chi(q)$, all solutions of the original equation (5.8) are also solutions of the derived equation (5.14).*

Proof. The derived equation is written in the form

$$\chi(q)\left[\tilde{a}(q)y_k - \tilde{b}(q)u_k\right] = 0_k\,, \tag{5.18}$$

where $0_k = \begin{bmatrix} 0 & \cdots & 0 \end{bmatrix}'$ for all $k \geq 0$. Obviously, the vectors u_k, y_k satisfy (5.18) for all $k \geq 0$, when Equation (5.8) holds for all of them and all $k \geq 0$. ∎

Remark 5.3. The inverse statement to Lemma 5.2 in general is not true. Indeed, let $\{v\}$ be a solution of the equation

$$\chi(q)v_k = 0_k\,.$$

Then any solution of the equation

$$\tilde{a}(q)y_k = \tilde{b}(q)u_k + v_k \tag{5.19}$$

for all possible v_k presents a solution of the derived Equation (5.14), but only for $v_k = 0_k$ it is a solution of the original equation. It is easy to show that Relation (5.19) contains all solutions of the derived equation.

7. Consider the important special case, when in (5.14) the matrix $\chi(q)$ is unimodular. In this case, the transition from the original equation (5.8) to the derived equation (5.14) means manipulating the system (5.3) by operations of the following types:

a) Exchange the places of two equations.
b) Multiply an equation by a non-zero constant.
c) Add one equation to any other equation that was multiplied before by an arbitrary polynomial $f(q)$.

In what follows, Equations (5.8) and (5.14) are called *equivalent by the unimodular matrix* $\chi(q)$. The reasons for using this terminology arise from the next lemma.

Lemma 5.4. *The solution sets of the equivalent equations (5.8) and (5.14) coincide.*

Proof. Let Equations (5.8) and (5.14) be equivalent, and $\mathcal{R}$, $\mathcal{R}_x$ are their solution sets. Lemma 5.2 implies $\mathcal{R} \subset \mathcal{R}_x$. On the other side, Equations (5.8) are gained from Equations (5.14) by multiplying them from left by $\chi^{-1}(q)$. Then also Lemma 5.2 implies $\mathcal{R}_x \subset \mathcal{R}$, thus $\mathcal{R} = \mathcal{R}_x$. ∎

8. Assume in (5.14) a unimodular matrix $\chi(q)$. Introduce the notation

$$\chi(q)\tilde{a}(q) = \tilde{a}_\varrho(q)\,, \qquad \chi(q)\tilde{b}(q) = \tilde{b}_\varrho(q)\,. \tag{5.20}$$

Then the derived equation (5.14) can be written in the form

$$\tilde{a}_\varrho(q)y_k = \tilde{b}_\varrho(q)u_k\,. \tag{5.21}$$

From Section 1.6, it is known that under supposition (5.11), the matrix $\chi(q)$ can always be selected in such a way that the matrix $\tilde{a}_\varrho(q)$ becomes row reduced. In this case, Equation (5.21) also is said to be row reduced. Let $\tilde{a}_1(q), \ldots, \tilde{a}_n(q)$ be the rows of the matrix $\tilde{a}_\varrho(q)$. As before, denote

$$\alpha_i = \deg \tilde{a}_i(q)\,, \quad (i = 1, \ldots, n)\,.$$

If under these conditions, Equation (5.21) is row reduced, then independently of the concrete shape of the matrix $\chi(q)$, the quantities

$$\alpha_l = \sum_{i=1}^{n} \alpha_i\,, \quad \alpha_{\max} = \deg \tilde{a}_\varrho(q) = \max_{1\le i\le n} \{\alpha_i\}$$

take their minimal values in the set of equivalent equations to the original equation (5.8).

Example 5.5. Consider the anomalous process

$$\begin{aligned} y_{1,k+4} + 2y_{1,k+2} + y_{1,k+1} + 2y_{2,k+2} + y_{2,k} &= u_{k+3} + 2u_{k+1} \\ y_{1,k+3} + y_{1,k+1} + y_{1,k} + 2y_{2,k+1} &= u_{k+2} + u_k\,. \end{aligned} \tag{5.22}$$

In the present case, we have

$$\tilde{a}(q) = \begin{bmatrix} q^4+2q^2+q & 2q^2+1 \\ q^3+q+1 & 2q \end{bmatrix} = \tilde{a}_0 q^4 + \tilde{a}_1 q^3 + \tilde{a}_2 q^2 + \tilde{a}_3 q + \tilde{a}_4\,,$$

$$\tilde{b}(q) = \begin{bmatrix} q^3+2q \\ q^2+1 \end{bmatrix} = \tilde{b}_1 q^3 + \tilde{b}_2 q^2 + \tilde{b}_3 q + \tilde{b}_4\,,$$

where

$$\tilde{a}_0 = \begin{bmatrix} 1\,0 \\ 0\,0 \end{bmatrix}, \quad \tilde{a}_1 = \begin{bmatrix} 0\,0 \\ 1\,0 \end{bmatrix}, \quad \tilde{a}_2 = \begin{bmatrix} 2\,2 \\ 0\,0 \end{bmatrix}, \quad \tilde{a}_3 = \begin{bmatrix} 1\,0 \\ 1\,2 \end{bmatrix}, \quad \tilde{a}_4 = \begin{bmatrix} 0\,1 \\ 1\,0 \end{bmatrix},$$

$$\tilde{b}_1 = \begin{bmatrix} 1 \\ 0 \end{bmatrix}, \quad \tilde{b}_2 = \begin{bmatrix} 0 \\ 1 \end{bmatrix}, \quad \tilde{b}_3 = \begin{bmatrix} 2 \\ 0 \end{bmatrix}, \quad \tilde{b}_4 = \begin{bmatrix} 0 \\ 1 \end{bmatrix}.$$

Choose

$$\chi(q) = \begin{bmatrix} 1 & -q \\ -q & q^2+1 \end{bmatrix}.$$

So, we generate the derived matrices

$$\tilde{a}_\varrho(q) = \chi(q)\tilde{a}(q) = \begin{bmatrix} q^2 & 1 \\ q+1 & q \end{bmatrix}, \quad \tilde{b}_\varrho(q) = \chi(q)\tilde{b}(q) = \begin{bmatrix} q \\ 1 \end{bmatrix},$$

where the matrix $\tilde{a}_\varrho(q)$ is row reduced. Applying this, Equations (5.22) might be expressed equivalently by

$$\begin{aligned} y_{1,k+2} + y_{2,k} &= u_{k+1} \\ y_{1,k+1} + y_{1,k} + y_{2,k+1} &= u_k\,. \end{aligned}$$

□

5.2 Transfer Matrices and Causality of LTI Processes

1. For non-singular processes (5.8) under Condition (5.11), the rational matrix

$$\tilde{w}(q) = \tilde{a}^{-1}(q)\tilde{b}(q) \tag{5.23}$$

is defined, which is called the *transfer matrix (-function)* of the forward model. The next lemma indicates an important property of the transfer matrix.

Lemma 5.6. *The transfer matrices of the original equation (5.8) and of the derived equation (5.21) coincide.*

Proof. Suppose

$$\tilde{w}_\varrho(q) = \tilde{a}_\varrho^{-1}(q)\tilde{b}_\varrho(q)\,. \tag{5.24}$$

Then applying (5.20), we get

$$\tilde{w}_\varrho(q) = \tilde{a}_\varrho^{-1}(q)\tilde{b}_\varrho(q) = \tilde{a}^{-1}(q)\tilde{b}(q) = \tilde{w}(q)\,.$$

■

Corollary 5.7. *The transfer functions of equivalent forward models coincide.* ■

2. From the above said emerges that any forward model (5.8) is uniquely assigned to a transfer matrix. The reverse statement is obviously wrong. Therefore the question arises, how is the set of forward models structured, that possess a given transfer matrix? The next theorem gives the answer.

Theorem 5.8. *Let the rational $n \times m$ matrix $\tilde{w}(q)$ be given and*

$$\tilde{w}(q) = \tilde{a}_0^{-1}\tilde{b}_0(q)$$

be an ILMFD. Then the set of all forward models of LTI processes possessing this transfer matrix is determined by the relations

$$\tilde{a}(q) = \psi(q)\tilde{a}_0(q), \quad \tilde{b}(q) = \psi(q)\tilde{b}_0(q)\,, \tag{5.25}$$

where $\psi(q)$ is any non-singular polynomial matrix.

Proof. The right side of Relation (5.24) presents a certain LMFD of the rational matrix $\tilde{w}(q)$. Hence by the properties of LMFDs considered in Section 2.4, we conclude that the set of all pairs $(\tilde{a}(q), \tilde{b}(q))$ according to the transfer matrix $\tilde{w}(q)$ is determined by Relations (5.25). ■

Corollary 5.9. *A forward model of the LTI processes (5.8) is called* controllable, *if the pair $(\tilde{a}(q), \tilde{b}(q))$ is irreducible. Hence Theorem 5.8 is formulated in the following way: Let the forward model defined by the pair $(\tilde{a}(q), \tilde{b}(q))$ be controllable. Then the set of all forward models with transfer function (5.23) coincides with the set of all derived forward models.* ■

3. The LTI process (5.8) is called *weakly causal*, *strictly causal* or *causal*, if its transfer matrix (5.23) is proper, strictly proper or at least proper, respectively. From the content of Section 2.6, it emerges that the LTI process (5.8), (5.9) is causal, if there exists the finite limit

$$\lim_{q \to \infty} \tilde{w}(q) = w_0 \,. \tag{5.26}$$

Besides, when $w_0 = O_{nm}$ holds, the process is strictly causal. When the limit (5.26) becomes infinite, the process is named *non-causal.*

Theorem 5.10. *For the process (5.8), (5.9) to be causal, the condition*

$$\ell \geq s \tag{5.27}$$

is necessary. For strictly causality the inequality

$$\ell > s$$

must be valid.

Proof. Let us have the transfer matrix $\tilde{w}(q)$ in the standard form (2.21)

$$\tilde{w}(q) = \frac{\tilde{N}(q)}{\tilde{d}(q)} \,.$$

When this matrix is at least proper, then $\deg \tilde{N}(q) \leq \deg \tilde{d}(q)$ becomes true. Besides, Corollary 2.23 delivers for any LMFD (5.23) $\deg \tilde{a}(q) \geq \deg \tilde{b}(q)$, which is equivalent to (5.27). For strict causality, we conclude analogously. ∎

Remark 5.11. For further investigations, we optionally consider causal processes. Thus, in Equations (5.8), (5.9) always $\ell \geq s$ is assumed.

Remark 5.12. The conditions of Theorem 5.10 are in general not sufficient, as it is illustrated by the following example.

Example 5.13. Assume the LTI process (5.8) with

$$\tilde{a}(q) = \begin{bmatrix} q^3 & 1 \\ q+1 & q+2 \end{bmatrix}, \quad \tilde{b}(q) = \begin{bmatrix} 1 \\ q^2 \end{bmatrix}. \tag{5.28}$$

In this case, we have $\ell = \deg \tilde{a}(q) = 3$, $s = \deg \tilde{b}(q) = 2$. At the same time, we receive

$$\tilde{w}(q) = \frac{\begin{bmatrix} -q^2 + q + 2 \\ q^5 - q - 1 \end{bmatrix}}{q^4 + 2q^3 - q - 1} \,.$$

Hence the process (5.28) is non-causal. □

4. If Equation (5.8) is row reduced, the causality question for the processes (5.8) can be answered without constructing the transfer matrix.

Theorem 5.14. *Let Equation (5.8) be row reduced, and α_i be the degree of the i-th row of the matrix $\tilde{a}(q)$ and β_i be the degree of the i-th row of the matrix $\tilde{b}(q)$. Then the following statements are true:*

a) For the weak causality of the process, it is necessary and sufficient that the conditions

$$\alpha_i \geq \beta_i\,, \quad (i = 1, \dots, n) \tag{5.29}$$

are true, where at least for one $1 \leq i \leq n$ in (5.29) the equality sign has to be taken place.

b) For the strict causality of the process, the fulfilment of the inequalities

$$\alpha_i > \beta_i\,, \quad (i = 1, \dots, n) \tag{5.30}$$

is necessary and sufficient.

c) When for at least one $1 \leq i \leq n$

$$\alpha_i < \beta_i$$

becomes true, then the process is non-causal.

Proof. The proof emerges immediately from Theorem 2.24. ∎

5.3 Normal LTI Processes

1. This section considers LTI processes of the form

$$\tilde{a}_0 y_{k+\ell} + \ldots + \tilde{a}_\ell y_k = \tilde{b}_0 u_{k+\ell} + \ldots + \tilde{b}_\ell u_k\,, \quad (k = 0, 1, \ldots) \tag{5.31}$$

under the supposition

$$\det \tilde{a}_0 \neq 0\,. \tag{5.32}$$

Some important properties of normal LTI processes will be formulated, which emerge from Relation (5.32).

Theorem 5.15. *For the weak causality of the normal processes (5.31), the fulfillment of*

$$\tilde{b}_0 \neq O_{nm} \tag{5.33}$$

is necessary and sufficient.

For the strict causality of the normal processes (5.31), the fulfillment of

$$\tilde{b}_0 = O_{nm} \tag{5.34}$$

is necessary and sufficient.

Proof. From (5.32) it follows that the matrix $\tilde{a}(q)$ for a normal process is row reduced, and we have

$$\alpha_1 = \alpha_2 = \ldots = \alpha_n = \ell\,.$$

If (5.33) takes place, then in Condition (5.29) the equality sign stands for at least one $1 \leq i \leq n$. Therefore, as a consequence of Theorem 5.14, the process is weakly causal. If however, (5.34) takes place, then Condition (5.30) is true and the process is strictly causal. ∎

2. Let the vector input sequence

$$\{u\} = \{u_0, u_1, \ldots\} \tag{5.35}$$

be given, and furthermore assume any ensemble of ℓ constant vectors of dimension $n \times 1$

$$\bar{y}_0\,,\ \bar{y}_1\,,\ \ldots,\ \bar{y}_{\ell-1}\,. \tag{5.36}$$

In what follows, the vectors (5.36) are called *initial values.*

Theorem 5.16. *For any input sequence (5.35) and any ensemble of initial values (5.36), there exists a unique solution of the normal equation (5.31)*

$$\{y\} = \{y_0, y_1, \ldots, y_{\ell-1}, y_\ell, \ldots\}$$

satisfying the initial conditions

$$y_i = \bar{y}_i\,, \qquad (i = 1, \ldots, \ell - 1)\,. \tag{5.37}$$

Proof. Assume that the vectors y_i, $(i = 0, 1, \ldots, \ell - 1)$ satisfy Condition (5.37). Since Condition (5.32) is fulfilled, Equation (5.31) might be written in the shape

$$y_{k+\ell} = \bar{a}_1 y_{k+\ell-1} + \ldots + \bar{a}_\ell y_k + \bar{b}_0 u_{k+\ell} + \bar{b}_1 u_{k+\ell-1} + \ldots + \bar{b}_\ell u_k\,, \quad (k = 0, 1, \ldots), \tag{5.38}$$

where

$$\bar{a}_i = -\tilde{a}_0^{-1}\tilde{a}_i, \quad (i = 1, 2, \ldots, \ell); \qquad \bar{b}_i = \tilde{a}_0^{-1}\tilde{b}_i, (i = 0, 1, \ldots, \ell)\,.$$

For $k = 0$ from (5.38), we obtain

$$y_\ell = \bar{a}_1 y_{\ell-1} + \ldots + \bar{a}_\ell y_0 + \bar{b}_0 u_\ell + \bar{b}_1 u_{\ell-1} + \ldots + \bar{b}_\ell u_0\,. \tag{5.39}$$

Hence for a known input sequence (5.35) and given initial values (5.36), the vector y_ℓ is uniquely determined. For $k = 1$ from (5.38), we derive

$$y_{\ell+1} = \bar{a}_1 y_\ell + \ldots + \bar{a}_\ell y_1 + \bar{b}_0 u_{\ell+1} + \bar{b}_1 u_\ell + \ldots + \bar{b}_\ell u_1\,.$$

Thus with the help of (5.35), (5.36) and (5.39), the vector $y_{\ell+1}$ is uniquely calculated. Obviously, this procedure can be uniquely continued for all $k > 0$. As a result, in a unique way the sequence

$$\{y\} = \{\bar{y}_0, \dots, \bar{y}_{\ell-1}, y_\ell, \dots\}$$

is generated, that is a solution of Equation (5.31) and fulfills the initial conditions (5.37). ∎

Remark 5.17. It follows from the proof of Theorem 5.16 that for weakly causal normal processes for given initial conditions, the vector y_k of the solution $\{y\}$ is determined by the values of the input sequence $u_0, u_1, \dots, u_k$. If the process, however, is strictly causal, then the vector y_k is determined by the vectors $u_0, u_1, \dots, u_{k-1}$.

3.

Theorem 5.18. *Let the input (5.2) be a Taylor sequence (see Appendix A). Then all solutions of Equation (5.31) are Taylor sequences.*

Proof. Using (5.8), (5.9), the polynomial matrices

$$\begin{aligned} a(\zeta) &= \zeta^\ell \tilde{a}(\zeta^{-1}) = \tilde{a}_0 + \tilde{a}_1\zeta + \dots \tilde{a}_\ell \zeta^\ell , \\ b(\zeta) &= \zeta^\ell \tilde{b}(\zeta^{-1}) = \tilde{b}_0 + \tilde{b}_1\zeta + \dots \tilde{b}_\ell \zeta^\ell \end{aligned} \tag{5.40}$$

are considered. Condition (5.32) implies

$$\det a(0) = \det \tilde{a}_0 \neq 0 . \tag{5.41}$$

Under the assumed conditions, there exists the ζ-transform of the input sequence

$$u^0(\zeta) = \sum_{i=0}^{\infty} u_i \zeta^i . \tag{5.42}$$

Consider the vector

$$\begin{aligned} y^0(\zeta) = a^{-1}(\zeta) & \left[\tilde{a}_0\bar{y}_0 + \zeta(\tilde{a}_0\bar{y}_1 + \tilde{a}_1\bar{y}_0) + \dots \right. \\ & \left. + \zeta^{\ell-1}(\tilde{a}_0\bar{y}_{\ell-1} + \tilde{a}_1\bar{y}_{\ell-2} + \dots + \tilde{a}_{\ell-1}\bar{y}_0)\right] \\ + a^{-1}(\zeta) & \Big(\tilde{b}_0 \left[u^0(\zeta) - u_0 - \zeta u_1 - \dots - \zeta^{\ell-1} u_{\ell-1}\right] \\ & + \zeta\tilde{b}_1 \left[u^0(\zeta) - u_0 - \zeta u_1 - \dots - \zeta^{\ell-2} u_{\ell-2}\right] + \dots + \zeta^\ell \tilde{b}_\ell u^0(\zeta)\Big) , \end{aligned} \tag{5.43}$$

where $u^0(\zeta)$ is the convergent series (5.42). Since the vector $u^0(\zeta)$ is analytical in the point $\zeta = 0$ and Condition (5.41) is valid, the right side of (5.43) is analytical in $\zeta = 0$, and consequently

$$y^0(\zeta) = \sum_{i=0}^{\infty} y_i \zeta^i \tag{5.44}$$

also defines a convergent series. For determining the coefficients of Expansion (5.44), substitute this equation on the left side of (5.43). Thus by taking advantage of (5.40), we come out with

$$
\begin{aligned}
(\tilde{a}_0 + \tilde{a}_1\zeta + \ldots + \tilde{a}_\ell\zeta^\ell)\sum_{i=0}^{\infty} y_i\zeta^i = \tilde{a}_0\bar{y}_0 + \zeta(\tilde{a}_0\bar{y}_1 + \tilde{a}_1\bar{y}_0) + \ldots \\
+\zeta^{\ell-1}(\tilde{a}_0\bar{y}_{\ell-1} + \tilde{a}_1\bar{y}_{\ell-2} + \ldots + \tilde{a}_{\ell-1}\bar{y}_0) \\
+\tilde{b}_0\left[u^0(\zeta) - u_0 - \zeta u_1 - \ldots - \zeta^{\ell-1}u_{\ell-1}\right] \\
+\zeta\tilde{b}_1\left[u^0(\zeta) - u_0 - \zeta u_1 - \ldots - \zeta^{\ell-2}u_{\ell-2}\right] + \ldots + \zeta^\ell\,\tilde{b}_\ell u^0(\zeta)\,,
\end{aligned}
\tag{5.45}
$$

which holds for all sufficiently small $|\zeta|$. Notice that the coefficients of the matrices $\tilde{b}_i$, $(i = 0, \ldots, \ell)$ on the right side of (5.45) are proportional to ζ^i. Hence comparing the coefficients for ζ^i, $(i = 0, \ldots, \ell - 1)$ on both sides of (5.45) yields

$$
\begin{aligned}
\tilde{a}_0 y_0 &= \tilde{a}_0\bar{y}_0 \\
\tilde{a}_1 y_0 + \tilde{a}_0 y_1 &= \tilde{a}_1\bar{y}_0 + \tilde{a}_0\bar{y}_1 \\
\vdots \qquad & \qquad \vdots \\
\tilde{a}_{\ell-1}y_0 + \tilde{a}_{\ell-1}y_1 + \ldots + \tilde{a}_0 y_\ell &= \tilde{a}_{\ell-1}\bar{y}_0 + \tilde{a}_{\ell-2}\bar{y}_1 + \ldots \tilde{a}_0\bar{y}_{\ell-1}\,.
\end{aligned}
\tag{5.46}
$$

With regard to (5.41), we generate from (5.46)

$$
y_i = \bar{y}_i\,, \quad (i = 0, 1, \ldots, \ell - 1)\,. \tag{5.47}
$$

Using (5.47) and (5.46), Relation (5.45) is written as

$$
\begin{aligned}
\tilde{a}_0\sum_{i=\ell}^{\infty} y_i\zeta^i + \zeta\tilde{a}_1\sum_{i=\ell-1}^{\infty} y_i\zeta^i\!+\!. + \zeta^\ell\,\tilde{a}_\ell\sum_{i=0}^{\infty} y_i\zeta^i \\
= \tilde{b}_0\sum_{i=\ell}^{\infty} u_i\zeta^i + \zeta\tilde{b}_1\sum_{i=\ell-1}^{\infty} u_i\zeta^i + \ldots + \zeta^\ell\,\tilde{b}_\ell\sum_{i=0}^{\infty} u_i\zeta^i\,.
\end{aligned}
$$

Dividing both sides of the last equation by ζ^ℓ yields the relation

$$
\begin{aligned}
\tilde{a}_0\sum_{i=0}^{\infty} y_{i+\ell}\,\zeta^i + \tilde{a}_1\sum_{i=0}^{\infty} y_{i+\ell-1}\zeta^i\!+\!. + \tilde{a}_\ell\sum_{i=0}^{\infty} y_i\zeta^i \\
= \tilde{b}_0\sum_{i=0}^{\infty} u_{i+\ell}\,\zeta^i + \tilde{b}_1\sum_{i=0}^{\infty} u_{i+\ell-1}\zeta^i + \ldots + \tilde{b}_\ell\sum_{i=0}^{\infty} u_i\zeta^i\,.
\end{aligned}
$$

A comparison of the coefficients of equal powers of ζ on both sides produces

$$\tilde{a}_0 y_{k+\ell} + \tilde{a}_1 y_{k+\ell-1} + \ldots + \tilde{a}_\ell\, y_k = \tilde{b}_0 u_{k+\ell} + \tilde{b}_1 u_{k+\ell-1} + \ldots + \tilde{b}_\ell\, u_k, \qquad (k = 0, 1, \ldots)\,.$$

Bringing this face to face with (5.31) and taking advantage of (5.47), we conclude that the coefficients of the expansion (5.44) build a solution of Equation (5.31), the initial conditions of which satisfy (5.37) for any initial vectors (5.36). But owing to Theorem 5.16, every ensemble of initial values (5.36) uniquely corresponds to a solution. Hence we discover that for any initial vectors (5.36), the totality of coefficients of the expansion (5.44) exhaust the whole solution set of the normal equation (5.31). Thus in case of convergence of the ζ-transforms (5.42), all solutions of the normal equation (5.31) are Taylor sequences. ∎

Corollary 5.19. *When the input is a Taylor sequence $\{u\}$, it emerges from the proof of Theorem 5.18 that the right side of Relation (5.43) defines the ζ-transform of the general solution of the normal equation (5.31).* ∎

4. From Theorem 5.18 and its Corollary, as well as from the relations between the z-transforms and ζ-transforms, it arises that for a Taylor input sequence $\{u\}$, any solution $\{y\}$ of the normal equation (5.31) possesses the z-transform

$$y^*(z) = \sum_{k=0}^{\infty} y_k z^{-k}\,.$$

Applying (A.8), after transition in (5.31) to the z-transforms, we arrive at

$$\begin{aligned}\tilde{a}_0 \left[z^\ell y^*(z) - z^\ell y_0 - \ldots - z y_{\ell-1}\right] + \tilde{a}_1 \left[z^{\ell-1} y^*(z) - z^{\ell-1} y_0 - \ldots - z y_{\ell-2}\right] \\ + \ldots + \tilde{a}_\ell\, y^*(z) = \tilde{b}_0 \left[z^\ell u^*(z) - z^\ell u_0 - \ldots - z u_{\ell-1}\right] \\ + \tilde{b}_1 \left[z^{\ell-1} u^*(z) - z^{\ell-1} u_0 - \ldots - z u_{\ell-2}\right] + \ldots + \tilde{b}_\ell\, u^*(z)\,.\end{aligned}$$

Using (5.23) after rearrangement, this is represented in the form

$$\begin{aligned}y^*(z) = \tilde{w}(z) u^*(z) + z^\ell(\tilde{a}_0 y_0 - \tilde{b}_0 u_0) + z^{\ell-1}(\tilde{a}_0 y_1 + \tilde{a}_1 y_0 - \tilde{b}_0 u_1 - \tilde{b}_1 u_0) + \ldots \\ + z(\tilde{a}_0 y_{\ell-1} + \tilde{a}_1 y_{\ell-2} + \ldots + \tilde{a}_{\ell-1} y_0 - \tilde{b}_0 u_{\ell-1} - \tilde{b}_1 u_{\ell-2} - \ldots - u_0)\,.\end{aligned}$$

The initial vectors $y_0^0, \ldots, y_{\ell-1}^0$ have to be selected in such a way that the relations

$$\begin{aligned}\tilde{a}_0 y_0^0 &= \tilde{b}_0 u_0 \\ \tilde{a}_1 y_0^0 + \tilde{a}_0 y_1^0 &= \tilde{b}_1 u_0 + \tilde{b}_0 u_1 \\ \vdots \qquad & \qquad \vdots \\ \tilde{a}_{\ell-1} y_0^0 + \ldots + \tilde{a}_0 y_{\ell-1}^0 &= \tilde{b}_{\ell-1} u_0 + \ldots + \tilde{b}_0 u_{\ell-1}\end{aligned} \tag{5.48}$$

hold. Owing to $\det \tilde{a}_0 \neq 0$, the system (5.48) uniquely determines the totality of initial vectors $y_0^0, \dots, y_{\ell-1}^0$. Taking these vectors as initial values, we conclude that the solution $\{y^0\}$, which is configured to the initial values $y_0^0, \dots, y_{\ell-1}^0$, possesses the z-transform

$$y_0^*(z) = \tilde{w}(z)u^*(z)\,. \tag{5.49}$$

In what follows, those solution of Equation (5.31) having the transform (5.49) is called the solution with *vanishing initial energy.* As a result of the above considerations, the following theorem is formulated.

Theorem 5.20. *For the normal equation (5.31) and any Taylor input sequence $\{u\}$, there exists the solution with vanishing initial energy $\{y^0\}$, which has the z-transform (5.49). The initial conditions of this solution are uniquely determined by the system of equations (5.48).* ∎

5. The Taylor matrix sequence

$$\{H\} = \{H_0, H_1, \dots, \}, \quad (i = 0, 1, \dots)$$

for which the equation

$$H^*(z) = \sum_{i=0}^{\infty} H_i z^{-i} = \tilde{w}(z)$$

holds, is called the *weighting sequence* of the normal process (5.31).

Based on the above reasons arising in the proof of Theorem 5.20, we are able to show that the weighting sequence $\{H\}$ is the solution the matrix difference equation

$$\tilde{a}_0 H_{k+\ell} + \dots + \tilde{a}_\ell H_k = \tilde{b}_0 U_{k+\ell} + \dots + \tilde{b}_\ell U_k\,, \quad (k = 0, 1, \dots) \tag{5.50}$$

for the matrix input

$$\{U\} = \{I_m, O_{mm}, O_{mm}, \dots\} \tag{5.51}$$

with the solution of the equations

$$\begin{aligned} \tilde{a}_0 H_0 &= \tilde{b}_0 \\ \tilde{a}_1 H_0 + \tilde{a}_0 H_1 &= \tilde{b}_1 \\ \vdots \qquad &\ \ \vdots \quad \vdots \\ \tilde{a}_{\ell-1} H_0 + \dots + \tilde{a}_0 H_{\ell-1} &= \tilde{b}_{\ell-1} \end{aligned}$$

as initial values $H_0, \dots, H_{\ell-1}$. Notice that due to (5.51) for $k > 0$, Equation (5.50) converts into the homogeneous equation

$$\tilde{a}_0 H_{k+\ell} + \dots + \tilde{a}_\ell H_k = O_{nm}\,, \quad (k = 0, 1, \dots)\,.$$

5.4 Anomalous LTI Processes

1. Conformable with the above introduced concepts, the non-singular causal LTI process, described by the equations

$$\tilde{a}(q)y_k = \tilde{b}(q)u_k \tag{5.52}$$

with

$$\begin{aligned} \tilde{a}(q) &= \tilde{a}_0 q^\ell + \tilde{a}_1 q^{\ell-1} + + \ldots + \tilde{a}_\ell \,, \\ \tilde{b}(q) &= \tilde{b}_0 q^\ell + \tilde{b}_1 q^{\ell-1} + + \ldots + \tilde{b}_\ell \end{aligned} \tag{5.53}$$

is called *anomalous*, when

$$\det \tilde{a}_0 = 0 \,. \tag{5.54}$$

From a mathematical point of view, anomalous processes, which include descriptor systems [34], are provided with a number of properties that distinguish them fundamentally from normal processes. Especially, notice the following:

a) While a normal process, which is described by (5.52), (5.53), will always be causal, an anomalous process, described by (5.52)–(5.54), might be non-causal, as shown in Example 5.13.
b) The successive procedure in Section 5.3 on the basis of (5.38), that was used for calculating the sequence is not applicable for anomalous processes. Indeed, denote

$$d_i = -\tilde{a}_1 y_{k+\ell+i-1} - \ldots - \tilde{a}_\ell y_{k+i} + \tilde{b}_0 y_{k+i} + \ldots + \tilde{b}_\ell u_{k+i} \,,$$

then Equation (5.52) is written as difference equation

$$\tilde{a}_0 y_{\ell+i} = d_i \,, \quad (i = 0, 1, \ldots)\,. \tag{5.55}$$

However due to (5.54), Equations (5.55) possess either no solution or infinitely many solutions. In both cases, a unique resolution of (5.55) is impossible and the successive calculation of the output sequence breaks down.
c) For a normal equation (5.52), the totality of initial vectors (5.36), configured according to (5.37), can be prescribed arbitrarily. For causal anomalous processes (5.52)–(5.54), the solution in general exists only for certain initial conditions that are bound by additional equations, which also include values of the input sequence.

Example 5.21. Consider the anomalous process

$$\begin{aligned} y_{1,k+1} + 3y_{1,k} + 2y_{2,k+1} &= x_k \\ y_{1,k+1} + 2y_{2,k+1} + y_{2,k} &= 2x_k \,. \end{aligned} \qquad (k = 0, 1, \ldots) \tag{5.56}$$

In this case, we configure

$$\tilde{a}(q) = \begin{bmatrix} q+3 & 2q \\ q & 2q+1 \end{bmatrix}, \quad \tilde{b}(q) = \begin{bmatrix} 1 \\ 2 \end{bmatrix}.$$

By means of (5.23), we find the transfer matrix

$$\tilde{w}(q) = \frac{\begin{bmatrix} -2q+1 \\ q+6 \end{bmatrix}}{7q+3}.$$

This matrix is proper and thus the process (5.56) is causal. For $k = 0$ from (5.56), we obtain

$$\begin{aligned} y_{1,1} + 2y_{2,1} &= x_0 - 3y_{1,0} \\ y_{1,1} + 2y_{2,1} &= 2x_0 - y_{2,0}\,. \end{aligned} \tag{5.57}$$

Equations (5.57) are consistent under the condition

$$x_0 - 3y_{1,0} = 2x_0 - y_{2,0} \tag{5.58}$$

which makes the system (5.57) to

$$y_{1,1} + 2y_{2,1} = x_0 - 3y_{1,0} = 2x_0 - 2y_{2,0}$$

that has infinitely many solutions. □

With respect to the above said, it is clear that dealing with anomalous processes (5.52)–(5.54) needs special attention. The present section presents adequate investigations.

2.

Lemma 5.22. *If the input of a causal anomalous process is a Taylor-sequence, then all solutions of Equation (5.52) are Taylor sequences.*

Proof. Without loss of generality, we assume that Equation (5.52) is row reduced, so that utilising (1.21) gives

$$\tilde{a}(q) = \operatorname{diag}\{q^{\alpha_1}, \dots, q^{\alpha_n}\}\tilde{A}_0 + \tilde{a}_1(q)\,, \tag{5.59}$$

where the degree of the i-th row of the matrix $\tilde{a}_1(q)$ is lower than α_i and $\det \tilde{A}_0 \neq 0$. Suppose $\deg \tilde{a}(q) = \alpha_{\max}$. Select

$$\chi(q) = \operatorname{diag}\{q^{\alpha_{\max}-\alpha_1}, \dots, q^{\alpha_{\max}-\alpha_n}\}$$

and consider the derived equations (5.14), which with the help of (5.59) takes the form

$$\left[\tilde{A}_0 q^{\alpha_{\max}} + \chi(q)\tilde{a}_1(q)\right] y_k = \chi(q)\tilde{b}(q)u_k\,. \tag{5.60}$$

As is easily seen, Equation (5.60) is normal under the given suppositions. Therefore, owing to Theorem 5.18 for Taylor input sequence $\{u\}$, all solutions of Equation (5.60) are Taylor sequences. But due to Lemma 5.2, all solutions of the original equation (5.52) are also solutions of the derived equation (5.60), thus Lemma 5.22 is proven. ■

3. Lemma 5.22 motivates a construction procedure for the solution set of Equation (5.52) according to its initial conditions. For this reason, the process equations (5.52) are written as a system of scalar equations of the shape (5.3):

$$\sum_{p=1}^{n} a_{ip}^{(0)} y_{p,k+\alpha_i} + \ldots + \sum_{p=1}^{n} a_{ip}^{(\alpha_i)} y_{p,k} = \sum_{r=1}^{m} b_{ir}^{(0)} u_{r,k+\alpha_i} + \ldots + \sum_{r=1}^{m} b_{ir}^{(\alpha_i)} u_{r,k} \tag{5.61}$$
$$(p = 1, \ldots, n;\ r = 1, \ldots, m;\ \ k = 0, 1, \ldots),$$

where due to the row reducibility, the condition

$$\det \left[a_{ip}^{(0)} \right] = \det \tilde{A}_0 \neq 0\,. \tag{5.62}$$

Without loss of generality, we suppose

$$\alpha_i \geq \alpha_{i+1}\,, \quad (i = 1, \ldots, n-1), \quad \alpha_n > 0\,,$$

because this can always be obtained by rearrangement. Passing formally from (5.61) to the z-transforms, we obtain

$$\begin{aligned} &\sum_{p=1}^{n} a_{ip}^{(0)} \left[z^{\alpha_i} y_p^*(z) - z^{\alpha_i} \bar{y}_{p,0} - \ldots - z \bar{y}_{p,\alpha_i - 1} \right] + \\ &\quad + \sum_{p=1}^{n} a_{ip}^{(1)} \left[z^{\alpha_i - 1} y_p^*(z) - z^{\alpha_i - 1} \bar{y}_{p,0} - \ldots - z \bar{y}_{p,\alpha_i - 2} \right] + \ldots \\ &\quad \ldots + \sum_{p=1}^{n} a_{ip}^{(\alpha_i)} y_p^*(z) = \tilde{B}_i(z)\,, \end{aligned}$$

where $\bar{y}_{p,0}, \ldots, \bar{y}_{p,\alpha_i - 1}$ are the initial values, and $\tilde{B}_i(z)$ is a polynomial in z, which depends on the coefficients $b_{ir}^{(j)}$ and the excitation $\{u\}$. Substituting here ζ^{-1} for z, we obtain the equations for the ζ-transforms

$$\begin{aligned} &\sum_{p=1}^{n} a_{ip}^{(0)} \left[y_p^0(\zeta) - \bar{y}_{p,0} - \ldots - \zeta^{\alpha_i - 1} \bar{y}_{p,\alpha_i - 1} \right] + \\ &\quad + \sum_{p=1}^{n} a_{ip}^{(1)} \left[y_p^0(\zeta) - \bar{y}_{p,0} - \ldots - z^{\alpha_i - 2} \bar{y}_{p,\alpha_i - 2} \right] + \ldots \\ &\quad \ldots + \zeta^{\alpha_i} \sum_{p=1}^{n} a_{ip}^{(\alpha_i)} y_p^0(z) = B_i(\zeta)\,, \end{aligned} \tag{5.63}$$

where $B_i(\zeta)$ is a polynomial in ζ. The solution is put up as a set of power series

$$y_p^0(\zeta) = \sum_{k=0}^{\infty} y_{p,k} \zeta^k\,, \tag{5.64}$$

which exists due to Condition (5.62). Besides, the condition

$$y_{p,k} = \bar{y}_{p,k} \tag{5.65}$$

has to be fulfilled for all $\bar{y}_{p,k}$, that are practically configured by the left side of (5.63). Inserting (5.64) on the left side of (5.63), and comparing the coefficients of ζ^k, $(k = 0, 1, \dots)$ on both sides, a system of successive linear equations for the quantities $y_{p,k}$, $(k = 0, 1, \dots)$ is created, which due to (5.62) is always solvable. In order to meet Condition (5.65), we generate the totality of linear relations, that have to be fulfilled between the quantities $\bar{y}_{p,k}$ and the first values of the input sequence $\{u\}$. These conditions determine the set of initial conditions $\bar{y}_{p,k}$, for which the wanted solution of Equation (5.61) exists. Since with respect to Lemma 5.22, all solutions of Equation (5.61) (whenever they exist) possess ζ-transforms, the suggested procedure always delivers the wanted result.

Example 5.23. Investigate the row reduced anomalous system of equations

$$\begin{aligned} y_{1,k+2} + y_{1,k} + 2y_{2,k+1} &= u_{k+1} \\ y_{1,k} + y_{2,k+1} &= u_k \end{aligned} \qquad (k = 0, 1, \dots) \tag{5.66}$$

of the form (5.61). In the present case, we have $\alpha_1 = 2,\ \alpha_2 = 1$ and

$$a_{ip}^{(0)} = \begin{bmatrix} 1 & 0 \\ 0 & 1 \end{bmatrix}.$$

A formal pass to the z-transforms yields

$$\begin{aligned} z^2 y_1^*(z) - z^2\bar{y}_{1,0} - z\bar{y}_{1,1} + y_1^*(z) + 2zy_2^*(z) - 2z\bar{y}_{2,0} &= zu^*(z) - zu_0 \\ y_1^*(z) + zy_2^*(z) - z\bar{y}_{2,0} &= u^*(z) \end{aligned}$$

so, we gain

$$\begin{bmatrix} y_1^*(z) \\ y_2^*(z) \end{bmatrix} = \begin{bmatrix} z^2+1 & 2 \\ 1 & z \end{bmatrix}^{-1} \begin{bmatrix} z^2\bar{y}_{1,0} + z\bar{y}_{1,1} + 2z\bar{y}_{2,0} + zu^*(z) - zu_0 \\ z\bar{y}_{2,0} + u^*(z) \end{bmatrix}. \tag{5.67}$$

Substituting ζ^{-1} for z gives

$$\begin{bmatrix} y_1^0(\zeta) \\ y_2^0(\zeta) \end{bmatrix} = \begin{bmatrix} 1+\zeta^2 & 2\zeta \\ \zeta & 1 \end{bmatrix}^{-1} \begin{bmatrix} \bar{y}_{1,0} + \zeta\bar{y}_{1,1} + 2\zeta\bar{y}_{2,0} + \zeta u^0(\zeta) - \zeta u_0 \\ \bar{y}_{2,0} + \zeta u^0(z) \end{bmatrix}, \tag{5.68}$$

where the conditions

$$y_1^0(\zeta) = y_1^*(\zeta^{-1}), \quad y_2^0(\zeta) = y_2^*(\zeta^{-1}), \quad u^0(\zeta) = u^*(\zeta^{-1})$$

were used. Since by supposition the input $\{u\}$ is a Taylor sequence, the expansion

$$u^0(\zeta) = \sum_{k=0}^{\infty} u_k \zeta^k$$

converges. Thus, the right side of (5.68) is analytical in the point $\zeta = 0$. Hence the pair of convergent expansions

$$y_1^0(\zeta) = \sum_{k=0}^{\infty} y_{1,k}\zeta^k, \quad y_2^0(\zeta) = \sum_{k=0}^{\infty} y_{2,k}\zeta^k \tag{5.69}$$

exists uniquely. From (5.68), we obtain

$$\begin{aligned} y_1^0(\zeta) - \bar{y}_{1,0} - \zeta\bar{y}_{1,1} + \zeta^2 y_1^0(\zeta) + 2\zeta y_2^0(\zeta) - 2\zeta\bar{y}_{2,0} &= \zeta u^0(\zeta) - \zeta u_0 \\ \zeta y_1^0(\zeta) + y_2^0(\zeta) - \bar{y}_{2,0} &= \zeta u^0(z)\,. \end{aligned} \tag{5.70}$$

Now we insert (5.69) into (5.70) and set equal those terms on both sides, which do not depend on ζ. Thus, we receive

$$y_{1,0} = \bar{y}_{1,0}\,, \quad y_{2,0} = \bar{y}_{2,0}\,, \tag{5.71}$$

and for the term with ζ, we get

$$y_{1,1} = \bar{y}_{1,1}\,. \tag{5.72}$$

When (5.71) and (5.72) hold, then in the first row of (5.70) the terms of zero and first degree in ζ neutralise each other, respectively, and in the second equation the absolute terms cancel each other. Altogether, Equations (5.70) under Conditions (5.71), (5.72) might be written in the shape

$$\begin{aligned} \sum_{k=2}^{\infty} y_{1,k}\zeta^k + \zeta^2 \sum_{k=0}^{\infty} y_{1,k}\zeta^k + 2\zeta \sum_{k=1}^{\infty} y_{2,k}\zeta^k &= \zeta \sum_{k=1}^{\infty} u_k\zeta^k \\ \zeta \sum_{k=0}^{\infty} y_{1,k}\zeta^k + \sum_{k=1}^{\infty} y_{2,k}\zeta^k &= \zeta \sum_{k=0}^{\infty} u_k\zeta^k\,. \end{aligned}$$

Canceling the first equation by ζ^2, and the second by ζ, we find

$$\begin{aligned} \sum_{k=0}^{\infty} y_{1,k+2}\zeta^k + \sum_{k=0}^{\infty} y_{1,k}\zeta^k + 2\sum_{k=0}^{\infty} y_{2,k+1}\zeta^k &= \sum_{k=0}^{\infty} u_{k+1}\zeta^k \\ \sum_{k=0}^{\infty} y_{1,k}\zeta^k + \sum_{k=0}^{\infty} y_{2,k+1}\zeta^k &= \sum_{k=0}^{\infty} u_k\zeta^k\,. \end{aligned}$$

Comparing the coefficients of the powers ζ^k, $(k = 0, 1, \dots)$, we conclude that for any selection of the constants $y_{1,0} = \bar{y}_{1,0}$, $y_{2,0} = \bar{y}_{2,0}$, $y_{1,1} = \bar{y}_{1,1}$ the coefficients of the expansion (5.69) present a solution of Equation (5.66) which satisfies the initial conditions (5.71), (5.72). As result of the above analysis, the following facts are ascertained:

a) The general solution of Equation (5.66) is determined by the initial conditions $\bar{y}_{1,0}$, $\bar{y}_{2,0}$, $\bar{y}_{1,1}$, which can be chosen arbitrarily.
b) The right side of Relation (5.67) presents the z-transform of the general solution of Equation (5.66).
c) The right side of Relation (5.68) presents the ζ-transform of the general solution of Equation (5.66). □

Example 5.24. Investigate the row reduced anomalous system of equations

$$\begin{aligned} y_{1,k+2} + y_{1,k} + 2y_{2,k+2} &= u_{k+1} \\ y_{1,k+1} + y_{2,k+1} &= u_k \end{aligned} \qquad (k = 0, 1, \dots) \tag{5.73}$$

of the form (5.61). In the present case, we have $\alpha_1 = 2, \ \alpha_2 = 1$ and

$$a_{i1}^{(0)} = \begin{bmatrix} 1 & 2 \\ 1 & 1 \end{bmatrix}, \qquad \det a_{i1}^{(0)} \neq 0\,. \tag{5.74}$$

By formal pass to z-transforms, we find

$$\begin{bmatrix} y_1^*(z) \\ y_2^*(z) \end{bmatrix} = \begin{bmatrix} z^2+1 & 2z^2 \\ z & z \end{bmatrix}^{-1} \begin{bmatrix} z^2\bar{y}_{1,0} + z\bar{y}_{1,1} + 2z^2\bar{y}_{2,0} + 2z\bar{y}_{2,1} + zu^*(z) - zu_0 \\ z\bar{y}_{1,0} + z\bar{y}_{2,0} + u^*(z) \end{bmatrix}. \tag{5.75}$$

Although the values of the numbers α_1 and α_2 are the same as in Example 5.23, the right side of Relation (5.75) now depends on four values $\bar{y}_{1,0}$, $\bar{y}_{2,0}$, $\bar{y}_{1,1}$ and $\bar{y}_{2,1}$. Substitute $z = \zeta^{-1}$, so, as in the preceding example, the relations

$$\begin{bmatrix} y_1^0(\zeta) \\ y_2^0(\zeta) \end{bmatrix} = \begin{bmatrix} 1+\zeta^2 & 2 \\ 1 & 1 \end{bmatrix}^{-1} \begin{bmatrix} \bar{y}_{1,0} + \zeta\bar{y}_{1,1} + 2\bar{y}_{2,0} + +2\zeta\bar{y}_{2,1} + \zeta u^0(z) - \zeta u_0 \\ \bar{y}_{1,0} + \bar{y}_{2,0} + \zeta u^0(\zeta) \end{bmatrix} \tag{5.76}$$

take place. The right side of (5.76) is analytical in the point $\zeta = 0$. Hence there exists uniquely a pair of convergent expansions (5.69), which are the Taylor series of the right side of (5.76). Thus from (5.76), we obtain

$$\begin{aligned} y_1^0(\zeta) - \bar{y}_{1,0} - \zeta\bar{y}_{1,1} + \zeta^2 y_1^0(\zeta) + 2\left[y_2^0(\zeta) - \bar{y}_{2,0} - \zeta\bar{y}_{2,1}\right] &= \zeta u^0(\zeta) - \zeta u_0 \\ y_1^0(\zeta) - \bar{y}_{1,0} + y_2^0(\zeta) - \bar{y}_{2,0} &= \zeta u^0(\zeta)\,. \end{aligned} \tag{5.77}$$

Equating on both sides the terms not depending on ζ in (5.9), we find

$$\begin{aligned} (y_{1,0} - \bar{y}_{1,0}) + 2(y_{2,0} - \bar{y}_{2,0}) &= 0 \\ (y_{1,0} - \bar{y}_{1,0}) + (y_{2,0} - \bar{y}_{2,0}) &= 0\,, \end{aligned}$$

so we gain with the help of (5.74)

$$y_{1,0} = \bar{y}_{1,0}\,, \quad y_{2,0} = \bar{y}_{2,0}\,. \tag{5.78}$$

Comparing the coefficients for ζ, we find

$$\begin{aligned} y_{1,1} - \bar{y}_{1,1} + 2(y_{2,1} - \bar{y}_{2,1}) &= 0 \\ y_{1,1} + y_{2,1} &= u_0 \end{aligned}$$

which might be composed in the form

$$\begin{aligned} (y_{1,1} - \bar{y}_{1,1}) + 2(y_{2,1} - \bar{y}_{2,1}) &= 0 \\ (y_{1,1} - \bar{y}_{1,1} + (y_{2,1} - \bar{y}_{2,1} &= -\bar{y}_{1,1} - \bar{y}_{2,1} + u_0\,. \end{aligned}$$

Recall (5.74) and recognise that the equations

$$y_{1,1} = \bar{y}_{1,1}\,, \quad y_{2,1} = \bar{y}_{2,1} \tag{5.79}$$

hold, if and only if the condition

$$\bar{y}_{1,1} + \bar{y}_{2,1} = u_0 \tag{5.80}$$

is satisfied. The last equation is a consequence of the second equation in (5.73) for $k = 0$. Suppose (5.80), then also Relations (5.78) and (5.79) are fulfilled, such that Relations (5.77) might be comprised to the equations

$$\begin{aligned} \sum_{k=2}^{\infty} y_{1,k}\zeta^k + \zeta^2 \sum_{k=0}^{\infty} y_{1,k}\zeta^k + 2\sum_{k=2}^{\infty} y_{2,k}\zeta^k &= \zeta \sum_{k=1}^{\infty} u_k\zeta^k \\ \sum_{k=1}^{\infty} y_{1,k}\zeta^k + \sum_{k=1}^{\infty} y_{2,k}\zeta^k &= \zeta \sum_{k=0}^{\infty} u_k\zeta^k\,. \end{aligned}$$

Reducing the first equation by ζ^2, and the second one by ζ, we find

$$\begin{aligned} \sum_{k=0}^{\infty} y_{1,k+2}\zeta^k + \sum_{k=0}^{\infty} y_{1,k}\zeta^k + 2\sum_{k=0}^{\infty} y_{2,k+2}\zeta^k &= \sum_{k=0}^{\infty} u_{k+1}\zeta^k \\ \sum_{k=0}^{\infty} y_{1,k+1}\zeta^k + \sum_{k=0}^{\infty} y_{2,k+1}\zeta^k &= \sum_{k=0}^{\infty} u_k\zeta^k\,. \end{aligned}$$

Comparing the coefficients for the powers ζ^k, $(k = 0, 1, \dots)$, we conclude that for any selection of the constants $\bar{y}_{1,0}$, $\bar{y}_{2,0}$ and the quantities $\bar{y}_{1,1}$, $\bar{y}_{2,1}$, which are connected by Relation (5.80), there exists a solution of Equation (5.73) satisfying the initial conditions (5.78), (5.79). As result of the above analysis, the following facts are ascertained:

a) The general solution of Equation (5.73) is determined by the quantities $\bar{y}_{1,0}$, $\bar{y}_{2,0}$, $\bar{y}_{1,1}$ and $\bar{y}_{2,1}$, where the first two are free selectable and the other two are bound by Relation (5.80).
b) When (5.80) holds, the right side of Relation (5.75) presents the z-transform of the general solution of Equation (5.73).
c) When (5.80) holds, the right side of Relation (5.76) presents the ζ-transform of the general solution of Equation (5.73). □

4. In order to find the solutions of the equation system (5.61) for admissible initial conditions, various procedures can be constructed. For this purpose, the discovered expressions for the z- and the ζ-transforms are very helpful. A further suitable approach consists in the direct solution of the derived normal equation (5.60). Using this approach, the method of successive approximation (5.38) is applicable.

Example 5.25. Under the conditions of Example 5.23 substitute k by $k+1$ in the second equation of (5.66). Then we obtain the derived normal system of equations

$$\begin{aligned} y_{1,k+2} + y_{1,k} + 2y_{2,k+1} &= u_{k+1} \\ y_{1,k+1} + y_{2,k+1} &= u_{k+1}\,, \end{aligned} \qquad (k = 0, 1, \dots)$$

that might be written in the form

$$\begin{aligned} y_{1,k+2} &= -y_{1,k} - 2y_{2,k+1} + u_{k+1} \\ y_{2,k+2} &= -y_{1,k+1} + u_{k+1} \end{aligned} \qquad (k = 0, 1, \dots)\,. \tag{5.81}$$

Specify the initial conditions as

$$y_{1,0}\,, \quad y_{2,0}\,, \quad y_{1,1}\,, \quad y_{2,1} = u_0 - y_{1,0}\,,$$

then the wanted solution is generated directly from (5.81). □

Example 5.26. A corresponding consideration of Equation (5.73) leads to the normal system

$$\begin{aligned} y_{1,k+2} + 2y_{2,k+2} &= -y_{1,k} + u_{k+1} \\ y_{1,k+2} + y_{2,k+2} &= \qquad\quad u_{k+1} \end{aligned} \qquad (k = 0, 1, \dots)$$

with the initial conditions

$$y_{1,0}\,, \quad y_{2,0}\,, \quad y_{1,1}\,, \quad y_{2,1} = u_0 - y_{1,1}\,.$$ □

5. Although in general, the solution of a causal anomalous equation (5.52) does exist only over the set of admissible initial conditions, for such an anomalous system always exists the solution for vanishing initial energy.

Theorem 5.27. *For the causal anomalous process (5.52) with Taylor input sequence $\{u\}$, there always exists the solution $\{y_0\}$, the z-transform $y_0^*(z)$ of which is determined by the relation*

$$y_0^*(z) = \tilde{w}(z)u^*(z)\,,$$

where

$$\tilde{w}(z) = \tilde{a}^{-1}(z)\tilde{b}(z)$$

is the assigned transfer matrix. The ζ*-transform of the solution* $\{y_0\}$ *has the view*

$$y_0^0(\zeta) = w(\zeta)u^0(\zeta) \tag{5.82}$$

with

$$w(\zeta) = \tilde{w}(\zeta^{-1}) = \tilde{a}^{-1}(\zeta^{-1})\tilde{b}(\zeta^{-1})\,. \tag{5.83}$$

Proof. Without loss of generality, we assume that Equation (5.52) is row reduced and in (5.59) $\det \tilde{A}_0 \neq 0$. In this case, the right side of (5.82) is analytical in the point $\zeta = 0$. Thus, there uniquely exists the Taylor series expansion

$$y_0^0(\zeta) = \sum_{k=0}^{\infty} \tilde{y}_k\zeta^k \tag{5.84}$$

that converges for sufficiently small $|\zeta|$. From (5.83) and (5.53) using (5.40), we get

$$w(\zeta) = a^{-1}(\zeta)b(\zeta)\,, \tag{5.85}$$

where

$$a(\zeta) = \tilde{a}_0 + \tilde{a}_1\zeta + \ldots + \tilde{a}_\ell\,\zeta^\ell, \qquad b(\zeta) = \tilde{b}_0 + \tilde{b}_1\zeta + \ldots + \tilde{b}_\ell\,\zeta^\ell\,. \tag{5.86}$$

Applying (5.84)–(5.86) and (5.82), we obtain

$$\left(\tilde{a}_0 + \tilde{a}_1\zeta + \ldots + \tilde{a}_\ell\,\zeta^\ell\right)\sum_{k=0}^{\infty} \tilde{y}_k\zeta^k = \left(\tilde{b}_0 + \tilde{b}_1\zeta + \ldots + \tilde{b}_\ell\,\zeta^\ell\right)\sum_{k=0}^{\infty} \tilde{u}_k\zeta^k\,. \tag{5.87}$$

By comparison of the coefficients for ζ^k, $(k = 0, 1, \ldots, \ell - 1)$, we find

$$\begin{aligned}
\tilde{a}_0\tilde{y}_0 &= \tilde{b}_0 u_0 \\
\tilde{a}_1\tilde{y}_0 + \tilde{a}_0 y_1 &= \tilde{b}_1\tilde{u}_0 + \tilde{b}_0 u_1 \\
\vdots \qquad &\vdots \qquad \vdots \\
\tilde{a}_{\ell-1}\tilde{y}_0 + \ldots + \tilde{a}_0 y_{\ell-1} &= \tilde{b}_{\ell-1}\tilde{u}_0 + \ldots + \tilde{b}_0 u_{\ell-1}\,.
\end{aligned} \tag{5.88}$$

With the aid of (5.88), Equation (5.87) is easily brought into the form

$$\begin{aligned}
&\tilde{a}_0\sum_{k=\ell}^{\infty} \tilde{y}_k\zeta^k + \zeta\tilde{a}_1\sum_{k=\ell-1}^{\infty} \tilde{y}_k\zeta^k + \ldots + \zeta^\ell\tilde{a}_\ell\sum_{k=0}^{\infty} \tilde{y}_k\zeta^k \\
&\qquad = \tilde{b}_0\sum_{k=\ell}^{\infty} \tilde{u}_k\zeta^k + \zeta\tilde{b}_1\sum_{k=\ell-1}^{\infty} \tilde{u}_k\zeta^k + \ldots + \zeta^\ell\tilde{b}_\ell\sum_{k=0}^{\infty} \tilde{u}_k\zeta^k\,.
\end{aligned}$$

Cancellation on both sides by ζ^ℓ yields

$$\begin{aligned}
&\tilde{a}_0 \sum_{k=0}^{\infty} \tilde{y}_{k+\ell}\, \zeta^k + \tilde{a}_1 \sum_{k=0}^{\infty} \tilde{y}_{k+\ell-1} \zeta^k + \ldots + \tilde{a}_\ell \sum_{k=0}^{\infty} \tilde{y}_k \zeta^k \\
&\qquad = \tilde{b}_0 \sum_{k=0}^{\infty} \tilde{u}_{k+\ell}\, \zeta^k + \tilde{b}_1 \sum_{k=0}^{\infty} \tilde{u}_{k+\ell-1} \zeta^k + \ldots + \tilde{b}_\ell \sum_{k=0}^{\infty} \tilde{u}_k \zeta^k \,.
\end{aligned}$$

Comparing the coefficients of the powers ζ^k, $(k = 0, 1, \ldots)$ on both sides, we realise that the coefficients of Expansion (5.84) satisfy Equation (5.52) for all $k \geq 0$. ∎

Remark 5.28. Since in the anomalous case $\det \tilde{a}_0 = 0$, Relations (5.88) do not allow to determine the initial conditions that are assigned to the solution with vanishing initial energy. For the determination of these initial conditions, the following procedure is possible. Using (5.59), we obtain

$$w(\zeta) = A^{-1}(\zeta) B(\zeta)\,, \tag{5.89}$$

where

$$\begin{aligned}
A(\zeta) &= \operatorname{diag}\{\zeta^{\alpha_1}, \ldots \zeta^{\alpha_n}\} \tilde{a}(\zeta^{-1}) = \tilde{A}_0 + \tilde{A}_1 \zeta + \ldots + \tilde{A}_\ell\, \zeta^\ell\,, \\
B(\zeta) &= \operatorname{diag}\{\zeta^{\alpha_1}, \ldots \zeta^{\alpha_n}\} \tilde{b}(\zeta^{-1}) = \tilde{B}_0 + \tilde{B}_1 \zeta + \ldots + \tilde{B}_\ell\, \zeta^\ell\,.
\end{aligned} \tag{5.90}$$

With the help of (5.82), (5.89) and (5.90), we derive

$$\left(\tilde{A}_0 + \tilde{A}_1 \zeta + \ldots + \tilde{A}_\ell\, \zeta^\ell\right) \sum_{k=0}^{\infty} y_k \zeta^k = \left(\tilde{B}_0 + \tilde{B}_1 \zeta + \ldots + \tilde{B}_\ell\, \zeta^\ell\right) \sum_{k=0}^{\infty} u_k \zeta^k \,.$$

By comparing the coefficients, we find

$$\begin{aligned}
\tilde{A}_0 \tilde{y}_0 &= \tilde{B}_0 u_0 \\
\tilde{A}_1 \tilde{y}_0 + \tilde{A}_0 \tilde{y}_1 &= \tilde{B}_1 u_0 + \tilde{B}_0 u_1 \\
\vdots \qquad & \vdots \qquad \vdots \\
\tilde{A}_{\ell-1} \tilde{y}_0 + \ldots + \tilde{A}_0 \tilde{y}_{\ell-1} &= \tilde{B}_{\ell-1} u_0 + \ldots + \tilde{B}_0 u_{\ell-1}\,.
\end{aligned} \tag{5.91}$$

Since per construction $\det \tilde{A}_0 \neq 0$, Equations (5.91) provide to determine the vectors $\tilde{y}_0, \ldots, \tilde{y}_{\ell-1}$.

Example 5.29. Find the initial conditions for the solution with vanishing initial energy for Equations (5.73). Notice that in this example

$$\tilde{a}(z) = \begin{bmatrix} z^2 + 1 & 2 \\ z & z \end{bmatrix}, \quad \tilde{b}(z) = \begin{bmatrix} z \\ 1 \end{bmatrix}$$

is assigned. From this and (5.90), we obtain

$$A(\zeta) = \begin{bmatrix} 1+\zeta^2 & 2\zeta^2 \\ 1 & 1 \end{bmatrix}, \quad B(\zeta) = \begin{bmatrix} \zeta \\ \zeta \end{bmatrix},$$

that means

$$\begin{aligned} \tilde{A}_0 &= \begin{bmatrix} 1 & 2 \\ 1 & 1 \end{bmatrix}, \quad \tilde{A}_1 = \begin{bmatrix} 0 & 0 \\ 0 & 0 \end{bmatrix}, \quad \tilde{A}_2 = \begin{bmatrix} 1 & 0 \\ 0 & 0 \end{bmatrix}, \\ \tilde{B}_0 &= \begin{bmatrix} 0 \\ 0 \end{bmatrix}, \quad \tilde{B}_0 = \begin{bmatrix} 1 \\ 1 \end{bmatrix}, \quad \tilde{B}_0 = \begin{bmatrix} 0 \\ 0 \end{bmatrix}. \end{aligned} \tag{5.92}$$

Applying (5.92) and (5.91), we get

$$\tilde{A}_0\tilde{y}_0 = \tilde{B}_0 u_0$$

i.e. $\tilde{y}_0 = O_{21}$. Thus, the second equation in (5.91) takes the form

$$\tilde{A}_0\tilde{y}_1 = \tilde{B}_1 u_0$$

with the consequence

$$\tilde{y}_1 = \tilde{A}_0^{-1}\tilde{B}_1 u_0 = \begin{bmatrix} u_0 \\ 0 \end{bmatrix}.$$

Hence the solution with vanishing initial energy is determined by the initial conditions

$$\bar{y}_{1,0} = 0, \quad \bar{y}_{2,0} = 0, \quad \bar{y}_{1,1} = u_0\,, \quad \bar{y}_{2,1} = 0\,.$$

Here, Relation (5.80) is satisfied. □

6. For anomalous causal LTI processes, in the same way as for normal processes, we introduce the concept of the weighting sequence

$$\{H\} = \{H_0, H_1, \dots\}\,. \tag{5.93}$$

The weighting sequence is a matrix sequence, whose z-transform $H^*(z)$ is determined by

$$H^*(z) = \tilde{w}(z)\,. \tag{5.94}$$

From (5.94) it follows that the weighting sequence (5.93) might be seen as the matrix solution of Equation (5.52) under vanishing initial energy for the special input

$$\{U\} = \{I_m, O_{mm}, O_{mm}, \dots\}\,,$$

because the z-transform of this input amounts to

$$U^*(z) = I_m\,.$$

Passing in (5.94) to the variable ζ, we obtain an equation for the ζ-transform

$$H^0(\zeta) = w(\zeta)\,.$$

Since the right side is analytical in $\zeta = 0$, there exists the convergent expansion

$$H^0(\zeta) = \sum_{k=0}^{\infty} H_k \zeta^k .$$

Applying Relations (5.85), (5.86), we obtain from the last two equations

$$(\tilde{a}_0 + \tilde{a}_1\zeta + \ldots + \tilde{a}_\ell\,\zeta^\ell\,) \sum_{k=0}^{\infty} H_k\zeta^k = \tilde{b}_0 + \tilde{b}_1\zeta + \ldots + \tilde{b}_\ell\,\zeta^\ell . \tag{5.95}$$

By comparison of the coefficients at ζ^i, $(i = 0, \ldots, \ell)$ on both sides, we find

$$\begin{aligned} \tilde{a}_0 H_0 &= \tilde{b}_0 \\ \tilde{a}_1 H_0 + \tilde{a}_0 H_1 &= \tilde{b}_1 \\ \vdots \qquad & \vdots\ \vdots \\ \tilde{a}_\ell H_0 + \ldots + \tilde{a}_0 H_\ell &= \tilde{b}_\ell . \end{aligned} \tag{5.96}$$

When (5.96) is fulfilled, the terms with ζ^i, $(i = 0, \ldots, \ell)$ on both sides of (5.95) neutralise each other, hence this equation might be written as

$$\tilde{a}_0 \sum_{k=\ell+1}^{\infty} H_k\zeta^k + \zeta\tilde{a}_1 \sum_{k=\ell}^{\infty} H_k\zeta^k + \ldots + \zeta^\ell\tilde{a}_\ell \sum_{k=1}^{\infty} H_k\zeta^k = O_{nm} .$$

Canceling both sides by $\zeta^{\ell+1}$ results in

$$\tilde{a}_0 \sum_{k=0}^{\infty} H_{k+\ell+1}\zeta^k + \tilde{a}_1 \sum_{k=0}^{\infty} H_{k+\ell}\zeta^k + \ldots + \tilde{a}_\ell \sum_{k=0}^{\infty} H_{k+1}\zeta^k = O_{nm} .$$

If we make the coefficients at all powers of ζ on the left side equal to zero, then we find

$$\tilde{a}_0 H_{k+\ell+1} + \tilde{a}_1 H_{k+\ell} + \ldots + \tilde{a}_\ell H_{k+1} = O_{nm} , \quad (k = 0, 1, \ldots) ,$$

which is equivalently expressed by

$$\tilde{a}_0 H_{k+\ell} + \tilde{a}_1 H_{k+\ell-1} + \ldots + \tilde{a}_\ell H_k = O_{nm} , \quad (k = 1, 2, \ldots) .$$

From this is seen that for $k \geq 1$, the elements of the weighting sequence satisfy the homogeneous equation, which is derived from (5.52) for $\{u\} = O_{m1}$. Notice that for $\det \tilde{a}_0 = 0$, the determination of the matrices H_i, $(i = 0, 1, \ldots, \ell)$ is not possible with the help of (5.96). To overcome this difficulty, Relation (5.89) is recruited. So instead of (5.95), we obtain the result

$$\left(\tilde{A}_0 + \tilde{A}_1\zeta + \ldots + \tilde{A}_\ell\,\zeta^\ell\right) \sum_{k=0}^{\infty} H_k\zeta^k = \tilde{B}_0 + \tilde{B}_1\zeta + \ldots + \tilde{B}_\ell\,\zeta^\ell ,$$

where a corresponding formula to (5.96) arises:

$$\begin{aligned} \tilde{A}_0 H_0 &= \tilde{B}_0 \\ \tilde{A}_1 H_0 + \tilde{A}_0 H_1 &= \tilde{B}_1 \\ \vdots \qquad & \quad \vdots \; \vdots \\ \tilde{A}_\ell H_0 + \ldots + \tilde{A}_0 H_\ell &= \tilde{B}_\ell \,. \end{aligned} \tag{5.97}$$

Owing to $\det \tilde{A}_0 \neq 0$, the matrices H_i, $(i = 0, 1, \ldots, \ell)$ can be determined. If (5.97) is valid, we create the recursion formula

$$\tilde{A}_0 H_{k+\ell+1} = -\tilde{A}_1 H_{k+\ell} - \ldots - \tilde{A}_\ell H_{k+1} \,, \quad (k = 0, 1, \ldots) \,.$$

Thus with the help of the initial conditions (5.97), the weighting sequence $\{H\}$ can be calculated.

Example 5.30. Under the conditions of Example 5.29 and applying (5.97) and (5.92), we find

$$H_0 = \begin{bmatrix} 0 \\ 0 \end{bmatrix}, \quad H_1 = \begin{bmatrix} 1 \\ 0 \end{bmatrix}, \quad H_2 = \begin{bmatrix} 0 \\ 0 \end{bmatrix}.$$

The further elements of the weighting sequence are calculated by means of the recursion formula

$$\tilde{A}_0 H_{k+3} = -\tilde{A}_2 H_{k+1} \,, \quad (k = 0, 1, \ldots)$$

or

$$H_{k+3} = \begin{bmatrix} 1 & 0 \\ -1 & 0 \end{bmatrix} H_{k+1} \,.$$

□

5.5 Forward and Backward Models

1. Suppose the causal LTI process

$$\tilde{a}(q) y_k = \tilde{b}(q) u_k \,, \tag{5.98}$$

where

$$\begin{aligned} \tilde{a}(q) &= \tilde{a}_0 q^\ell + \ldots + \tilde{a}_\ell \,, \quad (n \times n) \,, \\ \tilde{b}(q) &= \tilde{b}_0 q^\ell + \ldots + \tilde{b}_\ell \,, \quad (n \times m) \,. \end{aligned} \tag{5.99}$$

As before, Equation (5.98) is designated as a *forward model* of the LTI process. Select a unimodular matrix $\chi(q)$, such that the matrix $\tilde{a}_\varrho(q) = \chi(q)\tilde{a}(q)$ in (5.20) becomes row reduced and consider the equivalent equation

$$\tilde{a}_\varrho(q)y_k = \tilde{b}_\varrho(q)u_k\,, \tag{5.100}$$

where $\tilde{b}_\varrho(q) = \chi(q)\tilde{b}(q)$. Let α_i be the degree of the i-th row of the matrix $\tilde{a}_\varrho(q)$, then we have

$$\tilde{a}_\varrho(q) = \operatorname{diag}\{q^{\alpha_1}, \ldots, q^{\alpha_n}\}\left(\tilde{A}_0 + \tilde{A}_1 q^{-1} + \ldots + \tilde{A}_\ell q^{-\ell}\right),$$

where $\det \tilde{A}_0 \neq 0$. Then the equation of the form [1]

$$a(\zeta)y_k = b(\zeta)u_k \tag{5.101}$$

with

$$a(\zeta) = A_0 + A_1\zeta + \ldots + A_\ell\,\zeta^\ell = \operatorname{diag}\{\zeta^{\alpha_1}, \ldots, \zeta^{\alpha_n}\}\,\tilde{a}_\varrho(\zeta^{-1}) \tag{5.102}$$

and

$$b(\zeta) = B_0 + B_1\zeta + \ldots + B_\ell\,\zeta^\ell = \operatorname{diag}\{\zeta^{\alpha_1}, \ldots, \zeta^{\alpha_n}\}\,\tilde{b}_\varrho(\zeta^{-1}) \tag{5.103}$$

is called the associated *backward model* of the LTI process. From (5.29), we recognise that $b(\zeta)$ is a polynomial matrix.

2. Hereinafter, the matrix

$$\tilde{w}(q) = \tilde{a}^{-1}(q)\tilde{b}(q) = \tilde{a}_\varrho^{-1}(q)\tilde{b}_\varrho(q) \tag{5.104}$$

is called the transfer matrix of the forward model, and the matrix

$$w(\zeta) = a^{-1}(\zeta)b(\zeta) = \tilde{a}^{-1}(\zeta^{-1})\tilde{b}(\zeta^{-1}) = \tilde{a}_\varrho^{-1}(\zeta^{-1})\tilde{b}_\varrho(\zeta^{-1}) \tag{5.105}$$

is the transfer matrix of the backward model. From (5.104) and (5.105), we take the reciprocal relations

$$w(\zeta) = \tilde{w}(\zeta^{-1}), \qquad \tilde{w}(q) = w(q^{-1})\,. \tag{5.106}$$

The matrix $\tilde{a}(q)$ is named as before the *eigenoperator* of the forward model and the matrix $a(\zeta)$ is the eigenoperator of the backward model. As seen from (5.102), the eigenoperator $a(\zeta)$ of the backward model is independent of the shape of the matrix $\tilde{b}(q)$ in (5.98). Obviously, the matrices $a(\zeta)$ and $b(\zeta)$ in (5.101) are not uniquely determined. Nevertheless, as we realise from (5.105), the transfer matrix $w(\zeta)$ is not affected. Moreover, later on we will prove that the structural properties of the matrix $a(\zeta)$ also do not depend on the special procedure for its construction.

[1] In (5.101) for once ζ means the operator q^{-1}. A distinction from the complex variable of the ζ-transformation is not made, because the operator q^{-1}, due to the mentioned difficulties, will not be used later on.

Example 5.31. Consider the forward model

$$\begin{aligned} 3y_{1,k+2} + y_{2,k+3} + y_{2,k} &= u_{k+2} + u_k \\ 2y_{1,k+1} + y_{2,k+2} &= u_{k+1}\,. \end{aligned}$$

In the present case, we have

$$\tilde{a}(q) = \begin{bmatrix} 3q^2 & q^3+1 \\ 2q & q^2 \end{bmatrix}, \quad \tilde{b}(q) = \begin{bmatrix} q^2+1 \\ q \end{bmatrix}.$$

Thus, the transfer matrix of the forward model $\tilde{w}(q)$ emerge as

$$\tilde{w}(q) = \frac{\begin{bmatrix} q-1 \\ q^2-2 \end{bmatrix}}{q^3-2}$$

and the corresponding LTI process is strictly causal. Select the unimodular matrix

$$\chi(q) = \begin{bmatrix} 1 & -q \\ 0 & 1 \end{bmatrix},$$

so we obtain

$$\tilde{a}_\varrho(q) = \chi(q)\tilde{a}(q) = \begin{bmatrix} q^2 & 1 \\ 2q & q^2 \end{bmatrix}, \quad \tilde{b}_\varrho(q) = \chi(q)\tilde{b}(q) = \begin{bmatrix} 1 \\ q \end{bmatrix}.$$

Thus, the matrices $a(\zeta)$, $b(\zeta)$ of the associated backward model take the shape

$$\begin{aligned} a(\zeta) &= \begin{bmatrix} 1 & \zeta^2 \\ 2\zeta & 1 \end{bmatrix} = \begin{bmatrix} 1 & 0 \\ 0 & 1 \end{bmatrix} + \begin{bmatrix} 0 & 0 \\ 2 & 0 \end{bmatrix}\zeta + \begin{bmatrix} 0 & 1 \\ 0 & 0 \end{bmatrix}\zeta^2, \\ b(\zeta) &= \begin{bmatrix} \zeta^2 \\ \zeta \end{bmatrix} = \begin{bmatrix} 0 \\ 1 \end{bmatrix}\zeta + \begin{bmatrix} 1 \\ 0 \end{bmatrix}\zeta^2. \end{aligned}$$

□

3.

Lemma 5.32. *For the causality of the processes (5.98), it is necessary and sufficient that the transfer function $w(\zeta)$ is analytical in the point $\zeta = 0$. For the strict causality of the process (5.98), the fulfillment of the equation*

$$w(0) = O_{nm}$$

is necessary and sufficient.

Proof. Necessity: If the process (5.98) is causal, then the matrix $\tilde{w}(q)$ is at least proper and thus analytical in the point $q = \infty$. Hence Matrix (5.105) is analytical in the point $\zeta = 0$. If the process is strictly causal, then $\tilde{w}(\infty) = O_{nm}$ is valid and we obtain the claim. Thus, the necessity is shown. We realise that the condition is also *sufficient*, when we reverse the steps of the proof. ∎

Corollary 5.33. *The process (5.98) is strictly causal, if and only if the equation*

$$w(\zeta) = \zeta w_1(\zeta)$$

holds with a matrix $w_1(\zeta)$*, which is analytical in the point* $\zeta = 0$. ■

4. It is shown that the concepts of forward and backward models are closely connected with the properties of the z- and ζ-transforms of the solution for Equation (5.98). Indeed, suppose a causal process, then it was shown above that for a Taylor input sequence $\{u\}$, independently of the fact whether the process is normal or anomalous, Equation (5.98) always possesses the solution with vanishing initial energy, and its z-transform $y^*(z)$ satisfies the equation

$$\tilde{a}(z)y^*(z) = \tilde{b}(z)u^*(z) \tag{5.107}$$

that formally coincides with (5.98). In what follows, Relation (5.107) is also called a forward model of the process (5.98). From (5.107), we receive

$$y^*(z) = \tilde{w}(z)u^*(z) = \tilde{a}^{-1}(z)\tilde{b}(z)u^*(z)\,.$$

Substituting here ζ^{-1} for z, we obtain

$$y^0(\zeta) = w(\zeta)u^0(\zeta)\,,$$

where $y^0(\zeta)$, $u^0(\zeta)$ are the ζ-transforms of the process output for vanishing initial energy and the input sequence, respectively. Moreover, $w(\zeta)$ is the transfer matrix of the backward model (5.105). Owing to (5.105), the last equation might be presented in the form

$$a(\zeta)y^0(\zeta) = b(\zeta)u^0(\zeta) \tag{5.108}$$

which coincides with the associated backward model (5.101).

5. The forward model (5.107) is called *controllable*, if the pair $(\tilde{a}(z), \tilde{b}(z))$ is irreducible, *i.e.* for all finite z

$$\operatorname{rank} \tilde{R}_h(z) = \operatorname{rank}\begin{bmatrix} \tilde{a}(z) & \tilde{b}(z) \end{bmatrix} = n$$

is true. Analogously, the backward model (5.108) is called controllable, if for all finite ζ

$$\operatorname{rank} R_h = \operatorname{rank}\begin{bmatrix} a(\zeta) & b(\zeta) \end{bmatrix} = n$$

is true. We will derive some general properties according to the controllability of forward and backward models.

6.

Lemma 5.34. *If the forward model (5.107) is controllable, then the associated backward model (5.108) is also controllable.*

Proof. Let the model (5.107) be controllable. Then the row reduced model (5.100) is also controllable and hence for all finite z

$$\operatorname{rank}\left[\,\tilde{a}_\varrho(z)\;\tilde{b}_\varrho(z)\,\right] = n\,.$$

Let $z_0 = 0, z_1, \ldots, z_q$ be the distinct eigenvalues of the matrix $\tilde{a}(z)$. Then the matrix $\tilde{a}_\varrho(z)$ has the same eigenvalues. From Formula (5.102), we gain with respect to $\det \tilde{A}_0 \neq 0$ that the set of eigenvalues of the matrix $a(\zeta)$ contains the quantities $\zeta_i = z_i^{-1}$, $i = 1, \ldots, q$. From the above rank condition for all $1 \leq i \leq q$, we obtain

$$\operatorname{rank}\left[\,a(\zeta_i)\;b(\zeta_i)\,\right] = \operatorname{rank}\left(\operatorname{diag}\left\{z_i^{-\alpha_i}, \ldots, z_i^{-\alpha_n}\right\}\left[\,\tilde{a}_\varrho(z_i)\;\tilde{b}_\varrho(z_i)\,\right]\right) = n\,.$$

Thus due to Lemma 1.42, the claim emerges. ∎

7. Let the forward model (5.107) be controllable and equation

$$\tilde{w}(z) = C(zI_p - A)^{-1}B + D \tag{5.109}$$

should describe a minimal standard realisation of the transfer matrix $\tilde{w}(z)$. Then with the help of (5.106), we find

$$w(\zeta) = \zeta C(I_p - \zeta A)^{-1}B + D\,. \tag{5.110}$$

The expression on the right side is called a *minimal standard realisation* of the transfer matrix of the associated backward model. Besides, the rational matrix

$$w_0(\zeta) = \zeta C(I_p - \zeta A)^{-1}B$$

might be seen as the transfer matrix of the PMD

$$\tau_0(\zeta) = (I_p - \zeta A, \zeta B, C)\,.$$

Lemma 5.35. *Under the named suppositions, the PMD $\tau_0(\zeta)$ is minimal, i.e. the pairs $(I_p - \zeta A, \zeta B)$ and $[I_p - \zeta A, C]$ are irreducible.*

Proof. Let $z_0 = 0, z_1, \ldots, z_q$ be the eigenvalues of the matrix A in (5.109). Then the eigenvalues of the matrix $I_p - \zeta A$ turn out to be the numbers $\zeta_1 = z_1^{-1}, \ldots, \zeta_q = z_q^{-1}$. Since the representation (5.109) is minimal, for all finite z, it follows

$$\operatorname{rank}\left[\,zI_p - A \quad B\,\right] = n, \quad \operatorname{rank}\begin{bmatrix} zI_p - A \\ C \end{bmatrix} = n\,.$$

Thus, for all $i = 1, \ldots, q$

$$\operatorname{rank}\begin{bmatrix} I_p - \zeta_i A & \zeta_i B \end{bmatrix} = \operatorname{rank}\begin{bmatrix} z_i I_p - A & B \end{bmatrix} = n\,,$$

$$\operatorname{rank}\begin{bmatrix} I_p - \zeta_i A \\ C \end{bmatrix} = \operatorname{rank}\begin{bmatrix} I_p - \zeta_i A \\ \zeta_i C \end{bmatrix} = \operatorname{rank}\begin{bmatrix} z_i I_p - A \\ C \end{bmatrix} = n\,.$$

These conditions together with Lemma 1.42 imply the irreducibility of the pairs

$$(I_p - \zeta A, \zeta B), \quad [I_p - \zeta A, C]\,.$$

■

Lemma 5.36. *Let $z_0 = 0, z_1, \ldots, z_q$ be the different eigenvalues of the matrix A and $a_1(z), \ldots, a_\rho(z)$ be the totality of its invariant polynomials different from one, having the shape*

$$\begin{aligned} a_1(z) &= z^{\mu_{01}}(z - z_1)^{\mu_{11}} \cdots (z - z_q)^{\mu_{q1}} \\ &\;\;\vdots \qquad\quad \vdots \qquad\quad \vdots \qquad\quad \vdots \\ a_\rho(z) &= z^{\mu_{0\rho}}(z - z_1)^{\mu_{1\rho}} \cdots (z - z_q)^{\mu_{q\rho}}\,, \end{aligned} \tag{5.111}$$

where

$$\mu_{si} \geq \mu_{s,i-1}\,, \quad (s = 0, \ldots q,\ i = 2, \ldots, \rho), \quad \sum_{s=0}^{q}\sum_{i=1}^{\rho} \mu_{si} = p\,.$$

Then the totality of invariant polynomials different from one of the matrix $I_p - \zeta A$ consists of the ρ polynomials $\alpha_1(\zeta), \ldots, \alpha_\rho(\zeta)$ having the shape

$$\begin{aligned} \alpha_1(\zeta) &= (\zeta - z_1^{-1})^{\mu_{11}} \cdots (\zeta - z_q^{-1})^{\mu_{q1}} \\ &\;\;\vdots \qquad\quad \vdots \qquad\quad \vdots \qquad\quad \vdots \\ \alpha_\rho(\zeta) &= (\zeta - z_1^{-1})^{\mu_{1\rho}} \cdots (\zeta - z_q^{-1})^{\mu_{q\rho}}\,. \end{aligned} \tag{5.112}$$

Proof. Firstly, we consider the matrix A as Jordan-Block (1.76) of dimension $\mu \times \mu$

$$A = J_\mu(a) = \begin{bmatrix} a & 1 & \ldots & 0 & 0 \\ 0 & a & \ddots & 0 & 0 \\ \vdots & \vdots & \ddots & \ddots & \vdots \\ 0 & 0 & \ldots & a & 1 \\ 0 & 0 & \ldots & 0 & a \end{bmatrix}. \tag{5.113}$$

For $a \neq 0$, we obtain

$$I_\mu - \zeta J_\mu(a) = \begin{bmatrix} 1 - \zeta a & -\zeta & \ldots & 0 & 0 \\ 0 & 1 - \zeta a & \ddots & 0 & 0 \\ \vdots & \vdots & \ddots & \ddots & \vdots \\ 0 & 0 & \ldots & 1 - \zeta a & -\zeta \\ 0 & 0 & \ldots & 0 & 1 - \zeta a \end{bmatrix}. \tag{5.114}$$

Besides,

$$\det[I_\mu - \zeta J_\mu(a)] = (1-\zeta a)^\mu \tag{5.115}$$

and Matrix (5.114) possesses only the eigenvalue $\zeta = a^{-1}$ of multiplicity μ. For $\zeta = a^{-1}$ from (5.114), we receive

$$I_\mu - a^{-1} J_\mu(a) = \begin{bmatrix} 0 & -a^{-1} & 0 & \dots & 0 \\ 0 & 0 & -a^{-1} & \dots & 0 \\ \vdots & \vdots & \vdots & \ddots & \vdots \\ 0 & 0 & 0 & \dots & -a^{-1} \\ 0 & 0 & 0 & \dots & 0 \end{bmatrix} . \tag{5.116}$$

Obviously, $\operatorname{rank}[I_\mu - a^{-1} J_\mu(a)] = \mu - 1$ is true. Thus, owing to Theorem 1.28, Matrix (5.114) possesses an elementary divisor $(\zeta - a^{-1})^\mu$. For $a = 0$ from (5.114), we obtain

$$I_\mu - \zeta J_\mu(0) = \begin{bmatrix} 1 & -\zeta & \dots & 0 & 0 \\ 0 & 1 & \ddots & 0 & 0 \\ \vdots & \vdots & \ddots & \ddots & \vdots \\ 0 & 0 & \dots & 1 & -\zeta \\ 0 & 0 & \dots & 0 & 1 \end{bmatrix} . \tag{5.117}$$

Obviously

$$\det[I_\mu - \zeta J_\mu(0)] = 1 ,$$

thus Matrix (5.117) is unimodular and has no elementary divisor.

Now, consider the general case and A is expressed in the Jordan form

$$A = U \operatorname{diag}\left\{ J_{\mu_{01}}(0), J_{\mu_{11}}(z_1), \dots, J_{\mu_{q\rho}}(z_q) \right\} U^{-1}$$

where U is a certain non-singular matrix. Hence

$$I_p - \zeta A = U \operatorname{diag}\left\{ I_{\mu_{01}} - \zeta J_{\mu_{01}}(0), I_{\mu_{11}} - \zeta J_{\mu_{11}}(z_1), \dots, I_{\mu_{q\rho}} - \zeta J_{\mu_{q\rho}}(z_q) \right\} U^{-1}. \tag{5.118}$$

According to Lemma 1.25, the set of elementary divisors of the block-diagonal matrix (5.118) consists of the unification of the sets of elementary divisors of its diagonal blocks. As follows from (5.113)–(5.117), no elementary divisor of Matrix (5.118) is assigned to the eigenvalue zero and a non-zero eigenvalues z_k corresponds to the totality of elementary divisors

$$(\zeta - z_k^{-1})^{\mu_{k1}}, \dots, (\zeta - z_k^{-1})^{\mu_{k\rho}} ,$$

from which directly follows (5.112). ∎

8. The next theorem establishes the connection between the eigenoperators of controllable forward and backward models.

Theorem 5.37. *Suppose the forward and backward models (5.107) and (5.108) be controllable, and the sequence of the invariant polynomials different from 1 of the matrix* $\tilde{a}(z)$ *should have the form (5.111). Then the sequence of invariant polynomials different from 1 of the matrix* $a(\zeta)$ *has the form (5.112).*

Proof. The proof is divided into several steps.

a) When the right side of (5.109) defines a minimal standard realisation of the transfer matrix of a controllable forward model, then the sequences of the invariant polynomials different from 1 of the matrices $\tilde{a}(z)$ and $zI_p - A$ coincide. Thus, we get

$$\det \tilde{a}(z) \approx \det(zI_p - A)\,.$$

This statement directly emerges from the content of Section 2.4.

b) In the same way, we conclude that, when the right side of (5.110) presents a minimal standard realisation of the transfer matrix of the controllable backward model (5.108), then

$$\det a(\zeta) \approx \det(I_p - \zeta A)\,.$$

Thus, the sequences of invariant polynomials different from 1 of the matrices $a(\zeta)$ and $I_p - \zeta A$ coincide, because the pairs $(I_p - \zeta A, \zeta B)$ and $[I_p - \zeta A, C]$ are irreducible.

c) Owing to Lemma 5.36, the set of invariant polynomials of the matrices $zI_p - A$ and $I_p - \zeta A$ are connected by Relations (5.111) and (5.112), such that with respect to a) and b) analogue connections also exist between the sets of invariant polynomials of the matrices $\tilde{a}(z)$ and $a(\zeta)$. ∎

Corollary 5.38. *The last equivalence implies*

$$\det a(0) \neq 0\,.$$

Hence the eigenoperator of a controllable backward models does not possess a zero eigenvalue. ∎

Corollary 5.39. *Denote*

$$\det \tilde{a}(z) = \tilde{\Delta}(z), \quad \det a(\zeta) = \Delta(\zeta) \tag{5.119}$$

and let $\deg \tilde{\Delta}(z) = p$. *Then*

$$\Delta(\zeta) \approx \zeta^p\, \tilde{\Delta}(\zeta^{-1})\,. \tag{5.120}$$

Proof. Assume (5.111), then with

$$\mu_i = \sum_{s=1}^{\rho} \mu_{is}, \quad \sum_{i=1}^{q} \mu_i = p$$

and taking advantage of (5.112), we can write

$$\tilde{\Delta}(z) \approx \tilde{a}_1(\zeta)\cdots\tilde{a}_\rho(z) = z^{\mu_0}(z-z_1)^{\mu_1}\cdots(z-z_q)^{\mu_q},$$
$$\Delta(\zeta) \approx \alpha_1(\zeta)\cdots\alpha_\rho(\zeta) = (\zeta - z_1^{-1})^{\mu_1}\cdots(\zeta - z_q^{-1})^{\mu_q}.$$

Thus, (5.120) is ensured. ∎

In the following, the polynomials (5.119) are referred to as the *characteristic polynomial* of the forward or of the backward model, respectively.

Remark 5.40. For an arbitrary polynomial $f(\zeta)$ with $\deg f(\zeta) = p$, the polynomial

$$f^*(\zeta) = \zeta^p f(\zeta^{-1})$$

usually is designated as the *reciprocal* to $f(\zeta)$, [14]. Thus Relation (5.120) might be written in the form

$$\Delta(\zeta) \approx \tilde{\Delta}^*(\zeta)$$

i.e. the characteristic polynomial of the controllable backward model is equivalent to the reciprocal characteristic polynomial of the controllable forward model.

9. The next assertion can be interpreted as completion of Theorem 5.37.

Theorem 5.41. *Let $\tilde{a}(z)$ be a non-singular $n \times n$ polynomial matrix and $\chi(z)$ is a unimodular matrix, such that $\tilde{a}_\varrho(z) = \chi(z)\tilde{a}(z)$ becomes row reduced. Furthermore, let α_i be the degree of the i-th row of the matrix $\tilde{a}_\varrho(z)$ and build $a(\zeta)$ with the help of (5.102). Then for the matrix $a(\zeta)$ all assertions of Theorem 5.37 and its corollaries are true.*

Proof. Consider the controllable forward model

$$\tilde{a}(z)y^*(z) = \chi^{-1}(z)u^*(z).$$

Multiplying this from left by $\chi(z)$, we obtain the row reduced causal model

$$\tilde{a}_\varrho(z)y^*(z) = I_n u^*(z)$$

and hence both models are causal. Passing from the last model to the associated backward model by applying Relations (5.102), (5.103), we get

$$a(\zeta)y^0(\zeta) = b(\zeta)u^0(\zeta)$$

and all assertions of Theorem 5.41 emerge from Theorem 5.37 and its corollaries. ∎

10. As follows from the above shown, the set of eigenoperators of the associated backward model does not depend on the matrix $\tilde{b}(z)$ in (5.107) and it can be found by Formula (5.102). The reverse statement is in general not true. Therefore, the transition from a backward model to the associated forward model has to be considered separately.

a) For a given controllable backward model

$$a(\zeta)y^0(\zeta) = b(\zeta)u^0(\zeta)\,, \tag{5.121}$$

the transfer function of the associated controllable forward model can be designed with the help of the ILMFD

$$\tilde{a}^{-1}(z)\tilde{b}(z) = a^{-1}(z^{-1})b(z^{-1})\,.$$

b) The following lemma provides a numerical well posed method.

Lemma 5.42. *Let a controllable backward model (5.121) be given, where the matrix $a(\zeta)$ has the form (5.102) and* $\det A_0 \neq 0$. *Furthermore, let γ_i be the degree of the i-th row of the matrix*

$$R_h(\zeta) = \begin{bmatrix} a(\zeta) & b(\zeta) \end{bmatrix}\,.$$

Introduce the polynomial matrices

$$\begin{aligned} \tilde{a}(z) &= \operatorname{diag}\{z^{\gamma_1}, \ldots, z^{\gamma_n}\}\, a(z^{-1})\,, \\ \tilde{b}(z) &= \operatorname{diag}\{z^{\gamma_1}, \ldots, z^{\gamma_n}\}\, b(z^{-1})\,. \end{aligned} \tag{5.122}$$

Then under the condition

$$\operatorname{rank}\begin{bmatrix} \tilde{a}(0) & \tilde{b}(0) \end{bmatrix} = n\,, \tag{5.123}$$

the pair $(\tilde{a}(z), \tilde{b}(z))$ defines an associated controllable forward model.

Proof. The proof follows the reasoning for Lemma 5.34. ∎

Example 5.43. Consider the controllable backward model (5.121) with

$$a(\zeta) = \begin{bmatrix} 1 & 2\zeta \\ 1+\zeta & 1 \end{bmatrix}, \quad b(\zeta) = \begin{bmatrix} \zeta \\ \zeta^2+1 \end{bmatrix}. \tag{5.124}$$

In this case, we have $\gamma_1 = 1$, $\gamma_2 = 2$ and the matrices (5.122) take the form

$$\tilde{a}(z) = \begin{bmatrix} z & 2 \\ z^2+z & z^2 \end{bmatrix}, \quad \tilde{b}(z) = \begin{bmatrix} 1 \\ 1+z^2 \end{bmatrix}.$$

Here Condition (5.123) is satisfied. Thus, the matrices define an associated controllable forward model. □

11. As just shown for a known eigenoperator of the forward model $\tilde{a}(z)$, the set of all eigenoperators of the associated controllable backward models can be generated. When with the aid of Formula (5.102), one eigenoperator $a_0(\zeta)$ has been designed, then the set of all such operators is determined by the relation

$$a(\zeta) = \psi(\zeta)a_0(\zeta),$$

where $\psi(\zeta)$ is any unimodular matrix. The described procedure does not depend on the input operator $\tilde{b}(z)$. However, the reverse pass from an eigenoperator of a controllable backward model $a(\zeta)$ to the eigenoperator $\tilde{a}(z)$ in general requires additional information about the input operator $b(\zeta)$. In this connection, we ask for general rules for the transition from the matrix $a(\zeta)$ to the matrix $\tilde{a}(z)$.

Theorem 5.44. *Let the two controllable backward models*

$$\begin{aligned} a(\zeta)y^0(\zeta) &= b_1(\zeta)u^0(\zeta), \\ a(\zeta)x^0(\zeta) &= b_2(\zeta)v^0(\zeta) \end{aligned} \tag{5.125}$$

be given, where $a(\zeta)$, $b_1(\zeta)$ and $b_2(\zeta)$ are polynomial matrices of dimensions $n \times n$, $n \times m$ and $n \times \ell$, respectively. Furthermore, let $\tilde{a}_1(z)$ and $\tilde{a}_2(z)$ be the eigenoperators of the controllable forward models associated to (5.125). Then the relations

$$\tilde{a}_1(z) = \beta_1(z)\tilde{a}_0(z), \quad \tilde{a}_2(z) = \beta_2(z)\tilde{a}_0(z)$$

are true, where $\beta_1(z)$ and $\beta_2(z)$ are nilpotent polynomial matrices, i.e. they only possess the eigenvalue zero, and the $n \times n$ polynomial matrix $\tilde{a}_0(z)$ can be chosen independently on the matrices $b_1(\zeta)$ and $b_2(\zeta)$, it is only committed by the matrix $a(\zeta)$.

Proof. Consider the transfer matrices of the models (5.125)

$$w_1(\zeta) = a^{-1}(\zeta)b(\zeta), \quad w_2(\zeta) = a^{-1}(\zeta)b_2(\zeta). \tag{5.126}$$

Since the matrices (5.126), roughly speaking are not strictly proper, they could be written as

$$w_1(\zeta) = \bar{w}_1(\zeta) + d_1(\zeta), \quad w_2(\zeta) = \bar{w}_2(\zeta) + d_2(\zeta), \tag{5.127}$$

where $d_1(\zeta)$ and $d_2(\zeta)$ are polynomial matrices, and the matrices $\bar{w}_1(\zeta)$ and $\bar{w}_2(\zeta)$ are strictly proper. But the right sides of Relations (5.126) are ILMFD, so Lemma 2.15 delivers that the relations

$$\bar{w}_1(\zeta) = a^{-1}(\zeta)\bar{b}_1(\zeta), \qquad \bar{w}_2(\zeta) = a^{-1}(\zeta)\bar{b}_2(\zeta), \tag{5.128}$$

where

$$\bar{b}_1(\zeta) = b_1(\zeta) - a(\zeta)d_1(\zeta), \qquad \bar{b}_2(\zeta) = b_2(\zeta) - a(\zeta)d_2(\zeta)$$

determine an ILMFD of Matrix (5.128). Let us have the minimal standard realisation

$$\bar{w}_1(\zeta) = C(\zeta I_q - G)^{-1} B_1 \,. \tag{5.129}$$

In (5.129) the matrix G is non-singular, because of $\det a(0) \neq 0$. Moreover, $q = \operatorname{Mdeg} w_1(\zeta)$ is valid. Thus, we get from (5.128)

$$\bar{w}_2(\zeta) \underset{l}{\prec} \bar{w}_1(\zeta)$$

and owing to Theorem 2.56, the matrix $\bar{w}_2(\zeta)$ allows the representation

$$\bar{w}_2(\zeta) = C(\zeta I_q - G)^{-1} B_2 \,. \tag{5.130}$$

Since the right sides of (5.128) are ILMFD,

$$\operatorname{Mdeg} \bar{w}_1(\zeta) = \operatorname{Mdeg} \bar{w}_2(\zeta) = \deg \det a(\zeta) \,.$$

Hence the right side of (5.130) is a minimal standard realisation of the matrix $\bar{w}_2(\zeta)$. Inserting (5.129) and (5.130) into (5.127), we arrive at

$$\begin{aligned} w_1(\zeta) &= C(\zeta I_q - G)^{-1} B_1 + d_1(\zeta) \,, \\ w_2(\zeta) &= C(\zeta I_q - G)^{-1} B_2 + d_2(\zeta) \,, \end{aligned}$$

where the matrices C and G do not depend on the matrices $b_1(\zeta)$ and $b_2(\zeta)$ configured in (5.125). Substituting now z^{-1} for ζ, we obtain the transfer matrices of the forward models

$$\begin{aligned} \tilde{w}_1(z) &= -C(zI_q - G^{-1})^{-1} G^{-2} B_1 - CG^{-1} B_1 + d_1(z^{-1}) \,, \\ \tilde{w}_2(z) &= -C(zI_q - G^{-1})^{-1} G^{-2} B_2 - CG^{-1} B_2 + d_2(z^{-1}) \,, \end{aligned} \tag{5.131}$$

where the realisations $(G^{-1}, G^{-2}B_1, C)$ and $(G^{-1}, G^{-2}B_2, C)$ turn out to be minimal, because the realisations (G, B_1, C) and (G, B_2, C) are minimal. Build the ILMFD

$$C(zI_q - G^{-1})^{-1} = \tilde{a}_0(z) \tilde{b}_0(z) \,. \tag{5.132}$$

The matrix $\tilde{a}_0(z)$ does not depend on the matrices $b_1(\zeta)$ or $b_2(z)$ in (5.126), because the matrices C and G do not. Besides, $\tilde{a}_0(z)$ has no eigenvalues equal to zero, because G^{-1} is regular. Using (5.132) from (5.131), we gain

$$\begin{aligned} \tilde{w}_1(z) &= \tilde{a}_0^{-1}(z) \left[-b_0(z) G^{-2} B_1 - \tilde{a}_0(z) C G^{-1} B_1 + \tilde{a}_0(z) d_1(z^{-1}) \right] , \\ \tilde{w}_2(z) &= \tilde{a}_0^{-1}(z) \left[-b_0(z) G^{-2} B_2 - \tilde{a}_0(z) C G^{-1} B_2 + \tilde{a}_0(z) d_2(z^{-1}) \right] . \end{aligned} \tag{5.133}$$

The matrices in the brackets possess poles only in the point $z = 0$. Thus, in the ILMFDs

$$\begin{aligned} -b_0(z) G^{-2} B_1 - \tilde{a}_0(z) C G^{-1} B_1 + \tilde{a}_0(z) d_1(z^{-1}) &= \beta_1^{-1}(z) q_1(z) \,, \\ -b_0(z) G^{-2} B_2 - \tilde{a}_0(z) C G^{-1} B_2 + \tilde{a}_0(z) d_2(z^{-1}) &= \beta_2^{-1}(z) q_2(z) \end{aligned}$$

the matrices $\beta_1(z)$ and $\beta_2(z)$ are nilpotent. Applying this and (5.133), as well as Corollary 2.19, we find out that the ILMFDs

$$\tilde{w}_1(z) = [\beta_1(z)\tilde{a}_0(z)]^{-1} q_1(z), \quad \tilde{w}_2(z) = [\beta_2(z)\tilde{a}_0(z)]^{-1} q_2(z)$$

exist, from which all assertions of the Theorem may be read. ∎

12. Sometimes in engineering literature, the pass from the original controllable forward model (5.98) to an associated backward model is made by procedures that are motivated by the SISO case. Then simply

$$a(\zeta) = \zeta^{\ell}\tilde{a}(\zeta^{-1}), \quad b(\zeta) = \zeta^{\ell}\tilde{b}(\zeta^{-1}) \tag{5.134}$$

is applied. It is easy to see that this procedure does not work, when $\det \tilde{a}_0 = 0$ in (5.98). In this case, we would get $\det a(0) = 0$, which is impossible for a controllable backward model. If however, in (5.99) $\det \tilde{a}_0 \neq 0$ takes place, *i.e.* the original process is normal, then Formula (5.134) delivers a controllable associated backward model.

13. In recent literature [69, 80, 115], the backward model is usually written in the form

$$a(q^{-1})y_k = b(q^{-1})u_k\,, \tag{5.135}$$

where q^{-1} is the right-shift operator that is inverse to the operator q. Per definition, we have

$$q^{-1}y_k = y_{k-1}\,, \quad q^{-1}u_k = u_{k-1}\,. \tag{5.136}$$

Example 5.45. The backward model corresponding to the matrices (5.124) is written with the notation (5.136) in the form

$$\begin{aligned} y_{1,k} + 2y_{2,k-1} &= u_{k-1} \\ y_{1,k} + y_{1,k-1} + y_{2,k} &= u_{k-2} + u_k\,. \end{aligned} \tag{5.137}$$

□

As was demonstrated in [14], a strict foundation for using the operator q^{-1} for a correct description of discrete LTI processes is connected with honest difficulties. The reason arises from the fact, that the operator q is only invertible over the set of two-sided unlimited sequences. If however, the equations of the LTI process (5.4) are only considered for $k \geq 0$, then the application of the operator q^{-1} needs special attention. From this point of view, the application of the ζ-transformation for investigating the properties of backward models seems more careful. Nevertheless, the description in the form (5.135) appears sometimes more comfortable, and it will also be used later on.

5.6 Stability of Discrete-time LTI Systems

1. The vector sequence $\{y\} = \{y_0, y_1, \ldots\}$ is called *stable*, if the inequality

$$\|y_k\| < c\theta^k, \quad (k = 0, 1, \ldots)$$

is true, where $\|\cdot\|$ is a certain norm for finite dimensional number vectors and c, θ are positive constants with $0 < \theta < 1$. If for the sequence $\{y\}$, such an estimate does not hold, then it is called *unstable*.

The homogeneous vector difference equation

$$\tilde{a}_0 y_{k+\ell} + \ldots + \tilde{a}_\ell \, y_k = O_{n1} \tag{5.138}$$

is called *stable*, if all of its solutions are stable sequences. Equations of the form (5.138), that are not stable, will be called *unstable*.

2. The next theorem establishes a criterion for the stability of Equation (5.138).

Theorem 5.46. *Suppose the $n \times n$ polynomial matrix*

$$\tilde{a}(z) = \tilde{a}_0(z) z^\ell + \ldots + \tilde{a}_\ell$$

be non-singular. Let z_i, $(i = 1, \ldots, q)$ be the eigenvalues of the matrix $\tilde{a}(z)$, i.e. the roots of the equation

$$\tilde{\Delta}(z) = \det \tilde{a}(z) = 0\,. \tag{5.139}$$

Then, for the stability of Equation (5.138), it is necessary and sufficient that

$$|z_j| < 1, \quad (j = 0, 1, \ldots, q)\,. \tag{5.140}$$

Proof. Sufficiency: Let $\chi(z)$ be a unimodular matrix, such that the matrix

$$\tilde{a}_\rho(z) = \chi(z)\tilde{a}(z)$$

is row reduced. Thus, the equivalent equation

$$\tilde{a}_\rho(z) y_k = O_{n1} \tag{5.141}$$

at the same time with Equations (5.137) is stable or unstable, and the equation $\det \tilde{a}_\rho(z) = 0$ possesses the same roots as Equation (5.139). Since the zero input is a Taylor sequence, owing to Lemma 5.22, all solutions of Equation (5.19) are Taylor sequences. Passing in Equation (5.141) to the z-transforms, we obtain the result that for any initial conditions, the transformed solution of Equation (5.141) has the shape

$$y^*(z) = \frac{R(z)}{\tilde{\Delta}(z)}\,,$$

where $R(z)$ is a polynomial vector. Besides under Condition (5.140), the inverse z-transformation formula [1, 123] ensures that all originals according to the transforms of (5.141) must be stable. Thus the sufficiency is shown.

Necessity: It is shown that, if Equation (5.139) has one root z_0 with $|z_0| \geq 1$, then Equation (5.138) is unstable. Let d be a constant vector, which is a solution of the equation

$$\tilde{a}(z_0)d = O_{n1} .$$

Then, we directly verify that

$$y_k = z_0^k d, \quad (k = 0, 1, \ldots)$$

is a solution of Equation (5.138). Besides due to $|z_0| \geq 1$, this sequence is unstable and hence Equation (5.138) is unstable. ∎

3. Let $a(\zeta)$ be the eigenoperator of the associated backward model designed by Formula (5.102). Then the homogeneous process equation might be written in form of the backward model

$$a(\zeta)y = a_0 y_k + a_1 y_{k-1} + \ldots + a_\ell \, y_{k-\ell} = O_{n1} \tag{5.142}$$

with $\det a_0 \neq 0$. Denote

$$\Delta(\zeta) = \det a(\zeta) ,$$

then the stability condition of Equation (5.142) may be formulated as follows.

Theorem 5.47. *For the stability of Equation (5.142), it is necessary and sufficient that the characteristic polynomial*

$$\det(a_0 + a_1\zeta + \ldots + a_\ell \, \zeta^\ell) = \det a(\zeta) = \Delta(\zeta) = 0$$

has no roots inside the unit disc or on its border.

Proof. The proof follows immediately from Theorems 5.41–5.46. ∎

Corollary 5.48. *As a special case, Equation (5.142) is stable, if* $\Delta(\zeta) =$ const. $\neq 0$, *i.e. if the matrix* $a(\zeta)$ *is unimodular.* ∎

4. Further on, the non-singular $n \times n$ polynomial matrices $\tilde{a}(z)$ and $a(\zeta)$ are called *stable*, if the conditions of Theorems 5.46, 5.47 are true for them. Matrices $\tilde{a}(z)$ and $a(\zeta)$ are named *unstable*, when they are not stable. The set of real stable polynomial matrices $\tilde{a}(z)$ and $a(\zeta)$ is denoted by $\tilde{\mathbb{R}}^+_{nn}[z]$ and $\mathbb{R}^+_{nn}[\zeta]$, respectively. For the sets of adequate scalar polynomials, we write $\tilde{\mathbb{R}}^+[z]$ and $\mathbb{R}^+[\zeta]$, respectively.

5. In the following considerations, the stability conditions for Equations (5.135) and (5.138) will be applied to explain the stability of the inverse matrices $\tilde{a}^{-1}(z)$ and $a^{-1}(\zeta)$.

In what follows, the rational matrix $\tilde{w}(z) \in \mathbb{R}_{nm}(z)$ is called *stable*, if its poles $z_1, \ldots, z_q$ satisfy Condition (5.140). The rational matrix $w(\zeta)$ is called stable, if it is free of poles inside or on the border of the unit disc. In the light of this definition, any polynomial matrix is a stable rational matrix. The sets of real stable matrices $\tilde{w}(z)$ and $w(\zeta)$ are denoted by $\tilde{\mathbb{R}}^{+}_{nm}(z)$ and $\mathbb{R}^{+}_{nm}(\zeta)$, respectively.

Rational matrices, which are not stable, are named *unstable*.

Theorem 5.49. *Equations (5.138) and (5.142) are stable, if and only if the rational matrices $\tilde{a}^{-1}(z)$ and $a^{-1}(\zeta)$ are stable.*

Proof. Applying Formula (2.114), we obtain the irreducible representation

$$\tilde{a}^{-1}(z) = \frac{\widetilde{\operatorname{adj} \tilde{a}(z)}}{d_{\tilde{a}\,\min}(z)}, \tag{5.143}$$

where $d_{\tilde{a}\,\min}(z)$ is the minimal polynomial of the matrix $\tilde{a}(z)$. Since the set of roots of the polynomial $d_{\tilde{a}\,\min}(z)$ contains all roots of the polynomial $\det \tilde{a}(z)$, the matrices $\tilde{a}(z)$ and $\tilde{a}^{-1}(z)$ are at the same time stable or unstable. The same can be said about the matrices $a(\zeta)$ and $a^{-1}(\zeta)$. ∎

6. Hitherto, the forward model (5.107) and the backward model (5.108) are called *stable*, when the matrices $\tilde{a}(z)$ and $a(\zeta)$ are stable. For the considered class of systems, this definition is *de facto* equivalent to the asymptotic stability in the sense of Lyapunov.

Theorem 5.50. *Let the forward model (5.107) and the associated backward model (5.108) be controllable. Then for the stability of the corresponding models, it is necessary and sufficient that their transfer matrices $\tilde{w}(z)$ resp. $w(\zeta)$ are stable.*

Proof. Using (5.104) and (5.143), we obtain

$$\tilde{w}(z) = \frac{\widetilde{\operatorname{adj} \tilde{a}(z)}\, \tilde{b}(z)}{d_{\tilde{a}\,\min}(z)}.$$

Under the made suppositions, this matrix is irreducible, and this fact arises from Theorem 2.42. Thus, the matrices $\tilde{a}(z)$ and $\tilde{w}(z)$ are either both stable or both unstable. This fact proves Theorem 5.50 for forward models. The proof for backward models runs analogously. ∎

5.7 Closed-loop LTI Systems of Finite Dimension

1. The input signal $\{u\}$ of the process in Fig. 5.1 is now separated into two components. The first component is still denoted by $\{u\}$ and contains the directly controllable quantities called as the *control input*. Besides the control input, additional quantities effect the process L, that depend on external factors. In Fig. 5.2, these quantities are assigned by the sequence $\{g\}$ called as the *disturbance input*. The forward model of this process might be represented by the equation

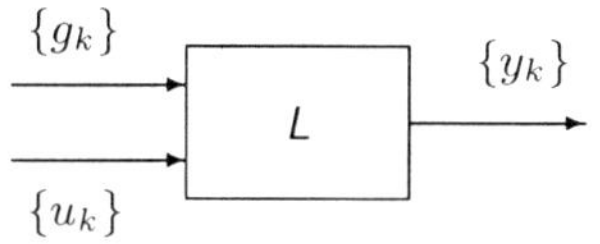

Fig. 5.2. Process with two inputs

$$\tilde{a}(q)y_k = \tilde{b}(q)u_k + \tilde{f}(q)g_k\,, \quad (k = 0, 1, \ldots)\,, \tag{5.144}$$

where $\tilde{a}(q) \in \mathbb{R}_{nn}[q]$, $\tilde{b}(z) \in \mathbb{R}_{nm}[q]$, $\tilde{f}(q) \in \mathbb{R}_{n\ell}[q]$. In future, we will only consider non-singular processes, for which $\det \tilde{a}(q) \not\equiv 0$ is true. When this condition is ensured, the rational matrices

$$\tilde{w}(q) = \tilde{a}^{-1}(q)\tilde{b}(q), \quad \tilde{w}_g(q) = \tilde{a}^{-1}(q)\tilde{f}(q) \tag{5.145}$$

are explained, and they will be called the *control* and *disturbance transfer matrix*, respectively.

For the further investigations, we always suppose the following assumptions:

A1 The matrix $\tilde{w}_g(q)$ is at least proper, *i.e.* the process is causal with respect to the input $\{g\}$.

A2 The matrix $\tilde{w}(q)$ is strictly proper, *i.e.* the process is strictly causal with respect to the input $\{u\}$. This assumption is motivated by the following reasons:

  a) In further considerations, only such kind of models will occur.
  b) This assumption enormously simplifies the answer to the question about the causality of the controller.
  c) It can be shown that, when the matrix $\tilde{w}(q)$ is only proper, then the closed-loop system contains *de facto* algebraic loops, which cannot appear in real sampled-data control systems [19].

The process (5.144) is called *controllable by the control input*, if for all finite q

$$\operatorname{rank} R_h(q) = \operatorname{rank}\left[\, \tilde{a}(q) \;\; \tilde{b}(q) \,\right] = n\,, \tag{5.146}$$

and it is named *controllable by the disturbance input*, if for all finite q

$$\operatorname{rank} R_g(q) = \operatorname{rank} \begin{bmatrix} \tilde{a}(q) & \tilde{f}(q) \end{bmatrix} = n\,.$$

2. To impart the process (5.144) appropriate dynamical properties, a controller R is fed back, what results in the structure shown in Fig. 5.3. The

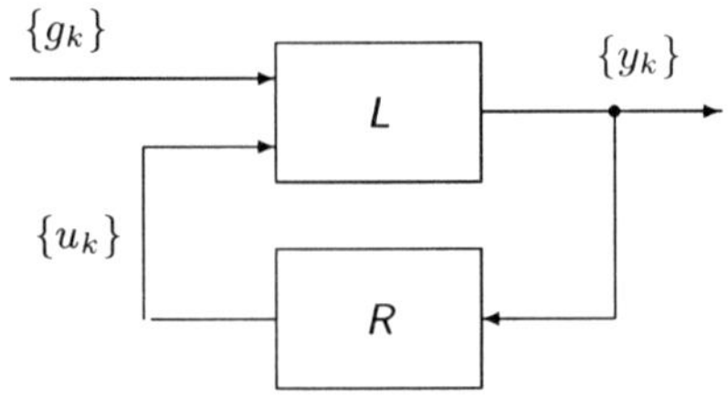

Fig. 5.3. Controlled process

controller R itself is an at least causal discrete-time LTI object, which is given by the forward model

$$\tilde{\alpha}(q)u_k = \tilde{\beta}(q)y_k$$

with $\tilde{\alpha}(q) \in \mathbb{R}_{mm}[q]$, $\tilde{\beta}(q) \in \mathbb{R}_{mn}[q]$.

Together with Equation (5.144), this performs a model of the closed-loop system:

$$\begin{aligned} \tilde{a}(q)y_k - \tilde{b}(q)u_k &= \tilde{f}(q)g_k \\ -\tilde{\beta}(q)y_k + \tilde{\alpha}(q)u_k &= O_{m1}\,. \end{aligned} \tag{5.147}$$

3. The dynamical properties of the closed-loop system (5.147) are characterised by the polynomial matrix

$$\tilde{Q}_l(q, \tilde{\alpha}, \tilde{\beta}) = \begin{bmatrix} \tilde{a}(q) & -\tilde{b}(q) \\ -\tilde{\beta}(q) & \tilde{\alpha}(q) \end{bmatrix}, \tag{5.148}$$

which is named the (left) *characteristic matrix* of the forward model of the closed-loop system. A wide class of control problems might be expressed purely algebraic.

Abstract control problem. For a given pair $(\tilde{a}(q), \tilde{b}(q))$ with strictly proper transfer matrix $\tilde{w}(q)$, find the set of pairs $(\tilde{\alpha}(q), \tilde{\beta}(q))$ such that the matrix

$$w_d(q) = \tilde{\alpha}(q)^{-1}\tilde{\beta}(q)$$

is at least proper and the characteristic matrix (5.148) adopt certain prescribed properties. Besides, the closed-loop system (5.147), has to be causal.

4. For the solution of many control problems, it is suitable to use the associated backward model additionally to the forward model (5.144) of the process. We will give a general approach for the design of such models, which suppose the controllability of the process by the control input. For this reason, we write (5.144) in the form

$$y_k = \tilde{w}(q)u_k + \tilde{w}_g(q)g_k \,.$$

Substituting here ζ^{-1} for q, we obtain

$$y_k = w(\zeta)u_k + w_g(\zeta)g_k \,, \tag{5.149}$$

where

$$w(\zeta) = \tilde{w}(\zeta^{-1}) \,, \quad w_g(\zeta) = \tilde{w}_g(\zeta^{-1}) \,.$$

When we have the ILMFD

$$w(\zeta) = a^{-1}(\zeta)b_0(\zeta) \,, \tag{5.150}$$

then (5.149) might be written as

$$a(\zeta)y_k = b_0(\zeta)u_k + a(\zeta)w_g(\zeta)g_k \,. \tag{5.151}$$

For the further arrangements the next property is necessary.

Lemma 5.51. *Let the process (5.144) be controllable by the control input. Then the matrix*

$$b_g(\zeta) = a(\zeta)w_g(\zeta) \tag{5.152}$$

turns out to be a polynomial.

Proof. Due to supposition (5.146), the first relation in (5.145) is an ILMFD of the matrix $\tilde{w}(q)$. Thus from (5.145), we obtain

$$\tilde{w}_g(q) \underset{l}{\prec} \tilde{w}(q) \,, \tag{5.153}$$

because the polynomial $\tilde{a}(q)$ reduces the matrix $\tilde{w}_g(q)$. Starting with the minimal standard realisation

$$\tilde{w}(q) = C(qI_p - A)^{-1}B + D \,,$$

where A, B, C, D are constant matrices of appropriate dimension, we find from (5.153) with the help of Theorem 2.56, that the matrix $\tilde{w}_g(q)$ allows the representation

$$\tilde{w}_g(q) = C(qI_p - A)^{-1}B_g + D_g \,, \tag{5.154}$$

where B_g and D_g are constant matrices of the dimensions $p \times \ell$ and $n \times \ell$, respectively. Substituting ζ^{-1} for q, we obtain the standard realisation of the matrix $w(\zeta)$:

$$w(\zeta) = \zeta C(I_p - \zeta A)^{-1}B + D\,,$$

which is minimal due to Lemma 5.35, *i.e.* the pairs $(I_p - \zeta A, \zeta B)$ and $[I_p - \zeta A, C]$ are irreducible. Thus, when we build the ILMFD

$$C(I_p - \zeta A)^{-1} = a_1^{-1}(\zeta)b_1(\zeta)\,, \tag{5.155}$$

then the right side of the formula

$$w(\zeta) = a_1^{-1}(\zeta)\left[\zeta\, b_1(\zeta)B + a_1(\zeta)D\right]$$

turns out as an ILMFD of the matrix $w(\zeta)$. Hence the right side of (5.150) is also an ILMFD of the matrix $w(\zeta)$, such that

$$a(\zeta) = \psi(\zeta)a_1(\zeta)$$

is valid with a unimodular matrix $\psi(\zeta)$. From (5.154), we find

$$w_g(\zeta) = \tilde{w}_g(\zeta^{-1}) = \zeta C(I_p - \zeta A)^{-1}B_g + D_g\,.$$

Therefore, using (5.155), we realise that

$$a(\zeta)w_g(\zeta) = \zeta\psi(\zeta)\left[b_1(\zeta)B_g + a_1(\zeta)D_g\right] = b_g(\zeta)$$

is a polynomial matrix. ■

Inserting (5.152) into (5.151), we obtain the wanted backward model of the form

$$a(\zeta)y_k = b_0(\zeta)u_k + b_g(\zeta)g_k\,.$$

4. Due to the supposed strict causality of the process with respect to the control input, the conditions

$$\det a(0) \neq 0, \quad b_0(0) = O_{nm} \tag{5.156}$$

hold. That's why for further considerations, the associated backward model of the process is denoted in the form

$$a(\zeta)y_k = \zeta b(\zeta)u_k + b_g(\zeta)g_k\,, \tag{5.157}$$

where the first condition in (5.156) is ensured. Starting with the backward model of the process (5.157), the controller is attempted in the form

$$\alpha(\zeta)u_k = \beta(\zeta)y_k \tag{5.158}$$

with

$$\det\alpha(0) \neq 0\,.$$

When we put this together with (5.157), we obtain the backward model of the closed-loop system

$$\begin{aligned} a(\zeta)y_k - \zeta b(\zeta)u_k &= b_g(\zeta)g_k \\ -\beta(\zeta)y_k + \alpha(\zeta)u_k &= O_{m1}\,. \end{aligned} \tag{5.159}$$

Besides, the characteristic matrix of the backward model of the closed-loop system $Q_l(\zeta\alpha, \beta)$ takes the form

$$Q_l(\zeta, \alpha, \beta) = \begin{bmatrix} a(\zeta) & -\zeta b(\zeta) \\ -\beta(\zeta) & \alpha(\zeta) \end{bmatrix}. \tag{5.160}$$

Introduce the extended output vector

$$\Upsilon_k = \begin{bmatrix} y_k \\ u_k \end{bmatrix},$$

so Equations (5.159) might be written in form of the backward model

$$Q_l(\zeta, \alpha, \beta)\Upsilon_k = B(\zeta)g_k\,, \quad B(\zeta) = \begin{bmatrix} b_g(\zeta) \\ O_{m\ell} \end{bmatrix}.$$

5. In analogy to the preceding investigations, consider the case when the process is described by a PMD of the form

$$\tau_0 = (a(\zeta), \zeta b(\zeta), c(\zeta)) \in \mathbb{R}_{npm}[\zeta]$$

with $\det a(0) \neq 0$. Then the backward model of the closed-loop system might be presented in the shape

$$\begin{aligned} a(\zeta)x_k &= \zeta b(\zeta)u_k + b_g(\zeta)g_k \\ y_k &= c(\zeta)x_k \\ \alpha(\zeta)u_k &= \beta(\zeta)y_k\,. \end{aligned} \tag{5.161}$$

In this case, the characteristic matrix of the closed-loop system $Q_\tau(\zeta, \alpha, \beta)$ takes the form

$$Q_\tau(\zeta, \alpha, \beta) = \begin{bmatrix} a(\zeta) & O_{pn} & -\zeta b(\zeta) \\ -c(\zeta) & I_n & O_{nm} \\ O_{mp} & -\beta(\zeta) & \alpha(\zeta) \end{bmatrix}. \tag{5.162}$$

Introduce here the extended vector

$$\Upsilon_{\tau k} = \begin{bmatrix} x_k \\ y_k \\ u_k \end{bmatrix},$$

so (5.161) might be written as backward model

$$Q_\tau(\zeta, \alpha, \beta)\Upsilon_{\tau k} = B_\tau(\zeta)g_k\,, \quad B(\zeta) = \begin{bmatrix} b_g(\zeta) \\ O_{n\ell} \\ O_{m\ell} \end{bmatrix}.$$

5.8 Stability and Stabilisation of the Closed Loop

1. The following investigations for stability and stabilisation refer to closed-loop systems and will be done with the backward models (5.159) and (5.161), which are preferred in the whole book. In what follows, an arbitrary controller $(\alpha(\zeta), \beta(\zeta))$ is said to be *stabilising*, if the closed-loop system with it is stable.

The polynomial

$$\Delta(\zeta) = \det Q_l(\zeta, \alpha, \beta) \tag{5.163}$$

is called the *characteristic polynomial* of the system (5.159), and the polynomial

$$\Delta_\tau(\zeta) = \det Q_\tau(\zeta, \alpha, \beta)$$

the characteristic polynomial of the system (5.161).

Theorem 5.52. *For the stability of the systems (5.159) or (5.161), it is necessary and sufficient that the characteristic polynomials $\Delta(\zeta)$ or $\Delta_\tau(\zeta)$, respectively, are stable.*

Proof. The proof immediately follows from Theorems 5.46 and 5.47. ∎

Corollary 5.53. *Any stabilising controller for the processes (5.157) or (5.161) is causal, i.e.*

$$\det \alpha(0) \neq 0 . \tag{5.164}$$

Proof. When the system (5.159) is stable, due to Theorem 5.46, we have

$$\det Q_l(0, \alpha, \beta) = \Delta(0) \neq 0 ,$$

which with the aid of (5.160) yields

$$\det a(0) \det \alpha(0) \neq 0 ,$$

hence (5.164) is true. The proof for the system (5.161) runs analogously. ∎

Corollary 5.54. *Any stabilising controller $(\alpha(\zeta), \beta(\zeta))$ for the systems (5.159) or (5.161) possesses a transfer matrix*

$$w_d(\zeta) = \alpha^{-1}(\zeta)\beta(\zeta) ,$$

because from (5.164) immediately emerge that the matrix $\alpha(\zeta)$ is invertible. ∎

Corollary 5.55. *The stable closed-loop systems (5.159) and (5.161) possess the transfer matrices*

$$w_0(\zeta) = Q_l^{-1}(\zeta, \alpha, \beta)B(\zeta) ,$$

$$w_\tau(\zeta) = Q_\tau^{-1}(\zeta, \alpha, \beta)B_\tau(\zeta) ,$$

which are analytical in the point $\zeta = 0$. ∎

2. Let the LMFD

$$w(\zeta) = \zeta a_l^{-1}(\zeta) b_l(\zeta) \tag{5.165}$$

be given. Then using the terminology of Chapter 4, the pair $(a_l(\zeta), \zeta b_l(\zeta))$ is called a *left process model* . Besides, if the pair $(a_l(\zeta), \zeta b_l(\zeta))$ is irreducible, then the left process model is named *controllable.* If we have at the same time the RMFD

$$w(\zeta) = \zeta b_r(\zeta) a_r^{-1}(\zeta)\,, \tag{5.166}$$

then the pair $[a_r(\zeta), \zeta b_r(\zeta)]$ is called a *right process model.* The right process model is named *controllable*, when the pair $[a_r(\zeta), \zeta b_r(\zeta)]$ is irreducible.

Related to the above, the concept of controllability for left and right models of the controllers might be introduced. If the LMFD and RMFD

$$w_d(\zeta) = \alpha_l^{-1}(\zeta)\beta_l(\zeta) = \beta_r(\zeta)\alpha_r^{-1}(\zeta)$$

exist, then the pairs $(\alpha_l(\zeta), \beta_l(\zeta))$ and $[\alpha_r(\zeta), \beta_r(\zeta)]$ are left and right models of the controller, respectively. As above, we introduce the concepts of controllable left and right controller models. The matrices

$$\begin{aligned} Q_l(\zeta, \alpha_l, \beta_l) &= \begin{bmatrix} a_l(\zeta) & -\zeta b_l(\zeta) \\ -\beta_l(\zeta) & \alpha_l(\zeta) \end{bmatrix}, \\ Q_r(\zeta, \alpha_r, \beta_r) &= \begin{bmatrix} \alpha_r(\zeta) & \zeta b_r(\zeta) \\ \beta_r(\zeta) & a_r(\zeta) \end{bmatrix} \end{aligned} \tag{5.167}$$

are called the *left and right characteristic matrices*, respectively.

Lemma 5.56. *Let* $(a_l(\zeta), \zeta b_l(\zeta))$, $[a_r(\zeta), \zeta b_r(\zeta)]$ *as well as* $(\alpha_l(\zeta), \beta_l(\zeta))$, $[\alpha_r(\zeta), \beta_r(\zeta)]$ *be irreducible left and right models of the process or controller, respectively. Then*

$$\det Q_l(\zeta, \alpha_l, \beta_l) \approx \det Q_r(\zeta, \alpha_r, \beta_r)\,. \tag{5.168}$$

Proof. Applying the general formulae (4.76) and (5.167), we easily find

$$\begin{aligned} \det Q_l(\zeta, \alpha_l, \beta_l) &= \det a_l \det \alpha_l \det \left[I_n - \zeta a_l^{-1}(\zeta) b_l(\zeta) \alpha_l^{-1}(\zeta)\beta_l(\zeta)\right], \\ \det Q_r(\zeta, \alpha_r, \beta_r) &= \det a_r \det \alpha_r \det \left[I_n - \zeta b_r(\zeta) a_r^{-1}(\zeta) \beta_r(\zeta)\alpha_r^{-1}(\zeta)\right]. \end{aligned} \tag{5.169}$$

Due to the supposed irreducibility, we obtain for the left and right models

$$\det a_l(\zeta) \approx \det a_r(\zeta), \quad \det \alpha_l(\zeta) \approx \det \alpha_r(\zeta)\,.$$

Moreover, the expressions in the brackets of (5.169) coincide, that's why (5.168) is true. ∎

From Lemma 5.56, it arises that the design problems for left and right models of stabilising controllers are in principal equivalent.

3. A number of statements is listed, concerning the stability of the closed-loop system and the design of the set of stabilising controllers.

Theorem 5.57. *Let the process be controllable by the control input, and Relations (5.165) and (5.166) should determine controllable left and right IMFDs. Then a necessary and sufficient condition for the fact, that the pair* $(\alpha_l(\zeta), \beta_l(\zeta))$ *is a left model of a stabilising controller, is that the matrices* $\alpha_l(\zeta)$ *and* $\beta_l(\zeta)$ *satisfy the relation*

$$\alpha_l(\zeta)a_r(\zeta) - \zeta\beta_l(\zeta)b_r(\zeta) = D_l(\zeta)\,, \tag{5.170}$$

where $D_l(\zeta)$ *is any stable polynomial matrix. For the pair* $[\alpha_r(\zeta), \beta_r(\zeta)]$ *to be a right model of a stabilising controller, it is necessary and sufficient that the matrices* $\alpha_r(\zeta)$ *and* $\beta_r(\zeta)$ *fulfill the relation*

$$a_l(\zeta)\alpha_r(\zeta) - \zeta b_l(\zeta)\beta_r(\zeta) = D_r(\zeta)\,, \tag{5.171}$$

where $D_r(\zeta)$ *is any stable polynomial matrix.*

Proof. Relation (5.170) will be shown.
Necessity: Let the polynomial matrices $\alpha_{0r}(\zeta)$, $\beta_{0r}(\zeta)$ satisfy the equation

$$a_l(\zeta)\alpha_{0r}(\zeta) - \zeta b_l(\zeta)\beta_{0r}(\zeta) = I_n\,.$$

Then owing to Lemma 4.4, the matrix

$$Q_r(\zeta, \alpha_{0r}, \beta_{0r}) = \begin{bmatrix} \alpha_{0r}(\zeta) & \zeta b_r(\zeta) \\ \beta_{0r}(\zeta) & a_r(\zeta) \end{bmatrix}$$

is unimodular. Besides, we obtain

$$Q_l(\zeta, \alpha_l, \beta_l)Q_r(\zeta, \alpha_{0r}, \beta_{0r}) = \begin{bmatrix} I_n & O_{nm} \\ M_l(\zeta) & D_l(\zeta) \end{bmatrix}, \tag{5.172}$$

where $D_l(\zeta)$ and $M_l(\zeta)$ are polynomial matrices and in addition (5.170) is fulfilled. Per construction, the sets of eigenvalues of the matrices $Q_l(\zeta, \alpha_l, \beta_l)$ and $D_l(\zeta)$ coincide. Thus, the stability of the matrix $Q_l(\zeta, \alpha_l, \beta_l)$ implies the stability of the matrix $D_l(\zeta)$.
Sufficiency: Take the steps of the proof in reverse order to realise that the conditions are sufficient.
Relation (5.171) is shown analogously. ■

Theorem 5.58. *For the rational* $m \times n$ *matrix* $w_d(\zeta)$ *to be the transfer matrix of a stabilising controller for the system (5.159), where the process is completely controllable, it is necessary and sufficient that it allows a representation of the form*

$$w_d(\zeta) = F_1^{-1}(\zeta)F_2(\zeta) = G_2(\zeta)G_1^{-1}(\zeta)\,, \tag{5.173}$$

where $F_1(\zeta)$, $F_2(\zeta)$ and $G_1(\zeta)$, $G_2(\zeta)$ are stable rational matrices satisfying

$$\begin{aligned} F_1(\zeta)a_r(\zeta) - \zeta F_2(\zeta)b_r(\zeta) &= I_m\,, \\ a_l(\zeta)G_1(\zeta) - \zeta b_l(\zeta)G_2(\zeta) &= I_n\,. \end{aligned} \tag{5.174}$$

Proof. The first statement in (5.174) will be shown.
Necessity: Let $(\alpha_l(\zeta), \beta_l(\zeta))$ be a left model of a stabilising controller. Then the matrices of this pair satisfy Relation (5.170) for a certain stable matrix $D_l(\zeta)$. Besides, we convince that the matrices

$$F_1(\zeta) = D_l^{-1}(\zeta)\alpha_l(\zeta), \quad F_2(\zeta) = D_l^{-1}(\zeta)\beta_l(\zeta)$$

are stable and satisfy Relations (5.173) and (5.174).
Sufficiency: Let the matrices $F_1(\zeta)$ and $F_2(\zeta)$ be stable and Relation (5.174) be satisfied. Then the rational matrix

$$F(\zeta) = \begin{bmatrix} F_1(\zeta) \; F_2(\zeta) \end{bmatrix}$$

is stable. Consider the ILMFD

$$F(\zeta) = a_F^{-1}(\zeta)b_F(\zeta) = a_F^{-1}(\zeta) \overset{\quad\; m \qquad\; n}{\begin{bmatrix} d_1(\zeta) & d_2(\zeta) \end{bmatrix}}\, m\,,$$

where the matrix $a_F(\zeta)$ is stable and

$$d_1(\zeta) = a_F(\zeta)F_1(\zeta), \quad d_2(\zeta) = a_F(\zeta)F_2(\zeta)$$

are polynomial matrices. Due to

$$d_1(\zeta)a_r(\zeta) - \zeta d_2(\zeta)b_r(\zeta) = a_F(\zeta)\,,$$

the pair $(d_1(\zeta), d_2(\zeta))$, owing to Theorem 5.57, is a stabilising controller with the transfer function

$$w_d(\zeta) = d_1^{-1}(\zeta)d_2(\zeta) = F_1^{-1}(\zeta)F_2(\zeta)\,.$$

Thus, the first statement in (5.174) is proven.
The second statement in (5.174) can be shown analogously. ∎

Theorem 5.59. *Let the pair $(a_l(\zeta), \zeta b_l(\zeta))$ be irreducible and $(\alpha_{0l}(\zeta), \beta_{0l}(\zeta))$ should be an arbitrary basic controller, such that the matrix $Q_l(\zeta, \alpha_{0l}, \beta_{0l})$ becomes unimodular. Then the set of all stabilising left controllers $(\alpha_l(\zeta), \beta_l(\zeta))$ for the system (5.159) is determined by the relations*

$$\begin{aligned} \alpha_l(\zeta) &= D_l(\zeta)\alpha_{0l}(\zeta) - \zeta M_l(\zeta)b_l(\zeta)\,, \\ \beta_l(\zeta) &= D_l(\zeta)\beta_{0l}(\zeta) - M_l(\zeta)a_l(\zeta)\,, \end{aligned}$$

where $D_l(\zeta)$, $M_l(\zeta)$ are any polynomial matrices, but $D_l(\zeta)$ has to be stable.

Proof. The proof immediately emerges from Theorem 4.21. ■

Theorem 5.60. *Let the pairs* $(a_l(\zeta), \zeta b_l(\zeta))$ *and* $(\alpha_l(\zeta), \beta_l(\zeta))$ *be irreducible and the matrix* $Q_l^{-1}(\zeta, \alpha_l, \beta_l)$ *be represented in the form*

$$Q_l^{-1}(\zeta, \alpha_l, \beta_l) = \begin{array}{c} \begin{array}{cc} \quad n & \quad m \end{array} \\ \begin{bmatrix} V_1(\zeta) & q_{12}(\zeta) \\ V_2(\zeta) & q_{21}(\zeta) \end{bmatrix} \end{array} \begin{array}{l} n \\ m \end{array} . \tag{5.175}$$

Then a necessary and sufficient condition for $(\alpha_l(\zeta), \beta_l(\zeta))$ *to be a stabilising controller is the fact that the matrices* $V_1(\zeta)$ *and* $V_2(\zeta)$ *are stable.*

Proof. The necessity of the conditions of the theorem emerges immediately from Theorem 5.49.
Sufficiency: Let the ILMFD (5.165) and IRMFD (5.166) exist and $(\alpha_{0l}(\zeta), \beta_{0l}(\zeta))$, $(\alpha_{0r}(\zeta), \beta_{0r}(\zeta))$ should be dual left and right basic controllers. Then observing (5.172), we get

$$Q_l(\zeta, \alpha_l, \beta_l) = N_l(\zeta) Q_l(\zeta, \alpha_{0l}, \beta_{0l}) , \tag{5.176}$$

where

$$N_l(\zeta) = \begin{bmatrix} I_n & O_{nm} \\ M_l(\zeta) & D_l(\zeta) \end{bmatrix} . \tag{5.177}$$

Inverting the matrices in Relation (5.176), we arrive at

$$Q_l^{-1}(\zeta, \alpha_l, \beta_l) = Q_l^{-1}(\zeta, \alpha_{0l}, \beta_{0l}) N_l^{-1}(\zeta) .$$

With respect to the properties of dual controllers, we obtain

$$Q_l^{-1}(\zeta, \alpha_l, \beta_l) = Q_r(\zeta, \alpha_{0r}, \beta_{0r}) = \begin{bmatrix} \alpha_{0r}(\zeta) & \zeta b_r(\zeta) \\ \beta_{0r}(\zeta) & a_r(\zeta) \end{bmatrix} , \tag{5.178}$$

where from (5.177), we find

$$N_l^{-1}(\zeta) = \begin{bmatrix} I_n & O_{nm} \\ -D_l^{-1}(\zeta) M_l(\zeta) & D_l^{-1}(\zeta) \end{bmatrix} .$$

Applying this and (5.178), we obtain

$$Q_l^{-1}(\zeta) = \begin{bmatrix} \alpha_{0r}(\zeta) - \zeta b_r(\zeta)\phi(\zeta) & \zeta b_r(\zeta) D_l^{-1}(\zeta) \\ \beta_{0r}(\zeta) - a_r(\zeta)\phi(\zeta) & a_r(\zeta) D_l^{-1}(\zeta) \end{bmatrix} \tag{5.179}$$

with the notation

$$\phi(\zeta) = D_l^{-1}(\zeta) M_l(\zeta) . \tag{5.180}$$

Comparing (5.175) with (5.179), we produce

$$\begin{aligned} V_1(\zeta) &= \alpha_{0r}(\zeta) - \zeta b_r(\zeta)\phi(\zeta)\,, \\ V_2(\zeta) &= \beta_{0r}(\zeta) - a_r(\zeta)\phi(\zeta)\,, \end{aligned} \tag{5.181}$$

or equivalently

$$\begin{bmatrix} V_1(\zeta) \\ V_2(\zeta) \end{bmatrix} = Q_r(\zeta, \alpha_{0r}, \beta_{0r}) \begin{bmatrix} I_n \\ -\phi(\zeta) \end{bmatrix}.$$

From this equation and (5.178), we generate

$$\begin{bmatrix} I_n \\ -\phi(\zeta) \end{bmatrix} = Q_l(\zeta, \alpha_{0l}, \beta_{0l}) \begin{bmatrix} V_1(\zeta) \\ V_2(\zeta) \end{bmatrix},$$

thus we read

$$\phi(\zeta) = \beta_{0l}(\zeta)V_1(\zeta) - \alpha_{0l}(\zeta)V_2(\zeta)\,.$$

When the matrices $V_1(\zeta)$ and $V_2(\zeta)$ are stable, then the matrix $\phi(\zeta)$ is also stable.

Furthermore, notice that the pair $(\alpha_l(\zeta), \beta_l(\zeta))$ is irreducible, because the pair $(D_l(\zeta), M_l(\zeta))$ is also irreducible. Hence Equation (5.180) defines an ILMFD of the matrix $\phi(\zeta)$. But the matrix $\phi(\zeta)$ is stable and therefore, $D_l(\zeta)$ is also stable. Since also the matrix $Q_l^{-1}(\zeta, \alpha_l, \beta_l)$ is stable, the blocks in (5.179) must be stable. Hence owing to Theorem 5.49, it follows that the matrix $Q_l(\zeta, \alpha_l, \beta_l)$ is stable and consequently, the controller $(\alpha_l(\zeta), \beta_l(\zeta))$ is stabilising. ∎

Remark 5.61. In principle, we could understand the assertions of Theorem 5.60 as a corollary to Theorem 5.58. Nevertheless, in the proof we gain some important additional relations that will be used in the further disclosures.

Besides (5.181), an additional representation of the matrices $V_1(\zeta)$, $V_2(\zeta)$ will be used. For this purpose, notice that from (5.167) and (5.175), it emerges

$$\begin{aligned} a_l(\zeta)V_1(\zeta) - \zeta b_l(\zeta)V_2(\zeta) &= I_n\,, \\ -\beta_l(\zeta)V_1(\zeta) + \alpha_l(\zeta)V_2(\zeta) &= O_{nm}\,. \end{aligned}$$

Resolving these equations for the variables $V_1(\zeta)$ and $V_2(\zeta)$, we obtain

$$\begin{aligned} V_1(\zeta) &= \left[a_l(\zeta) - \zeta b_l(\zeta)w_d(\zeta)\right]^{-1}, \\ V_2(\zeta) &= w_d(\zeta)\left[a_l(\zeta) - \zeta b_l(\zeta)w_d(\zeta)\right]^{-1}, \end{aligned} \tag{5.182}$$

where $w_d(\zeta)$ is the transfer matrix of the controller. From (5.182), it follows directly

$$w_d(\zeta) = V_2(\zeta)V_1(\zeta)^{-1}\,. \tag{5.183}$$

4. On basis of Theorem 4.24, the stabilisation problem can be solved for the system (5.159) even in those cases, when the pair $(a(\zeta), \zeta b(\zeta))$ is reducible.

Theorem 5.62. *Suppose in (5.159)*

$$a(\zeta) = \lambda(\zeta)a_1(\zeta), \qquad \zeta b(\zeta) = \zeta\lambda(\zeta)b_1(\zeta)$$

with a latent matrix $\lambda(\zeta)$ and the irreducible pair $(a_1(\zeta), \zeta b_1(\zeta))$. Then, if the matrix $\lambda(\zeta)$ is unstable, the system (5.159) never can be stabilised by a feedback of the form (5.158), i.e. the process (5.157) is not stabilisable. However, if the matrix $\lambda(\zeta)$ is stable, then there exists for this process a set of stabilising controllers, i.e. the process is stabilisable. The corresponding set of stabilising controllers coincides with the set of stabilising controllers of the irreducible pair $(a_1(\zeta), \zeta b_1(\zeta))$. ∎

5. In analogy, following the reasoning of Section 4.8, the stabilisation problem for PMD processes is solved.

Theorem 5.63. *Suppose the strictly causal LTI process as a minimal PMD*

$$\tau_0(\zeta) = (a(\zeta), \zeta b(\zeta), c(\zeta))\,, \tag{5.184}$$

where $\det a(0) \not\equiv 0$. *For the transfer matrix*

$$w_\tau(\zeta) = \zeta c(\zeta)a^{-1}(\zeta)b(\zeta)\,,$$

there should exist the ILMFD

$$w_\tau(\zeta) = \zeta p^{-1}(\zeta)q(\zeta)\,. \tag{5.185}$$

Then the stabilisation problem

$$\det Q_\tau(\zeta, \alpha, \beta) \approx d^+(\zeta)\,,$$

where $Q_\tau(\zeta, \alpha, \beta)$ is Matrix (5.162), is solvable for any stable polynomial $d^+(\zeta)$. Besides, the set of stabilising controllers coincides with the set of stabilising controllers of the irreducible pair $(p(\zeta), \zeta q(\zeta))$ and can be designed on basis of Theorems 5.57 and 5.59. ∎

Theorem 5.64. *Let the strictly causal PMD process (5.184) be not minimal and the polynomial $\eta(\zeta)$ should be defined by the relation*

$$\eta(\zeta) = \frac{\det a(\zeta)}{\det p(\zeta)}\,,$$

where the matrix $p(\zeta)$ is determined by an ILMFD (5.185). Then for the stabilisability of the PMD process (5.184), it is necessary and sufficient that the polynomial $\eta(\zeta)$ is stable. If this condition is fulfilled, the set stabilising controllers of the original system coincides with the set of stabilising controllers of the irreducible pair $(p(\zeta), \zeta q(\zeta))$ in the ILMFD (5.185). ∎

6. The results in Sections 4.7 and 4.8, together with the design of the set of stabilising controllers allow to obtain at the same time information about the structure of the set of invariant polynomials for the characteristic matrices (5.160) and (5.162). For instance, in case of Theorem 5.57 or 5.63, the n invariant polynomials of the matrices (5.160) $a_1(\zeta), \dots, a_n(\zeta)$ are equal to 1, and the set of the remaining invariant polynomials $a_{n+1}(\zeta), \dots, a_{n+m}(\zeta)$ coincides with the set of invariant polynomials of the matrix $D_l(\zeta)$.

7. For practical applications, the question on insensitivity of the obtained solution of the stabilisation problem plays a great role. The next theorem supplies the answer to this question.

Theorem 5.65. *Let $(\alpha_l(\zeta), \beta_l(\zeta))$ be a stabilising controller for the strictly causal process $(a_l(\zeta), \zeta b_l(\zeta))$, and instead of the process $(a_l(\zeta), \zeta b_l(\zeta))$ there exists the disturbed strictly causal process*

$$(a_l(\zeta) + a_{l1}(\zeta), \zeta b_l(\zeta) + \zeta b_{l1}(\zeta)) \tag{5.186}$$

with

$$a_{l1}(\zeta) = \sum_{k=0}^{r} A_k \zeta^k, \quad b_{l1}(\zeta) = \sum_{k=0}^{r} B_k \zeta^k,$$

where $r \geq 0$ is an integer, and A_k, B_k, $(k = 0, \dots, r)$ are constant matrices. Suppose $\| \cdot \|$ be a certain norm for finite-dimensional number matrices. Then there exists a positive constant ϵ, such that for

$$\|A_k\| < \epsilon, \quad \|B_k\| < \epsilon, \qquad (k = 0, \dots, r), \tag{5.187}$$

the closed-loop system with the disturbed process (5.186) and the controller $(\alpha_l(\zeta), \beta_l(\zeta))$ remains stable.

Proof. The characteristic matrix of the closed-loop system with the disturbed process has the form

$$Q_{l1}(\zeta, \alpha_l, \beta_l)) = \begin{bmatrix} a_l(\zeta) + a_{l1}(\zeta) & -\zeta[b_l(\zeta) + b_{l1}(\zeta)] \\ -\beta_l(\zeta) & \alpha_l(\zeta) \end{bmatrix}.$$

Applying the sum theorem for determinants, we find

$$\det Q_{l1}(\zeta, \alpha_l, \beta_l) = \Delta_1(\zeta) = \Delta(\zeta) + \Delta_2(\zeta),$$

where

$$\Delta(\zeta) = \det \begin{bmatrix} a_l(\zeta) & -\zeta b_l(\zeta) \\ -\beta_l(\zeta) & \alpha_l(\zeta) \end{bmatrix} \tag{5.188}$$

is the characteristic polynomial of the undisturbed system, and $\Delta_2(\zeta)$ is a polynomial, the coefficients of which tend to zero for $\epsilon \to 0$. Denote

$$\min_{|\zeta|=1} |\Delta(\zeta)| = \delta. \tag{5.189}$$

Under our suppositions, $\delta > 0$ is true, because the polynomial $\Delta(\zeta)$ has no zeros on the unit circle. Attempt

$$\Delta_2(\zeta) = d_0\zeta^{\mu} + d_1\zeta^{\mu-1} + \ldots + d_{\mu} \, ,$$

where the coefficients d_i, $(i = 0, 1, \ldots, \mu)$ continuously depend on the elements of the matrices A_k, B_k, and all of them become zero, when for all $\kappa = 0, 1, \ldots, r$ $A_k = O_{nn}$, $B_k = O_{nm}$ is valid. Thus, there exists an ϵ, such that the inequalities

$$|d_i| < \frac{\delta}{\mu + 1} \, , \qquad (i = 0, 1, \ldots, \mu)$$

remain true, as long as Estimates (5.187) remain true. If (5.189) is fulfilled, we get

$$\max_{|\zeta|=1} |\Delta_2(\zeta)| < \delta \, .$$

Comparing this and (5.188), we realise that for any point of the unit circle $|\zeta| = 1$

$$|\Delta_2(\zeta)| < |\Delta(\zeta)| \, ,$$

and from the Theorem of Rouché [171], it arises that the polynomials $\Delta(\zeta)$ and $\Delta(\zeta) + \Delta_2(\zeta)$ have the same number of zeros inside the unit disc. Hence the stability of the polynomial $\Delta(\zeta)$ implies the stability of the polynomial $\Delta_1(\zeta)$. ∎

Remark 5.66. It can be shown that for the solution of the stabilisation problem in case of forward models of the closed-loop systems (5.147), an analogue statement with respect to the insensitivity of the solution of the stabilisation problem cannot be derived.

Part III

Frequency Methods for MIMO SD Systems

6

Parametric Discrete-time Models of Continuous-time Multivariable Processes

6.1 Response of Linear Continuous-time Processes to Exponential-periodic Signals

This section presents some auxiliary relations that are needed for the further disclosures.

1. Suppose the linear continuous-time process

$$y = w(p)x\,, \tag{6.1}$$

where $p = \frac{\mathrm{d}}{\mathrm{d}t}$ is the differential operator, $w(p) \in \mathbb{R}_{nm}(p)$ is a rational matrix and $x = x(t)$, $y = y(t)$ are vectors of dimensions $m \times 1$, $n \times 1$, respectively. The process is symbolically presented in Fig. 6.1. In the following, the matrix

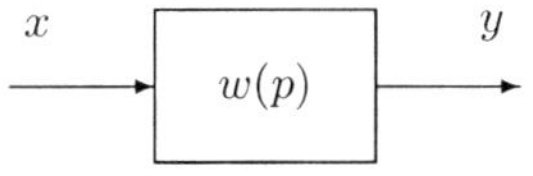

Fig. 6.1. Continuous-time process

$w(p)$ is called the *transfer matrix* of the continuous-time process (6.1).

2. The vectorial input signal $x(t)$ is called *exponential-periodic* (*exp.per.*), if it has the form

$$x(t) = \mathrm{e}^{st}x_T(t), \quad x_T(t) = x_T(t+T)\,, \tag{6.2}$$

where s is a complex number and $T > 0$ is a real constant. Here, s is designated as the *exponent* and T as the *period* of the exponential-periodic function $x(t)$. Further on, all components of the vector $x_T(t)$ are supposed to be of bounded variation.

Assuming an exp.per. input signal $x(t)$ (6.2), this section handles the existence problem for an exp.per. output signal of the processes (6.1), *i.e.*

$$y(t) = \mathrm{e}^{st} y_T(t), \quad y_T(t) = y_T(t+T) . \tag{6.3}$$

3.

Lemma 6.1. *Let the matrix $w(p)$ be given in the standard form*

$$w(p) = \frac{N(p)}{d(p)}$$

with $N(s) \in \mathbb{R}_{nm}[p]$ and the scalar polynomial

$$d(p) = (p - p_1)^{\mu_1} \cdots (p - p_q)^{\mu_q}, \qquad \mu_1 + \ldots + \mu_q = r . \tag{6.4}$$

Furthermore, suppose

$$x(t) \stackrel{\triangle}{=} x_s(t) = X\mathrm{e}^{st} , \tag{6.5}$$

where $X \in \mathbb{C}_{m1}$ is a constant vector and s is a complex number with

$$s \neq p_i , \quad (i = 1, \ldots, q) . \tag{6.6}$$

Then there exists a unique output of the form

$$y(t) \stackrel{\triangle}{=} y_s(t) = Y(s)\mathrm{e}^{st} \tag{6.7}$$

with a constant vector $Y(s) \in \mathbb{C}_{n1}$. Besides,

$$Y(s) = w(s)X$$

and

$$y_s(t) = w(s)X\mathrm{e}^{st} . \tag{6.8}$$

Proof. Suppose a certain ILMFD

$$w(s) = a_l^{-1}(s) b_l(s)$$

with $a_l(s) \in \mathbb{R}_{nn}[s]$, $b_l(s) \in \mathbb{R}_{nm}[s]$. Then Relation (6.1) is equivalent to the differential equation

$$a_l\left(\frac{\mathrm{d}}{\mathrm{d}t}\right) y = b_l\left(\frac{\mathrm{d}}{\mathrm{d}t}\right) x . \tag{6.9}$$

Relations (6.5) and (6.7) should hold, and the vectors $x_s(t)$ and $y_s(t)$ should determine special solutions of Equation (6.9). Due to

$$a_l\left(\frac{\mathrm{d}}{\mathrm{d}t}\right) y_s(t) = a(s)Y(s)\mathrm{e}^{st} ,$$

$$b_l\left(\frac{\mathrm{d}}{\mathrm{d}t}\right) x_s(t) = b(s)X\mathrm{e}^{st} ,$$

the condition

$$a_l(s)Y(s) = b_l(s)X \tag{6.10}$$

must be satisfied. Owing to the properties of ILMFDs, the eigenvalues of the matrix $a_l(s)$ turn out as the roots of the polynomial (6.4), but possibly with higher multiplicity. Thus, (6.6) implies $\det a_l(s) \neq 0$ and from (6.10) we derive

$$Y(s) = a_l^{-1}(s)b_l(s)X = w(s)X\,,$$

i.e. Formula (6.8) really determines the wanted solution.

Now, we prove that the found solution is unique. Beside of (6.5) and (6.7), let Equation (6.9) have an additional special solution of the form

$$x(t) = X\mathrm{e}^{st}, \quad y_{s1}(t) = Y_1(s)\mathrm{e}^{st}\,,$$

where $Y_1(s)$ is a constant vector. Then the difference

$$\varepsilon_s(t) \stackrel{\triangle}{=} y_s(t) - y_{s1}(t) = [Y(s) - Y_1(s)]\mathrm{e}^{st} \tag{6.11}$$

must be a non-vanishing solution of the equation

$$a_l\left(\frac{\mathrm{d}}{\mathrm{d}t}\right)\varepsilon_s(t) = 0\,. \tag{6.12}$$

Relation (6.12) represents a homogeneous system of linear difference equations with constant coefficients. This system may possess non-trivial solutions of the form (6.11) only when $\det a_l(s) = 0$. But this case is excluded by (6.6). Thus, $Y(s) = Y_1(s)$ is true, *i.e.* the solution of the form (6.7) is unique. ∎

4. The question about the existence of an exp.per. output signal with the same exponent and the same period is investigated.

Theorem 6.2. *Let the transfer function of the processes (6.1) be strictly proper, the input signal should have the form (6.2), and for all* k, $(k = 0, \pm 1, \ldots)$ *the relations*

$$s + k\mathrm{j}\omega \neq p_i, \quad (i = 1, \ldots, q), \quad \omega = 2\pi/T, \quad \mathrm{j} = \sqrt{-1} \tag{6.13}$$

should be valid. Then there exists a unique exp.per. output of the form (6.3) with

$$y_T(t) = \int_0^T \varphi_w(T, s, t - \tau)x_T(\tau)\,\mathrm{d}\tau\,, \tag{6.14}$$

where $\varphi_w(T, s, t)$ *is defined by the series*

$$\varphi_w(T, s, t) = \frac{1}{T}\sum_{k=-\infty}^{\infty} w(s + k\mathrm{j}\omega)\mathrm{e}^{k\mathrm{j}\omega t}\,. \tag{6.15}$$

Proof. The function $x_T(t)$ is represented as Fourier series

$$x_T(t) = \sum_{k=-\infty}^{\infty} x_k \mathrm{e}^{kj\omega t},$$

where

$$x_k = \frac{1}{T} \int_0^T x_T(\tau) \mathrm{e}^{-kj\omega\tau} \, \mathrm{d}\tau . \tag{6.16}$$

Then we obtain

$$x(t) = \sum_{k=-\infty}^{\infty} x_k \mathrm{e}^{(s+kj\omega)t} .$$

According to the linearity of the operator (6.1) and Condition (6.13), Lemma 6.1 yields

$$y(t) = \sum_{k=-\infty}^{\infty} w(s + kj\omega) x_k \mathrm{e}^{(s+kj\omega)t} = \mathrm{e}^{st} y_T(t) , \tag{6.17}$$

where

$$y_T(t) = \sum_{k=-\infty}^{\infty} w(s + kj\omega) x_k \mathrm{e}^{kj\omega t} . \tag{6.18}$$

Using (6.16), the last expression sounds

$$y_T(t) = \frac{1}{T} \sum_{k=-\infty}^{\infty} w(s + kj\omega) \int_0^T x_T(\tau) \mathrm{e}^{-kj\omega\tau} \, \mathrm{d}\tau \, \mathrm{e}^{kj\omega t} .$$

Under our suppositions, series (6.15) converges. Hence due to the general properties of Fourier series [171], the order of summation and integration could be exchanged. Thus, we obtain Formula (6.14).

It remains to show the uniqueness of the above generated exp.per. solution. Assume the existence of a second exp.per. output

$$y_1(t) = \mathrm{e}^{st} y_{1T}(t), \quad y_{1T}(t) = y_{1T}(t + T)$$

in addition to the solution (6.3). Then the difference $\varepsilon(t) = y(t) - y_1(t)$ is a solution of the homogeneous equation (6.12) with exponent s and period T. But, from (6.13) emerge that Equation (6.12) does not possess solutions different from zero. Thus, $\varepsilon(t) = 0$ and hence the exp.per. solutions (6.3) and (6.14) coincide. ∎

5. In the following, the series (6.15) is called the *displaced pulse frequency response*, which is abbreviated as *DPFR*. This notation has a physical interpretation. Let $\delta(t)$ be the Dirac impulse and

$$\delta_T(t) = \sum_{k=-\infty}^{\infty} \delta(t - kT)$$

is a periodic pulse sequence. Then, it is well known [159] that the function $\delta_T(t)$ could be developed in a generalised Fourier series

$$\delta_T(t) = \frac{1}{T} \sum_{k=-\infty}^{\infty} \mathrm{e}^{kj\omega t} .$$

For the response of the process (6.1) to the exp.per. input

$$x(t) = \mathrm{e}^{st}\delta_T(t) \tag{6.19}$$

recruit Formulae (6.17), (6.18) with $x_k = 1$, $(k = 0, \pm 1, \dots)$. Thus, we obtain

$$y(t) = \mathrm{e}^{st}\varphi_w(T, s, t) .$$

Hence the DPFR $\varphi_w(T, s, t)$ is related to the response of the process (6.1) to an exponentially modulated sequence of unit impulses (6.19).

6.2 Response of Open SD Systems to Exp.per. Inputs

1. In this section and further on by a *digital control unit DCU*, we understand a system with a structure as shown in Fig. 6.2.[1] If the digital control unit works as a controller, we also will call it a *digital controller*. Hereby, $y = y(t)$ and

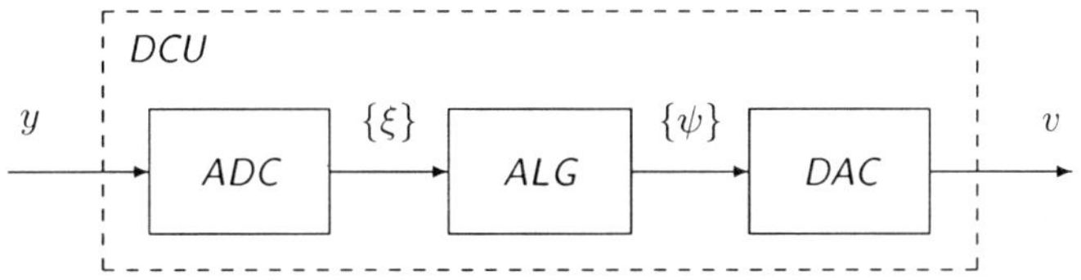

Fig. 6.2. Structure of a digital control unit

$v = v(t)$ are vectors of dimensions $m \times 1$ and $n \times 1$, respectively. Furthermore, $y(t)$ is assumed to be a continuous function of t.

In Fig. 6.2 *ADC* is the analog to digital converter, which converts a continuous-time input signal $y(t)$ into a discrete-time vector sequence $\{\xi\}$ with the elements ξ_k, $(k = 0, \pm 1, \dots)$, *i.e.*

$$\xi_k = y(kT) \stackrel{\triangle}{=} y_k, \quad (k = 0, \pm 1, \dots) . \tag{6.20}$$

[1] The concepts for the elements in a digital control system are not standardised in the literature.

The number $T > 0$, arising in (6.20), is named as the *sampling period* or the period of time quantisation.

The block ALG in Fig. 6.2 stands for the *control program* or the *control algorithm*. If confusion is excluded, also the short name *controller* is used. It calculates from the sequence $\{\xi\}$ a new sequence $\{\psi\}$ with elements ψ_k, $(k = 0, \pm 1, \dots)$. The ALG is a causal discrete LTI object, which is described for instance by its forward model

$$\tilde{\alpha}_0 \psi_{k+\ell} + \tilde{\alpha}_1 \psi_{k+\ell-1} + \ldots + \tilde{\alpha}_\ell \psi_k = \tilde{\beta}_0 \xi_{k+\ell} + \tilde{\beta}_1 \xi_{k+\ell-1} + \ldots + \tilde{\beta}_\ell \xi_k \quad (6.21)$$

or by the associated backward model

$$\alpha_0 \psi_k + \alpha_1 \psi_{k-1} + \ldots + \alpha_\ell \psi_{k-\rho} = \beta_0 \xi_k + \beta_1 \xi_{k-1} + \ldots + \beta_\rho \xi_{k-\rho} \,. \quad (6.22)$$

In (6.21) and (6.22) the $\tilde{\alpha}_i$, $\tilde{\beta}_i$ and α_i, β_i are constant real matrices of appropriate dimensions.

Finally in Fig. 6.2, the block DAC is the digital to analog converter, which transforms a discrete sequence $\{\psi\}$ into a continuous-time signal $v(t)$ by the relation

$$v(t) = m(t - kT)\psi_k, \quad kT < t < (k+1)T \,. \quad (6.23)$$

In (6.23), $m(t)$ is a given function on the interval $0 < t < T$, which is named as *form function*, because it establishes the shape of the control pulses [148]. In what follows, we always suppose that the function $m(t)$ is of bounded variation on the interval $0 \le t \le T$.

2. During the investigation of open and closed sampled-data systems, the transition of exp.per. signals through a digital control unit (6.20)-(6.23) plays an important role. Suppose the input of a digital control unit be the continuous-time signal

$$y(t) = \mathrm{e}^{st} y_T(t), \quad y_T(t) = y_T(t+T) \quad (6.24)$$

with the exponent s and the period T, which coincides with the time quantisation period. We search for an exp.per. output of the form

$$v(t) = \mathrm{e}^{st} v_T(s,t), \quad v_T(s,t) = v_T(s, t+T) \,. \quad (6.25)$$

At first, notice a special feature, when an exp.per. signal (6.24) is sent through a digital control unit. If (6.24) and (6.20) is valid, we namely obtain

$$\xi_k = \mathrm{e}^{ksT} \xi_0 \,, \quad \xi_0 = y_T(0).$$

The result would be the same, if instead of the input $y(t)$ the exponential signal

$$y_s(t) = \mathrm{e}^{st} y_T(0)$$

would be considered. The equivalence of the last two equations shows the so-called *stroboscopic property* of a digital control unit.

The awareness of the stroboscopic property makes it possible to connect the response of the digital control unit to an exp.per. excitation with its response to an exponential signal.

3. In connection with the above said, consider the design task for a solution of Equations (6.20)–(6.23) under the conditions

$$y(t) = \mathrm{e}^{st} y_0; \quad v(t) = \mathrm{e}^{st} v_T(t), \; v_T(t) = v_T(t+T) . \tag{6.26}$$

Assume at first

$$m(t) = 1, \quad 0 \le t < T . \tag{6.27}$$

Then from (6.23) and (6.25), we obtain

$$\mathrm{e}^{st} v_T(s,t) = \psi_k, \quad kT < t < (k+1)T . \tag{6.28}$$

Consider $t \to kT + 0$, so we find

$$\psi_k = \mathrm{e}^{ksT} g(s) , \tag{6.29}$$

where

$$g(s) = v_T(s, +0)$$

is an unknown vector function. The equality

$$\xi_k = y(kT) = \mathrm{e}^{ksT} y_0$$

emerges from (6.28), so after inserting this and (6.29) into (6.22), we receive

$$\left(\alpha_0 + \alpha_1 \mathrm{e}^{-sT} + \ldots + \alpha_\rho \mathrm{e}^{-\rho sT}\right) g(s) = \left(\beta_0 + \beta_1 \mathrm{e}^{-sT} + \ldots + \beta_\rho \mathrm{e}^{-\rho sT}\right) y_0$$

or the equivalent relation

$$\overset{\frown}{\alpha}(s) g(s) = \overset{\frown}{\beta}(s) y_0 , \tag{6.30}$$

where

$$\overset{\frown}{\alpha}(s) = \alpha_0 + \alpha_1 \mathrm{e}^{-sT} + \ldots + \alpha_\rho \mathrm{e}^{-\rho sT} ,$$

$$\overset{\frown}{\beta}(s) = \beta_0 + \beta_1 \mathrm{e}^{-sT} + \ldots + \beta_\rho \mathrm{e}^{-\rho sT}$$

are polynomial matrices in the variable e^{-sT}. Hereinafter, ’ $\frown$ ’ means that the corresponding function depends on e^{-sT}. For $\det \overset{\frown}{\alpha}(s) \not\equiv 0$ from (6.30), we obtain

$$g(s) = \overset{\frown}{w}_d(s) y_0 , \tag{6.31}$$

where

$$\overset{\frown}{w}_d(s) = \overset{\frown}{\alpha}^{-1}(s) \overset{\frown}{\beta}(s) .$$

From (6.29) and (6.31), it arises

$$\psi_k = \mathrm{e}^{ksT} \widehat{w}_d(s) y_0 \,.$$

Substituting this in (6.28), we arrive at

$$\mathrm{e}^{st} v_T(s,t) = \mathrm{e}^{ksT} \widehat{w}_d(s) y_0, \quad kT < t < (k+1)T \,,$$

which implies

$$v_T(s,t) = \widehat{w}_d(s) y_0 \mathrm{e}^{-s(t-kT)}, \quad kT < t < (k+1)T \,.$$

This formula is equivalent to

$$v_T(s,t) = \widehat{w}_d(s) y_0 \mathrm{e}^{-st}, \quad 0 < t < T, \quad v_T(s,t) = v_T(s,t+T) \,. \tag{6.32}$$

4. Using (6.32), we are able to obtain the general solution for the case, when $m(t)$ is an arbitrary given function on the interval $0 \le t \le T$. Then instead of (6.32), formula

$$v_T(s,t) = \widehat{w}_d(s) y_0 \mathrm{e}^{-st} m(t), \quad 0 < t < T, \quad v_T(s,t) = v_T(s,t+T) \tag{6.33}$$

comes up. As result of the above considerations, the following theorem was proven.

Theorem 6.3. *Let the input of the digital control unit (6.20)–(6.23) be the continuous-time signal (6.24). Furthermore, suppose*

$$\det \widehat{\alpha}(s) = \det \left(\alpha_0 + \alpha_1 \mathrm{e}^{-sT} + \ldots + \alpha_\rho \mathrm{e}^{-\rho sT} \right) \not\equiv 0 \,. \tag{6.34}$$

Then there exists an exp.per. solution of Equations (6.20)–(6.23) with the exponent s and the period T in the shape (6.33) ∎

Remark 6.4. It can be shown that under Supposition (6.34) the obtained exp.per. solution with the exponent s and the period T is unique.

Remark 6.5. The obtained solution does not depend on the vector $y(t)$ for $0 < t < T$, but only on $y_0 = y_T(0)$. The stroboscopic property expresses itself in this way.

5. Perform the Fourier expansion of the vector $v_T(s,t)$ as a function of t. At first, we detect

$$v_T(s,t) = \widehat{w}_d(s) y_0 \varphi_\mu(T,s,t) \,, \tag{6.35}$$

where $\varphi_\mu(T,s,t)$ is a scalar periodic function given by

$$\varphi_\mu(T,s,t) = \mathrm{e}^{-st} m(t), \quad 0 < t < T; \quad \varphi_\mu(T,s,t) = \varphi_\mu(T,s,t+T) \,. \tag{6.36}$$

Take the Fourier series

$$\varphi_\mu(T,s,t) = \sum_{k=-\infty}^{\infty} \varphi_k(s)\mathrm{e}^{kj\omega t}, \tag{6.37}$$

where

$$\varphi_k(s) = \frac{1}{T}\int_0^T \varphi_\mu(T,s,\tau)\mathrm{e}^{-kj\omega\tau}\,\mathrm{d}\tau\,.$$

Now, introduce the function

$$\mu(s) = \int_0^T \mathrm{e}^{-s\tau} m(\tau)\,\mathrm{d}\tau\,, \tag{6.38}$$

which is called the *transfer function of the form element.* Thus from (6.36) and (6.38), we obtain

$$\varphi_k(s) = \frac{1}{T}\int_0^T \mathrm{e}^{-(s+kj\omega)\tau} m(\tau)\,\mathrm{d}\tau = \frac{1}{T}\mu(s+kj\omega)\,,$$

so Formula (6.37) sounds

$$\varphi_\mu(T,s,t) = \frac{1}{T}\sum_{k=-\infty}^{\infty} \mu(s+kj\omega)\mathrm{e}^{kj\omega t}\,, \tag{6.39}$$

and it allows to establish the Fourier series of the vector (6.35).

Notice that in the special case (6.27), the transfer function of the form element (6.38) takes the form

$$\mu(s) = \frac{1-\mathrm{e}^{-sT}}{s}\,.$$

6. Let us consider now the more general question about the pass of an exp.per. signals through the open sampled-data system of Fig. 6.3, where *DCU* is a digital control unit described by Equations (6.20)–(6.23) and $L(p)$ is a continuous-times LTI process of the form (6.1) with the transfer function $w(p)$, that is at least proper. The problem amounts to the solution of the

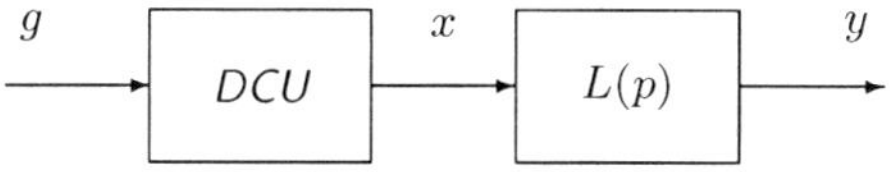

Fig. 6.3. Digital control unit with continuous-time process

general system of Equations (6.20)–(6.23) and (6.9), where the conditions

$$\begin{gathered} g(t) = \mathrm{e}^{st}g_T(t),\ g_T(t) = g_T(t+T);\quad x(t) = \mathrm{e}^{st}x_T(t),\ x_T(t) = x_T(t+T);\\ y(t) = \mathrm{e}^{st}y_T(t),\ y_T(t) = y_T(t+T) \end{gathered} \tag{6.40}$$

hold. In order to solve the just stated problem, we point out that owing to the stroboscopic property, we could restrict ourselves to exponential inputs of the form $g(s) = \mathrm{e}^{st} g_T(0)$. Hence instead of (6.40), the equivalent task with

$$g(t) = \mathrm{e}^{st} g_0\,, \quad x(t) = \mathrm{e}^{st} x_T(t), \quad y(t) = \mathrm{e}^{st} y_T(t) \tag{6.41}$$

might be considered. Further on assume that Conditions (6.13) and $\det \alpha_n(s) \not\equiv 0$ take place. Then with the aid of (6.32), we find

$$x(t) = \mathrm{e}^{st} \varphi_\mu(T, s, t) \widehat{w}_d(s) g_0\,. \tag{6.42}$$

The exp.per. signal (6.42) acts as input to the continuous-time process. With regard to (6.39), this input might be written in the form

$$x(t) = \frac{1}{T} \sum_{k=-\infty}^{\infty} \mu(s + k\mathrm{j}\omega) \mathrm{e}^{(s+k\mathrm{j}\omega)t} \widehat{w}_d(s) g_0\,. \tag{6.43}$$

Calculating the responses of the continuous-time process to the various parts of the input signal (6.43), we find

$$y(t) = \mathrm{e}^{st} \varphi_{w\mu}(T, s, t) \widehat{w}_d(s) g_0 \tag{6.44}$$

with

$$\varphi_{w\mu}(T, s, t) = \frac{1}{T} \sum_{k=-\infty}^{\infty} w(s + k\mathrm{j}\omega) \mu(s + k\mathrm{j}\omega) \mathrm{e}^{k\mathrm{j}\omega t}. \tag{6.45}$$

Denote

$$G(s) = w(s)\mu(s)\,, \tag{6.46}$$

so Formula (6.45) can be written as the DPFR

$$\varphi_{w\mu}(T, s, t) = \varphi_G(T, s, t)\,.$$

Matrix (6.46) is called the transfer matrix of the *modulated process.*

6.3 Functions of Matrices

1. Expansions of the forms (6.15) and (6.45) will play an important role in the following investigations. Above all, closed expressions for the sums of series as well as algebraic properties of these sums are needed. The solution of these problems succeeds by applying the theory of matrix functions. The present section discusses some elements of this theory, where the declarations orientate itself by [51]. The skilled reader may skip this and the next section.

Let A be a constant $p \times p$ matrix, so

$$A_\lambda = \lambda I_p - A$$

is the assigned characteristic matrix and

$$d_A(\lambda) = \det A_\lambda = \det(\lambda I_p - A)$$

is the characteristic polynomial of the matrix A. The minimal polynomial of the matrix A is denoted by $d_{A\,\min}(\zeta)$. Using (2.117), we can write

$$(\lambda I_p - A)^{-1} = \frac{\widetilde{\operatorname{adj}(\lambda I_p - A)}}{d_{A\,\min}(\lambda)}, \tag{6.47}$$

where the numerator is the monic adjoint matrix. Compatible with earlier results, Matrix (6.47) turns out to be strictly proper. Assume

$$d_{A\,\min}(\lambda) = (\lambda - \lambda_1)^{\nu_1} \cdots (\lambda - \lambda_q)^{\nu_q}, \quad \nu_1 + \ldots + \nu_q = r \leq p. \tag{6.48}$$

Then the partial fraction expansion (2.98) yields

$$\begin{aligned}(\lambda I_p - A)^{-1} = \frac{M_{11}}{\lambda - \lambda_1} + \frac{M_{12}}{(\lambda - \lambda_1)^2} + \ldots + \frac{M_{1,\nu_1}}{(\lambda - \lambda_1)^{\nu_1}} + \ldots \\ \ldots + \frac{M_{q1}}{\lambda - \lambda_q} + \frac{M_{q2}}{(\lambda - \lambda_q)^2} + \ldots + \frac{M_{q,\nu_q}}{(\lambda - \lambda_q)^{\nu_q}},\end{aligned} \tag{6.49}$$

where the $M_{ik} = N_{i,\nu_i - k+1}$ are constant matrices, and the N_{ik} are calculated by Formula (2.99). Since the fraction in (6.47) is irreducible, observing (2.100) produces

$$M_{i,\nu_i} \neq O_{pp} \quad (i = 1, \ldots, q).$$

2. Denote

$$Z_{ik} = \frac{M_{ik}}{(k-1)!}, \quad (i = 1, \ldots, q;\ k = 1, \ldots, \nu_i). \tag{6.50}$$

The constant matrices (6.50) are called *components* of the matrix A. Each root λ_i of the minimal polynomial (6.48) with the multiplicity ν_i corresponds to ν_i components

$$Z_{i1},\ Z_{i2},\ \ldots,\ Z_{i,\nu_i}\,.$$

The totality of all these matrices is named the *set of components* of the matrix A according to the eigenvalue λ_i. The total number of the components of the matrix A is equal to the degree of its minimal polynomial.

3. Some general properties of the components (6.50) are listed below [51].

a) If the matrix A is real, and λ_1, λ_2 are two conjugated complex eigenvalues, which arise in the minimal polynomial with the power ν, then the corresponding components Z_{1k}, Z_{2k} $(k = 1, \ldots, \nu)$ are conjugated complex.

b) The generating matrices (6.50) of any matrix are commutative, *i.e.*

$$Z_{ik} Z_{lm} = Z_{lm} Z_{ik} .$$

c) No component (6.50) is a zero matrix.
d) The components (6.50) of the matrix A are linearly independent, *i.e.* the equality

$$\sum_i \sum_k c_{ik} Z_{ik} = O_{pp}$$

with scalar constant c_{ik} implies $c_{ik} = 0$ for all i, k.
e) In future, the particular notation

$$Z_{i1} = Q_i , \quad (i = 1, \dots, q)$$

is used. The matrices Q_i are named the *projectors* of the matrix A. Some important properties of the projectors Q_i are advised now:

(i) The following relations hold:

$$Q_i Z_{ik} = Z_{ik} Q_i = Z_{ik}, \quad (k = 1, \dots, \nu_i) , \tag{6.51}$$

$$Q_i Z_{\rho k} = Z_{\rho k} Q_i = O_{pp}, \quad (i \neq \rho) . \tag{6.52}$$

(ii) From (6.51) for $k = 1$, we get

$$Q_i^2 = Q_i , \quad (i = 1, \dots, q) .$$

(iii) From (6.52) for $k = 1$, we find

$$Q_i Q_\rho = Q_\rho Q_i = O_{pp}, \quad (i \neq \rho) .$$

(iv) Moreover, it can be shown

$$\sum_{i=1}^{q} Q_i = I_p .$$

4. Let the matrix A possess the minimal polynomial (6.48), and $f(\lambda)$ should be a known scalar function. It is said that the function $f(\lambda)$ is defined over the spectrum of the matrix A, if the expressions

$$\begin{matrix} f(\lambda_1), & f'(\lambda_1), & \dots & f^{(\nu_1-1)}(\lambda_1) \\ f(\lambda_2), & f'(\lambda_2), & \dots & f^{(\nu_2-1)}(\lambda_2) \\ \vdots & \vdots & \dots & \vdots \\ f(\lambda_q), & f'(\lambda_q), & \dots & f^{(\nu_q-1)}(\lambda_q) \end{matrix} \tag{6.53}$$

make sense. The totality of the values (6.53) is addressed if we speak about the *values of the function* $f(\lambda)$ *on the spectrum* of the matrix A.

5. Let the function $f(\lambda)$ be given, that should take the values (6.53) on the spectrum of the matrix A. Moreover, a polynomial $h(\lambda)$ may fulfill the conditions

$$\begin{aligned} h(\lambda_1) &= f(\lambda_1), \; \dots \; h^{(\nu_1-1)}(\lambda_1) = f^{(\nu_1-1)}(\lambda_1) \\ h(\lambda_2) &= f(\lambda_2), \; \dots \; h^{(\nu_2-1)}(\lambda_2) = f^{(\nu_2-1)}(\lambda_2) \\ &\vdots \qquad \dots \qquad \vdots \\ h(\lambda_q) &= f(\lambda_q), \; \dots \; h^{(\nu_q-1)}(\lambda_q) = f^{(\nu_q-1)}(\lambda_q)\,. \end{aligned}$$

If these relations are true, then the polynomial $h(\lambda)$ is said to take the same values as the function $f(\lambda)$ on the spectrum of the matrix A, and we write

$$h(\Lambda_A) = f(\Lambda_A) \tag{6.54}$$

for this fact. If we have a polynomial $h(\lambda)$ that satisfies Conditions (6.54), then the matrix

$$f(A) \stackrel{\triangle}{=} h(A)$$

is established by definition as the *value of the function* $f(\lambda)$ for $\lambda = A$. Using this definition, the value of the function of a matrix does not depend on the concrete choice of the polynomial $h(\lambda)$ satisfying (6.54).

6. In particular, if all components (6.50) of the matrix A are known, then for a function $f(\lambda)$ defined on the spectrum of this matrix, the formula

$$f(A) = \sum_{i=1}^{q} \left[f(\lambda_i) Z_{i1} + f'(\lambda_i) Z_{i2} + \ldots + f^{(\nu_i-1)}(\lambda_i) Z_{i,\nu_i} \right] \tag{6.55}$$

is valid. Consider a scalar polynomial $h_{k\ell}$, which takes the following value on the spectrum of the matrix A:

$$\begin{aligned} &h_{k\ell}(\lambda_i) = h'_{k\ell}(\lambda_i) = \ldots = h_{k\ell}^{(\nu_i-1)}(\lambda_i) = 0, \quad (i = 1, \ldots, k-1, k+1, \ldots, q), \\ &h_{k\ell}(\lambda_k) = h'_{k\ell}(\lambda_k) = \ldots = h_{k\ell}^{(\ell-2)}(\lambda_k) = 0, \quad h_{k\ell}^{(\ell-1)}(\lambda_k) = 1\,, \\ &h_{k\ell}^{(\ell)}(\lambda_k) = \ldots = h_{k\ell}^{(\nu_k-1)}(\lambda_k) = 0\,. \end{aligned}$$

In this case, from (6.55) arise

$$h_{k\ell}(A) = Z_{k\ell}\,.$$

Thus we recognise that each component of the matrix A turns out to be a polynomial of this matrix. Applying Relation (6.50), we produce from (6.55) a representation of the matrix $f(A)$ by the coefficients of the partial fraction expansion (6.49)

$$f(A) = \sum_{i=1}^{q} \left[f(\lambda_i) M_{i1} + \frac{f'(\lambda_i)}{1!} M_{i2} + \ldots + \frac{f^{(\nu_i-1)}(\lambda_i)}{(\nu_i-1)!} M_{i,\nu_i} \right]. \tag{6.56}$$

7. The representation of the function of a matrix in the shape (6.55), (6.56) remains its sense, also in those cases, when the function $f(\lambda)$ is given by a sum of an (infinite) series, which converges on the spectrum of the matrix A. Suppose

$$f(\lambda) = \sum_{k=-\infty}^{\infty} u_k(\lambda) .$$

Then, if

$$f(\Lambda_A) = \sum_{k=-\infty}^{\infty} u_k(\Lambda_A)$$

is true, so

$$f(A) = \sum_{k=-\infty}^{\infty} u_k(A)$$

is established.

8. Functions of one and the same matrix are always commutative, *i.e.* if the functions $f_1(\lambda)$ and $f_2(\lambda)$ are defined on the spectrum of the matrix A, then

$$f_1(A)f_2(A) = f_2(A)f_1(A) .$$

9. In a number of problems, connected functions of matrices are considered. For example, let us have

$$f(\lambda) = F[f_1(\lambda), \dots, f_n(\lambda)] , \tag{6.57}$$

where the $f_i(\lambda)$ are known scalar functions. Now, if the functions $f(\lambda)$ and $f_i(\lambda)$, $(i = 1, \dots, n)$ are defined on the spectrum of A, then we obtain

$$f(A) = F[f_1(A), \dots, f_n(A)] .$$

Notice that the connected function (6.57) could be defined on the spectrum of A, even if some of the functions $f_i(\lambda)$, $(i = 1, \dots, n)$ are not defined on the spectrum of A.

Example 6.6. Consider the scalar function

$$f(\lambda) = \frac{\sin\lambda}{\lambda} ,$$

which might be written in the form

$$f(\lambda) = f_1(\lambda)f_2(\lambda), \qquad f_1(\lambda) = \sin\lambda, \quad f_2(\lambda) = \lambda^{-1} .$$

The function $f_2(\lambda)$ is only defined on the spectrum of non-singular matrices. Nevertheless, the function $f(\lambda)$ is an integral function, and thus defined on the spectrum of arbitrary matrices. Hence the matrix

$$f(A) = (\sin A)\, A^{-1} = A^{-1} \sin A$$

is defined for any matrix A and it could be constructed by (6.55) or (6.56). □

10. Let the $p\times p$ matrix A possess the eigenvalues $\lambda_1,\dots,\lambda_q$ with the multiplicities $\mu_1,\dots,\mu_q$, and the characteristic polynomial of the matrix A should have the shape

$$d_A(\lambda) = (\lambda-\lambda_1)^{\mu_1}\cdots(\lambda-\lambda_q)^{\mu_q}\,. \tag{6.58}$$

If under these conditions, the function $f(\lambda)$ is defined on the spectrum of the matrix A, then the characteristic polynomial of the matrix $f(A)$ has the form

$$d_f(\lambda) = (\lambda-f(\lambda_1))^{\mu_1}\cdots(\lambda-f(\lambda_q))^{\mu_q}\,.$$

Besides, if $f(\lambda_i)\neq 0$, $(i=1,\dots,q)$ is valid, then the matrix $f(A)$ only nonzero eigenvalues, *i.e.* $f(A)$ is non-singular. However, if $f(\lambda_i)=0$ for any i takes place, then the matrix $f(A)$ is singular.

11. Now, the important question about the structure of the set of elementary divisors and the invariant polynomials of the matrix $f(A)$ is investigated. Let the matrix A_λ have the elementary divisors

$$(\lambda-\lambda_1)^{\phi_1},\ \dots,\ (\lambda-\lambda_r)^{\phi_r}\,, \tag{6.59}$$

where among the numbers $\lambda_1,\dots,\lambda_r$, equal ones are allowed. Then the following assertion is true [51]: In those cases, where $\phi_i=1$, or $\phi_i>1$ and $f'(\lambda_i)\neq 0$, the elementary divisor $(\lambda-\lambda_i)^{\phi_i}$ of the matrix A corresponds to an elementary divisor

$$(\lambda-f(\zeta_i))^{\phi_i}$$

of the matrix $f(A)$. In case of $f'(\lambda_i)=0$, $\phi_i>1$ for the elementary divisor $(\lambda-\lambda_i)^{\phi_i}$, there exist more than one elementary divisors of the matrix $f(A)$.

12. Suppose again that the characteristic polynomial of the matrix A has the shape (6.58) and the sequence of its elementary divisors has the shape (6.59). It is said that the matrices A and $f(A)$ have the *same structure*, if among the numbers $f(\lambda_1),\dots,f(\lambda_q)$ are no equal ones and the sequence of elementary divisors of the matrix $f(A)$ possesses the analogue form to (6.59)

$$(\lambda-f(\lambda_1))^{\phi_1},\dots,(\lambda-f(\lambda_r))^{\phi_r}\,.$$

The above derived results are formulated in the next theorem.

Theorem 6.7. *The following two conditions are necessary and sufficient for the matrices A and f(A) to possess the same structure:*

a)

$$f(\lambda_i)\neq f(\lambda_k),\quad (i\neq k;\ i,k=1,\dots,q) \tag{6.60}$$

b) For all exponents ϕ_{i_ℓ} $(\ell=1,\dots,\rho<r)$ *with* $\phi_{i_\ell}>1$

$$f'(\lambda_{i_\ell})\neq 0,\quad (\ell=1,\dots,\rho)\,.$$

∎

Corollary 6.8. *Let the matrix A be cyclic, i.e. in (6.59)* $r=q$ *and* $\phi_i=\mu_i$ *are true. Then the matrix f(A) is also cyclic, if and only if Conditions a) and b) are true.* ∎

6.4 Matrix Exponential Function

1. Consider the scalar function

$$f(\lambda) = \mathrm{e}^{\lambda t},$$

where t is a real parameter. This function is defined on the spectrum of every matrix. Thus for any matrix A, formula (6.56) is applicable and we obtain

$$f(A) = \sum_{i=1}^{q} \left[\mathrm{e}^{\lambda_i t} M_{i1} + \frac{t\mathrm{e}^{\lambda_i t}}{1!} M_{i2} + \ldots + \frac{t^{(\nu_i - 1)}\mathrm{e}^{\lambda_i t}}{(\nu_i - 1)!} M_{i,\nu_i} \right], \tag{6.61}$$

where

$$f(\lambda_i) = \mathrm{e}^{\lambda_i t}, \quad f'(\lambda_i) = t\mathrm{e}^{\lambda_i t}, \quad \ldots, \quad f^{(\nu_i - 1)}(\lambda_i) = t^{\nu_i - 1}\mathrm{e}^{\lambda_i t}.$$

Matrix (6.61) is named the *exponential function* of the matrix A and it is denoted by

$$f(A) = \mathrm{e}^{At}.$$

2. A further important representation of the matrix e^{At} is obtained by applying the series expansion

$$\mathrm{e}^{\lambda t} = 1 + \lambda t + \frac{\lambda^2 t^2}{2!} + \ldots, \tag{6.62}$$

which converges for all λ, and consequently on any spectrum too. Inserting the matrix A instead of λ into (6.62), we receive

$$\mathrm{e}^{At} = I_p + At + \frac{A^2 t^2}{2!} + \ldots . \tag{6.63}$$

Particularly for $t = 0$, we get

$$\mathrm{e}^{At}\,|_{t=0} = I_p .$$

3. Differentiating (6.63) by t, we obtain

$$\frac{\mathrm{d}}{\mathrm{d}t}\left(\mathrm{e}^{At}\right) = A\left(I_p + At + \frac{A^2 t^2}{2!} + \ldots\right) = A\mathrm{e}^{At}.$$

4. Substituting the parameter $-\tau$ for t in (6.62), we receive

$$\mathrm{e}^{-A\tau} = I_p - A\tau + \frac{A^2 \tau^2}{2!} - + \ldots .$$

By multiplying this expansion with (6.62), we prove

$$\mathrm{e}^{At}\mathrm{e}^{-A\tau} = \mathrm{e}^{-A\tau}\mathrm{e}^{At} = I_p + A(t-\tau) + \frac{A^2(t-\tau)^2}{2!} + \ldots$$

hence

$$\mathrm{e}^{At}\mathrm{e}^{-A\tau} = \mathrm{e}^{-A\tau}\mathrm{e}^{At} = \mathrm{e}^{A(t-\tau)} .$$

For $\tau = t$, we find immediately

$$\mathrm{e}^{At}\mathrm{e}^{-At} = \mathrm{e}^{-At}\mathrm{e}^{At} = I_p$$

or

$$\left(\mathrm{e}^{At}\right)^{-1} = \mathrm{e}^{-At} .$$

5.

Theorem 6.9. *For a positive constant T, the matrices A and e^{AT} possess the same structure, if the eigenvalues $\lambda_1, \ldots, \lambda_q$ of A satisfy the conditions*

$$\mathrm{e}^{\lambda_i T} \neq \mathrm{e}^{\lambda_k T}, \quad (i \neq k;\ i,k = 1,\ldots,q) \tag{6.64}$$

or equivalently

$$\lambda_i - \lambda_k \neq \frac{2n\pi\mathrm{j}}{T} = n\mathrm{j}\omega\,, \quad (i \neq k;\ i,k = 1,\ldots,q)\,, \tag{6.65}$$

where n is an arbitrary integer and $\omega = 2\pi/T$.

Proof. Owing to $\mathrm{d}(\mathrm{e}^{\lambda T})/\,\mathrm{d}T = T\mathrm{e}^{\lambda T} \neq 0$ for an exponential function, Condition b) in Theorem 6.7 is always ensured. Therefore, the matrices A and e^{AT} have the same structure, if and only if Conditions (6.60) hold, which in the present case have the shape (6.64). Conditions (6.65) are obviously implications of (6.64). ∎

Corollary 6.10. *Let the matrix A be cyclic. Then, a necessary and sufficient condition for the matrix e^{AT} to become cyclic is the demand that Conditions (6.64) hold.* ∎

6. The next theorem is fundamental for the future declarations.

Theorem 6.11 ([71]). *Let the pair (A, B) controllable and the pair $[A, C]$ observable. Then under Conditions (6.64), (6.65), the pair (e^{AT}, B) is controllable and the pair $[\mathrm{e}^{AT}, C]$ is observable.* ∎

7. Since for any λ, t always $\mathrm{e}^{\lambda t} \neq 0$, the matrix e^{At} becomes non-singular for any finite t and any matrix A.

6.5 DPFR and DLT of Rational Matrices

1. Let $w(s) \in \mathbb{R}_{nm}(s)$ be a strictly proper rational matrix and

$$\varphi_w(T,s,t) = \frac{1}{T} \sum_{k=-\infty}^{\infty} w(s+kj\omega)e^{kj\omega t}, \quad \omega = \frac{2\pi}{T} \tag{6.66}$$

be its displaced pulse frequency response. As the *discrete Laplace transform (DLT)* of the matrix $w(s)$, we understand the sum of the series

$$\widehat{D}_w(T,s,t) = \frac{1}{T} \sum_{k=-\infty}^{\infty} w(s+kj\omega)e^{(s+kj\omega)t}, \quad -\infty < t < \infty . \tag{6.67}$$

The transforms (6.66) and (6.67) are closely connected:

$$\widehat{D}_w(T,s,t) = e^{st}\varphi_w(T,s,t) . \tag{6.68}$$

Per definition, we have

$$\varphi_w(T,s,t) = \varphi_w(T,s,t+T) .$$

In this section, we will derive closed formulae for the sums of the series (6.66) and (6.67).

2.

Lemma 6.12. *Let the matrix $w(s)$ be strictly proper and possess the partial fraction expansion*

$$w(s) = \sum_{i=1}^{q} \sum_{k=1}^{\mu_i} \frac{w_{ik}}{(s-s_i)^k} , \tag{6.69}$$

where the w_{ik} are constant matrices. Then the sum of the series (6.67) is determined by the formulae

$$\widehat{D}_w(T,s,t) = \widehat{\mathcal{D}}_w(T,s,t), \quad 0 < t < T , \tag{6.70}$$

$$\widehat{D}_w(T,s,t) = \widehat{\mathcal{D}}_w(T,s,t-\ell T)e^{\ell sT}, \quad \ell T < t < (\ell+1)T, \quad (\ell = 0, \pm 1, \dots), \tag{6.71}$$

where

$$\widehat{\mathcal{D}}_w(T,s,t) = \sum_{i=1}^{q} \sum_{k=1}^{\mu_i} \frac{w_{ik}}{(k-1)!} \left[\frac{\partial^{k-1}}{\partial \lambda^{k-1}} \frac{e^{\lambda t}}{1-e^{(\lambda-s)T}} \right]_{|\lambda=s_i} . \tag{6.72}$$

Proof. Substituting (6.69) into (6.67), we gain

$$\widehat{D}_w(T,s,t) = \sum_{i=1}^{q}\sum_{k=1}^{\mu_i} w_{ik}\left[\frac{1}{T}\sum_{m=-\infty}^{\infty}\frac{\mathrm{e}^{(s+mj\omega)t}}{(s+mj\omega-s_i)^k}\right]. \tag{6.73}$$

Appendix B yields

$$\frac{1}{T}\sum_{m=-\infty}^{\infty}\frac{\mathrm{e}^{(s+mj\omega)t}}{(s+mj\omega-a)^k} = \frac{1}{(k-1)!}\left[\frac{\partial^{k-1}}{\partial\lambda^{k-1}}\frac{\mathrm{e}^{\lambda t}}{1-\mathrm{e}^{(\lambda-s)T}}\right]_{|\lambda=a}.$$

Inserting this into (6.73), we obtain Formulae (6.70) and (6.72). We recognise Formula (6.71) as follows. Let $\ell T < t < (\ell+1)T$, so we conclude

$$\begin{aligned}\widehat{D}_w(T,s,t) &= \varphi_w(T,s,t)\mathrm{e}^{st} = \varphi_w(T,s,t-\ell T+\ell T)\mathrm{e}^{s(t-\ell T)}\mathrm{e}^{\ell sT}\\ &= \left[\varphi_w(T,s,t-\ell T)\mathrm{e}^{s(t-\ell T)}\right]\mathrm{e}^{\ell sT} = \widehat{\mathcal{D}}_w(T,s,t-\ell T)\mathrm{e}^{\ell sT}.\end{aligned}$$

∎

3.

Lemma 6.13. *Let A be a constant $p\times p$ matrix and*

$$w_A(s) = (sI_p - A)^{-1}. \tag{6.74}$$

Then, the following formulae hold:

$$\begin{aligned}\widehat{D}_{w_A}(T,s,t) &= \widehat{\mathcal{D}}_{w_A}(T,s,t)\\ &= \left(I_p-\mathrm{e}^{-sT}\mathrm{e}^{AT}\right)^{-1}\mathrm{e}^{At} = \mathrm{e}^{At}\left(I_p-\mathrm{e}^{-sT}\mathrm{e}^{AT}\right)^{-1}, \quad 0<t<T,\end{aligned} \tag{6.75}$$

$$\begin{aligned}\widehat{D}_{w_A}(T,s,t) &= \left(I_p-\mathrm{e}^{-sT}\mathrm{e}^{AT}\right)^{-1}\mathrm{e}^{A(t-\ell T)}\mathrm{e}^{\ell sT}\\ &= \mathrm{e}^{A(t-\ell T)}\mathrm{e}^{\ell sT}\left(I_p-\mathrm{e}^{-sT}\mathrm{e}^{AT}\right)^{-1}, \quad \ell T<t<(\ell+1)T.\end{aligned} \tag{6.76}$$

Proof. Consider the partial fraction expansion of the form (6.49)

$$w_A(s) = (\lambda I_p - A)^{-1} = \sum_{i=1}^{q}\sum_{k=1}^{\nu_i}\frac{M_{ik}}{(\lambda-\lambda_i)^k}. \tag{6.77}$$

Here, due to (6.50)

$$M_{ik} = (k-1)!\,Z_{ik}\,, \tag{6.78}$$

where Z_{ik} are the components of the matrix A. Substituting

$$w_{ik} = (k-1)!\,Z_{ik}$$

into (6.72) gives

$$\widehat{\mathcal{D}}_{w_A}(T,s,t) = \sum_{i=1}^{q}\sum_{k=1}^{\nu_i} Z_{ik}\left[\frac{\partial^{k-1}}{\partial\lambda^{k-1}}\frac{\mathrm{e}^{\lambda t}}{1-\mathrm{e}^{(\lambda-s)T}}\right]_{|\lambda=s_i} . \tag{6.79}$$

Introduce the scalar function

$$\widehat{f}(\lambda,t) = \mathrm{e}^{\lambda t}\left(1-\mathrm{e}^{-\lambda T}\mathrm{e}^{-sT}\right)^{-1},$$

so Relation (6.79) can be presented in the form

$$\widehat{\mathcal{D}}_{w_A}(T,s,t) = \sum_{i=1}^{q}\sum_{k=1}^{\nu_i} Z_{ik}\left[\frac{\partial^{k-1}}{\partial\lambda^{k-1}}\widehat{f}(\lambda,t)\right]_{|\lambda=s_i} .$$

Comparing this with (6.55), we find

$$\widehat{\mathcal{D}}_{w_A}(T,s,t) = \widehat{f}(A,t),$$

which is equivalent to (6.75). Relation (6.76) follows immediately from (6.71) and (6.75). ∎

4. On basis of the stated facts, the next theorem is easily derived.

Theorem 6.14. *Let the matrix $w(s)$ have the standard realisation*

$$w(s) = C(sI_p - A)^{-1}B . \tag{6.80}$$

Then the following formulae take place:

$$\begin{aligned}\widehat{D}_w(T,s,t) = \widehat{\mathcal{D}}_w(T,s,t) &= C\left(I_p - \mathrm{e}^{-sT}\mathrm{e}^{AT}\right)^{-1}\mathrm{e}^{At}B\\ &= C\mathrm{e}^{At}\left(I_p - \mathrm{e}^{-sT}\mathrm{e}^{AT}\right)^{-1}B, \qquad 0<t<T,\end{aligned} \tag{6.81}$$

$$\begin{aligned}\widehat{D}_w(T,s,t) &= \mathrm{e}^{\ell sT}C\left(I_p - \mathrm{e}^{-sT}\mathrm{e}^{AT}\right)^{-1}\mathrm{e}^{A(t-\ell T)}B,\\ &\qquad\qquad \ell T < t < (\ell+1)T .\end{aligned} \tag{6.82}$$

Proof. Insert (6.80) into (6.67) and obtain

$$\begin{aligned}\widehat{D}_w(T,s,t) &= C\left\{\frac{1}{T}\sum_{k=-\infty}^{\infty}\left[(s+kj\omega)I_p - A\right]^{-1}\mathrm{e}^{(s+kj\omega)t}\right\}B\\ &= C\widehat{D}_{w_A}(T,s,t)B ,\end{aligned}$$

so Relations (6.81), (6.82) directly emerge from (6.75), (6.76). ∎

Corollary 6.15. *Since the left sides of Relations (6.81) and (6.82) only depend on the transfer matrix $w(s)$, also the right sides of Formulae (6.81), (6.82) does not depend on the concrete choice of the realisation (A,B,C) configured by (6.80). Therefore, we are able to state that for two realisations (A,B,C) and (A_1,B_1,C_1), which define one and the same transfer matrix, the equality*

$$C\left(I_p - \mathrm{e}^{-sT}\mathrm{e}^{AT}\right)^{-1}\mathrm{e}^{At}B = C_1\left(I_q - \mathrm{e}^{-sT}\mathrm{e}^{A_1T}\right)^{-1}\mathrm{e}^{A_1t}B_1$$

holds. ∎

5. From the above formulae by using the relation

$$\varphi_w(T,s,t) = \widehat{D}_w(T,s,t)\mathrm{e}^{-st}\,,$$

we obtain closed formulae for the DPFR $\varphi_w(T,s,t)$.

6.6 DPFR and DLT for Modulated Processes

1. In the present section, the properties of the DPFR (6.45)

$$\varphi_G(T,s,t) = \varphi_{w\mu}(T,s,t) = \frac{1}{T}\sum_{k=-\infty}^{\infty} w(s+kj\omega)\mu(s+kj\omega)\mathrm{e}^{kj\omega t} \tag{6.83}$$

as well as of the DLT

$$\widehat{D}_G(T,s,t) = \widehat{D}_{w\mu}(T,s,t) = \frac{1}{T}\sum_{k=-\infty}^{\infty} w(s+kj\omega)\mu(s+kj\omega)\mathrm{e}^{(s+kj\omega)t} \tag{6.84}$$

will be investigated, which are connected by the simple relation

$$\widehat{D}_{w\mu}(T,s,t) = \varphi_{w\mu}(T,s,t)\mathrm{e}^{st}\,.$$

The properties of the DPFR imply

$$\varphi_{w\mu}(T,s,t) = \varphi_{w\mu}(T,s,t+T)\,.$$

Now, closed expressions for the sums of the series (6.83), (6.84) will be derived.

2.

Lemma 6.16. *Suppose the strictly proper matrix $w(s) \in \mathbb{R}_{nm}(s)$ of the shape (6.69). Then, the sum of the series (6.84) converges for all $s \neq s_i + kj\omega$, ($i = 1,\dots,q$; $k = 0,\pm 1,\dots$), the sum depends continuously on t and is determined by the formula [148]*

$$\widehat{D}_{w\mu}(T,s,t) = \widehat{\mathcal{D}}_{w\mu}(T,s,t), \quad 0 \le t \le T\,, \tag{6.85}$$

where the matrix $\widehat{\mathcal{D}}_{w\mu}(T,s,t)$ is bound by any one of the both equivalent relations

$$\widehat{\mathcal{D}}_{w\mu}(T,s,t) = \sum_{i=1}^{q}\sum_{k=1}^{\mu_i} \frac{w_{ik}}{(k-1)!}\left[\frac{\partial^{k-1}}{\partial\lambda^{k-1}}\,\frac{\mu(\lambda)\mathrm{e}^{\lambda t}}{\mathrm{e}^{(s-\lambda)T}-1}\right]_{|\lambda=s_i} + h_{w\mu}(t)\,, \tag{6.86}$$

$$\widehat{\mathcal{D}}_{w\mu}(T,s,t) = \sum_{i=1}^{q}\sum_{k=1}^{\mu_i} \frac{w_{ik}}{(k-1)!}\left[\frac{\partial^{k-1}}{\partial\lambda^{k-1}}\,\frac{\mu(\lambda)\mathrm{e}^{\lambda t}}{1-\mathrm{e}^{(\lambda-s)T}}\right]_{|\lambda=s_i} + h^{*}_{w\mu}(t)\,, \tag{6.87}$$

where

$$h_{w\mu}(t) = \int_0^t h_w(t-\tau)m(\tau)\,\mathrm{d}\tau, \quad h^*_{w\mu}(t) = -\int_t^T h_w(t-\tau)m(\tau)\,\mathrm{d}\tau$$

and

$$h_w(t) = \sum_{i=1}^{q}\sum_{k=1}^{\mu_i} \frac{w_{ik}}{(k-1)!}\left[\frac{\partial^{k-1}}{\partial\lambda^{k-1}}\,\mathrm{e}^{\lambda t}\right]_{|\lambda=s_i}. \tag{6.88}$$

Formulae (6.86), (6.87) is extended onto the whole $t-$axis by the relation

$$\widehat{\mathcal{D}}_{w\mu}(T,s,t) = \widehat{\mathcal{D}}_{w\mu}(T,s,t-\ell T)\mathrm{e}^{\ell sT}, \quad \ell T < t < (\ell+1)T\,.$$

Proof. Placing (6.69) into (6.84) gives

$$\widehat{D}_{w\mu}(T,s,t) = \sum_{i=1}^{q}\sum_{k=1}^{\mu_i} w_{ik}\left[\frac{1}{T}\sum_{m=-\infty}^{\infty}\frac{\mu(s+m\mathrm{j}\omega)\mathrm{e}^{(s+m\mathrm{j}\omega)t}}{(s+m\mathrm{j}\omega-s_i)^k}\right].$$

From Appendix B, it follows

$$\begin{aligned}&\frac{1}{T}\sum_{m=-\infty}^{\infty}\frac{\mu(s+m\mathrm{j}\omega)\mathrm{e}^{(s+m\mathrm{j}\omega)t}}{(s+m\mathrm{j}\omega-a)^k}\\ &\quad= \frac{1}{(k-1)!}\frac{\partial^{k-1}}{\partial\lambda^{k-1}}\left[\frac{\mu(\lambda)\mathrm{e}^{\lambda t}}{\mathrm{e}^{(s-\lambda)T}-1}\right]_{|\lambda=a} + \frac{1}{(k-1)!}\frac{\partial^{k-1}}{\partial\lambda^{k-1}}\left[\mathrm{e}^{\lambda t}\right]_{|\lambda=a},\end{aligned}$$

which results in Formulae (6.85), (6.86). Relation (6.87) is shown in the same way. The continuity with respect to t is a consequence of the properties of the above series. ∎

3. In the particular case, when $w(s) = w_A(s)$, where the matrix $w_A(s)$ is from (6.74), the following relations hold.

Lemma 6.17. *In case of (6.74), Formulae (6.86), (6.87) can be represented by the both equivalent formulae*

$$\widehat{\mathcal{D}}_{w_A\mu}(T,s,t) = h_\mu(A,t) + \mathrm{e}^{At}\mu(A)\left(\mathrm{e}^{sT}\mathrm{e}^{-AT} - I_p\right)^{-1}, \tag{6.89}$$

$$\widehat{\mathcal{D}}_{w_A\mu}(T,s,t) = h^*_\mu(A,t) + \mathrm{e}^{At}\mu(A)\left(I_p - \mathrm{e}^{-sT}\mathrm{e}^{AT}\right)^{-1}, \tag{6.90}$$

where the notations

$$h_\mu(t) = \int_0^t \mathrm{e}^{A(t-\tau)}m(\tau)\,\mathrm{d}\tau, \quad h^*_\mu(t) = -\int_t^T \mathrm{e}^{A(t-\tau)}m(\tau)\,\mathrm{d}\tau$$

and

$$\mu(A) = \int_0^T \mathrm{e}^{-A\tau}m(\tau)\,\mathrm{d}\tau$$

were used.

Proof. At first, we obtain from (6.77), (6.78) and (6.88)

$$h_{w_A}(t) = \sum_{i=1}^{q} \sum_{k=1}^{\nu_i} Z_{ik} \left[\frac{\partial^{k-1}}{\partial \lambda^{k-1}} \mathrm{e}^{\lambda t} \right]_{|\lambda = s_i} \tag{6.91}$$

that with the aid of (6.55) supplies

$$h_{w_A}(t) = \mathrm{e}^{At} .$$

Using (6.91) and (6.77), Formulae (6.86) and (6.87) can be given the form

$$\widehat{\mathcal{D}}_{w_A \mu}(T, s, t) = \sum_{i=1}^{q} \sum_{k=1}^{\nu_i} Z_{ik} \frac{\partial^{k-1}}{\partial \lambda^{k-1}} \left[\frac{\mathrm{e}^{\lambda t} \mu(\lambda)}{\mathrm{e}^{(s-\lambda)T} - 1} + \int_0^t \mathrm{e}^{\lambda(t-\tau)} m(\tau) \, \mathrm{d}\tau \right]_{|\lambda = s_i} , \tag{6.92}$$

$$\widehat{\mathcal{D}}_{w_A \mu}(T, s, t) = \sum_{i=1}^{q} \sum_{k=1}^{\nu_i} Z_{ik} \frac{\partial^{k-1}}{\partial \lambda^{k-1}} \left[\frac{\mathrm{e}^{\lambda t} \mu(\lambda)}{1 - \mathrm{e}^{(\lambda - s)T}} - \int_t^T \mathrm{e}^{\lambda(t-\tau)} m(\tau) \, \mathrm{d}\tau \right]_{|\lambda = s_i} . \tag{6.93}$$

Explain the scalar function

$$\begin{aligned} f_\mu(\lambda, t) &= \frac{\mathrm{e}^{\lambda t} \mu(\lambda)}{\mathrm{e}^{(s-\lambda)T} - 1} + \int_0^t \mathrm{e}^{\lambda(t-\tau)} m(\tau) \, \mathrm{d}\tau \\ &= \frac{\mathrm{e}^{\lambda t} \mu(\lambda)}{1 - \mathrm{e}^{(\lambda - s)T}} - \int_t^T \mathrm{e}^{\lambda(t-\tau)} m(\tau) \, \mathrm{d}\tau , \end{aligned} \tag{6.94}$$

then Formulae (6.92) and (6.93) can be written as

$$\widehat{\mathcal{D}}_{w_A \mu}(T, s, t) = \sum_{i=1}^{q} \sum_{k=1}^{\nu_i} Z_{ik} \frac{\partial^{k-1}}{\partial \lambda^{k-1}} \left[f_\mu(\lambda, t) \right]_{|\lambda = s_i} .$$

According to (6.55), we win for this expression the compact form

$$\widehat{\mathcal{D}}_{w_A \mu}(T, s, t) = f_\mu(A, t) . \tag{6.95}$$

If in Formula (6.94) the argument λ is substituted by the matrix A, Formulae (6.89) and (6.90) are achieved. ∎

4. Consider Relations (6.89) and (6.90) in more detail for the important special case of a zero-order hold (6.27). Thus it appears that

$$\mu(s) = \mu_0(s) = \frac{1 - \mathrm{e}^{-sT}}{s} .$$

Owing to this equation and (6.27) from Formula (6.94), we gain the equivalent expressions

$$f_\mu(\lambda,t) = \frac{\mathrm{e}^{\lambda t}-1}{\lambda} + \frac{1-\mathrm{e}^{-\lambda T}}{\lambda}\,\frac{\mathrm{e}^{\lambda t}}{\mathrm{e}^{(s-\lambda)T}-1}\,,$$

$$f_\mu(\lambda,t) = \frac{\mathrm{e}^{\lambda(t-T)}-1}{\lambda} + \frac{1-\mathrm{e}^{-\lambda T}}{\lambda}\,\frac{\mathrm{e}^{\lambda t}}{1-\mathrm{e}^{(\lambda-s)T}}\,.$$

Passing in these equations to functions of matrices according to (6.95), we find in the present case

$$\widehat{\mathcal{D}}_{w_A\mu}(T,s,t) = \\ A^{-1}\left(\mathrm{e}^{At}-I_p\right) + A^{-1}\left(I_p-\mathrm{e}^{-AT}\right)^{-1}\mathrm{e}^{At}\left(\mathrm{e}^{sT}\mathrm{e}^{-AT}-I_p\right)^{-1}, \tag{6.96}$$

$$\widehat{\mathcal{D}}_{w_A\mu}(T,s,t) = \\ A^{-1}\left(\mathrm{e}^{A(t-T)}-I_p\right) + A^{-1}\left(I_p-\mathrm{e}^{-AT}\right)^{-1}\mathrm{e}^{At}\left(I_p-\mathrm{e}^{-sT}\mathrm{e}^{AT}\right)^{-1}. \tag{6.97}$$

Remark 6.18. At glance, Formulae (6.96) and (6.97) seem to be meaningful only for a regular matrix A, but this is a fallacy. Applying the recognitions from Section 6.3, it is easily shown that the scalar functions

$$h_{\mu_0}(\lambda,t) = \frac{\mathrm{e}^{\lambda t}-1}{\lambda}\,,\quad h^*_{\mu_0}(\lambda,t) = \frac{\mathrm{e}^{\lambda(t-T)}-1}{\lambda}\,,\quad \mu_0(\lambda) = \frac{1-\mathrm{e}^{-\lambda T}}{\lambda} \tag{6.98}$$

are integral functions in the argument λ. In particular for $\lambda = 0$

$$\begin{aligned} h_{\mu_0}(0,t) &= t, \quad & h^*_{\mu_0}(0,t) &= t-T, \quad & \mu_0(0) &= T\,,\\ h'_{\mu_0}(0,t) &= \frac{t^2}{2}\,, \quad & h^{*\prime}_{\mu_0}(0,t) &= \frac{(t-T)^2}{2}\,, \quad & \mu'_0(0) &= \frac{T^2}{2}\,, \;\ldots\,. \end{aligned} \tag{6.99}$$

Thus, the functions (6.96), (6.97) are defined for any matrices A, among them are singular ones too.

Example 6.19. Suppose

$$A = \begin{bmatrix} 1 & 0 \\ 2 & 0 \end{bmatrix}.$$

In this case, we have

$$\lambda I_2 - A = \begin{bmatrix} \lambda-1 & 0 \\ -2 & \lambda \end{bmatrix}$$

and $\det(\lambda I_2 - A) = \lambda(\lambda-1)$. Thus the eigenvalues of the matrix A are the numbers $\lambda_1 = 1$, $\lambda_2 = 0$. Furthermore, we obtain

$$(\lambda I_2 - A)^{-1} = \frac{\begin{bmatrix} \lambda & 0 \\ 2 & \lambda-1 \end{bmatrix}}{\lambda(\lambda-1)}\,.$$

The partial fraction expansion gives

$$(\lambda I_2 - A)^{-1} = \frac{\begin{bmatrix} 1 & 0 \\ 2 & 0 \end{bmatrix}}{\lambda - 1} + \frac{\begin{bmatrix} 0 & 0 \\ -2 & 1 \end{bmatrix}}{\lambda}$$

i.e., the components of A possess the form

$$Z_1 = \begin{bmatrix} 1 & 0 \\ 2 & 0 \end{bmatrix}, \quad Z_2 = \begin{bmatrix} 0 & 0 \\ -2 & 1 \end{bmatrix}.$$

For any function $f(\lambda)$ defined for $\lambda = 1$ and $\lambda = 0$, we obtain from (6.55)

$$f(A) = Z_1 f(\lambda_1) + Z_2 f(\lambda_2).$$

Particularly, from (6.98) and (6.99), we receive for the function $h_{\mu_0}(\lambda, t)$

$$h_{\mu_0}(1, t) = \mathrm{e}^t - 1, \quad h_{\mu_0}(0, t) = t$$

and hence

$$h_{\mu_0}(A, t) = \left(\mathrm{e}^t - 1\right) \begin{bmatrix} 1 & 0 \\ 2 & 0 \end{bmatrix} + t \begin{bmatrix} 0 & 0 \\ -2 & 1 \end{bmatrix} = \begin{bmatrix} \mathrm{e}^t - 1 & 0 \\ 2(\mathrm{e}^t - 1) - 2t & t \end{bmatrix}.$$

In the same way, we can convince ourself that

$$h^*_{\mu_0}(1, t) = \mathrm{e}^{t-T} - 1, \quad h^*_{\mu_0}(0, t) = t - T$$

and

$$h^*_{\mu_0}(A, t) = \begin{bmatrix} \mathrm{e}^{t-T} - 1 & 0 \\ 2(\mathrm{e}^{t-T} - 1) - 2(t - T) & t - T \end{bmatrix}.$$

Analogously, we find

$$\mu_0(A) = \begin{bmatrix} 1 - \mathrm{e}^{-T} & 0 \\ 2(1 - \mathrm{e}^{-T}) - 2T & T \end{bmatrix}.$$

□

5. The following results generalise the preceding considerations.

Theorem 6.20. *Suppose the matrix $w(s)$ in the standard form (6.80). Then the formula*

$$\widehat{D}_{w\mu}(T, s, t) = \widehat{\mathcal{D}}_{w\mu}(T, s, t) = C\widehat{\mathcal{D}}_{w_A\mu}(T, s, t)B, \quad 0 \le t \le T \qquad (6.100)$$

is valid, where the matrix $\widehat{\mathcal{D}}_{w_A\mu}(T, s, t)$ is determined by Formulae (6.89), (6.90). This formula is extended onto the whole time axis by

$$\widehat{D}_{w\mu}(T, s, t) = C\widehat{\mathcal{D}}_{w_A\mu}(T, s, t - \ell T)B\mathrm{e}^{\ell sT}, \quad \ell T < t < (\ell + 1)T. \qquad (6.101)$$

■

Corollary 6.21. *Let (A, B, C) and $(\tilde{A}, \tilde{B}, \tilde{C})$ be any realisations of the matrix $w(s)$, then*

$$C\widehat{D}_{w_A\mu}(T, s, t)B = \tilde{C}\widehat{D}_{w_{\tilde{A}}\mu}(T, s, t)\tilde{B}.$$

■

6. It is shown that a series of the form (6.84) is also convergent in the case, when $w(s)$ is only proper. Indeed, in this case we write

$$w(s) = w_1(s) + w_0 \,, \tag{6.102}$$

where the matrix $w_1(s)$ is strictly proper and

$$w_0 = \lim_{s\to\infty} w(s)\,.$$

When (6.102) holds, then (6.84) implies

$$\widehat{D}_{w\mu}(T,s,t) = \widehat{D}_{w_1\mu}(T,s,t) + w_0 \frac{1}{T}\sum_{k=-\infty}^{\infty} \mu(s+kj\omega)\mathrm{e}^{(s+kj\omega)t}\,. \tag{6.103}$$

The first term on the right side of (6.103) can be calculated by Formulae (6.100) and (6.101). In order to calculate the second term, we observe that

$$\frac{1}{T}\sum_{k=-\infty}^{\infty} \mu(s+kj\omega)\mathrm{e}^{(s+kj\omega)t} = \varphi_\mu(T,s,t)\mathrm{e}^{st}\,,$$

where the function $\varphi_\mu(T,s,t)$ is defined by (6.36). Thus, we obtain

$$\frac{1}{T}\sum_{k=-\infty}^{\infty} \mu(s+kj\omega)\mathrm{e}^{(s+kj\omega)t} = \widehat{D}_\mu(T,s,t) = m(t-\ell T)\mathrm{e}^{\ell sT}, \quad \ell T < t < (\ell+1)T\,.$$

6.7 Parametric Discrete Models of Continuous Processes

1. Substituting

$$\mathrm{e}^{-sT} = \zeta \tag{6.104}$$

into (6.81) results in a rational matrix of the argument ζ :

$$\begin{aligned} \mathcal{D}_w(T,\zeta,t) &= \widehat{\mathcal{D}}_w(T,s,t)\mid_{\mathrm{e}^{-sT}=\zeta} \\ &= C(I_p - \zeta\mathrm{e}^{AT})^{-1}\mathrm{e}^{At}B\,, \quad 0 < t < T. \end{aligned} \tag{6.105}$$

Matrix (6.105) is called the *parametric discrete model* of the matrix (of the process) $w(s)$. A list of general properties of the matrix $\mathcal{D}_w(T,\zeta,t)$ will now be derived from Relation (6.105).

Since the matrix e^{AT} is regular, we conclude from (6.105) that the matrix $\mathcal{D}_w(T,\zeta,t)$ is strictly proper for all t. Besides, for $w(s) \in \mathbb{R}_{nm}(s)$ also $\mathcal{D}_w(T,\zeta,t) \in \mathbb{R}_{nm}(\zeta)$ is true.

Moreover, Corollary 6.15 ensures that the parametric discrete model (6.105) does not depend on the concrete choice of the realisation (A,B,C) configured in (6.80).

2. We introduce a new concept to prepare the next theorem.
Suppose the monic polynomial

$$r(s) = (s - s_1)^{\mu_1} \cdots (s - s_q)^{\mu_q}.$$

Then the monic polynomial

$$r^d(\zeta) = \left(\zeta - e^{-s_1 T}\right)^{\mu_1} \cdots \left(\zeta - e^{-s_q T}\right)^{\mu_q}$$

is called the *discretisation* of the polynomial $r(s)$.

Theorem 6.22. *Relation (6.80) should define a minimal standard realisation of the matrix $w(s)$, and the eigenvalues s_i, $(i = 1, \ldots, q)$ of the matrix A should satisfy the relations*

$$e^{s_i T} \neq e^{s_k T}, \quad (i \neq k;\ i, k = 1, \ldots, q) \tag{6.106}$$

which are called conditions for non-pathological behavior. *Then the PMD*

$$\tau_d(\zeta, t) = \left(I_p - \zeta e^{AT}, e^{At} B, C\right) \tag{6.107}$$

is minimal for any t.

Proof. Observing Theorem 6.11, we realise that under the conditions for non-pathological behavior (6.106), the minimality of the representation (6.80) ensures that the pair (e^{AT}, B) is controllable and the pair $[e^{AT}, C]$ is observable. Since the matrix e^{At} for all t is commutative with the matrix e^{AT}, and in addition $\det e^{At} \neq 0$ is true, Theorem 1.53 provides the pair $(e^{AT}, e^{At}B)$ to be controllable. As a consequence, the controllability of the pair $(e^{AT}, e^{At}B)$ and the observability of the pair $[e^{AT}, C]$ together with Lemma 5.34 ensure the irreducibility of the pairs $(I_p - \zeta e^{AT}, e^{At}B)$ and $[I_p - \zeta e^{AT}, C]$. But this means nothing else but the PMD (6.107) is minimal. ∎

3.

Theorem 6.23. *Let under the conditions of Theorem 6.22 the sequence of invariant polynomials $a_1(s), \ldots, a_p(s)$ of the matrix $sI_p - A$ have the form*

$$\begin{aligned} a_1(s) &= (s - s_1)^{\mu_{11}} \cdots (s - s_q)^{\mu_{1q}} \\ \vdots\ \ &\quad\ \vdots \qquad\qquad \vdots \\ a_p(s) &= (s - s_1)^{\mu_{p1}} \cdots (s - s_q)^{\mu_{pq}}. \end{aligned} \tag{6.108}$$

Then for the invariant polynomials $\alpha_1(\zeta), \ldots, \alpha_p(\zeta)$ of the matrix $I_p - \zeta e^{AT}$, the relations

$$\alpha_i(\zeta) = a_i^d(\zeta), \quad (i = 1, \ldots, q) \tag{6.109}$$

are true, that means, each invariant polynomial $\alpha_i(\zeta)$ is the discretisation of the corresponding invariant polynomial $a_i(s)$.

Proof. Due to the suppositions, the matrices A and e^{AT} have the same structure, Thus, if (6.106) and (6.107) are fulfilled, the sequence $\tilde{\alpha}_1(z), \ldots, \tilde{\alpha}_p(z)$ of the invariant polynomials of the matrix $zI_p - \mathrm{e}^{AT}$ has the shape

$$\begin{aligned}\tilde{\alpha}_1(z) &= \left(z - \mathrm{e}^{s_1 T}\right)^{\mu_{11}} \cdots \left(z - \mathrm{e}^{s_q T}\right)^{\mu_{1q}} \\ \vdots \quad & \quad \vdots \qquad\qquad \vdots \\ \tilde{\alpha}_p(z) &= \left(z - \mathrm{e}^{s_1 T}\right)^{\mu_{p1}} \cdots \left(z - \mathrm{e}^{s_q T}\right)^{\mu_{pq}} .\end{aligned}$$

But then owing to Lemma 5.36, the sequence of invariant polynomials of the matrix $I_p - \zeta \mathrm{e}^{AT}$ has the form

$$\begin{aligned}\alpha_1(\zeta) &= \left(\zeta - \mathrm{e}^{-s_1 T}\right)^{\mu_{11}} \cdots \left(\zeta - \mathrm{e}^{-s_q T}\right)^{\mu_{1q}} \\ \vdots \quad & \quad \vdots \qquad\qquad \vdots \\ \alpha_p(\zeta) &= \left(\zeta - \mathrm{e}^{-s_1 T}\right)^{\mu_{p1}} \cdots \left(\zeta - \mathrm{e}^{-s_q T}\right)^{\mu_{pq}} ,\end{aligned} \tag{6.110}$$

which is equivalent to (6.109). ∎

Corollary 6.24. *Let $\psi_w(s)$ and $\psi_{\mathcal{D}}(\zeta)$ be the McMillan denominators of the matrices $w(s)$ and $\mathcal{D}_w(T, \zeta, t)$, respectively. Then under the conditions of Theorem 6.22*

$$\psi_{\mathcal{D}}(\zeta) = \psi_w^d(\zeta) \tag{6.111}$$

and hence

$$\operatorname{Mdeg} w(s) = \operatorname{Mdeg} \mathcal{D}_w(T, \zeta, t) . \tag{6.112}$$

Proof. If (6.108) and (6.110) hold, we obtain

$$\begin{aligned}\psi_w(s) &= a_1(s) \cdots a_p(s) , \\ \psi_{\mathcal{D}}(\zeta) &= \alpha_1(\zeta) \cdots \alpha_p(\zeta)\end{aligned}$$

and Equations (6.111) and (6.112) result from (6.109). ∎

Corollary 6.25. *If under the conditions of Theorem 6.22, the matrix $w(s)$ is normal, then for any t the matrix $\mathcal{D}_w(T, \zeta, t)$ is also normal.*

Proof. If the matrix $w(s)$ is normal, then in any minimal standard representation (6.80) the matrix A is cyclic. Regarding to Corollary 6.10, also e^{AT} becomes cyclic. Besides, the minimality of the PMD (6.107) and Theorem 3.17 ensure that Matrix (6.105) is normal. ∎

4.

Theorem 6.26. *Let the strictly proper rational matrices $w(s)$ and $w_1(s)$ of sizes $n \times m$ and $m \times r$, respectively, be related by*

$$w_1(s) \underset{l}{\prec} w(s) \tag{6.113}$$

and the poles $s_1, \dots, s_q$ of the matrix $w(s)$ should satisfy the conditions for non-pathological behavior (6.106). Then

$$\mathcal{D}_{w_1}(T, \zeta, t) \underset{l}{\prec} \mathcal{D}_w(T, \zeta, t) . \tag{6.114}$$

Proof. Let us have the minimal standard realisation

$$w(s) = C(sI_p - A)^{-1}B .$$

Then Theorem 2.56 and (6.113) imply the existence of a representation of the form

$$w_1(s) = C(sI_p - A)^{-1}B_1 .$$

Passing to the parametric discrete-time models, we receive

$$\begin{aligned} \mathcal{D}_w(T, \zeta, t) &= C \left(I_p - \zeta e^{AT}\right)^{-1} e^{At} B , \\ \mathcal{D}_{w_1}(T, \zeta, t) &= C \left(I_p - \zeta e^{AT}\right)^{-1} e^{At} B_1 . \end{aligned}$$

Consider now the ILMFD

$$C \left(I_p - \zeta e^{AT}\right)^{-1} = a^{-1}(\zeta) b(\zeta) .$$

Then the minimality of the PMD (6.107) together with Lemma 2.9 ensures that the right side of the relation

$$\mathcal{D}_w(T, \zeta, t) = a^{-1}(\zeta) \left[b(\zeta) e^{At} B\right] \tag{6.115}$$

is an ILMFD of the matrix $\mathcal{D}_w(T, \zeta, t)$ for any t. Besides, the product

$$a(\zeta)\mathcal{D}_{w_1}(T, \zeta, t) = b(\zeta) e^{At} B_1$$

turns out to be a polynomial matrix, *i.e.* we obtain (6.114). ∎

Remark 6.27. As a consequence from (6.115) we find out that for a parametric discrete model $\mathcal{D}_w(T, \zeta, t)$, there always exists an ILMFD (6.115), in which the left divisor $a(\zeta)$ does not depend on t.

5.

Theorem 6.28. *Let the rational matrices* $F(s)$, $G(s)$ *of size* $n\times\rho$ *and* $\rho\times m$, *respectively, be given. Suppose the matrices* $F(s)$ *and*

$$L(s) = F(s)G(s)$$

be strictly proper and the poles of the matrices $F(s)$ *and* $G(s)$ *should satisfy together the conditions for non-pathological behavior (6.106). Moreover, let us have the ILMFD*

$$L(s) = a_0^{-1}(s)b_0(s), \qquad F(s) = a_1^{-1}(s)b_1(s),$$
$$G(s) = a_2^{-1}(s)b_2(s), \qquad b_1(s)a_2^{-1}(s) = a_3^{-1}(s)b_3(s)$$

and $a_{i\kappa}(s)$, $(i = 1,\ldots,n_\kappa;\ \kappa = 0,1,2)$ *should be the totality of invariant polynomials of the matrices* $a_0(s)$, $a_1(s)$, $a_2(s)$. *Then owing to the validity of*

$$F(s) \underset{l}{\prec} L(s) = F(s)G(s),$$

the following relations hold:

a)

$$\mathcal{D}_F(T,\zeta,t) \underset{l}{\prec} \mathcal{D}_L(T,\zeta,t) = \mathcal{D}_{FG}(T,\zeta,t). \tag{6.116}$$

b) If we have the ILMFDs

$$\mathcal{D}_{FG}(T,\zeta,t) = \alpha_0^{-1}(\zeta)\beta_0(\zeta,t), \quad \mathcal{D}_F(T,\zeta,t) = \alpha_1^{-1}(\zeta)\beta_1(\zeta,t), \tag{6.117}$$

then

$$\alpha_0(\zeta) = \alpha_2(\zeta)\alpha_1(\zeta), \tag{6.118}$$

where $\alpha_2(\zeta)$ *is an* $n\times n$ *polynomial matrix. Besides, if* $\alpha_{i\kappa}(\zeta)$, $(i = 1,\ldots,n;\ \kappa = 0,1)$ *are the sequences of invariant polynomials of the matrices* $\alpha_0(\zeta)$ *and* $\alpha_1(\zeta)$, *respectively, then*

$$\alpha_{i\kappa}(\zeta) = a_{i\kappa}^d(\zeta), \quad (i = 1,\ldots,n;\ \kappa = 0,1). \tag{6.119}$$

c) Denote

$$\Delta_i(s) = \det a_i(s), \quad (i = 0,1,2),$$

then the conditions

$$\det \alpha_i(\zeta) = \Delta_i^d(\zeta), \quad (i = 0,1,2) \tag{6.120}$$

are satisfied.

d) The following relation holds:

$$\alpha_0(\zeta)\mathcal{D}_F(T,\zeta,t) = \alpha_2(\zeta)\beta_1(\zeta,t). \tag{6.121}$$

e) If the matrices $L(s)$ *and* $F(s)$ *are normal, then the matrix* $\alpha_2(\zeta)$ *is simple.*

Proof. a) Relation (6.116) is a consequence of Theorem 6.26.

b) Relations (6.118) and (6.119) follow immediately from (6.116), the concept of subordination, and Theorems 6.23, 6.26.

c) Formulae (6.120) are consequences of Theorem 6.23.

d) Formula (6.121) results from (6.118), because we have the ILMFD (6.117).

e) If the matrices $L(s)$ and $F(s)$ are normal, then owing to Corollary 6.25, the matrices $\mathcal{D}_{FG}(T,\zeta,t)$ and $\mathcal{D}_F(T,\zeta,t)$ are normal. Thus according to Theorem 3.1, it follows that the matrices $\alpha_0(\zeta)$ and $\alpha_1(\zeta)$ are simple. Hence the matrix $\alpha_2(\zeta)$ is simple, because the product of two latent matrices becomes simple if and only if both factors are simple matrices. ∎

Remark 6.29. Recall that by transposition, analogue statements as in Theorems 6.26 and 6.28 can be formulated in case of subordination from right and according to IRMFD.

6.8 Parametric Discrete Models of Modulated Processes

1. Substituting the closed expressions (6.89), (6.90) into (6.100), we obtain the equivalent formulae

$$\widehat{\mathcal{D}}_{w\mu}(T,s,t) = C\left[h_\mu(A,t) + \mathrm{e}^{At}\mu(A)\left(\mathrm{e}^{sT}\mathrm{e}^{-AT} - I_p\right)^{-1}\right]B\,,$$

$$\widehat{\mathcal{D}}_{w\mu}(T,s,t) = C\left[h^*_\mu(A,t) + \mathrm{e}^{At}\mu(A)\left(I_p - \mathrm{e}^{-sT}\mathrm{e}^{AT}\right)^{-1}\right]B\,.$$

After replacing the complex variable in these relations by (6.104), we find the rational matrix

$$\widehat{\mathcal{D}}_{w\mu}(T,s,t)\mid_{\mathrm{e}^{-sT}=\zeta} = \mathcal{D}_{w\mu}(T,\zeta,t)\,,$$

which is determined by the expressions

$$\begin{aligned} \mathcal{D}_{w\mu}(T,\zeta,t) &= C\left[h_\mu(A,t) + \zeta\mathrm{e}^{A(t+T)}\mu(A)\left(I_p - \zeta\mathrm{e}^{AT}\right)^{-1}\right]B\,, \\ \mathcal{D}_{w\mu}(T,\zeta,t) &= C\left[h^*_\mu(A,t) + \mathrm{e}^{At}\mu(A)\left(I_p - \zeta\mathrm{e}^{AT}\right)^{-1}\right]B\,. \end{aligned} \tag{6.122}$$

The matrix $\mathcal{D}_{w\mu}(T,\zeta,t)$ is called the *parametric discrete model of the modulated process* $w(s)\mu(s)$. Clearly, $w(s) \in \mathbb{R}_{nm}(s)$ implies $\mathcal{D}_{w\mu}(T,\zeta,t) \in \mathbb{R}_{nm}(\zeta)$.

2. Starting from Relations (6.122), some important properties of the matrix $\mathcal{D}_{w\mu}(T,\zeta,t)$ will be established.

a) At first (6.122) provides that there exists the finite limit

$$\lim_{\zeta\to\infty} \mathcal{D}_{w\mu}(T,\zeta,t) = Ch^*_\mu(A,t)B\,.$$

Thus for all t, the matrix $\mathcal{D}_{w\mu}(T,\zeta,t)$ is at least proper.

b) As in Corollary 6.21, we recognise that the right sides of Formulae (6.122) do not depend on the concrete choice of the realisation (A, B, C) in the standard realisation (6.80).

3. Bring the first formula of (6.122) into the form

$$\mathcal{D}_{w\mu}(T, \zeta, t) = Ch_{\mu}(A, t)B + R(T, \zeta, t)\,,$$

where $R(T, \zeta, t)$ is the rational matrix

$$R(T, \zeta, t) = \zeta C\mathrm{e}^{A(t+T)}\mu(A)\left(I_p - \zeta\mathrm{e}^{AT}\right)^{-1}B\,.$$

The right side could be seen as the transfer matrix of the PMD

$$\tau_m(\zeta, t) = (I_p - \zeta\mathrm{e}^{AT}, \zeta\mathrm{e}^{A(t+T)}\mu(A)B, C)\,, \tag{6.123}$$

which is called the *parametric discrete model of the modulated process* $w(s)\mu(s)$.

Theorem 6.30. *Let the standard realisation (6.80) be minimal and the eigenvalues* $s_1, \dots, s_q$ *of the matrix* A *should satisfy the conditions for non-pathological behavior*

$$\mathrm{e}^{s_i T} \neq \mathrm{e}^{s_k T}, \quad (i \neq k;\ i, k = 1, \dots, q) \tag{6.124}$$

and

$$\mu(s_i) \neq 0, \quad (i = 1, \dots, q)\,. \tag{6.125}$$

Then the PMD (6.123) is minimal for any t.

Proof. With attention to Theorem 6.11, we realise that for satisfied Conditions (6.124), the pair (e^{AT}, B) is controllable and the pair $[\mathrm{e}^{AT}, C]$ is observable. Thus, the pair $(\mathrm{e}^{AT}, \mathrm{e}^{A(t+T)}\mu(A)B)$ is also controllable, which is a consequence of Theorem 1.53, because the matrices e^{AT} and $\mathrm{e}^{A(t+T)}\mu(A)$ are regular due to (6.125). Since the matrix e^{AT} is also non-singular, from the controllability of the pair $(\mathrm{e}^{AT}, \mathrm{e}^{A(t+T)}\mu(A)B)$ and the observability of the pair $[\mathrm{e}^{AT}, C]$, we derive that the pairs $(I_p - \zeta\mathrm{e}^{AT}, \zeta\mathrm{e}^{A(t+T)}B)$ and $[I_p - \zeta\mathrm{e}^{AT}, C]$ are irreducible. But this means, the PMD (6.123) is minimal. ∎

Corollary 6.31. *Under the conditions of Theorem 6.30 for the matrix* $\mathcal{D}_{w\mu}(T, \zeta, t)$, *there exists an ILMFD*

$$\mathcal{D}_{w\mu}(T, \zeta, t) = \alpha^{-1}(\zeta)\beta(\zeta, t)\,, \tag{6.126}$$

where the matrix $\alpha(\zeta)$ *depends neither on* t *nor on the function* $m(t)$.

Proof. Take the ILMFD

$$C(I_p - \zeta\mathrm{e}^{AT})^{-1} = \alpha^{-1}(\zeta)\beta_1(\zeta)\,.$$

Then owing to Lemma 2.9, the expression

$$\mathcal{D}_{w\mu}(T, \zeta, t) = \alpha^{-1}(\zeta)\left[\zeta\beta_1(\zeta)\mathrm{e}^{A(t+T)}\mu(A)B + \alpha(\zeta)Ch_{\mu}(A, t)B\right]$$

defines an ILMFD of the matrix $\mathcal{D}_{w\mu}(T, \zeta, t)$. ∎

4. Henceforth, the entirety of Relations (6.124) and (6.125) will be called the *strict conditions for non-pathological behavior*. It is shown that in case of a zero-order hold (6.27), where $m(t) = 1$, the strict conditions for non-pathological behavior (6.124), (6.125) coincide with the ordinary Conditions (6.124). Indeed, if (6.124) is true, then none of the eigenvalues $s_1, \dots, s_q$ will have the form $2\pi n\mathrm{j}/T$, where n is an integer different from zero. Thus for all i, $(1 \le i \le q)$ the relations

$$\mu_0(s_i) = \frac{1 - \mathrm{e}^{-s_i T}}{s_i} \neq 0$$

are valid, this means, Conditions (6.124) imply (6.125).

5. The properties of the matrix $\mathcal{D}_{w\mu}(T, \zeta, t)$ will be characterized by a number of theorems. The proofs of these theorems are repetitions of the proofs of the corresponding statements in Section 6.7, with the only distinction that we always have to replace the conditions for non-pathological behavior (6.124) by the strict conditions for non-pathological behavior (6.124), (6.125). Therefore, hereinafter all statements of this section will be given without proofs.

Theorem 6.32. *Let $\psi_w(s)$ and $\psi_{\mathcal{D}\mu}(\zeta)$ be the McMillan denominators of the matrices $w(s)$ and $\mathcal{D}_{w\mu}(T, \zeta, t)$, respectively. Then for fulfilled strict conditions for non-pathological behavior, the equation*

$$\psi_{\mathcal{D}\mu}(\zeta) = \psi_w^d(\zeta)$$

holds, and thus

$$\operatorname{Mdeg} w(s) = \operatorname{Mdeg} \mathcal{D}_{w\mu}(T, \zeta, t)\,.$$

∎

Theorem 6.33. *If Conditions (6.124), (6.125) are fulfilled and the standard realisation (6.80) is minimal, where the matrix A is cyclic, then the matrix $\mathcal{D}_{w\mu}(T, \zeta, t)$ is normal.* ∎

Theorem 6.34. *Let the matrices $w(s)$ and $w_1(s)$ of size $n \times m$ and $n \times r$, respectively, be at least proper and related by*

$$w_1(s) \underset{l}{\prec} w(s)\,.$$

Moreover, the poles $s_1, \dots, s_q$ of the matrix $w(s)$ satisfy the strict conditions for non-pathological behavior (6.125), (6.126). Then

$$\mathcal{D}_{w_1\mu}(T, \zeta, t) \underset{l}{\prec} \mathcal{D}_{w\mu}(T, \zeta, t)\,.$$

∎

Theorem 6.35. *Suppose the rational matrices $F(s)$, $G(s)$ of size $n \times \rho$, $\rho \times m$. Assume the matrices $F(s)$ and*

$$L(s) = F(s)G(s)$$

be at least proper, and the poles of the matrices $F(s)$ and $G(s)$ as a whole satisfy the strict conditions for non-pathological behavior (6.125), (6.126). Moreover, the ILMFD

$$L(s) = a_0^{-1}(s)b_0(s), \qquad F(s) = a_1^{-1}(s)b_1(s)\,,$$
$$G(s) = a_2^{-1}(s)b_2(s), \qquad b_1(s)a_2^{-1}(s) = a_3^{-1}(s)b_3(s)$$

should exist, and let $a_{i\lambda}(s)$, $(i = 1, \dots, n;\ \lambda = 0, 1, 2)$ be the totality of the invariant polynomials of the matrices $a_0(s)$, $a_1(s)$, $a_2(s)$. Then from the validity of

$$F(s) \underset{l}{\prec} L(s) = F(s)G(s)$$

the following relations emerge:

a)
$$\mathcal{D}_{F\mu}(T, \zeta, t) \underset{l}{\prec} \mathcal{D}_{FG\mu}(T, \zeta, t) = \mathcal{D}_{L\mu}(T, \zeta, t)\,.$$

b) Suppose the ILMFDs

$$\mathcal{D}_{L\mu}(T, \zeta, t) = \alpha_0^{-1}(\zeta)\beta_0(\zeta, t), \quad \mathcal{D}_{F\mu}(T, \zeta, t) = \alpha_1^{-1}(\zeta)\beta_1(\zeta, t)\,,$$

then

$$\alpha_0(\zeta) = \alpha_2(\zeta)\alpha_1(\zeta)\,,$$

where $\alpha_2(\zeta)$ is an $n \times n$ polynomial matrix. Besides, if $\alpha_{i\kappa}(\zeta)$, $(i = 1, \dots, n;\ \kappa = 0, 1)$ are the sequences of invariant polynomials of the matrices $\alpha_0(\zeta)$ and $\alpha_1(\zeta)$, then the relations

$$\alpha_{i\kappa}(\zeta) = a_{i\kappa}^d(\zeta), \quad (i = 1, \dots, n;\ \kappa = 0, 1)$$

take place.

c) Denote

$$g_\kappa(s) = \det a_\kappa(s), \quad (\kappa = 0, 1, 2)\,,$$

then the conditions

$$\det \alpha_\kappa(\zeta) = g_\kappa^d(\zeta), \quad (\kappa = 0, 1, 2)$$

are fulfilled.

d) The relation

$$\alpha_0(\zeta)\mathcal{D}_{F\mu}(T, \zeta, t) = \alpha_2(\zeta)\beta_1(\zeta, t)$$

consists, where the matrix $\beta_1(\zeta, t)$ is a polynomial in ζ for all t.

e) If the matrices $L(s)$ and $F(s)$ are normal, then the matrices $\mathcal{D}_{L\mu}(T, \zeta, t)$ and $\mathcal{D}_{F\mu}(T, \zeta, t)$ are also normal and the matrices $\alpha_0(\zeta)$, $\alpha_1(\zeta)$, $\alpha_2(\zeta)$ are simple. ■

Remark 6.36. Notice that by transposing, all above statements could be formulated in case of subordination from right and corresponding IRMFDs.

6.9 Reducibility of Parametric Discrete Models

1. As follows from Formulae (6.105) and (6.122), the above performed parametric discrete models of continuous-time and of modulated processes may be represented for $0 < t < T$ in the form

$$\mathcal{D}(\zeta, t) = \frac{N_{\mathcal{D}}(\zeta, t)}{\Delta(\zeta)}, \tag{6.127}$$

where $\Delta(\zeta)$ is a scalar polynomial and $N_{\mathcal{D}}(\zeta, t)$ is a polynomial matrix in ζ for any fixed t. A rational fraction (matrix) of the form (6.127) is called *irreducible*, when there does not exist a representation

$$\mathcal{D}(\zeta, t) = \frac{N_{\mathcal{D}1}(\zeta, t)}{\Delta_1(\zeta)}$$

with $\deg \Delta_1(\zeta) < \deg \Delta(\zeta)$. However, if this still becomes true, then the fraction (6.127) is named *reducible.*

2. Let $F(\zeta)$ be a matrix of the argument ζ. The already known replacement

$$\overset{\frown}{F}(s) = F(\zeta)\,|_{\zeta = \mathrm{e}^{-sT}}$$

will be investigated systematically. Obviously, the reverse relation holds:

$$F(\zeta) = \overset{\frown}{F}(s)\,|_{\mathrm{e}^{-sT} = \zeta} \; .$$

3. In particular, if the matrix $\mathcal{D}(\zeta, t)$ has the shape (6.127), then the matrix

$$\overset{\frown}{\mathcal{D}}(s, t) = \frac{\overset{\frown}{N}_{\mathcal{D}}(s, t)}{\overset{\frown}{\Delta}(s)}$$

is called *rational-periodic (rat.per.).* A rat.per. matrix is named (ir)reducible, if Matrix (6.127) is (ir)reducible.

4. Proceeding from the just introduced concepts, the question arise, wether the rat.per. matrix (6.81) is reducible for $0 < t < T$. With respect to

$$\overset{\frown}{D}_w(T, s, t) = \overset{\frown}{\mathcal{D}}_w(T, s, t) = C\left(I - \mathrm{e}^{-sT}\mathrm{e}^{AT}\right)^{-1}\mathrm{e}^{At}B\,, \tag{6.128}$$

the question on reducibility of Matrix (6.128) leads to the question on reducibility of the rational matrix

$$\mathcal{D}_w(T, \zeta, t) = C\left(I - \zeta\mathrm{e}^{AT}\right)^{-1}\mathrm{e}^{At}B\,, \tag{6.129}$$

which can be represented in the form (6.127) with

$$N_{\mathcal{D}}(\zeta, t) = C\,\mathrm{adj}\left(I - \zeta\mathrm{e}^{AT}\right)\mathrm{e}^{At}B\,,$$

$$\Delta(\zeta) = \det(I - \zeta\mathrm{e}^{AT})\,.$$

Theorem 6.37. *Let the realisation (A, B, C) be simple, i.e. the pair (A, B) is controllable, the pair $[A, C]$ is observable, and the matrix A is cyclic. Furthermore the conditions for non-pathological behavior (6.106) should be satisfied. Then Matrix (6.129) (and therefore also Matrix (6.128)) is irreducible.*

Proof. Owing to Theorem 6.22, the PMD

$$\tau_d(\zeta, t) = (I - \zeta \mathrm{e}^{AT}, \mathrm{e}^{At}B, C)$$

is minimal for any t. Besides, from Corollary 6.10 emerge that the matrices e^{AT} and e^{-AT} are cyclic. Thus, the matrix $I - \zeta \mathrm{e}^{AT} = -\mathrm{e}^{AT}(\zeta I - \mathrm{e}^{-AT})$ is simple and its minimal polynomial is equivalent to its characteristic polynomial. Hence by Theorem 2.47, Matrix (6.129) is irreducible for all t. This implies that the matrix

$$\widehat{\mathcal{D}}_w(T, s, t) = \frac{C \operatorname{adj}\left(I - \mathrm{e}^{-sT}\mathrm{e}^{AT}\right) \mathrm{e}^{At}B}{\det\left(I - \mathrm{e}^{-sT}\mathrm{e}^{AT}\right)}$$

is irreducible too. ∎

Remark 6.38. From the above proof, we read that under the conditions of Theorem 6.37 for any t, the right side of our last equation is irreducible.

Remark 6.39. If any one of the suppositions in Theorem 6.37 is violated, then we have reducibility in the above sense.

5. By analogy, we can handle the question on reducibility of the rat.per. matrix $\widehat{\mathcal{D}}_{w\mu}(T, s, t)$, which is written in the form

$$\widehat{\mathcal{D}}_{w\mu}(T, s, t) = \\ = \frac{C\mathrm{e}^{At}\mu(A) \operatorname{adj}\left(I - \mathrm{e}^{-sT}\mathrm{e}^{AT}\right) B + Ch_\mu^*(A, t)B \det(I - \mathrm{e}^{-sT}\mathrm{e}^{AT})}{\det(I - \mathrm{e}^{-sT}\mathrm{e}^{AT})}. \tag{6.130}$$

Here the next theorem takes place.

Theorem 6.40. *Let the pair (A, B) be controllable, the pair $[A, C]$ observable and the matrix A cyclic, and let the strict conditions for non-pathological behavior (6.124), (6.125) be fulfilled. Then, Matrix (6.130) is irreducible for any t (and also irreducible in the above sense).*

Proof. Same proof as for Theorem 6.37. ∎

Remark 6.41. If one of the conditions in Theorem 6.40 is violated, then Matrix (6.130) turns out to be reducible.

6. The introduced definitions of this section can be extended to wider classes of functions, which we will meet in further declarations.

Let the matrix $N_{\mathcal{D}}(\zeta, t)$ in (6.127) be an integral function of the argument ζ for any t. Then the fraction (6.127) is called *(ir)reducible*, if there does (not) exist a root ζ_0 of the polynomial $\Delta(\zeta)$ with

$$N_{\mathcal{D}}(\zeta_0, t) = O\,.$$

7

Mathematical Description, Stability and Stabilisation of the Standard Sampled-data System in Continuous Time

7.1 The Standard Sampled-data System

1. As a standard sampled-data system, we mean a system having the structure shown in Fig. 7.1. Here, $x = x(t)$, $y = y(t)$, $z = z(t)$, and $u = u(t)$ are

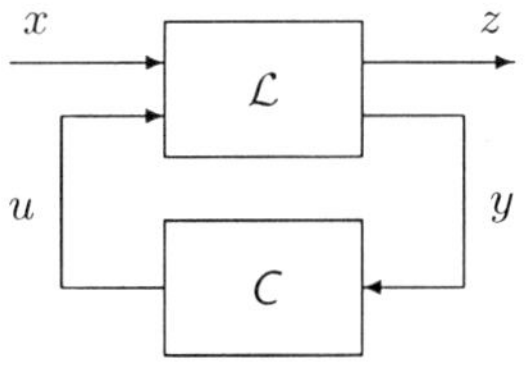

Fig. 7.1. Standard sampled-data system

column-vectors of dimensions $\ell \times 1$, $n \times 1$, $r \times 1$, and $m \times 1$, respectively. The block $\mathcal{L}$ is a generalised continuous-time LTI plant described by the following equations:

$$\begin{bmatrix} z \\ y \end{bmatrix} = w(p) \begin{bmatrix} x \\ u \end{bmatrix}, \tag{7.1}$$

where $p = \mathrm{d}/\mathrm{d}t$ and $w(p)$ is a real rational block transfer matrix of the form

$$w(p) = \begin{array}{c} \quad\ \ \ell \qquad\ \ m \\ \begin{bmatrix} K(p) & L(p) \\ M(p) & N(p) \end{bmatrix} \end{array} \begin{array}{c} \\ r \\ n \end{array}, \tag{7.2}$$

where the letters outside the matrix indicate the dimensions of the corresponding blocks. Henceforth, it is assumed that the matrix $N(p)$ is strictly proper and $L(p)$ is at least proper. The restrictions imposed on the matrices $K(p)$ and $M(p)$ will depend on the problem under consideration. Using (7.1) and (7.2), we can write the equations of the plant in the operator form

$$\begin{aligned} z &= K(p)x + L(p)u \\ y &= M(p)x + N(p)u\,. \end{aligned} \tag{7.3}$$

2. In addition to the continuous-time plant $\mathcal{L}$, the standard system contains a multivariable digital controller C described by the block-diagram shown in Fig. 7.2. According to Section 6.2, the equations of the ADC are

$$\xi_k = y(kT)\,, \tag{7.4}$$

where the input signal $y(t)$ is assumed to be continuous. The digital controller ALG can be described either by the forward model (6.21)

$$\tilde{\alpha}_0\psi_{k+\rho} + \ldots + \tilde{\alpha}_\rho\psi_k = \tilde{\beta}_0\xi_{k+\rho} + \ldots + \tilde{\beta}_\rho\xi_k \tag{7.5}$$

or by the associated backward model (6.22)

$$\alpha_0\psi_k + \ldots + \alpha_\rho\psi_{k-\rho} = \beta_0\xi_k + \ldots + \beta_\rho\xi_{k-\rho}\,, \tag{7.6}$$

where $\tilde{\alpha}_i$, $\tilde{\beta}_i$, α_i, and β_i are constant real matrices of compatible dimensions and $\det\alpha_0 \neq 0$. Moreover, the equations for the DAC have the form

$$u(t) = m(t-kT)\psi_k\,, \quad kT < t < (k+1)T. \tag{7.7}$$

Taken in the aggregate, Equations (7.3)–(7.7) form a system of equations, which hereinafter will be called the *operator model of the standard sampled-data system.* This model can be associated with the expanded structure shown in Fig. 7.2.

7.2 Equation Discretisation for the Standard SD System

1. Hereinafter, in addition to the assumptions taken in Section 7.1, we suppose that the matrices $K(p)$ and $M(p)$ are strictly proper. In this case the matrix $w(p)$ is at least proper and admits a multitude of standard realisations of the form

$$w(p) = C(pI_\chi - A)^{-1}B + D \tag{7.8}$$

with

$$D = \begin{bmatrix} O_{r\ell} & D_L \\ O_{n\ell} & O_{nm} \end{bmatrix}$$

and constant matrices A, B, C, and D_L. Separating the blocks of suitable dimensions, we receive

$$C = \overset{\chi}{\begin{bmatrix} C_1 \\ C_2 \end{bmatrix}}\begin{matrix} r \\ n \end{matrix}\,, \qquad B = \begin{matrix} \ell & m \\ \end{matrix}\! \begin{bmatrix} B_1 & B_2 \end{bmatrix}\,\chi\,.$$

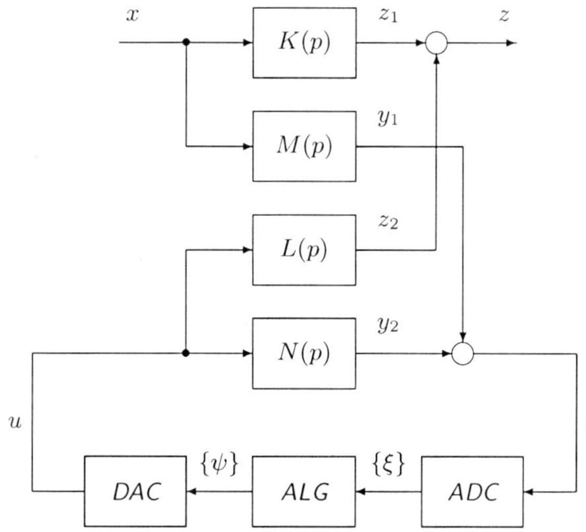

Fig. 7.2. Operator model of the standard sampled-data system

Then (7.8) yields

$$w(p) = \begin{bmatrix} C_1(pI_\chi - A)^{-1}B_1 & C_1(pI_\chi - A)^{-1}B_2 + D_L \\ C_2(pI_\chi - A)^{-1}B_1 & C_2(pI_\chi - A)^{-1}B_2 \end{bmatrix}.$$

Comparing this transfer matrix with (7.2), we find

$$\begin{aligned} K(p) &= C_1(pI_\chi - A)^{-1}B_1\,, \qquad L(p) = C_1(pI_\chi - A)^{-1}B_2 + D_L\,, \\ M(p) &= C_2(pI_\chi - A)^{-1}B_1\,, \qquad N(p) = C_2(pI_\chi - A)^{-1}B_2\,. \end{aligned} \tag{7.9}$$

The standard realisation (7.8) can be associated with the state equations of the plant

$$\begin{aligned} \frac{\mathrm{d}v}{\mathrm{d}t} &= Av + B_1x + B_2u \\ z &= C_1v + D_Lu, \qquad y = C_2v\,. \end{aligned} \tag{7.10}$$

Taken in the aggregate, the state equations (7.10) and the equations of the digital controllers (7.4)–(7.7) form a system of differential-difference equations, which will be called a *continuous-time model* of the standard sampled-data system.

2. Let us show that the continuous-time model (7.10), (7.4)–(7.7) can be transformed into an equivalent system of difference equations. With this aim in view, we integrate the first equation of (7.10) over the interval $kT \le t \le (k+1)T$, where k is any integer. So, we receive

$$v(t) = e^{A(t-kT)}v_k + \int_{kT}^{t} e^{A(t-\tau)} B_2 u(\tau)\, d\tau + \int_{kT}^{t} e^{A(t-\tau)} B_1 x(\tau)\, d\tau$$

with the notation $v_k = v(kT)$. Using (7.7), this equation can be written as

$$v(t) = e^{A(t-kT)}v_k + \int_{kT}^{t} e^{A(t-\tau)} m(\tau - kT)\, d\tau B_2 \psi_k + \int_{kT}^{t} e^{A(t-\tau)} B_1 x(\tau)\, d\tau\,. \tag{7.11}$$

Assuming

$$t = kT + \varepsilon, \quad \tau = kT + \nu, \quad 0 \le \varepsilon \le T, \quad 0 \le \nu \le T$$

after some transformations in (7.11), we get

$$v_k(\varepsilon) = e^{A\varepsilon}v_k + \int_0^{\varepsilon} e^{A(\varepsilon-\nu)} m(\nu)\, d\nu\, B_2\psi_k + \int_0^{\varepsilon} e^{A(\varepsilon-\nu)} B_1 x(kT+\nu)\, d\nu \tag{7.12}$$

with the notation

$$v_k(\varepsilon) \triangleq v(kT + \varepsilon)\,. \tag{7.13}$$

Moreover, using (7.10), we obtain

$$y_k(\varepsilon) \triangleq y(kT + \varepsilon) = C_2 v_k(\varepsilon), \quad 0 \le \varepsilon \le T\,, \tag{7.14}$$

$$\begin{aligned} z_k(\varepsilon) &\triangleq z(kT + \varepsilon) = C_1 v_k(\varepsilon) + D_L m(\varepsilon)\psi_k\,, \\ u_k(\varepsilon) &\triangleq u(kT + \varepsilon) = m(\varepsilon)\psi_k\,. \end{aligned} \qquad 0 < \varepsilon < T \tag{7.15}$$

Taken in the aggregate, the difference equations (7.12) and the equations of the digital controller (7.4)–(7.7) will be called a *parametric discrete model* of the standard sampled-data system. It can easily be shown that the continuous-time model (7.10), (7.4)–(7.7) and the parametric discrete model (7.12), (7.4)–(7.7) of the standard sampled-data system are equivalent, *i.e.*, if the set of functions $y(t)$, $z(t)$, and $u(t)$ and the sequence $\{\psi_k\}$ satisfy the equations of the continuous-time model, then the set of sequences $\{v_k(\varepsilon)\}$, $\{z_k(\varepsilon)\}$, $\{u_k(\varepsilon)\}$, and $\{\psi_k\}$ are a solution of the parametric discrete model, and *vice versa.*

3. Henceforth, we assume that the function $m(t)$ is piecewise smooth. Then under the given assumptions, the solutions $v(t)$ and $y(t)$ are continuous with respect to t. Therefore, assuming $\varepsilon = T$ in (7.12) and (7.14), we find

$$\begin{aligned} v_{k+1} &= e^{AT}v_k + e^{AT}\mu(A)B_2\psi_k + g_k \\ y_k &= C_2 v_k \end{aligned} \tag{7.16}$$

with the notation

$$g_k = \int_0^T e^{A(T-\nu)} B_1 x(kT + \nu)\, d\nu\,,$$

$$\mu(A) = \int_0^T e^{-A\nu} m(\nu)\, d\nu . \tag{7.17}$$

Then we supplement (7.16) with the equations of the digital controller as forward model (7.5)

$$\tilde{\alpha}_0 \psi_{k+\rho} + \ldots + \tilde{\alpha}_\rho \psi_k = \tilde{\beta}_0 y_{k+\rho} + \ldots + \tilde{\beta}_\rho y_k . \tag{7.18}$$

Taken in the aggregate, Equations (7.16)–(7.18) form a system of difference equations, which will be called a *discrete forward model* of the standard sampled-data system.

4. Without loss of generality, we assume that Equation (7.18) is row reduced. Then, it can be shown that the system (7.16)–(7.18) is row reduced. A proof of this fact will be given in Section 8.8. In this case, using Formulae (5.102), (5.103), and (7.6), we can obtain a *discrete backward model* of the standard sampled-data system

$$\begin{aligned} v_k &= e^{AT} v_{k-1} + e^{AT} \mu(A) B_2 \psi_{k-1} + g_{k-1} \\ y_k &= C_2 v_k \\ \alpha_0 \psi_k + \ldots + \alpha_\rho \psi_{k-\rho} &= \beta_0 y_k + \ldots + \beta_\rho y_{k-\rho} . \end{aligned} \tag{7.19}$$

It can be easily shown that the discrete backward model (7.19) together with (7.12) is equivalent to the original continuous-time model (7.10), (7.4)–(7.7), *i.e.*, if a set of sequences $\{v_k(\varepsilon)\}$, $y_k(\varepsilon)\}$, $\{z_k(\varepsilon)\}$ satisfies Equations (7.12), (7.19), then the functions

$$\begin{aligned} v(t) &= v_k(t - kT), \quad y(t) = y_k(t - kT), \qquad kT \le t \le (k+1)T \\ u(t) &= u_k(t - kT), \quad z(t) = z_k(t - kT), \qquad kT < t < (k+1)T \end{aligned}$$

determine a solution of the continuous-time models and *vice versa*.

7.3 Parametric Transfer Matrix (PTM)

1. The approach to a mathematical description of the standard system described in Sections 7.1 and 7.2 makes it possible to analyse the processes either in continuous time t, or in discrete time $t_k = kT$. These methods are similar to the description of continuous-time and discrete LTI systems by means of differential and difference equations, respectively. In this section we introduce a novel characteristic of the standard sampled-data system, describing its properties in the frequency domain. In this sense, this characteristic is a counterpart of the classical concept of transfer function (matrix) used in the theory of LTI systems.

2. Consider the following auxiliary problem. Assume in Fig. 7.3

$$x(t) = e^{st} I_\ell \,, \tag{7.20}$$

where s is a complex parameter. We search for a solution of the operator model (7.3)–(7.7) such that all signals in Fig. 7.3 are exponential periodic functions with exponent s and period equal to the sampling period T. In particular, this means

$$y(t) = y_T(s,t)e^{st}, \quad z(t) = z_T(s,t)e^{st}, \quad u(t) = u_T(s,t)e^{st}\,, \tag{7.21}$$

where

$$y_T(s,t) = y_T(s,t+T), \quad z_T(s,t) = z_T(s,t+T), \quad u_T(s,t) = u_T(s,t+T) \tag{7.22}$$

are matrices of dimensions $n \times \ell$, $r \times \ell$, and $m \times \ell$, respectively. The matrices (7.22) hereinafter will be called the *parametric transfer matrices (PTM)* of the standard sampled-data system from the input x to the outputs y, z, and u, respectively.

In this section, we present a formal method for constructing the PTM (7.22) on the basis of the stroboscopic property.

3. Henceforth for the PTM (7.22), we shall use the following special notation:

$$y_T(s,t) = w_{yx}(s,t), \quad z_T(s,t) = w_{zx}(s,t), \quad u_T(s,t) = w_{ux}(s,t)\,.$$

Let us begin with the matrix $w_{yx}(s,t)$. First of all, we notice that from the strict properness of the matrix $N(s)$ and Fig. 7.2, it follows that the PTM $w_{yx}(s,t)$ is continuous in t. Therefore, using the stroboscopic property, it can be assumed that the input of the ADC is acted upon by the exponential matrix signal

$$\tilde{y}(s,t) = w_{yx}(s,0)e^{st}\,.$$

Consider the open-loop system shown in Fig. 7.3. Using (6.41)–(6.45), we find

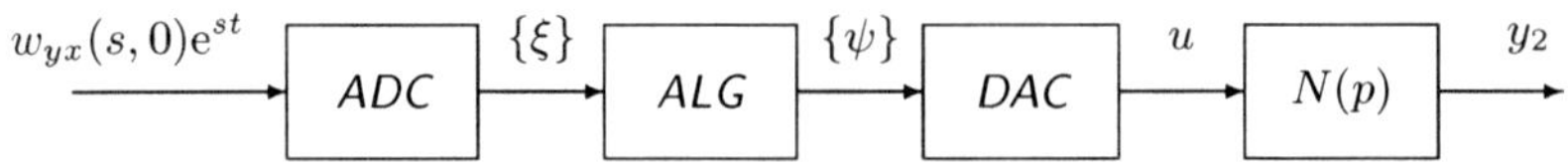

Fig. 7.3. Digital controller and continuous-time plant

the exp.per. output

$$y_2(t) = \varphi_{N\mu}(T,s,t)\widehat{w}_d(s) w_{yx}(s,0)e^{st}\,, \tag{7.23}$$

where

$$\varphi_{N\mu}(T,s,t) = \frac{1}{T}\sum_{k=-\infty}^{\infty} N(s+kj\omega)\mu(s+kj\omega)e^{kj\omega t}$$

and

$$\widehat{w}_d(s) = \widehat{\alpha}^{-1}(s)\widehat{\beta}(s)\,,$$

where

$$\begin{aligned}\widehat{\alpha}(s) &= \alpha_0 + \alpha_1 e^{-sT} + \ldots + \alpha_\rho e^{-\rho sT}\,,\\ \widehat{\beta}(s) &= \beta_0 + \beta_1 e^{-sT} + \ldots + \beta_\rho e^{-\rho sT}\,.\end{aligned} \tag{7.24}$$

From Fig. 7.2, it follows that

$$y_1(t) = M(s)e^{st}\,,$$

so we obtain

$$y(t) = y_1(t) + y_2(t) = \varphi_{N\mu}(T,s,t)\widehat{w}_d(s)w_{yx}(s,0)e^{st} + M(s)e^{st}\ .$$

Equating the expressions for $y(t)$ here and in (7.21), we get

$$w_{yx}(s,t) = \varphi_{N\mu}(T,s,t)\widehat{w}_d(s)w_{yx}(s,0) + M(s)\,. \tag{7.25}$$

Since the matrix $\varphi_{N\mu}(T,s,t)$ is continuous with respect to t, we can take $t=0$, so that

$$w_{yx}(s,0) = \widehat{D}_{N\mu}(T,s,0)\widehat{w}_d(s)w_{yx}(s,0) + M(s)\,, \tag{7.26}$$

where we used the fact that, due to (6.66) and (6.67), the following equality holds:

$$\varphi_{N\mu}(T,s,0) = \frac{1}{T}\sum_{k=-\infty}^{\infty} N(s+kj\omega)\mu(s+kj\omega) = \widehat{D}_{N\mu}(T,s,0)\,. \tag{7.27}$$

From (7.26), it follows that

$$w_{yx}(s,0) = \left[I_n - \widehat{D}_{N\mu}(T,s,0)\widehat{w}_d(s)\right]^{-1} M(s)\,. \tag{7.28}$$

Substituting (7.28) into (7.25), we find the required PTM

$$w_{yx}(s,t) = \varphi_{N\mu}(T,s,t)\widehat{R}_N(s)M(s) + M(s)$$

with the notation

$$\widehat{R}_N(s) = \widehat{w}_d(s)\left[I_n - \widehat{D}_{N\mu}(T,s,0)\widehat{w}_d(s)\right]^{-1}\,.$$

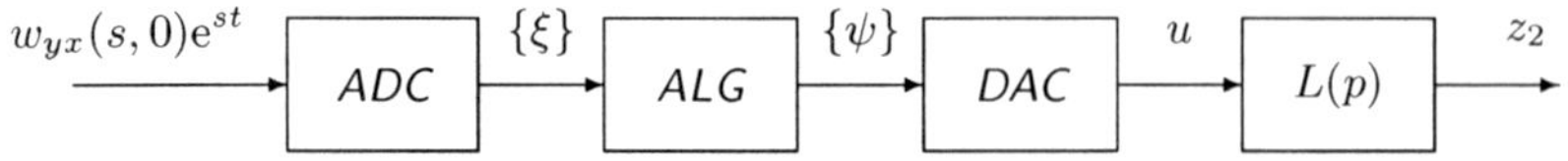

Fig. 7.4. Digital control unit and continuous-time plant

4. In order to find the PTM $w_{zx}(s,t)$, let us consider the open-loop system shown in Fig. 7.4. Similarly to (7.23), we construct the exp.per. output

$$z_2(t) = \varphi_{L\mu}(T,s,t)\widehat{w}_d(s)w_{yx}(s,0)e^{st}\,,$$

where

$$\varphi_{L\mu}(T,s,t) = \frac{1}{T}\sum_{k=-\infty}^{\infty} L(s+kj\omega)\mu(s+kj\omega)e^{kj\omega t}\,. \tag{7.29}$$

From Fig. 7.2, it follows that

$$z_1(t) = K(s)e^{st}\,.$$

Then with regard to (7.28), we obtain

$$z(t) = z_1(t) + z_2(t) = \varphi_{L\mu}(T,s,t)\widehat{R}_N(s)M(s)e^{st} + K(s)e^{st}\ .$$

Using (7.21), we obtain the required PTM from the input x to the output z

$$w_{zx}(s,t) = \varphi_{L\mu}(T,s,t)\widehat{R}_N(s)M(s) + K(s)\,. \tag{7.30}$$

5. Similar calculations provide also

$$w_{ux}(s,t) = \varphi_{\mu}(T,s,t)\widehat{R}_N(s)M(s)\,,$$

where $\varphi_\mu(T,s,t)$ is the function (6.39).

6. It should be noted that the standard system shown in Fig. 7.2 is fairly general. It can describe any sampled-data system containing, except for continuous-time LTI units, a single digital controller (7.4)–(7.7). Nevertheless, systems encountered in applications often are not given in the standard form. Then for obtaining the latter one, some structural transformations are needed. At the same time, there exists another way to construct the standard system for a given structure. For this purpose, we assume that the input of the system at hand is acted upon by an exponential signal (7.20), and all signals in the system are exponential periodic with exponent s and period T. Then using the stroboscopic property, the exp.per. system output $z(t)$ can always be found in the form

$$z(t) = e^{st} w_{zx}(s,t), \quad w_{zx}(s,t) = w_{zx}(s,t+T),$$

where $w_{zx}(s,t)$ is the PTM from the input x to the output z. Comparing this expression for $w_{zx}(s,t)$ with the general formula (7.30), we can always find the matrices $K(s)$, $L(s)$, $M(s)$, and $N(s)$ associated with the equivalent standard system.

Example 7.1. To illustrate the above approach, we consider the single-loop system shown in Fig. 7.5, where the $n \times m$ matrix $G(p)$ is strictly proper. To

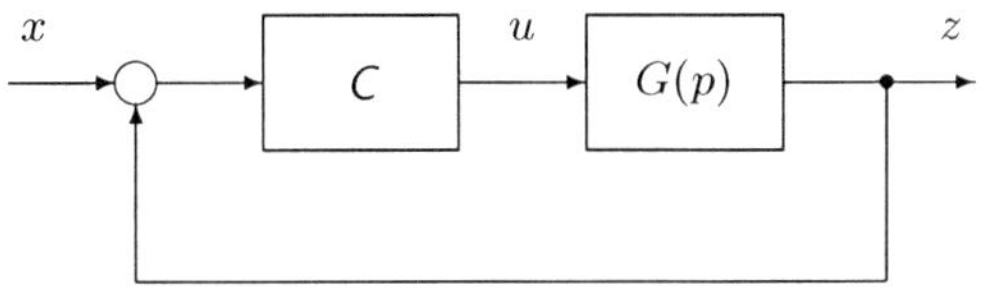

Fig. 7.5. Single-loop digital system

find the PTM $w_{zx}(s,t)$, we take

$$x(t) = e^{st} I_n, \quad z(t) = w_{zx}(s,t) e^{st}, \quad w_{zx}(s,t) = w_{zx}(s,t+T), \tag{7.31}$$

where the matrix $w_{zx}(s,t)$ is continuous with respect to t. Then using the stroboscopic property, consider the open-loop system shown in Fig. 7.6. The

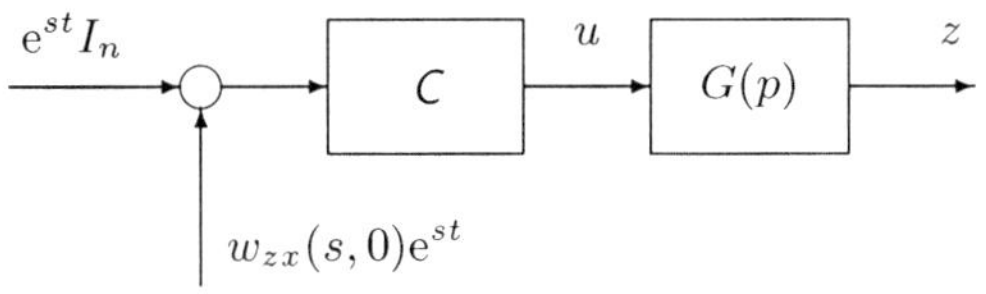

Fig. 7.6. Open-loop system for Fig. 7.5

exp.per. output of this open-loop system has the form

$$z(t) = \varphi_{G\mu}(T,s,t)\widehat{w}_d(s) e^{st} + \varphi_{G\mu}(T,s,t)\widehat{w}_d(s) w_{zx}(s,0) e^{st}.$$

Equating this expression for $z(t)$ with (7.31), we find

$$w_{zx}(s,t) = \varphi_{G\mu}(T,s,t)\widehat{w}_d(s) w_{zx}(s,0) + \varphi_{G\mu}(T,s,t)\widehat{w}_d(s). \tag{7.32}$$

Assuming $t = 0$ in (7.32), with the help of (7.27), we obtain

$$w_{zx}(s,0) = \left[I_n - \widehat{D}_{G\mu}(T,s,0)\widehat{w}_d(s) \right]^{-1} \widehat{D}_{G\mu}(T,s,0)\widehat{w}_d(s).$$

Substitution this into (7.32) yields

$$w_{zx}(s,t) = \varphi_{G\mu}(T,s,t)\widehat{w}_d(s)\left\{\left[I_n - \widehat{D}_{G\mu}(T,s,0)\widehat{w}_d(s)\right]^{-1}\widehat{D}_{G\mu}(T,s,0)\widehat{w}_d(s) + I_n\right\}.$$

After simplification, we receive

$$w_{zx}(s,t) = \varphi_{G\mu}(T,s,t)\widehat{R}_G(s)\,, \tag{7.33}$$

where $\widehat{R}_G(s)$ is constructed similarly to (7.42). Comparing (7.33) with (7.30), we find that in this example

$$K(s) = O_{nn}\,,\quad M(s) = I_n\,,\quad N(s) = L(s) = G(s)$$

i.e., Matrix (7.2) has the form

$$w(p) = \begin{bmatrix} O_{nn} & G(p) \\ I_n & G(p) \end{bmatrix}.$$

Multiplying both sides of (7.33) by e^{st}, with respect to (6.68), we find

$$\widehat{w}_d(s,t) \stackrel{\triangle}{=} e^{st}w_{zx}(s,t) = \widehat{D}_{G\mu}(T,s,t)\widehat{w}_d(s)\left[I_n - \widehat{D}_{G\mu}(T,s,0)\widehat{w}_d(s)\right]^{-1},$$

where

$$\widehat{D}_{G\mu}(T,s,t) = \frac{1}{T}\sum_{k=-\infty}^{\infty} G(s+kj\omega)\mu(s+kj\omega)e^{(s+kj\omega)t}$$

is the DLT of the matrix $G(s)\mu(s)$. It can be easily verified that for $t = \varepsilon$, $0 \le \varepsilon \le T$, the matrix $\widehat{w}_d(s,\varepsilon)$ determines the transfer matrix of the discrete system in Fig. 7.5 in the sense of the modified discrete Laplace transformation [177]. □

7. At the same time, if $M(s) = O_{n\ell}$, then the standard sampled-data system with the input x and output z reduces to the continuous-time LTI system

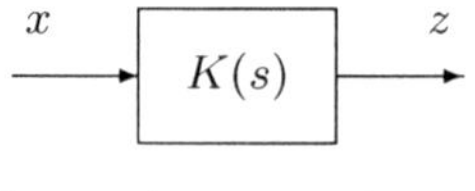

Fig. 7.7. Continuous-time system as a standard sampled-data system

shown in Fig. 7.7. The general Formula (7.30) yields

$$w(s,t) = K(s)\,.$$

Therefore, in the special cases of a continuous-time LTI system in Fig. 7.7 or a discrete-time system in Fig. 7.5, the PTM $w_{zx}(s,t)$ transforms into the classical frequency-domain descriptions for such systems.

8. The method for constructing the PTM, described in this section, is fairly general. In fact, it does not exploit the fact that the matrix $w(p)$ is rational. All the aforesaid still holds, when we assume that the matrices $K(p)$, $L(p)$, $M(p)$, $N(p)$ are transfer matrices of some linear stationary operators such that the series $\varphi_{L\mu}(T,s,t)$ and $\varphi_{N\mu}(T,s,t)$ converge and the latter sum is continuous with respect to t. As a special case, this method can be used for constructing the PTM for a standard system with pure-delay elements.

Example 7.2. Consider the system with delayed feedback shown in Fig. 7.8. Using the techniques described in detail in Chapter 9, we find the PTM

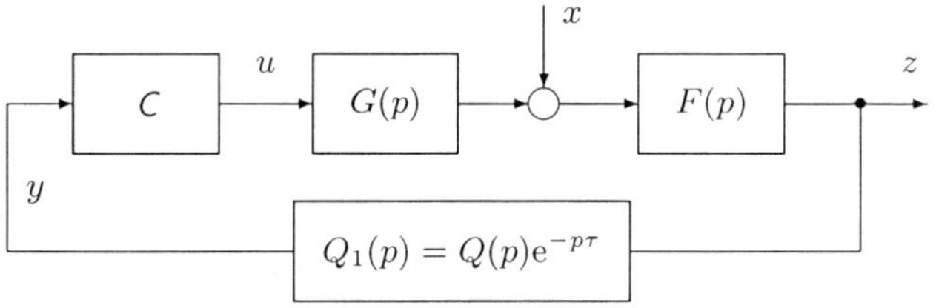

Fig. 7.8. Closed-loop sampled-data system with delayed feedback

$$w_{zx}(s,t) =$$
$$\varphi_{FG\mu}(T,s,t)\widetilde{w}_d(s)\left[I - \widetilde{D}_{QFG\mu}(T,s,-\tau)\widetilde{w}_d(s)\right]^{-1} Q(s)F(s)\mathrm{e}^{-s\tau} + F(s)$$

associated with the transfer matrix of the LTI plant

$$w(p) = \begin{bmatrix} F(p) & F(p)G(p) \\ Q(p)F(p)\mathrm{e}^{-p\tau} & Q(p)F(p)G(p)\mathrm{e}^{-p\tau} \end{bmatrix}.$$

□

7.4 PTM as Function of the Argument s

1. In this section, we construct a general representation for the PTM $w_{zx}(s,t)$ on the basis of the parametric discrete model of the standard sampled-data system. The general expression for the PTM makes it possible to completely characterise the set of singular points of the matrix $w_{zx}(s,t)$ as a function of the complex variable s.

2. The following theorem provides such a representation.

Theorem 7.3. *Suppose that the matrices $K(p)$ and $M(p)$ are strictly proper and the LTI plant of the standard sampled-data system is given by the state equations*

$$\begin{aligned}\frac{\mathrm{d}v}{\mathrm{d}t} &= Av + B_1 x + B_2 u \\ z = C_1 v + D_L u\,, &\qquad y = C_2 v\end{aligned} \tag{7.34}$$

with a $\chi \times \chi$ matrix A. Let the digital controller be given by the equations

$$\xi_k = y_k = y(kT) \tag{7.35}$$

$$\alpha_0 \psi_k + \ldots + \alpha_\rho \psi_{k-\rho} = \beta_0 \xi_k + \ldots + \beta_\rho \xi_{k-\rho} \tag{7.36}$$

$$u(t) = m(t - kT)\psi_k\,, \quad kT < t < (k+1)T\,. \tag{7.37}$$

Introduce the matrix

$$\widehat{Q}(s, \widehat{\alpha}, \widehat{\beta}) = \begin{bmatrix} I_\chi - \mathrm{e}^{-sT}\mathrm{e}^{AT} & O_{\chi n} & -\mathrm{e}^{-sT}\mathrm{e}^{AT}\mu(A)B_2 \\ -C_2 & I_n & O_{nm} \\ O_{m\chi} & -\widehat{\beta}(s) & \widehat{\alpha}(s) \end{bmatrix}, \tag{7.38}$$

where $\widehat{\alpha}(s)$ and $\widehat{\beta}(s)$ are the matrices (7.24). Then for any s with

$$\det \widehat{Q}(s, \widehat{\alpha}, \widehat{\beta}) \neq 0\,, \tag{7.39}$$

there exists a unique solution of Equations (7.34)–(7.37) satisfying Conditions (7.21)–(7.22). Moreover, for $0 < t < T$, we have

$$\begin{aligned} w_{zx}(s,t) = \mathrm{e}^{-st}&\left[C_1 \mathrm{e}^{At} v_0(s) + C_1 \int_0^t \mathrm{e}^{A(t-\nu)} m(\nu)\,\mathrm{d}\nu\, B_2 \psi_0(s)\right] + \\ &+ C_1 \int_0^t \mathrm{e}^{(A - sI_\chi)(t-\nu)}\,\mathrm{d}\nu B_1 + D_L \mathrm{e}^{-st} m(t)\psi_0(s)\,, \end{aligned} \tag{7.40}$$

where the matrices $v_0(s)$ and $\psi_0(s)$ are given by the equation

$$\widehat{Q}(s, \widehat{\alpha}, \widehat{\beta}) \begin{bmatrix} v_0(s) \\ y_0(s) \\ \psi_0(s) \end{bmatrix} = R(s) \tag{7.41}$$

and the matrix $R(s)$ has the form

$$R(s) = \overset{\ell}{\begin{bmatrix} (sI_\chi - A)^{-1}(I_\chi - \mathrm{e}^{-sT}\mathrm{e}^{AT})B_1 \\ O_{n\ell} \\ O_{m\ell} \end{bmatrix}} \begin{matrix} \chi \\ n \\ m \end{matrix}\,. \tag{7.42}$$

Proof. The proof is given in several stages.

a) First of all, we construct the parametric discrete forward model of Equations (7.34)–(7.37) for the input (7.20). With this aim in view, let $x(t) = \mathrm{e}^{st} I_\ell$ in (7.12). As a result after integration, we obtain

$$v_k(\varepsilon) = \mathrm{e}^{A\varepsilon} v_k + \int_0^{\varepsilon} \mathrm{e}^{A(\varepsilon-\nu)} m(\nu)\,\mathrm{d}\nu\, B_2 \psi_k + \mathrm{e}^{ksT} G(s,\varepsilon) \tag{7.43}$$

with

$$G(s,\varepsilon) = \mathrm{e}^{A\varepsilon} \int_0^{\varepsilon} \mathrm{e}^{(sI_\chi - A)\nu}\,\mathrm{d}\nu B_1 = (sI_\chi - A)^{-1}(\mathrm{e}^{s\varepsilon} I_\chi - \mathrm{e}^{A\varepsilon}) B_1 \,. \tag{7.44}$$

For $\varepsilon = T$ from (7.43), we find

$$v_{k+1} = \mathrm{e}^{AT} v_k + \mathrm{e}^{AT} \int_0^{T} \mathrm{e}^{-A\nu} m(\nu)\,\mathrm{d}\nu\, B_2 \psi_k + \mathrm{e}^{ksT} G(s,T)\,, \tag{7.45}$$

where

$$G(s,T) = (sI_\chi - A)^{-1}(\mathrm{e}^{sT} I_\chi - \mathrm{e}^{AT}) B_1 \,.$$

Combining (7.45) with the equations of the digital controller and using (7.17), we find the discrete backward model of the standard sampled-data system for the input (7.20)

$$\begin{aligned} v_k &= \mathrm{e}^{AT} v_{k-1} + \mathrm{e}^{AT} \mu(A) B_2 \psi_{k-1} + \mathrm{e}^{(k-1)sT} G(s,T) \\ y_k &= C_2 v_k \\ \alpha_0 \psi_k + \ldots + \alpha_\rho \psi_{k-\rho} &= \beta_0 y_k + \ldots + \beta_\rho y_{k-\rho} \,. \end{aligned} \tag{7.46}$$

The discrete model (7.46) together with (7.43) determines the backward parametric discrete model for the input (7.20).

b) If a solution of the continuous-time models (7.34)–(7.37) satisfies Conditions (7.21) and (7.22), then it is associated with discrete sequences

$$\begin{aligned} \{v(s)\} &= \{\ldots, v_{-1}(s), v_0(s), v_1(s), \ldots\} \\ \{y(s)\} &= \{\ldots, y_{-1}(s), y_0(s), y_1(s), \ldots\} \\ \{\psi(s)\} &= \{\ldots, \psi_{-1}(s), \psi_0(s), \psi_1(s), \ldots\} \end{aligned}$$

that determine a solution of Equations (7.46) and satisfy the conditions

$$v_k(s) = \mathrm{e}^{ksT} v_0(s), \quad y_k(s) = \mathrm{e}^{ksT} y_0(s), \quad \psi_k(s) = \mathrm{e}^{ksT} \psi_0(s)\,,$$

where $v_0(s)$, $y_0(s)$, $\psi_0(s)$ are unknown matrices to be found. Substituting these relations into (7.46), we obtain

$$\begin{aligned} v_0(s) &= \mathrm{e}^{-sT} \mathrm{e}^{AT} v_0(s) + \mathrm{e}^{-sT} \mathrm{e}^{AT} \mu(A) B_2 \psi_0(s) \\ &\quad + (sI_\chi - A)^{-1}(I_\chi - \mathrm{e}^{-sT} \mathrm{e}^{AT}) B_1 \\ y_0(s) &= C_2 v_0(s) \\ \widehat{\alpha}(s) \psi_0(s) &= \widehat{\beta}(s) y_0(s) \,. \end{aligned} \tag{7.47}$$

This system can be written in form of the linear system of equations (7.41) having a unique solution under Condition (7.39).

c) Now, we will prove (7.40). From the first equation in (7.47), we find

$$v_0(s) = \left(e^{sT}e^{-AT} - I_\chi\right)^{-1}\mu(A)B_2\psi_0(s) + (sI_\chi - A)^{-1}B_1\,. \tag{7.48}$$

Moreover, (7.47) yields

$$\psi_0(s) = \widehat{\alpha}^{-1}(s)\widehat{\beta}(s)y_0(s) = \widehat{w}_d(s)y_0(s)\,. \tag{7.49}$$

Substituting (7.49) into (7.48), we find

$$v_0(s) = \left(e^{sT}e^{-AT} - I_\chi\right)^{-1}\mu(A)B_2\widehat{w}_d(s)y_0(s) + (sI_\chi - A)^{-1}B_1\,. \tag{7.50}$$

Multiplying this from the left by C_2, we obtain

$$y_0(s) = C_2\left(e^{sT}e^{-AT} - I_\chi\right)^{-1}\mu(A)B_2\widehat{w}_d(s)y_0(s) + C_2(sI_\chi - A)^{-1}B_1\,. \tag{7.51}$$

But from (6.89), (6.100), and (7.9), it follows that

$$C_2\left(e^{sT}e^{-AT} - I_\chi\right)^{-1}\mu(A)B_2 = \varphi_{N\mu}(T,s,0) = \widehat{D}_{N\mu}(T,s,0)\,,$$
$$C_2(sI_\chi - A)^{-1}B_1 = M(s)\,.$$

Thus, Equation (7.51) can be written in the form

$$y_0(s) = \widehat{D}_{N\mu}(T,s,0)\widehat{w}_d(s)y_0(s) + M(s)\,,$$

whence

$$y_0(s) = \left[I_n - \widehat{D}_{N\mu}(T,s,0)\widehat{w}_d(s)\right]^{-1}M(s)\,.$$

Then using (7.49), we obtain

$$\psi_0(s) = \widehat{w}_d(s)\left[I_n - \widehat{D}_{N\mu}(T,s,0)\widehat{w}_d(s)\right]^{-1}M(s)\,. \tag{7.52}$$

For further transformations, we notice that for $0 \le \varepsilon = t \le T$, Equations (7.43) and (7.44) yield

$$v(t) = e^{At}v_0(s) + \int_0^t e^{A(t-\nu)}m(\nu)\,d\nu\,B_2\psi_0(s) + \int_0^t e^{A(t-\nu)}e^{s\nu}\,d\nu\,B_1\,.$$

Therefore, the PTM $w_{vx}(s,t)$ with respect to the output v is given for $0 \le t \le T$ by

$$\begin{aligned} w_{vx}(s,t) &= v(t)e^{-st} \\ &= e^{-st}\left[e^{At}v_0(s) + \int_0^t e^{A(t-\nu)}m(\nu)\,d\nu\,B_2\psi_0(s)\right] \\ &\quad + \int_0^t e^{(A-sI_\chi)(t-\nu)}\,d\nu\,B_1\,. \end{aligned} \tag{7.53}$$

Since for $0 < t < T$, we have

$$w_{zx}(s,t) = C_1 w_{vx}(s,t) + D_L \mathrm{e}^{-st} m(t)\psi_0(s)\,,$$

a substitution of (7.53) into this equation gives (7.40).

d) It remains to prove that for $0 < t < T$, Formula (7.40) can be reduced to the form (7.30). Obviously, for $0 < t < T$, we have

$$z(t) = C_1 v(t) + D_L m(t)\psi_0(s)$$

that gives, with respect to (7.43) and (7.44),

$$\begin{aligned} z(t) = C_1 &\left[\mathrm{e}^{At} v_0(s) + \int_0^t \mathrm{e}^{A(t-\nu)} m(\nu)\,\mathrm{d}\nu\, B_2 \psi_0(s)\right] + \\ &+ D_L m(t)\psi_0(s) + C_1 (sI_\chi - A)^{-1}(\mathrm{e}^{st} I_\chi - \mathrm{e}^{At}) B_1\,. \end{aligned} \tag{7.54}$$

Using (7.48) after simplification, we find

$$\begin{aligned} z(t) = C_1 &\left[\mathrm{e}^{At}\mu(A)\left(\mathrm{e}^{sT}\mathrm{e}^{-AT} - I_\chi\right)^{-1} + \int_0^t \mathrm{e}^{A(t-\nu)} m(\nu)\,\mathrm{d}\nu\right] B_2\,\psi_0(s) \\ &+ D_L m(t)\psi_0(s) + \mathrm{e}^{st} C_1 (sI_\chi - A)^{-1} B_1\,. \end{aligned}$$

Multiplying by e^{-st} and using the fact that for $0 < t < T$

$$\begin{aligned} \mathrm{e}^{-st} C_1 &\left[\mathrm{e}^{At}\mu(A)\left(\mathrm{e}^{sT}\mathrm{e}^{-AT} - I_\chi\right)^{-1} + \int_0^t \mathrm{e}^{A(t-\nu)} m(\nu)\,\mathrm{d}\nu\right] B_2 \\ &+ D_L \mathrm{e}^{-st} m(t) = \varphi_{L\mu}(T,s,t) \end{aligned}$$

and moreover,

$$C_1 (sI_\chi - A)^{-1} B_1 = K(s)\,,$$

we obtain

$$w_{zx}(s,t) = \varphi_{L\mu}(T,s,t)\psi_0(s) + K(s)\,.$$

Using (7.52), we get (7.30). ∎

3. Using Theorem 7.3, some general properties of the singular points of the PTM $w_{zx}(s,t)$ can be investigated.

Theorem 7.4. *Under the conditions of Theorem 7.3, the PTM $w_{zx}(s,t)$ given by (7.30) is a meromorphic function of the argument s, i.e., all its singular points are poles. The set of poles of $w_{zx}(s,t)$ belongs for any t to the set of roots of the equation*

$$\widehat{\Delta}(s) = \det \widehat{Q}(s, \widehat{\alpha}, \widehat{\beta}) = 0\,, \tag{7.55}$$

where $\widehat{Q}(s, \widehat{\alpha}, \widehat{\beta})$ is Matrix (7.38).

Proof. From (7.41), we have

$$\begin{bmatrix} v_0(s) \\ y_0(s) \\ \psi_0(s) \end{bmatrix} = \widehat{Q}^{-1}(s, \widehat{\alpha}, \widehat{\beta})R(s)\,. \tag{7.56}$$

The matrix $R(s)$ determined by (7.42) is an integral function of the argument s. This follows from the fact that the matrix

$$R_1(s, A) \triangleq (sI_\chi - A)^{-1}(I_\chi - e^{-sT}e^{AT})$$

is integral, because it is generated from the integral scalar function

$$R_1(s, a) \triangleq \frac{e^{-sT}(e^{sT} - e^{aT})}{s - a}\,,$$

where the scalar parameter a is changed for a matrix A. Therefore, from (7.56), it follows that the matrices $v_0(s)$, $y_0(s)$, and $\psi_0(s)$ are meromorphic functions, and their poles belong to the set of roots of Equation (7.55). From (7.54) for $0 < t < T$, we have

$$\begin{aligned} w_{zx}(s,t) &= e^{-st}C_1 e^{At}v_0(s) \\ &\quad + e^{-st}\left[C_1\int_0^t e^{A(t-\nu)}m(\nu)\,d\nu\,B_2 + D_L m(t)\right]\psi_0(s) \\ &\quad + e^{-st}C_1(sI_\chi - A)^{-1}(e^{st}I_\chi - e^{At})B_1\,. \end{aligned} \tag{7.57}$$

The coefficients for $v_0(s)$ and $\psi_0(s)$, as well as the last term on the right side of (7.57) are integral functions of s. Therefore, the claim of the theorem follows for $0 < t < T$ from the already proved properties of the matrices $v_0(s)$ and $\psi_0(s)$. Since $w_{zx}(s,t) = w_{zx}(s, t+T)$, this result holds for all t. ∎

4. Introduce the polynomial matrices

$$\alpha(\zeta) = \widehat{\alpha}(s)\,|_{e^{-sT}=\zeta} = \alpha_0 + \alpha_1\zeta + \ldots + \alpha_\rho\zeta^\rho\,,$$

$$\beta(\zeta) = \widehat{\beta}(s)\,|_{e^{-sT}=\zeta} = \beta_0 + \beta_1\zeta + \ldots + \beta_\rho\zeta^\rho\,,$$

$$Q(\zeta, \alpha, \beta) = \begin{bmatrix} I_\chi - \zeta e^{AT} & O_{\chi n} & -\zeta e^{AT}\mu(A)B_2 \\ -C_2 & I_n & O_{nm} \\ O_{m\chi} & -\beta(\zeta) & \alpha(\zeta) \end{bmatrix} \tag{7.58}$$

and the polynomial

$$\Delta(\zeta) = \widehat{\Delta}(s)\,|_{e^{-sT}=\zeta} = \det Q(\zeta, \alpha, \beta)\,, \tag{7.59}$$

which is called the *characteristic polynomial* of the standard sampled-data system.

Theorem 7.5. *Let $\zeta_1, \dots, \zeta_q$ be all different roots of the polynomial $\Delta(\zeta)$. Then for any t, the set of poles of $w_{zx}(s,t)$ belongs to the set of the numbers*

$$s_{in} = \frac{1}{T}\ln\zeta_i + \frac{2n\pi \mathrm{j}}{T}, \quad (i = 1, \dots, q;\; n = 0, \pm 1, \dots). \tag{7.60}$$

Proof. The matrix $\widehat{Q}(s, \widehat{\alpha}, \widehat{\beta})$ in (7.38) depends only on e^{-sT}. Therefore, substituting in (7.38) ζ for e^{-sT}, we obtain the polynomial matrix (7.58). The poles of the matrix $Q^{-1}(\zeta, \alpha, \beta)$ coincide with the roots of the polynomial $\Delta(\zeta)$, which are related to the poles of the matrices $v_0(s)$, $y_0(s)$, and $\psi_0(s)$ by Equations (7.60). Due to (7.57), the same is valid for the poles of the PTM $w_{zx}(s,t)$. ∎

Remark 7.6. The standard sampled-data system (7.3)–(7.7) is associated with a set of continuous-time models with different realisations of the matrix $w(p)$. Nevertheless, as follows from (7.30), the PTM $w_{zx}(s,t)$ is independent of the choice of this realisation, because for a given matrix $w(p)$ it is uniquely determined by the matrices $K(p)$, $L(p)$, $M(p)$, and $N(p)$ appearing in (7.2). Hence it follows, in particular, that the set of poles of the PTM $w_{zx}(s,t)$ is independent of the choice of the realisation (A, B, C) in the standard form (7.8). Therefore, without loss of generality, it can be assumed that the realisation (A, B, C) is minimal, *i.e.*, the pair (A, B) is controllable and the pair $[A, C]$ is observable. This situation is similar in case of an LTI system, where the uncontrollable and unobservable parts do not change the transfer function of a system.

7.5 Internal Stability of the Standard SD System

1. In this section, we investigate the stability behavior of the standard sampled-data system, assuming that the continuous-time plant is given by the state equations (7.34). In this case, combining the equations of the digital controller with (7.34), we can write a continuous-times model of the standard sampled-data system in the form

$$\begin{aligned} \frac{\mathrm{d}v}{\mathrm{d}t} &= Av + B_1 x + B_2 u \\ z &= C_1 v + D_L u\,, \quad y = C_2 v \end{aligned} \tag{7.61}$$

and

$$\alpha(\zeta)\psi_k = \beta(\zeta)y_k \tag{7.62}$$

$$u(t) = m(t - kT)\psi_k\,, \quad kT < t < (k+1)T\,, \tag{7.63}$$

where

$$\begin{aligned}\alpha(\zeta) &= \alpha_0 + \alpha_1\zeta + \ldots + \alpha_\rho\zeta^\rho\,,\\ \beta(\zeta) &= \beta_0 + \beta_1\zeta + \ldots + \beta_\rho\zeta^\rho\end{aligned} \tag{7.64}$$

are polynomial matrices.

2.

Definition 7.7. *The standard sampled-data system (7.61)–(7.64) will be called* internally stable, *if with* $x(t) = O_{\ell 1}$ *for any solution of Equations (7.61)–(7.64) and for* $t > 0$, $k > 0$, *the following estimations hold:*

$$\|v(t)\| < d_v e^{-\gamma t}, \quad \|u(t)\| < d_u e^{-\gamma t}, \quad \|\psi_k\| < d_\psi e^{-\gamma kT}\,, \tag{7.65}$$

where $\|\cdot\|$ *denotes any norm for number matrices,* d_v, d_u, d_ψ *and* γ *are positive constants, where* γ *is independent of the initial conditions.*

We note that under Conditions (7.65), the following estimates hold:

$$\|y(t)\| < d_y e^{-\gamma t}, \quad \|z(t)\| < d_z e^{-\gamma t} \tag{7.66}$$

with positive constants d_y and d_z. As follows from (7.57), the matrices C_1 and D_L do not influence the internal stability of the standard system (7.61)–(7.64).

In this section, we formulate some necessary and sufficient conditions for the internal stability of the standard sampled-data system. For brevity, we also shall use the term "stability", when we mean "internal stability".

If $x(t) = O_{\ell 1}$, the standard sampled-data system can be associated with the discrete backward model generated from (7.19) with $g_k = O_{\ell 1}$:

$$\begin{aligned} v_k &= e^{AT} v_{k-1} + e^{AT}\mu(A) B_2 \psi_{k-1}\\ y_k &= C_2 v_k\\ \alpha_0\psi_k + \ldots + \alpha_\rho\psi_{k-\rho} &= \beta_0 y_k + \ldots + \beta_\rho y_{k-\rho}\,. \end{aligned} \tag{7.67}$$

3.

Lemma 7.8. *For the standard sampled-data system (7.61)–(7.64) to be internally stable, a necessary and sufficient condition is that the discrete backward model (7.67) is stable.*

Proof. Necessity: Let the standard sampled-data system be stable. Then Estimates (7.65) and (7.66) hold. As a special case for $k > 0$, we have

$$\|v(kT)\| = \|v_k\| < d_v e^{-\gamma kT}, \; \|y(kT)\| = \|y_k\| < d_y e^{-\gamma kT}, \; \|\psi_k\| < d_\psi e^{-\gamma kT}\,. \tag{7.68}$$

With the notation $e^{-\gamma T} = \theta$, $|\theta| < 1$, we obtain

$$\|v_k\| < d_v\theta^k, \quad \|y_k\| < d_y\theta^k, \quad \|\psi_k\| < d_\psi\theta^k, \quad (k > 0)\,. \tag{7.69}$$

Since Conditions (7.69) hold for all solutions of Equations (7.67), the discrete model is stable by definition.

Sufficiency: Let the discrete model (7.67) be stable. Then, we have Inequalities (7.69), which can be written in the form (7.68). Due to (7.12) and (7.13), we have

$$v(kT+\varepsilon) = \mathrm{e}^{A\varepsilon} v_k + \int_0^{\varepsilon} \mathrm{e}^{A(\varepsilon-\nu)} m(\nu)\,\mathrm{d}\nu\, B_2 \psi_k$$

and, estimating by a norm, we obtain

$$\|v(kT+\varepsilon)\| \le L_1 \|v_k\| + L_2 \|\psi_k\|\,, \tag{7.70}$$

where

$$L_1 = \max_{0\le\varepsilon\le T} \|\mathrm{e}^{A\varepsilon}\|, \quad L_2 = \max_{0\le\varepsilon\le T} \left\| \int_0^{\varepsilon} \mathrm{e}^{A(\varepsilon-\nu)} m(\nu)\,\mathrm{d}\nu\, B_2 \right\|$$

are constants. From (7.70) and (7.68), it follows that

$$\|v(t)\| \le L\mathrm{e}^{-k\gamma T}, \quad kT \le t \le (k+1)T, \quad (k=0,1,\dots)\,, \tag{7.71}$$

where

$$L = L_1 d_v + L_2 d_\psi$$

is a constant. From (7.71), the following estimate can easily be derived:

$$\|v(t)\| \le L\mathrm{e}^{-\gamma T}\mathrm{e}^{-\gamma t}, \quad t > 0\,.$$

This relation and (7.61) yield (7.66). ∎

4. Necessary and sufficient conditions for the internal stability of the system (7.61)–(7.64) are given by the following theorem.

Theorem 7.9. *A necessary and sufficient condition for the standard sampled-data system (7.61)–(7.64) to be internally stable is that all eigenvalues of the matrix*

$$\widehat{Q}(s,\widehat{\alpha},\widehat{\beta}) = \begin{bmatrix} I_\chi - \mathrm{e}^{-sT}\mathrm{e}^{AT} & O_{\chi n} & -\mathrm{e}^{-sT}\mathrm{e}^{AT}\mu(A)B_2 \\ -C_2 & I_n & O_{nm} \\ O_{m\chi} & -\widehat{\beta}(s) & \widehat{\alpha}(s) \end{bmatrix}$$

lie in the open left half-plane, or equivalently, the polynomial matrix

$$Q(\zeta,\alpha,\beta) = \begin{bmatrix} I_\chi - \zeta\mathrm{e}^{AT} & O_{\chi n} & -\zeta\mathrm{e}^{AT}\mu(A)B_2 \\ -C_2 & I_n & O_{nm} \\ O_{m\chi} & -\beta(\zeta) & \alpha(\zeta) \end{bmatrix} \tag{7.72}$$

has to be stable, i.e., it is free of eigenvalues in the closed unit disk.

Proof. Due to Lemma 7.8, the standard sampled-data system is stable iff the discrete model (7.67) is stable. The latter can be written in the form of a homogeneous backward-difference equation

$$Q(\zeta,\alpha,\beta)\begin{bmatrix} v_k \\ y_k \\ \psi_k \end{bmatrix} = O\,.$$

Then from Theorem 5.47, it follows that the discrete model (7.67) is stable iff Matrix (7.72) is stable. The claim regarding Matrix (7.72) follows from the equality

$$\widehat{Q}(s,\widehat{\alpha},\widehat{\beta}) = Q(\zeta,\alpha,\beta)\,|_{\zeta=\mathrm{e}^{-sT}}\,. \qquad \blacksquare$$

Corollary 7.10. *Any controller ensuring under the given assumptions the internal stability of the closed-loop system is causal, i.e.,* $\det\alpha_0 \neq 0$, *because for* $\det\alpha_0 = 0$, *Matrix (7.72) is unstable.* $\blacksquare$

Corollary 7.11. *Theorem 7.9 can be formulated in an alternative way: a necessary and sufficient condition for the standard sampled-data system to be stable is that its characteristic polynomial (7.59) must be stable.* $\blacksquare$

7.6 Polynomial Stabilisation of the Standard SD System

1. Hereinafter, without loss of generality, we assume $D_L = O_{rm}$. Then, the matrix $w(p)$ is strictly proper and admits a set of realisations (A, B, C).

Definition 7.12. *A realisation* (A, B, C) *will be called* stabilisable, *if there exists a controller* $(\alpha(\zeta), \beta(\zeta))$, *such that Matrix (7.72) is stable. Such a controller will be called a* stabilising controller *for the realisation* (A, B, C).

In this section, we present solutions to the following problems:

a) Construction of the set of stabilisable realisations;
b) Construction of the set of stabilising controllers for a stabilisable realisation (A, B, C).

2. Firstly, we consider the above mentioned problems for the closed loop incorporated in the standard system in Fig. 7.3. This loop is shown in Fig. 7.9, where the control signal u is related to y by (7.4)–(7.7).

Let $(\tilde{A}, \tilde{B}_2, \tilde{C}_2)$ be a realisation of the matrix $N(p)$ in form of the state equations

$$\frac{\mathrm{d}\tilde{v}}{\mathrm{d}t} = \tilde{A}\tilde{v} + \tilde{B}_2 u\,, \quad y = \tilde{C}_2\tilde{v} \tag{7.73}$$

with an $\eta\times 1$ state vector $\tilde{v}$ and constant matrices $\tilde{A}$, $\tilde{B}_2$, and $\tilde{C}_2$ of dimensions $\eta\times\eta$, $\eta\times m$, and $n\times\eta$, respectively. Then, the closed loop will be called stable, if for the system of equations (7.73) and (7.62)–(7.64), estimates similar to (7.65) hold:

$$\|\tilde{v}(t)\| < d_{\tilde{v}}\mathrm{e}^{-\gamma t}, \quad \|u(t)\| < d_u\mathrm{e}^{-\gamma t}, \quad \|\psi_k\| < d_\psi\mathrm{e}^{-\gamma kT}, \quad t>0,\ k>0\,.$$

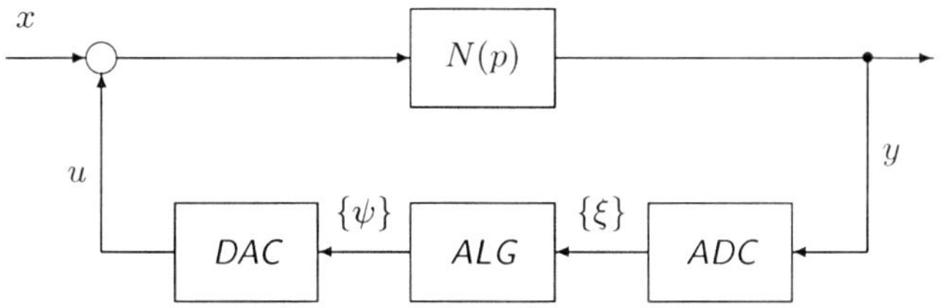

Fig. 7.9. Closed loop of the standard sampled-data system

3. The following theorem presents a solution to the stabilisation problem for the closed loop, when the continuous-time plant is given by its minimal realisation.

Theorem 7.13. *Let $(\tilde{A}_\nu, \tilde{B}_{2\nu}, \tilde{C}_{2\nu})$ be a minimal realisation of dimension (n, λ, m), and the rational matrix $w_N(\zeta)$ be determined by*

$$w_N(\zeta) \stackrel{\triangle}{=} \mathcal{D}_{N\mu}(T, \zeta, 0) = \zeta \tilde{C}_{2\nu} e^{\tilde{A}_\nu T} \mu(\tilde{A}_\nu)(I_\lambda - \zeta e^{\tilde{A}_\nu T})^{-1} \tilde{B}_{2\nu} \,,$$

where

$$\mu(\tilde{A}_\nu) = \int_0^T e^{-\tilde{A}_\nu \tau} m(\tau) \, d\tau \,.$$

Let us have an ILMFD

$$w_N(\zeta) = \zeta a_N^{-1}(\zeta) b_N(\zeta) \tag{7.74}$$

with $a_N(\zeta) \in R_{nn}[\zeta]$ and $b_N(\zeta) \in R_{nm}[\zeta]$. Then the relation

$$\varrho_N(\zeta) \stackrel{\triangle}{=} \frac{\det(I_\eta - \zeta e^{\tilde{A}_\nu T})}{\det a_N(\zeta)} \tag{7.75}$$

is a polynomial. For an arbitrary choice of the minimal realisation $(\tilde{A}_\nu, \tilde{B}_{2\nu}, \tilde{C}_{2\nu})$ and the matrices $a_N(\zeta)$, $b_N(\zeta)$, the polynomials (7.75) are equivalent, i.e., they are equal up to a constant factor. A necessary and sufficient condition for the set of minimal realisations of the matrix $N(p)$ to be stabilisable is that the polynomial (7.75) is stable, i.e., is free of roots inside the closed unit disk. If the polynomial $\varrho_N(\zeta)$ is stable and $(a_N(\zeta), \zeta b_N(\zeta))$ is an arbitrary pair forming the ILMFD (7.74), then the set of all controllers $(\alpha(\zeta), \beta(\zeta))$ stabilising the minimal realisations of the matrix $N(p)$ is defined as the set of all pairs ensuring the stability of the matrix

$$Q_N(\zeta, \alpha, \beta) = \begin{bmatrix} a_N(\zeta) & -\zeta b_N(\zeta) \\ -\beta(\zeta) & \alpha(\zeta) \end{bmatrix} . \tag{7.76}$$

Proof. Using an arbitrary minimal realisation $(\tilde{A}_\nu, \tilde{B}_{2\nu}, \tilde{C}_{2\nu})$ and (7.62)–(7.64), we arrive at the problem of investigating the stability of the system

$$\tilde{y} = \tilde{C}_{2\nu}\tilde{v}\,, \quad \frac{\mathrm{d}\tilde{v}}{\mathrm{d}t} = \tilde{A}_\nu \tilde{v} + \tilde{B}_{2\nu} u$$

$$\alpha(\zeta)\psi_k = \beta(\zeta) u_k$$

$$u(t) = m(t - kT)\psi_k\,, \quad kT < t < (k+1)T\,.$$

As follows from Theorem 7.9, a necessary and sufficient condition for the stability of this system is that the matrix

$$\tilde{Q}_\nu(\zeta, \alpha, \beta) = \begin{bmatrix} I_\lambda - \zeta \mathrm{e}^{\tilde{A}_\nu T} & O_{\lambda n} & -\zeta \mathrm{e}^{\tilde{A}_\nu T}\mu(\tilde{A}_\nu)\tilde{B}_{2\nu} \\ -\tilde{C}_{2\nu} & I_n & O_{nm} \\ O_{m\lambda} & -\beta(\zeta) & \alpha(\zeta) \end{bmatrix} \tag{7.77}$$

is stable. Hence the set of the pairs of stabilising polynomials $(\alpha(\zeta), \beta(\zeta))$ coincides with the set of stabilising pairs for the nonsingular PMD

$$\tilde{\tau}_N(\zeta) = \left(I_\lambda - \zeta \mathrm{e}^{\tilde{A}_\nu T}, \zeta \mathrm{e}^{\tilde{A}_\nu T}\mu(\tilde{A}_\nu)\tilde{B}_{2\nu}, \tilde{C}_{2\nu} \right)\,. \tag{7.78}$$

Then the claim of the theorem for a given realisation $(\tilde{A}_\nu, \tilde{B}_{2\nu}, \tilde{C}_{2\nu})$ and the pair $(a_N(\zeta), \zeta b_N(\zeta))$ follows from Theorem 5.64. It remains to prove that the set of stabilising controllers does not depend on the choice of the realisation $(\tilde{A}_\nu, \tilde{B}_{2\nu}, \tilde{C}_{2\nu})$ and the pair $(a_N(\zeta), \zeta b_N(\zeta))$. With this aim in view, we notice that from the formulae of Section 6.8, it follows that

$$w_N(\zeta) = \left[\frac{1}{T} \sum_{k=-\infty}^{\infty} N(s + k\mathrm{j}\omega)\mu(s + k\mathrm{j}\omega) \right] \Bigg|_{\mathrm{e}^{-sT} = \zeta}\,. \tag{7.79}$$

Since all minimal realisations are equivalent, the matrix $w_N(\zeta)$ is independent of the choice of the realisation $(\tilde{A}_\nu, \tilde{B}_{2\nu}, \tilde{C}_{2\nu})$. Hence the set of pairs $(a_N(\zeta), \zeta b_N(\zeta))$ is also independent of the choice of this realisation. The same proposition holds for the set of stabilising controllers. ∎

4. A more complete result can be obtained under the assumption that the poles of the matrix $N(p)$ satisfy the strict conditions for non-pathological behavior (6.124) and (6.125).

Theorem 7.14. *Let the eigenvalues $s_1, \ldots, s_q$ of the matrix $sI_\lambda - \tilde{A}_\nu$ satisfy the strict conditions for non-pathological behavior*

$$s_i - s_k \neq \frac{2n\pi\mathrm{j}}{T}\,, \quad (i \neq k;\; i, k = 1, \ldots, q;\; n = 0, \pm 1, \ldots)\,, \tag{7.80}$$

$$\mu(s_i) \neq 0\,, \quad (i = 1, \ldots, q)\,. \tag{7.81}$$

Then all minimal realisations $(\tilde{A}_\nu, \tilde{B}_{2\nu}, \tilde{C}_{2\nu})$ are stabilisable.

Proof. If (7.80) and (7.81) hold, due to Theorem 6.30, the PMD (7.78) is minimal. Then for the ILMFD

$$\tilde{C}_{2\nu}\left(I_\lambda - \zeta e^{\tilde{A}_\nu T}\right) = a_1^{-1}(\zeta)b_1(\zeta)\,,$$

we have

$$\det a_1(\zeta) \approx \det\left(I_\lambda - \zeta e^{\tilde{A}_\nu T}\right)\,.$$

Hence from Lemma 2.9, it follows that for any ILMFD (7.74)

$$a_N(\zeta) = \phi(\zeta)a_1(\zeta)$$

is true with a unimodular matrix $\phi(\zeta)$. Therefore, in this case due to (7.75), the polynomial

$$\varrho_N(\zeta) = \text{const.} \neq 0$$

is stable. Then the claim follows from Theorem 7.13. ∎

5. A general criterion for the stabilisability of the closed loop is given by the following theorem.

Theorem 7.15. *Let $(\tilde{A}, \tilde{B}_2, \tilde{C}_2)$ of dimension n, η, m be any realisation of the matrix $N(p)$, and $(\tilde{A}_\nu, \tilde{B}_{2\nu}, \tilde{C}_{2\nu})$ be one of its minimal realisation with dimension n, λ, m such that $\eta > \lambda$. Then the function*

$$r(s) = \frac{\det(sI_\eta - \tilde{A})}{\det(sI_\lambda - \tilde{A}_\nu)} \tag{7.82}$$

is a polynomial. If, in addition, the minimal realisation $(\tilde{A}_\nu, \tilde{B}_{2\nu}, \tilde{C}_{2\nu})$ is not stabilisable, so is the realisation $(\tilde{A}, \tilde{B}_2, \tilde{C}_2)$. If the minimal realisation $(\tilde{A}_\nu, \tilde{B}_{2\nu}, \tilde{C}_{2\nu})$ is stabilisable, then for the stabilisability of the realisation $(\tilde{A}, \tilde{B}_2, \tilde{C}_2)$, it is necessary and sufficient that all roots of the polynomial (7.82) be in the open left half-plane. Under this condition, the set of stabilising controllers $(\alpha(\zeta), \beta(\zeta))$ is independent of the realisation $(\tilde{A}, \tilde{B}_2, \tilde{C}_2)$ and is determined by the stability condition for the matrix (7.76).

Proof. Under the given assumptions, the PMDs

$$\begin{aligned} \tau_N(s) &= (sI_\eta - \tilde{A}, \tilde{B}_2, \tilde{C}_2)\,, \\ \tau_{N\nu}(s) &= (sI_\lambda - \tilde{A}_\nu, \tilde{B}_{2\nu}, \tilde{C}_{2\nu}) \end{aligned} \tag{7.83}$$

are equivalent, *i.e.* their transfer matrices coincide. Moreover, since the PMD $\tau_{N\nu}(s)$ is minimal, Relation (7.82) is a polynomial by Lemma 2.48. Let us have

$$\begin{aligned} \det(sI_\eta - \tilde{A}) &= (s - s_1)^{\kappa_1}\cdots(s - s_q)^{\kappa_q}, \quad \kappa_1 + \ldots + \kappa_q = \eta\,, \\ \det(sI_\lambda - \tilde{A}_\nu) &= (s - s_1)^{\ell_1}\cdots(s - s_q)^{\ell_q}, \quad \ell_1 + \ldots + \ell_q = \lambda\,, \end{aligned}$$

where $\kappa_i \geq \ell_i$, $(i = 1, \ldots, q)$. Let $\ell_i < \kappa_i$ for $i = 1, \ldots, \gamma$ and $\ell_i = \kappa_i$ for $i = \gamma + 1, \ldots, q$. Then from (7.82), we obtain

$$r(s) = (s - s_1)^{m_1} \cdots (s - s_\gamma)^{m_\gamma}, \tag{7.84}$$

where $m_i = \kappa_i - \ell_i$, $(i = 1, \ldots, \gamma)$. Moreover, since the PMDs (7.83) are equivalent, using (7.79), we receive

$$\begin{aligned} \zeta \tilde{C}_2 \left(I_\eta - \zeta \mathrm{e}^{\tilde{A}T} \right)^{-1} \mathrm{e}^{\tilde{A}T} \mu(\tilde{A}) \tilde{B}_2 &= \zeta \tilde{C}_{2\nu} \left(I_\lambda - \zeta \mathrm{e}^{\tilde{A}_\nu T} \right)^{-1} \mathrm{e}^{\tilde{A}_\nu T} \mu(\tilde{A}_\nu) \tilde{B}_{2\nu} \\ &= w_N(\zeta) \,. \end{aligned} \tag{7.85}$$

From (7.85), it follows that the PMDs

$$\tau_d(\zeta) = \left(I_\eta - \zeta \mathrm{e}^{\tilde{A}T}, \zeta \mathrm{e}^{\tilde{A}T} \mu(\tilde{A}) \tilde{B}_2, \tilde{C}_2 \right), \tag{7.86}$$

$$\tau_{d\nu}(\zeta) = \left(I_\lambda - \zeta \mathrm{e}^{\tilde{A}_\nu T}, \zeta \mathrm{e}^{\tilde{A}_\nu T} \mu(\tilde{A}_\nu) \tilde{B}_{2\nu}, \tilde{C}_{2\nu} \right) \tag{7.87}$$

are equivalent. Then

$$\begin{aligned} \det(I_\eta - \zeta \mathrm{e}^{\tilde{A}T}) &= (1 - \zeta \mathrm{e}^{s_1 T})^{\kappa_1} \cdots (1 - \zeta \mathrm{e}^{s_q T})^{\kappa_q}, \\ \det(I_\lambda - \zeta \mathrm{e}^{\tilde{A}_\nu T}) &= (1 - \zeta \mathrm{e}^{s_1 T})^{\ell_1} \cdots (1 - \zeta \mathrm{e}^{s_q T})^{\ell_q}, \end{aligned}$$

and the relation

$$\varrho_1(\zeta) = \frac{\det(I_\eta - \zeta \mathrm{e}^{\tilde{A}T})}{\det(I_\lambda - \zeta \mathrm{e}^{\tilde{A}_\nu T})} = (1 - \zeta \mathrm{e}^{s_1 T})^{m_1} \cdots (1 - \zeta \mathrm{e}^{s_\gamma T})^{m_\gamma} \tag{7.88}$$

is a polynomial. Consider the characteristic matrix (7.77) for the PMD (7.86)

$$\tilde{Q}(\zeta, \alpha, \beta) = \begin{bmatrix} I_\eta - \zeta \mathrm{e}^{\tilde{A}T} & O_{\eta n} & -\zeta \mathrm{e}^{\tilde{A}T} \mu(\tilde{A}) \tilde{B}_2 \\ -\tilde{C}_2 & I_n & O_{nm} \\ O_{m\eta} & -\beta(\zeta) & \alpha(\zeta) \end{bmatrix}.$$

Using Equation (4.71) for this and Matrix (7.77), and taking account of (7.85), we find

$$\begin{aligned} \det \tilde{Q}(\zeta, \alpha, \beta) &= \det(I_\eta - \zeta \mathrm{e}^{\tilde{A}T}) \det[\alpha(\zeta) - \beta(\zeta) \tilde{w}_N(\zeta)], \\ \det \tilde{Q}_\nu(\zeta, \alpha, \beta) &= \det(I_\lambda - \zeta \mathrm{e}^{\tilde{A}_\nu T}) \det[\alpha(\zeta) - \beta(\zeta) \tilde{w}_N(\zeta)]. \end{aligned}$$

Hence with (7.88), it follows

$$\det \tilde{Q}(\zeta, \alpha, \beta) = \varrho_1(\zeta) \det \tilde{Q}_\nu(\zeta, \alpha, \beta) \,.$$

If the minimal realisation $(\tilde{A}_\nu, \tilde{B}_{2\nu}, \tilde{C}_{2\nu})$ is not stabilisable, then the matrix $\tilde{Q}_\nu(\zeta, \alpha, \beta)$ is unstable for any controller $(\alpha(\zeta), \beta(\zeta))$. Due to the last equation, the matrix $\tilde{Q}(\zeta,\alpha,\beta)$ is also unstable. If the polynomial $\varrho_1(\zeta)$ is not stable, then the matrix $\tilde{Q}(\zeta\,\alpha,\beta)$ is also unstable, independently of the choice of the controller. Finally, if the polynomial $\varrho_1(\zeta)$ is stable, then the matrix $\tilde{Q}(\zeta,\alpha,\beta)$ is stable or unstable together with the matrix $\tilde{Q}_\nu(\zeta,\alpha,\beta)$.

As a conclusion, we note that from (7.88), it follows that the polynomial $\varrho_1(\zeta)$ is stable iff in (7.84), we have

$$\operatorname{Re} s_i < 0, \quad (i = 1, \dots, \gamma)\,.$$

This completes the proof. ∎

6. Using the above results, we can consider the stabilisation problem for the complete standard sampled-data system.

Theorem 7.16. *Let the continuous-time plant of the standard sampled-data system be given by the state equations (7.61) with a* $\chi \times \chi$ *matrix* A*. Let also* $(\tilde{A}_\nu, \tilde{B}_{2\nu}, \tilde{C}_{2\nu})$ *with dimension* (n, λ, m) *be any minimal realisation of the matrix* $N(p)$*. Then, we have* $\chi \geq \lambda$ *and the function*

$$r(s) = \frac{\det(sI_\chi - A)}{\det(sI_\lambda - \tilde{A}_\nu)} \tag{7.89}$$

is a polynomial. Moreover, if the minimal realisation $(\tilde{A}_\nu, \tilde{B}_{2\nu}, \tilde{C}_{2\nu})$ *is not stabilisable, then also the standard sampled-data system with the plant (7.61) is not stabilisable. If the minimal realisation* $(\tilde{A}_\nu, \tilde{B}_{2\nu}, \tilde{C}_{2\nu})$ *is stabilisable, then for stabilisability of the standard sampled-data system, it is necessary and sufficient that all roots* s_i *of the polynomial (7.89) lie in the open left half-plane. Under this condition, the set of stabilising controllers for the standard sampled-data system coincides with the set of stabilising controllers for the minimal realisation* $(\tilde{A}_\nu, \tilde{B}_{2\nu}, \tilde{C}_{2\nu})$.

Proof. Using (7.61) and (7.62)–(7.64) and assuming $x(t) = O_{\ell 1}$, we can represent the standard sampled-data system in the form

$$\begin{aligned} &y = C_2 v\,, \quad \frac{dv}{dt} = Av + B_2 u \\ &\alpha(\zeta)\psi_k = \beta(\zeta) y_k \\ &u(t) = m(t - kT)\psi_k\,, \quad kT < t < (k+1)T\,. \end{aligned} \tag{7.90}$$

that should be completed with the output equation

$$z(t) = C_1 y(t) + D_L u(t)\,. \tag{7.91}$$

Since

$$C_2(pI_\chi - A)^{-1} B_2 = N(p)\,,$$

due to (7.9), Equations (7.90) can be considered as equations of the closed loop, where the continuous-time plant $N(p)$ is given in form of a realisation (A, B_2, C_2) of dimension n, χ, m, which is not minimal in the general case. Obviously, a necessary and sufficient condition for the stability of the system (7.90) and (7.91) is that the system (7.90) is stable. Hence it follows the conclusion that the stabilisation problem for the standard sampled-data system with the plant (7.61) is equivalent to the stabilisation problem for the closed loop, where the continuous-time plant is given as a realisation (A, B_2, C_2). Therefore, all claims of Theorem 7.16 are corollaries of Theorem 7.15. ■

7.7 Modal Controllability and the Set of Stabilising Controllers

1. Let the continuous-time plant of the standard sampled-data system be given by the state equations

$$\begin{aligned} \frac{dv}{dt} &= Av + B_1 x + B_2 u \\ z = C_1 v\,, &\quad y = C_2 v\,. \end{aligned} \tag{7.92}$$

Then, as follows from Theorems 7.13–7.16, in case of a stabilisable plant (7.92), the characteristic polynomial of the closed-loop standard sampled-data system $\Delta(\zeta)$ can be represented in the form

$$\Delta(\zeta) = \varrho(\zeta)\Delta_d(\zeta)\,, \tag{7.93}$$

where $\varrho(\zeta)$ is a stable polynomial, which is independent of the choice of the controller. Moreover, in (7.93)

$$\Delta_d(\zeta) \approx \det \begin{bmatrix} a_N(\zeta) & -\ b_N(\zeta) \\ -\beta(\zeta) & \alpha(\zeta) \end{bmatrix} = \det Q_N(\zeta, \alpha, \beta)\,, \tag{7.94}$$

where $(\alpha(\zeta), \beta(\zeta))$ is a discrete controller and the matrices $a_N(\zeta)$, $\zeta b_N(\zeta)$ define an ILMFD

$$w_N(\zeta) = \zeta C_2 (I - \zeta \mathrm{e}^{AT})^{-1} \mu(A) \mathrm{e}^{AT} B_2 = \zeta a_N^{-1}(\zeta) b_N(\zeta)\,. \tag{7.95}$$

From (7.93), it follows that the roots of the characteristic polynomial of the standard sampled-data system $\Delta(\zeta)$ can be split up into two groups. The first group (roots of the polynomial $\rho(\zeta)$) is determined only by the properties of the matrix $w(p)$ and is independent of the properties of the discrete controller. Hereinafter, these roots will be called *uncontrollable*. The second group of roots consists of those roots of the polynomial (7.94), which are determined by the matrix $w(p)$ and the controller $(\alpha(\zeta), \beta(\zeta))$. Since the pair $(a_N(\zeta), \zeta b_N(\zeta))$ is irreducible, the controller $(\alpha(\zeta), \beta(\zeta))$ can be chosen in such a way that the polynomial $\Delta_d(\zeta)$ is equal to any given (stable) polynomial. In this connection, the roots of the second group will be called *controllable*

2. The standard sampled-data system with the plant (7.92) will be called *modal controllable*, if all roots of its characteristic polynomial are controllable, *i.e.*, $\varrho(\zeta) = \text{const.} \neq 0$. Under the strict conditions for non-pathological behavior, necessary and sufficient conditions for the system to be modal controllable are given by the following theorem.

Theorem 7.17. *Let the poles of the matrix*

$$w(p) = \begin{bmatrix} K(p) & L(p) \\ M(p) & N(p) \end{bmatrix} = \begin{bmatrix} C_1 \\ C_2 \end{bmatrix} (pI_\chi - A)^{-1} \begin{bmatrix} B_1 & B_2 \end{bmatrix} + \begin{bmatrix} O_{r\ell} & D_L \\ O_{n\ell} & O_{nm} \end{bmatrix} \tag{7.96}$$

satisfy Conditions (7.80) and (7.81). Then, a necessary and sufficient condition for the standard sampled-data system to be modal controllable is that the matrix $N(p)$ dominates in the matrix $w(p)$.

Proof. Sufficiency: Without loss of generality, we take $D_L = O_{rm}$ and assume that the standard representation is minimal. Let the matrix $N(p)$ dominate in Matrix (7.96). Then due to Theorem 2.67, the realisation (A, B_2, C_2) on the right-hand side of (7.96) is minimal. Construct the discrete model $\mathcal{D}_{w\mu}(T,\zeta,t)$ of the matrix $w(p)\mu(p)$. Obviously, we have

$$\mathcal{D}_{w\mu}(T,\zeta,t) = \begin{bmatrix} \mathcal{D}_{K\mu}(T,\zeta,t) & \mathcal{D}_{L\mu}(T,\zeta,t) \\ \mathcal{D}_{M\mu}(T,\zeta,t) & \mathcal{D}_{N\mu}(T,\zeta,t) \end{bmatrix}.$$

Using the second formula in (6.122) and (7.96), we obtain

$$\mathcal{D}_{w\mu}(T,\zeta,t) = \mathcal{D}_1(\zeta,t) + \mathcal{D}_2(t)\,,$$

where

$$\mathcal{D}_1(\zeta,t) = -\begin{bmatrix} C_1 \\ C_2 \end{bmatrix} (\zeta I_\chi - \mathrm{e}^{-AT})^{-1} \mathrm{e}^{A(t-T)} \mu(A) \begin{bmatrix} B_1 & B_2 \end{bmatrix},$$

$$\mathcal{D}_2(t) = -\begin{bmatrix} C_1 \\ C_2 \end{bmatrix} \int_t^T \mathrm{e}^{A(t-\tau)} \mu(\tau)\,\mathrm{d}\tau \begin{bmatrix} B_1 & B_2 \end{bmatrix}.$$

By virtue of Theorem 6.30, the right-hand side of the first equation defines a minimal standard representation of the matrix $\mathcal{D}_1(\zeta,t)$. At the same time, the realisation $\left(\mathrm{e}^{AT}, \mathrm{e}^{A(t-T)}\mu(A)B_2, C_2\right)$ is also minimal. Therefore, we can take $A = \tilde{A}_\nu$ in (7.89). Hence $r_1(s) = \text{const.} \neq 0$ and $\varrho(\zeta) = \text{const.} \neq 0$, and the sufficiency has been proven.

The necessity of the conditions of the theorem is seen by reversing the above derivations. ∎

3. Under the stabilisability condition, the set of stabilising controllers is completely determined by the properties of the matrix $N(p)$, and is defined as the set of pairs $(\alpha(\zeta), \beta(\zeta))$ satisfying (7.94) for all possible stable polynomials $\Delta_d(\zeta)$. The form of Equation (7.94) coincides with (5.163), where the matrix $Q_l(\zeta,\alpha,\beta)$ is given by (5.160). Therefore, to describe the set of stabilising controllers, all the results of Section 5.8 can be used.

4. As a special case, the following propositions hold:

a) Let $(\alpha_0(\zeta), \beta_0(\zeta))$ be a controller, such that

$$\det \begin{bmatrix} a_N(\zeta) & -\zeta b_N(\zeta) \\ -\beta_0(\zeta) & \alpha_0(\zeta) \end{bmatrix} = \text{const.} \neq 0\,.$$

Then the set of all stabilising controllers for the stabilisable standard sampled-data system is given by

$$\begin{aligned} \alpha(\zeta) &= D_l(\zeta)\alpha_0(\zeta) - \zeta M_l(\zeta) b_N(\zeta)\,, \\ \beta(\zeta) &= D_l(\zeta)\beta_0(\zeta) - M_l(\zeta) a_N(\zeta)\,, \end{aligned}$$

where $D_l(\zeta)$ and $M_l(\zeta)$ are any polynomial matrices, but $D_l(\zeta)$ has to be stable. ∎

b) Together with the ILMFD (7.95), let us have an IRMFD

$$w_N(\zeta) = \zeta C_2 (I - \zeta \mathrm{e}^{AT})^{-1} \mathrm{e}^{AT} \mu(A) B_2 = \zeta b_r(\zeta) a_r^{-1}(\zeta)\,.$$

Then the set of stabilising controllers $(\alpha(\zeta), \beta(\zeta))$ for the standard sampled-data system coincides with the set of solutions of the Diophantine equation

$$\alpha(\zeta) a_r(\zeta) - \zeta \beta(\zeta) b_r(\zeta) = D_l(\zeta)\,,$$

where $D_l(\zeta)$ is any stable polynomial matrix. ∎

5. Any stabilising controller $(\alpha(\zeta), \beta(\zeta))$ for the standard sampled-data system fulfills $\det \alpha(0) \neq 0$, *i.e.*, the matrix $\alpha(\zeta)$ is invertible. Therefore, any stabilising controller has a transfer matrix

$$w_d(\zeta) = \alpha^{-1}(\zeta)\beta(\zeta)\,.$$

The following propositions hold:

c) The set of transfer matrices of all stabilising controllers for the standard sampled-data system can be written in the form

$$w_d(\zeta) = \left[\alpha_0(\zeta) - \zeta\phi(\zeta) b_N(\zeta)\right]^{-1} \left[\beta_0(\zeta) - \phi(\zeta) a_N(\zeta)\right],$$

where $\phi(\zeta)$ is any stable rational matrix of compatible dimension.

d) The rational matrix $w_d(\zeta)$ is associated with a stabilising controller for a stabilisable standard system, if and only if there exists any of the following representations:

$$w_d(\zeta) = F_1^{-1}(\zeta) F_2(\zeta), \quad w_d(\zeta) = G_2(\zeta) G_1^{-1}(\zeta)\,,$$

where the pairs of rational matrices $(F_1(\zeta), F_2(\zeta))$ and $[G_1(\zeta), G_2(\zeta)]$ are stable and satisfy the equations

$$\begin{aligned} F_1(\zeta) a_r(\zeta) - \zeta F_2(\zeta) b_r(\zeta) &= I_m\,, \\ a_l(\zeta) G_1(\zeta) - \zeta b_l(\zeta) G_2(\zeta) &= I_n\,. \end{aligned}$$

8

Analysis and Synthesis of SD Systems Under Stochastic Excitation

8.1 Quasi-stationary Stochastic Processes in the Standard SD System

1. Let the input of the standard sampled-data system be acted upon by a vector signal $x(t)$ that is modelled as a centered stochastic process with the *autocorrelation matrix*

$$K_x(\tau) = \mathsf{E}\left[x(t)x'(t+\tau)\right],$$

where $\mathsf{E}[\cdot]$ denotes the operator of mathematical expectation. Assume that the integral

$$\Phi_x(s) = \int_{-\infty}^{\infty} K_x(\tau)\mathrm{e}^{-s\tau}\,\mathrm{d}\tau\,,$$

which will be called the *spectral density* of the input signal, converges absolutely in some stripe $-\alpha_0 \le \operatorname{Re} s \le \alpha_0$, where α_0 is a positive number.

2. Let the block $L(p)$ in the matrix $w(p)$ (7.2) be at least proper and the remaining blocks be strictly proper. Let also the system (7.3)–(7.7) be internally stable. When the input of the standard sampled-data system is the above mentioned signal, after fading away of transient processes, the steady-state stochastic process $z_\infty(t)$ is characterised by the covariance matrix [143, 148]

$$K_z(t_1,t_2) = \frac{1}{2\pi\mathrm{j}}\int_{-\mathrm{j}\infty}^{\mathrm{j}\infty} w(-s,t_1)\Phi_x(s)w'(s,t_2)\mathrm{e}^{s(t_2-t_1)}\,\mathrm{d}s\,, \qquad (8.1)$$

where $w(s,t) = w(s,t+T)$ is the PTM of the system. Hereinafter, the stochastic process $z_\infty(t)$ with the correlation matrix (8.1) will be called *quasi-stationary*. As follows from (8.1), the covariance matrix $K_z(t_1,t_2)$ depends separately on each of its arguments t_1 and t_2 rather than on their difference. Therefore, the quasi-stationary output $z_\infty(t)$ is a non-stationary stochastic process. Since

$$w(-s,t_1) = w(-s,t_1+T), \quad w'(s,t_2) = w'(s,t_2+T), \tag{8.2}$$

we have

$$K_z(t_1,t_2) = K_z(t_1+T, t_2+T).$$

Stochastic processes satisfying this condition will be called *periodically non-stationary*, or shortly periodical. Using this term, we state that the steady-state (quasi-stationary) response of a stable standard sampled-data system to a stationary input signal is a periodically non-stationary stochastic process.

3. The scalar function

$$d_z(t) = \operatorname{trace} K_z(t,t) \tag{8.3}$$

will be called the *variance* of the quasi-stationary output. Here 'trace' denotes the trace of a matrix defined as the sum of its diagonal elements. From (8.1) for $t_1 = t_2 = t$ and (8.3), we find

$$d_z(t) = \frac{1}{2\pi \mathrm{j}} \int_{-\mathrm{j}\infty}^{\mathrm{j}\infty} \operatorname{trace}\left[w(-s,t)\Phi_x(s)w'(s,t)\right] \mathrm{d}s\,. \tag{8.4}$$

Then using (8.2), we obtain

$$d_z(t) = d_z(t+T),$$

i.e., the variance of the quasi-stationary output is a periodic function of its argument t. For matrices A, B of compatible dimensions, the relation

$$\operatorname{trace}(AB) = \operatorname{trace}(BA) \tag{8.5}$$

is well known. Thus in addition to (8.4), the following equivalent relations hold:

$$d_z(t) = \frac{1}{2\pi \mathrm{j}} \int_{-\mathrm{j}\infty}^{\mathrm{j}\infty} \operatorname{trace}\left[\Phi_x(s)w'(s,t)w(-s,t)\right] \mathrm{d}s\,, \tag{8.6}$$

$$d_z(t) = \frac{1}{2\pi \mathrm{j}} \int_{-\mathrm{j}\infty}^{\mathrm{j}\infty} \operatorname{trace}\left[w'(s,t)w(-s,t)\Phi_x(s)\right] \mathrm{d}s\,. \tag{8.7}$$

4. Assume in particular that $\Phi_x(s) = I_\ell$, *i.e.* the input signal is white noise with uncorrelated components. For this case, we denote

$$r_z(t) \triangleq d_z(t).$$

Then (8.6) and (8.4) yield

$$\begin{aligned} r_z(t) &= \frac{1}{2\pi \mathrm{j}} \int_{-\mathrm{j}\infty}^{\mathrm{j}\infty} \operatorname{trace}\left[w'(s,t)w(-s,t)\right] \mathrm{d}s\,, \\ &= \frac{1}{2\pi \mathrm{j}} \int_{-\mathrm{j}\infty}^{\mathrm{j}\infty} \operatorname{trace}\left[w(-s,t)w'(s,t)\right] \mathrm{d}s\,. \end{aligned}$$

Substituting here the variable $-s$ for s, we find also

$$r_z(t) = \frac{1}{2\pi \mathrm{j}} \int_{-\mathrm{j}\infty}^{\mathrm{j}\infty} \operatorname{trace}\left[w'(-s,t)w(s,t)\right] \mathrm{d}s\,,$$
$$= \frac{1}{2\pi \mathrm{j}} \int_{-\mathrm{j}\infty}^{\mathrm{j}\infty} \operatorname{trace}\left[w(s,t)w'(-s,t)\right] \mathrm{d}s\,.$$

5. A practical calculation of the variance $d_z(t)$ using Formulae (8.4), (8.6) and (8.7) causes some technical difficulties, because the integrands of these formulae are transcendent functions of the argument s. To solve the problem, it is reasonable to transform these integrals to those with finite integration limits. The corresponding equations, which stem from (8.6) and (8.7) have the form

$$d_z(t) = \frac{T}{2\pi \mathrm{j}} \int_{-\mathrm{j}\omega/2}^{\mathrm{j}\omega/2} \operatorname{trace} \widehat{U}_1(T,s,t)\,\mathrm{d}s = \frac{T}{2\pi \mathrm{j}} \int_{-\mathrm{j}\omega/2}^{\mathrm{j}\omega/2} \operatorname{trace} \widehat{U}_2(T,s,t)\,\mathrm{d}s\,, \tag{8.8}$$

where $\omega = 2\pi/T$ and

$$\widehat{U}_1(T,s,t) = \frac{1}{T} \sum_{k=-\infty}^{\infty} \Phi_x(s+k\mathrm{j}\omega)w'(s+k\mathrm{j}\omega,t)\underline{w}(s+k\mathrm{j}\omega,t)\,, \tag{8.9}$$

$$\widehat{U}_2(T,s,t) = \frac{1}{T} \sum_{k=-\infty}^{\infty} w'(s+k\mathrm{j}\omega,t)\underline{w}(s+k\mathrm{j}\omega,t)\Phi_x(s+k\mathrm{j}\omega)\,. \tag{8.10}$$

In (8.9) and (8.10) for any function $f(s)$, we denote

$$\underline{f}(s) \stackrel{\triangle}{=} f(-s)\,.$$

Moreover, for any function (matrix) $g(\zeta)$, we use as before the notation

$$\widehat{g}(s) \stackrel{\triangle}{=} g(\zeta)\,|_{\zeta=\mathrm{e}^{-sT}}\,.$$

Obviously, the following reciprocal relations hold:

$$\widehat{g}(s) = g(\zeta)\,|_{\zeta=\mathrm{e}^{-sT}}\,, \quad g(\zeta) = \widehat{g}(s)\,|_{\mathrm{e}^{-sT}=\zeta} \tag{8.11}$$

and per construction

$$\widehat{g}(s) = \widehat{g}(s+\mathrm{j}\omega)\,, \quad \omega = 2\pi/T\,. \tag{8.12}$$

As follows from [148] for a rational matrix $\Phi_x(s)$, the matrices (8.9) and (8.10) are rational matrices of the argument $\zeta = \mathrm{e}^{-sT}$. Therefore, to calculate the integrals (8.8), we could take profit from the technique described in [148]. There exists an alternative way to compute the integrals in (8.8). With this aim in view, we pass to the integration variable ζ in (8.8), such that

$$d_z(t) = \frac{1}{2\pi \mathrm{j}} \oint_\Gamma \operatorname{trace} U_1(T,\zeta,t) \frac{\mathrm{d}\zeta}{\zeta} = \frac{1}{2\pi \mathrm{j}} \oint_\Gamma \operatorname{trace} U_2(T,\zeta,t) \frac{\mathrm{d}\zeta}{\zeta}, \tag{8.13}$$

where, according to the notation (8.11),

$$U_i(T,\zeta,t) = \widehat{U}_i(T,s,t) \,|_{\mathrm{e}^{-sT}=\zeta}\,, \qquad (i = 1,2)$$

are rational matrices in ζ. The integration in (8.13) is performed along the unit circle Γ in positive direction (anti-clockwise). The integrals (8.13) can be easily computed using the residue theorem with account for the fact that all poles of the PTM of a stable standard sampled-data system lie in the open left half-plane. Other ways of calculating these integrals are described in [11] and [177].

6.

Example 8.1. Let us find the variance of the quasi-stationary output for the simple single-loop system shown in Fig. 8.1, where the forming element is a

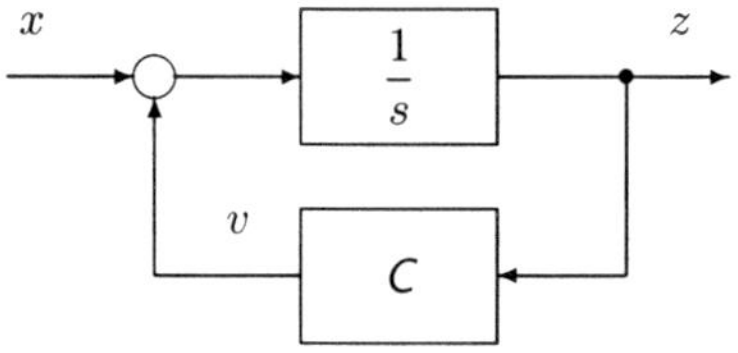

Fig. 8.1. Single sampled-data control loop

zero-order hold with transfer function

$$\mu(s) = \mu_0(s) = \frac{1 - \mathrm{e}^{-sT}}{s}$$

and $\Phi_x(s) = 1$. Using Formulae (7.40)-(7.42), it can be shown that in this case the PTM $w(s,t)$ can be written in the form

$$w(s,t) = \mathrm{e}^{-st}\left[v_0(s) + t\psi_0(s) + c(s,t)\right], \quad 0 \le t \le T \tag{8.14}$$

with

$$\begin{aligned} v_0(s) &= \frac{1-\mathrm{e}^{-sT}}{s} \frac{\widehat{\alpha}(s)}{\widehat{\Delta}(s)}, \\ \psi_0(s) &= \frac{1-\mathrm{e}^{-sT}}{s} \frac{\widehat{\beta}(s)}{\widehat{\Delta}(s)}, \\ c(s,t) &= \frac{\mathrm{e}^{st}-1}{s}, \end{aligned} \tag{8.15}$$

where

$$\widehat{\alpha}(s) = \alpha_0 + \alpha_1 e^{-sT} + \ldots + \alpha_\rho e^{-\rho sT}\,,$$
$$\widehat{\beta}(s) = \beta_0 + \beta_1 e^{-sT} + \ldots + \beta_\rho e^{-\rho sT} \tag{8.16}$$

and

$$\widehat{\Delta}(s) = \left(1 - e^{-sT}\right)\widehat{\alpha}(s) - T e^{-sT}\widehat{\beta}(s)\,. \tag{8.17}$$

Using (8.14)–(8.17) from (8.9) and (8.10) after fairly tedious calculations, it is found that for $0 \le t \le T$

$$\begin{aligned}
\widehat{U}_1(T,s,t) &= \widehat{U}_2(T,s,t) \\
&= T\frac{\widehat{\alpha}(s)\widehat{\alpha}(-s)}{\widehat{\Delta}(s)\widehat{\Delta}(-s)} + tT\left(\frac{\widehat{\alpha}(s)\widehat{\beta}(-s)}{\widehat{\Delta}(s)\widehat{\Delta}(-s)} + \frac{\widehat{\alpha}(-s)\widehat{\beta}(s)}{\widehat{\Delta}(s)\widehat{\Delta}(-s)}\right) \\
&\quad + t^2 T\frac{\widehat{\beta}(s)\widehat{\beta}(-s)}{\widehat{\Delta}(s)\widehat{\Delta}(-s)} + t\left(e^{-sT}\,\frac{\widehat{\alpha}(s)}{\widehat{\Delta}(s)} + e^{sT}\,\frac{\widehat{\alpha}(-s)}{\widehat{\Delta}(-s)}\right) \\
&\quad + t^2\left(e^{-sT}\,\frac{\widehat{\beta}(s)}{\widehat{\Delta}(s)} + e^{sT}\,\frac{\widehat{\beta}(-s)}{\widehat{\Delta}(-s)}\right) + t\,.
\end{aligned} \tag{8.18}$$

To derive Formula (8.18), we employed expressions for the sums of the following series:

$$\frac{1}{T}\sum_{k=-\infty}^{\infty} v_0(s+kj\omega)\underline{v}_0(s+kj\omega) = T\,\frac{\widehat{\alpha}(s)\widehat{\alpha}(-s)}{\widehat{\Delta}(s)\widehat{\Delta}(-s)}\,,$$

$$\frac{1}{T}\sum_{k=-\infty}^{\infty} \psi_0(s+kj\omega)\underline{\psi}_0(s+kj\omega) = T\,\frac{\widehat{\beta}(s)\widehat{\beta}(-s)}{\widehat{\Delta}(s)\widehat{\Delta}(-s)}\,,$$

$$\frac{1}{T}\sum_{k=-\infty}^{\infty} v_0(s+kj\omega)\underline{\psi}_0(s+kj\omega) = T\,\frac{\widehat{\alpha}(s)\widehat{\beta}(-s)}{\widehat{\Delta}(s)\widehat{\Delta}(-s)}\,,$$

$$\frac{1}{T}\sum_{k=-\infty}^{\infty} v_0(s+kj\omega)\underline{c}(s+kj\omega,t) = \frac{\widehat{\alpha}(s)}{\widehat{\Delta}(s)}\,e^{-sT}\,t\,,$$

$$\frac{1}{T}\sum_{k=-\infty}^{\infty} \psi_0(s+kj\omega)\underline{c}(s+kj\omega,t) = \frac{\widehat{\beta}(s)}{\widehat{\Delta}(s)}\,e^{-sT}\,t\,,$$

$$\frac{1}{T}\sum_{k=-\infty}^{\infty} c(s+kj\omega,t)\underline{c}(s+kj\omega,t) = t$$

and

$$
\begin{aligned}
\frac{1}{T}\sum_{k=-\infty}^{\infty}\frac{\mathrm{e}^{(s+kj\omega)t}}{(s+kj\omega)^2} &= \frac{(1-\mathrm{e}^{-sT})t+T\mathrm{e}^{-sT}}{(1-\mathrm{e}^{-sT})^2}\,, \\
\frac{1}{T}\sum_{k=-\infty}^{\infty}\frac{\mathrm{e}^{-(s+kj\omega)t}}{(s+kj\omega)^2} &= \frac{-(1-\mathrm{e}^{-sT})\mathrm{e}^{-sT}t+T\mathrm{e}^{-sT}}{(1-\mathrm{e}^{-sT})^2}\,,
\end{aligned}
\qquad 0\le t\le T\,.
$$

After substiting $\mathrm{e}^{-sT}=\zeta$ in (8.18), we find a rational function of the argument ζ, for which the integrals (8.13) can be calculated elementary. □

8.2 Mean Variance and $\mathcal{H}_2$-norm of the Standard SD System

1. Let $d_z(t)$ be the variance of the quasi-stationary output determined by anyone of Formulae (8.4), (8.6) or (8.7). Then the value

$$
\bar{d}_z \triangleq \frac{1}{T}\int_0^T d_z(t)\,\mathrm{d}t
$$

will be called the *mean variance* of the quasi-stationary output. Using here (8.6) and (8.7), we obtain

$$
\bar{d}_z = \frac{1}{2\pi\mathrm{j}}\int_{-\mathrm{j}\infty}^{\mathrm{j}\infty}\operatorname{trace}\left[\Phi_x(s)\tilde{w}_1(s)\right]\,\mathrm{d}s = \frac{1}{2\pi\mathrm{j}}\int_{-\mathrm{j}\infty}^{\mathrm{j}\infty}\operatorname{trace}\left[\tilde{w}_1(s)\Phi_x(s)\right]\,\mathrm{d}s\,, \tag{8.19}
$$

where

$$
\tilde{w}_1(s) \triangleq \frac{1}{T}\int_0^T w'(s,t)w(-s,t)\,\mathrm{d}t\,. \tag{8.20}
$$

2. When $\Phi_x(s)=I_\ell$, for the mean variance, we will use the special notation

$$
\bar{r}_z \triangleq \frac{1}{T}\int_0^T r_z(t)\,\mathrm{d}t\,.
$$

The value $\bar{r}_z$ is determined by the properties of the standard sampled-data system and does not depend on the properties of the exogenous excitations. Formulae for calculating $\bar{r}_z$ can be derived from (8.19) and (8.20) with $\Phi(s)=I_\ell$. In particular, assuming

$$
\tilde{w}(s) \triangleq \frac{1}{T}\int_0^T w'(-s,t)w(s,t)\,\mathrm{d}t\,, \tag{8.21}
$$

from the formulae in (8.19), we find

$$\bar{r}_z = \frac{1}{2\pi \mathrm{j}} \int_{-\mathrm{j}\infty}^{\mathrm{j}\infty} \operatorname{trace}\left[\tilde{w}(s)\right] \, \mathrm{d}s \,. \tag{8.22}$$

The value

$$\|\mathcal{S}\|_2 = +\sqrt{\bar{r}_z} \tag{8.23}$$

henceforth, will be called the *$\mathcal{H}_2$-norm* of the stable standard sampled-data system $\mathcal{S}$. Hence

$$\|\mathcal{S}\|_2^2 = \frac{1}{2\pi \mathrm{j}} \int_{-\mathrm{j}\infty}^{\mathrm{j}\infty} \operatorname{trace} \tilde{w}(s) \, \mathrm{d}s \,. \tag{8.24}$$

For further transformations, we write the right-hand side of (8.24) in the form

$$\|\mathcal{S}\|_2^2 = \frac{T}{2\pi \mathrm{j}} \int_{-\mathrm{j}\omega/2}^{\mathrm{j}\omega/2} \operatorname{trace} \widehat{\mathcal{D}}_{\tilde{w}}(T,s,0) \, \mathrm{d}s \,,$$

where $\omega = 2\pi/T$ and

$$\widehat{\mathcal{D}}_{\tilde{w}}(T,s,0) = \frac{1}{T} \sum_{k=-\infty}^{\infty} \tilde{w}(s + k\mathrm{j}\omega) \tag{8.25}$$

is a rational matrix in e^{-sT}. Using the substitution $\mathrm{e}^{-sT} = \zeta$, similarly to (8.13), we obtain

$$\|\mathcal{S}\|_2^2 = \frac{1}{2\pi \mathrm{j}} \oint_{\Gamma} \operatorname{trace} \mathcal{D}_{\tilde{w}}(T,\zeta,0) \frac{\mathrm{d}\zeta}{\zeta} \,. \tag{8.26}$$

Example 8.2. Under the conditions of Example 8.1, we get

$$\begin{aligned}
\mathcal{D}_{\tilde{w}}(T,\zeta,0) = {} & T\,\frac{\alpha(\zeta)\alpha(\zeta^{-1})}{\Delta(\zeta)\Delta(\zeta^{-1})} + \frac{T^2}{2}\left(\frac{\alpha(\zeta)\beta(\zeta^{-1})}{\Delta(\zeta)\Delta(\zeta^{-1})} + \frac{\alpha(\zeta^{-1})\beta(\zeta)}{\Delta(\zeta)\Delta(\zeta^{-1})}\right) \\
& + \frac{T^3}{3}\,\frac{\beta(\zeta)\beta(\zeta^{-1})}{\Delta(\zeta)\Delta(\zeta^{-1})} + \frac{T}{2}\left(\frac{\zeta\alpha(\zeta)}{\zeta\Delta(\zeta)} + \frac{\zeta^{-1}\alpha(\zeta^{-1})}{\Delta(\zeta^{-1})}\right) \\
& + \frac{T^2}{3}\left(\frac{\zeta\beta(\zeta)}{\Delta(\zeta)} + \frac{\zeta^{-1}\beta(\zeta^{-1})}{\Delta(\zeta^{-1})}\right) + \frac{T}{2}
\end{aligned}$$

and the integral (8.26) is calculated elementary. □

Remark 8.3. The $\mathcal{H}_2$-norm is defined directly by the PTM. This approach opens the possibility to define the $\mathcal{H}_2$-norm for any system possessing a PTM. Interesting results have been already published by the authors for the class of linear periodically time-varying systems [98, 100, 101, 88, 89]. In contrast to other approaches like [200, 32, 203, 28, 204], the norm computation over the PTM yields closed formulae and needs to evaluate matrices of only finite dimensions.

3. Let us find a general expression for the $\mathcal{H}_2$-norm of the standard sampled-data system (6.3)–(6.6). Using the assumptions of Section 8.1 and notation (8.11), the PTM of the system can be written in the form

$$w(s,t) = \varphi_{L\mu}(T,s,t)\widehat{R}_N(s)M(s) + K(s)\,. \tag{8.27}$$

Substituting $-s$ for s after transposition, we receive

$$\underline{w}'(s,t) = w'(-s,t) = \underline{M}'(s)\underline{\widehat{R}}'_N(s)\varphi_{L'\mu}(T,-s,t) + \underline{K}'(s)\,.$$

Multiplying the last two equations, we find

$$\begin{aligned}\underline{w}'(s,t)w(s,t) = {} & \underline{M}'(s)\underline{\widehat{R}}'_N(s)\varphi_{L'\mu}(T,-s,t)\varphi_{L\mu}(T,s,t)\widehat{R}_N(s)M(s)\\ & + \underline{M}'(s)\underline{\widehat{R}}'_N(s)\varphi_{L'\mu}(T,-s,t)K(s)\\ & + \underline{K}'(s)\varphi_{L\mu}(T,s,t)\widehat{R}_N(s)M(s) + \underline{K}'(s)K(s)\,.\end{aligned}$$

Using this in (8.21) yields

$$\begin{aligned}\tilde{w}(s) = {} & \underline{M}'(s)\underline{\widehat{R}}'_N(s)\widehat{D}_L(s)\widehat{R}_N(s)M(s) + \underline{M}'(s)\underline{\widehat{R}}'_N(s)\underline{Q}'_L(s)K(s)\\ & + \underline{K}'(s)Q_L(s)\widehat{R}_N(s)M(s) + \underline{K}'(s)K(s)\,,\end{aligned} \tag{8.28}$$

where

$$\begin{aligned}\widehat{D}_L(s) &= \frac{1}{T}\int_0^T \varphi_{L'\mu}(T,-s,t)\varphi_{L\mu}(T,s,t)\,\mathrm{d}t\,,\\ Q_L(s) &= \frac{1}{T}\int_0^T \varphi_{L\mu}(T,s,t)\,\mathrm{d}t = \frac{1}{T}L(s)\mu(s)\,,\\ \underline{Q}'_L(s) &= \frac{1}{T}\int_0^T \varphi_{L'\mu}(T,-s,t)\,\mathrm{d}t = \frac{1}{T}L'(-s)\mu(-s)\,.\end{aligned}$$

Using (8.28) in (8.24), we obtain

$$\begin{aligned}\|S\|_2^2 = \frac{1}{2\pi \mathrm{j}}\int_{-\mathrm{j}\infty}^{\mathrm{j}\infty} \operatorname{trace}\Big[& \underline{M}'(s)\underline{\widehat{R}}'_N(s)\widehat{D}_L(s)\widehat{R}_N(s)M(s)\\ & + \underline{M}'(s)\underline{\widehat{R}}'_N(s)\underline{Q}'_L(s)K(s)\\ & + \underline{K}'(s)Q_L(s)\widehat{R}_N(s)M(s) + \underline{K}'(s)K(s)\Big]\,\mathrm{d}s\,.\end{aligned} \tag{8.29}$$

All matrices in the integrand, except for the matrix $\widehat{R}_N(s)$, are determined by the transfer matrix $w(s)$ of the continuous plant and are independent of the transfer matrix $\widehat{w}_d(s)$ of the controller. Moreover, each transfer matrix

$\widehat{w}_d(s)$ of a stabilising controller is associated with a nonnegative value $\|S\|_2^2$. Therefore, the right-hand side of (8.29) can be considered as a functional defined over the set of transfer functions of stabilising controllers $\widehat{w}_d(s)$. Hence the following optimisation problem arises naturally.

> **$\mathcal{H}_2$-problem.** Let the matrix $w(p)$ in (7.2) be given, where the matrix $L(p)$ is at least proper and the remaining elements are strictly proper. Furthermore, the sampling period T and the impulse form $m(t)$ are fixed. Find the transfer function of a stabilising controller $\widehat{w}_d(s)$, which minimises the functional (8.29).

8.3 Representing the PTM in Terms of the System Function

1. Equation (8.29) is not fairly convenient for solving the $\mathcal{H}_2$-optimisation problem. As will be shown below, a representation of the $\mathcal{H}_2$-norm in terms of the so-called system function is more suitable for this purpose. To construct such a representation, we must at first write the PTM in terms of the system function. This topic is considered in the present section.

2. To simplify the further reading, we summarise some relations obtained above. Heed that the notation slightly differs from that in the previous exposition.

Using (8.11), we present the PTM of the standard sampled-data system (7.2)–(7.7) in the form (8.27), where the matrix $\varphi_{L\mu}(T, s, t)$ is determined by (7.29)

$$\widehat{R}_N(s) = \widehat{w}_d(s)\left[I_n - \widehat{D}_{N\mu}(T, s, 0)\widehat{w}_d(s)\right]^{-1} \tag{8.30}$$

with

$$\widehat{D}_{N\mu}(T, s, 0) = \varphi_{N\mu}(T, s, 0) = \frac{1}{T}\sum_{k=-\infty}^{\infty} N(s + kj\omega)\mu(s + kj\omega)$$

and

$$\widehat{w}_d(s) = \widehat{\alpha}_l^{-1}(s)\widehat{\beta}_l(s)\,, \tag{8.31}$$

where

$$\begin{aligned}\widehat{\alpha}_l(s) &= \alpha_0 + \alpha_1 e^{-sT} + \ldots + \alpha_\rho e^{-\rho sT}\,,\\ \widehat{\beta}_l(s) &= \beta_0 + \beta_1 e^{-sT} + \ldots + \beta_\rho e^{-\rho sT}\end{aligned}$$

are polynomial matrices in the variable $\zeta = e^{-sT}$. Moreover, $\mu(s)$ is the transfer function of the forming element (6.38). Matrix (8.31) will be called the transfer function of the controller.

3. The PTM (8.27) is associated with the rational matrix

$$w(p) = \begin{array}{c} \quad \ell \qquad m \\ \begin{bmatrix} K(p) & L(p) \\ M(p) & N(p) \end{bmatrix} \end{array} \begin{array}{c} \kappa \\ n \end{array}.$$

Henceforth as above, we assume that the matrix $L(p)$ is at least proper and the remaining elements are strictly proper. The above matrix can be associated with the state equations (7.10)

$$\frac{\mathrm{d}v}{\mathrm{d}t} = Av + B_1 x + B_2 u$$

$$z = C_1 v + D_L u\,, \quad y = C_2 v\,,$$

where A is a constant $\chi \times \chi$ matrix. Without loss of generality, we can assume that the pairs

$$(A, \begin{bmatrix} B_1 & B_2 \end{bmatrix}), \quad [A, \begin{bmatrix} C_1 \\ C_2 \end{bmatrix}]$$

are controllable and observable, respectively.

4. As follows from Theorems 7.3 and 7.4, the PTM (8.27) admits a representation of the form

$$w(s,t) = \frac{P_w(s,t)}{\widehat{\Delta}(s)}\,, \tag{8.32}$$

where $P_w(s,t) = P_w(s,t+T)$ is a $\kappa \times \ell$ matrix, whose elements are integer functions in s for all t and the function $\widehat{\Delta}(s)$ is given by

$$\widehat{\Delta}(s) = \det \widehat{Q}(s, \widehat{\alpha}_l, \widehat{\beta}_l)\,, \tag{8.33}$$

where $\widehat{Q}(s, \widehat{\alpha}_l, \widehat{\beta}_l)$ is a matrix of the form

$$\widehat{Q}(s, \widehat{\alpha}_l, \widehat{\beta}_l) = \begin{bmatrix} I_\chi - \mathrm{e}^{-sT}\mathrm{e}^{AT} & O_{\chi n} & -\mathrm{e}^{-sT}\mathrm{e}^{AT}\mu(A)B_2 \\ -C_2 & I_n & O_{nm} \\ O_{m\chi} & -\widehat{\beta}_l(s) & \widehat{\alpha}_l(s) \end{bmatrix}.$$

Assuming $\mathrm{e}^{-sT} = \zeta$ in (8.33), we find the characteristic polynomial

$$\Delta(\zeta) = \widehat{\Delta}(s)\,|_{\mathrm{e}^{-sT}=\zeta} = \det Q(\zeta, \alpha_l, \beta_l)\,, \tag{8.34}$$

where

$$Q(\zeta, \alpha_l, \beta_l) = \begin{bmatrix} I_\chi - \zeta\mathrm{e}^{AT} & O_{\chi n} & -\zeta\mathrm{e}^{AT}\mu(A)B_2 \\ -C_2 & I_n & O_{nm} \\ O_{m\chi} & -\beta_l(\zeta) & \alpha_l(\zeta) \end{bmatrix}$$

with

$$
\begin{aligned}
\alpha_l(\zeta) &= \alpha_0 + \alpha_1\zeta + \ldots + \alpha_\rho\zeta^\rho\,, \\
\beta_l(\zeta) &= \beta_0 + \beta_1\zeta + \ldots + \beta_\rho\zeta^\rho\,.
\end{aligned}
\tag{8.35}
$$

For brevity, we will refer to the matrices (8.35) as a controller and the matrix

$$
w_d(\zeta) = \widehat{w}_d(s)\,|_{e^{-T}=\zeta} = \alpha_l^{-1}(\zeta)\beta_l(\zeta)
$$

as well as $\widehat{w}_d(s)$ are transfer functions of this controller.

5. As was shown in Chapter 6, the following equations hold:

$$
\begin{aligned}
D_{N\mu}(T,\zeta,0) &= \widehat{D}_{N\mu}(T,s,0)\,|_{e^{-sT}=\zeta} \\
&= \left[\frac{1}{T}\sum_{k=-\infty}^{\infty} N(s+kj\omega)\mu(s+kj\omega)\right]|_{e^{-sT}=\zeta} \\
&= \zeta C_2(I_\chi - \zeta e^{AT})^{-1}e^{AT}\mu(A)B_2 = w_N(\zeta)\,.
\end{aligned}
$$

If this rational matrix is associated with an ILMFD

$$
w_N(\zeta) = D_{N\mu}(T,\zeta,0) = \zeta a_l^{-1}(\zeta)b_l(\zeta)\,, \tag{8.36}
$$

then the function

$$
\varrho(\zeta) = \frac{\det(I_\chi - \zeta e^{AT})}{\det a_l(\zeta)} \tag{8.37}
$$

is a polynomial, which is independent of the choice of the controller (8.35). The characteristic polynomial (8.34) has the form

$$
\Delta(\zeta) \approx \varrho(\zeta)\Delta_d(\zeta)\,, \tag{8.38}
$$

where $\Delta_d(\zeta)$ is a polynomial determined by the choice of the controller (8.35). Moreover, the standard system is stabilisable, if and only if the polynomial $\varrho(\zeta)$ is stable. The polynomial $\Delta_d(\zeta)$ appearing in (8.38) satisfies the relation

$$
\Delta_d(\zeta) \approx \det Q_N(\zeta,\alpha_l,\beta_l)\,,
$$

where

$$
Q_N(\zeta,\alpha,\beta) = \begin{bmatrix} a_l(\zeta) & -\zeta b_l(\zeta) \\ -\beta_l(\zeta) & \alpha_l(\zeta) \end{bmatrix} \tag{8.39}
$$

is a polynomial matrix (7.76). If the stabilisability conditions hold, then the set of stabilising controllers for the standard sampled-data system coincide with the set of controllers (8.35) with stable matrices (8.39).

6. Let $(\alpha_{0l}(\zeta), \beta_{0l}(\zeta))$ be a basic controller such that

$$\Delta_d(\zeta) \approx \det Q_N(\zeta, \alpha_{0l}, \beta_{0l}) = \text{const.} \neq 0 \,. \tag{8.40}$$

Then, as was proved before, the set of all causal stabilising controllers can be given by

$$\begin{aligned} \alpha_l(\zeta) &= D_l(\zeta)\alpha_{0l}(\zeta) - \zeta M_l(\zeta) b_l(\zeta)\,, \\ \beta_l(\zeta) &= D_l(\zeta)\beta_{0l}(\zeta) - M_l(\zeta) a_l(\zeta)\,, \end{aligned} \tag{8.41}$$

where $M_l(\zeta)$ and $D_l(\zeta)$ are polynomial matrices, the first of them can be chosen arbitrarily, while the second one must be stable. Then the transfer matrix of any stabilising controller for a given basic controller has a unique left representation of the form

$$w_d(\zeta) = \alpha_l^{-1}(\zeta)\beta_l(\zeta) = \left[\alpha_{0l}(\zeta) - \zeta\theta(\zeta) b_l(\zeta)\right]^{-1}\left[\beta_{0l}(\zeta) - \theta(\zeta) a_l(\zeta)\right],$$

where

$$\theta(\zeta) = D_l^{-1}(\zeta) M_l(\zeta)$$

is a stable rational matrix, which will hereinafter be called the *system function* of the standard sampled-data system.

7. Together with an ILMFD (8.36), let us have an IRMFD

$$w_N(\zeta) = D_{N\mu}(T, \zeta, 0) = \zeta b_r(\zeta) a_r^{-1}(\zeta) \tag{8.42}$$

and let $(\alpha_{0l}(\zeta), \beta_{0l}(\zeta))$ and $[\alpha_{0r}(\zeta), \beta_{0r}(\zeta)]$ be two dual basic controllers corresponding to the IMFDs (8.36) and (8.42). These controllers will be called *initial controllers*. Then the transfer matrix of a stabilising controller admits the right representation

$$w_d(\zeta) = \beta_r(\zeta)\alpha_r^{-1}(\zeta) \tag{8.43}$$

with

$$\begin{aligned} \alpha_r(\zeta) &= \alpha_{0r}(\zeta) D_r(\zeta) - \zeta b_r(\zeta) M_r(\zeta)\,, \\ \beta_r(\zeta) &= \beta_{0r}(\zeta) D_r(\zeta) - a_r(\zeta) M_r(\zeta)\,, \end{aligned} \tag{8.44}$$

where $D_r(\zeta)$ is a stable polynomial matrix and $M_r(\zeta)$ is an arbitrary polynomial matrix. Thus

$$M_r(\zeta) D_r^{-1}(\zeta) = \theta(\zeta)$$

and Equation (8.43) appears as

$$w_d(\zeta) = \left[\beta_{0r}(\zeta) - a_r(\zeta)\theta(\zeta)\right]\left[\alpha_{0r}(\zeta) - \zeta b_r(\zeta)\theta(\zeta)\right]^{-1} . \tag{8.45}$$

Moreover, according to (5.182), (5.183) and (5.181), we obtain

$$w_d(\zeta) = V_2(\zeta)V_1^{-1}(\zeta), \tag{8.46}$$

where

$$\begin{aligned} V_1(\zeta) &= \left[a_l(\zeta) - \zeta b_l(\zeta)w_d(\zeta)\right]^{-1} = \alpha_{0r}(\zeta) - \zeta b_r(\zeta)\theta(\zeta), \\ V_2(\zeta) &= w_d(\zeta)\left[a_l(\zeta) - \zeta b_l(\zeta)w_d(\zeta)\right]^{-1} = \beta_{0r}(\zeta) - a_r(\zeta)\theta(\zeta). \end{aligned} \tag{8.47}$$

8. Using (8.47), we can write the matrix $\overset{\frown}{R}_N(s)$ in (8.30) in terms of the system function $\theta(\zeta)$. Indeed, using (8.30), (8.36), and (8.47), we find

$$\begin{aligned} R_N(\zeta) = \overset{\frown}{R}_N(s)\,|_{e^{-sT}=\zeta} &= w_d(\zeta)\left[I_n - D_{N\mu}(T,\zeta,0)w_d(\zeta)\right]^{-1} \\ &= w_d(\zeta)\left[I_n - \zeta a_l^{-1}(\zeta)b_l(\zeta)w_d(\zeta)\right]^{-1} \\ &= w_d(\zeta)\left[a_l(\zeta) - \zeta b_l(\zeta)w_d(\zeta)\right]^{-1}a_l(\zeta) = V_2(\zeta)a_l(\zeta). \end{aligned} \tag{8.48}$$

From (8.47) and (8.48), we obtain

$$R_N(\zeta) = \beta_{0r}(\zeta)a_l(\zeta) - a_r(\zeta)\theta(\zeta)a_l(\zeta). \tag{8.49}$$

Hence

$$\overset{\frown}{R}_N(s) = R_N(\zeta)\,|_{\zeta=e^{-sT}} = \overset{\frown}{\beta}_{0r}(s)\overset{\frown}{a}_l(s) - \overset{\frown}{a}_r(s)\overset{\frown}{\theta}(s)\overset{\frown}{a}_l(s). \tag{8.50}$$

Substituting (8.50) into (8.27), we obtain

$$w(s,t) = \psi(s,t)\overset{\frown}{\theta}(s)\xi(s) + \eta(s,t), \tag{8.51}$$

where

$$\begin{aligned} \psi(s,t) &= -\varphi_{L\mu}(T,s,t)\overset{\frown}{a}_r(s), \\ \xi(s) &= \overset{\frown}{a}_l(s)M(s), \\ \eta(s,t) &= \varphi_{L\mu}(T,s,t)\overset{\frown}{\beta}_{0r}(s)\overset{\frown}{a}_l(s)M(s) + K(s). \end{aligned} \tag{8.52}$$

Equation (8.51) will hereinafter be called a *representation of the PTM in terms of the system function.* The matrices (8.52) will be called the *coefficients of this representation.*

9. Below, we will prove several propositions showing that the coefficients (matrices) (8.52) should be calculated with account for a number of important cancellations.

Theorem 8.4. *The poles of the matrix $\eta(s,t)$ belong to the set of roots of the function $\widehat{\varrho}(s) = \varrho(\mathrm{e}^{-sT})$, where $\rho(\zeta)$ is the polynomial given by (8.37).*

Proof. Assume $D_l(\zeta) = I$ and $M_l(\zeta) = O_{mn}$ in (8.41), *i.e.* we choose the initial controller $(\alpha_{0l}(\zeta), \beta_{0l}(\zeta))$. In this case, $\widehat{\theta}(s) = O_{mn}$ and

$$w(s,t) = \eta(s,t)\,.$$

Since the controller $(\alpha_{0l}(\zeta), \beta_{0l}(\zeta))$ is a basic controller, we have (8.40) and from (8.38), it follows

$$\Delta(\zeta) \approx \varrho(\zeta)\,.$$

Assuming this, from (8.32) we get

$$w(s,t) = \eta(s,t) = \frac{P_\eta(s,t)}{\widehat{\varrho}(s)}\,, \tag{8.53}$$

where the matrix $P_\eta(s,t)$ is an integral function of the argument s. The claim of the theorem follows from (8.53). ∎

Corollary 8.5. *Let the standard sampled-data system be modal controllable, i.e.* $\widehat{\varrho}(s) = \text{const.} \neq 0$. *Then the matrix $\eta(s,t)$ is an integral function in s.*

Theorem 8.6. *For any polynomial matrix $\theta(\zeta)$, the set of poles of the matrix*

$$G(s,t) = \psi(s,t)\widehat{\theta}(s)\xi(s) \tag{8.54}$$

belongs to the set of roots of the function $\widehat{\varrho}(s)$.

Proof. Let $\widehat{\theta}(s) = \theta(\zeta)\,|_{\zeta=\mathrm{e}^{-sT}}$ with a polynomial matrix $\theta(\zeta)$. Then for any ILMFD

$$\theta(\zeta) = D_l^{-1}(\zeta)M_l(\zeta)\,,$$

the matrix $D_l(\zeta)$ is unimodular. Therefore, due to Theorem 4.1, the controller (8.41) is a basic controller. Hence we have $\Delta_d(\zeta) = const \neq 0$. In this case, $\widehat{\Delta}_d(s) = \text{const.} \neq 0$ and (8.38) yields

$$\Delta(\zeta) \approx \varrho(\zeta)\,.$$

From this relation and (8.32), we obtain

$$w(s,t) = \frac{\tilde{P}_w(s,t)}{\widehat{\varrho}(s)}\,,$$

where the matrix $\tilde{P}_w(s,t)$ is an integral function in s. Using (8.53) and the last equation, we obtain

$$G(s,t) = w(s,t) - \eta(s,t) = \frac{P_G(s,t)}{\widehat{\varrho}(s)}\,, \tag{8.55}$$

where the matrix $P_G(s,t)$ is an integral function in s. ∎

Corollary 8.7. *If the standard sampled-data system is modal controllable, then the matrix $G(s,t)$ is an integral function of the argument s for any polynomial $\theta(\zeta)$.* ∎

10. In principle for any $\theta(\zeta)$, the right-hand side of (8.54) can be cancellee by a function $\widehat{\varrho}_1(s)$, where $\varrho_1(\zeta)$ is a polynomial independent of t. In this case after cancellation, we obtain an expression similar to (8.55):

$$G(s,t) = \frac{P_{Gm}(s,t)}{\widehat{\varrho}_m(s)}, \tag{8.56}$$

where $\deg \varrho_m(\zeta) < \deg \varrho(\zeta)$. If $\deg \varrho_m(\zeta)$ has the minimal possible value independent of the choice of $\theta(\zeta)$, the function (8.56) will be called *globally irreducible.*

Using (8.52), we can represent (8.56) in the form

$$-\varphi_{L\mu}(T,s,t)\widehat{a}_r(s)\widehat{\theta}(s)\widehat{a}_l(s)M(s) = \frac{P_{Gm}(s,t)}{\widehat{\varrho}_m(s)} = G(s,t)\,.$$

Multiplying this by e^{st}, we find

$$\mathrm{e}^{st}G(s,t) \triangleq G_1(s,t) = -\widehat{D}_{L\mu}(T,s,t)\widehat{a}_r(s)\widehat{\theta}(s)\widehat{a}_l(s)M(s)\,.$$

Hence

$$\begin{aligned}\frac{1}{T}\sum_{k=0}^{\infty} G_1(s+kj\omega,t)\mathrm{e}^{(s+kj\omega)t} &= -\widehat{D}_{L\mu}(T,s,t)\widehat{a}_r(s)\widehat{\theta}(s)\widehat{a}_l(s)\widehat{D}_M(T,s,t)\\ &= \frac{\widehat{N}_{G1}(s,t)}{\widehat{\varrho}_m(s)},\end{aligned} \tag{8.57}$$

where $N_{G1}(\zeta,t)$ is a polynomial matrix in ζ for any $0 < t < T$.

11. The following propositions prove some further cancellations in the calculation of the matrices $\psi(s,t)$ and $\xi(s)$ appearing in (8.52).

Theorem 8.8. *For $0 < t < T$, let us have the irreducible representations*

$$\widehat{D}_{L\mu}(T,s,t)\widehat{a}_r(s) = \frac{\widehat{N}_L(s,t)}{\widehat{\varrho}_L(s)}, \tag{8.58}$$

$$\widehat{a}_l(s)\widehat{D}_M(T,s,t) = \frac{\widehat{N}_M(s,t)}{\widehat{\varrho}_M(s)}, \tag{8.59}$$

where $N_L(\zeta,t)$ and $N_M(\zeta,t)$ are polynomial matrices in ζ, and $\varrho_M(\zeta)$ and $\varrho_M(\zeta)$ are scalar polynomials. Let also the fractions

$$\frac{N_L(\zeta,t)}{\varrho_M(\zeta)}, \quad \frac{N_M(\zeta,t)}{\varrho_L(\zeta)}$$

be irreducible and the fraction (8.57) be globally irreducible. Then the function

$$\gamma(\zeta) \stackrel{\triangle}{=} \frac{\varrho(\zeta)}{\varrho_L(\zeta)\varrho_M(\zeta)}$$

is a polynomial. Moreover,

$$\varrho_L(\zeta)\varrho_M(\zeta) \approx \varrho_m(\zeta)\,.$$

The proof of the theorem is preceded by two auxiliary claims.

Lemma 8.9. *Let A and B be constant matrices of dimensions $n \times m$ and $\ell \times \kappa$, respectively. Moreover, for any $m \times \ell$ matrix Ω, the equation*

$$A\Omega B = O_{n\kappa} \tag{8.60}$$

should hold. Then at least one of the matrices A, B is a zero matrix.

Proof. Assume the converse, namely, let us have (8.60) for any Ω, where A and B are both nonzero. Let the elements a_{ij} and b_{pq} of the matrices A and B be nonzero. Assume $\Omega = I_{jp}$, where I_{jp} is an $m \times \ell$ matrix having the single unit element at the cross of the j-th row and the p-th column, while all other elements are zero. It can be easily verified that the product $AI_{jp}B$ is nonzero. This contradiction proves the Lemma. ∎

Lemma 8.10. *Let us have two irreducible rational $n \times m$ and $\ell \times \kappa$ matrices in the standard form*

$$A(\lambda) = \frac{N_A(\lambda)}{d_A(\lambda)}, \quad B(\lambda) = \frac{N_B(\lambda)}{d_B(\lambda)}. \tag{8.61}$$

Let also the fractions

$$\frac{N_A(\lambda)}{d_B(\lambda)}, \quad \frac{N_B(\lambda)}{d_A(\lambda)} \tag{8.62}$$

be irreducible. For any $m \times \ell$ polynomial matrix $\Omega(\lambda)$, let us have

$$\frac{N_A(\lambda)}{d_A(\lambda)}\Omega(\lambda)\frac{N_B(\lambda)}{d_B(\lambda)} = \frac{N(\lambda)}{d_0(\lambda)}, \quad \deg d_0(\lambda) = \delta\,, \tag{8.63}$$

where $d_0(\lambda)$ is a fixed polynomial and $N(\lambda)$ is a polynomial matrix. Then the function

$$\tilde{\gamma}(\lambda) \stackrel{\triangle}{=} \frac{d_0(\lambda)}{d_A(\lambda)d_B(\lambda)} \tag{8.64}$$

is a polynomial. Moreover, if the right-hand side of (8.63) is globally irreducible, then

$$d_{AB}(\lambda) \stackrel{\triangle}{=} d_A(\lambda)d_B(\lambda) \approx d_0(\lambda)\,. \tag{8.65}$$

Proof. Assume that the function (8.64) is not a polynomial. If $p(\lambda)$ is a GCD of the the polynomials $d_0(\lambda)$ and $d_{AB}(\lambda)$, then

$$d_0(\lambda) = p(\lambda)d_1(\lambda), \quad d_{AB}(\lambda) = p(\lambda)d_2(\lambda),$$

where the polynomials $d_1(\lambda)$ and $d_2(\lambda)$ are coprime and $\deg d_2(\lambda) > 0$. Substituting these equations into (8.63), we obtain

$$N_A(\lambda)\Omega(\lambda)N_B(\lambda) = N(\lambda)\frac{d_2(\lambda)}{d_1(\lambda)}.$$

Let λ_0 be a root of the polynomial $d_2(\lambda)$. Then for $\lambda = \lambda_0$, the equality

$$N_A(\lambda_0)\Omega(\lambda_0)N_B(\lambda_0) = O_{n\kappa}$$

can be written for any constant matrix $\Omega(\lambda_0)$. Then with the help of Lemma 8.9, it follows that at least one of the following two equations holds:

$$N_A(\lambda_0) = O_{nm} \quad \text{or} \quad N_B(\lambda_0) = O_{\ell\kappa}.$$

In this case, at least one of the rational matrices (8.61) or (8.62) appears to be reducible. This contradicts the assumptions. Thus, $\deg d_2(\lambda) = 0$ and $\tilde{\gamma}(\lambda)$ is a polynomial.

Now, let the right-hand side of (8.63) be globally irreducible. We show that in this case $\deg d_1(\lambda) = 0$ and we have (8.65). Indeed, if we assume the converse, we have $\deg d_0(\lambda) > \deg d_{AB}(\lambda)$. This contradicts the assumption that the right-hand side of (8.63) is globally irreducible. ∎

Proof (of Theorem 8.8). From (8.57)-(8.59) for $\mathrm{e}^{-sT} = \zeta$, we obtain

$$\frac{N_L(\zeta,t)}{\varrho_L(\zeta)}\theta(\zeta)\frac{N_M(\zeta,t)}{\varrho_M(\zeta)} = \frac{N_{G1}(\zeta)}{\varrho_m(\zeta)}.$$

Since here the polynomial matrix $\theta(\zeta)$ can be chosen arbitrarily, the claim of the theorem stems directly from Lemma 8.10. ∎

Corollary 8.11. *When under the conditions of Theorem 8.8, the right-hand side of (8.55) is globally irreducible, then we have*

$$\varrho_L(\zeta)\varrho_M(\zeta) \approx \varrho(\zeta). \tag{8.66}$$

∎

Corollary 8.12. *As follows from the above reasoning, the converse proposition is also valid: When under the conditions of Theorem 8.8, Equation (8.66) holds, then the representations (8.53) and (8.55) are globally irreducible.* ∎

Theorem 8.13. *Let the conditions of Theorem 8.8 hold. Then we have the irreducible representations*

$$\psi(s,t) = -\varphi_{L\mu}(T,s,t)\widehat{a}_r(s) = \frac{P_\psi(s,t)}{\widehat{\varrho}_L(s)}\,, \tag{8.67}$$

$$\xi(s) = \widehat{a}_l(s)M(s) = \frac{P_\xi(s)}{\widehat{\varrho}_M(s)}\,, \tag{8.68}$$

where the numerators are integral functions of the argument s.

Proof. Multiplying the first equation in (8.58) by e^{-st}, we obtain

$$\psi(s,t) = -\varphi_{L\mu}(T,s,t)\widehat{a}_r(s) = \frac{\mathrm{e}^{-st}\widehat{N}_L(s,t)}{\widehat{\varrho}_L(s)}\,.$$

The matrix $\mathrm{e}^{-st}\widehat{N}_L(s,t)$ is an integral function and the fraction on the right-hand side is irreducible, because the function e^{-st} has no zeros. Thus, (8.67) has been proven.

Further, multiplying (8.59) by e^{-st} and integrating by t, we find

$$\begin{aligned}\xi(s) &= \widehat{a}_l(s)M(s) = \\ &= \int_0^T \widehat{a}_l(s)\mathrm{e}^{-st}\widehat{D}_M(T,s,t)\,\mathrm{d}t = \frac{\int_0^T \mathrm{e}^{-st}\widehat{N}_M(s,t)\,\mathrm{d}t}{\widehat{\varrho}_M(s)}\,.\end{aligned}$$

The numerator of the latter expression is an integral function in s, *i.e.*, we have (8.68). It remains to prove that representation (8.68) is irreducible.

Assume the converse, *i.e.*, let the representation (8.68) be reducible. Then

$$\xi(s) = \widehat{a}_l(s)M(s) = \frac{P_{\xi 1}(s)}{\widehat{\varrho}_{M1}(s)}\,,$$

where $\deg \varrho_{M1}(\zeta) < \deg \varrho_M(\zeta)$. With respect to (8.59), we thus obtain an expression of the form

$$\widehat{a}_l(s)\widehat{D}_M(T,s,t) = \frac{1}{T}\sum_{k=-\infty}^{\infty} \xi(s+kj\omega)\mathrm{e}^{(s+kj\omega)t} = \frac{\widehat{N}_{M1}(s,t)}{\widehat{\varrho}_{M1}(s)}\,, \tag{8.69}$$

where $N_{M1}(\zeta,t)$ is a polynomial matrix in ζ for $0 \le t \le T$. This contradicts the irreducibility assumption of the right-hand side of (8.59). Hence (8.68) is irreducible. ∎

Corollary 8.14. *In case of modal controllability, Matrices (8.67), (8.68) and (8.69) are integral functions of the argument s for* $0 \le t \le T$. ∎

Corollary 8.15. *In a similar way, it can be proved that for an irreducible representation (8.58), the following irreducible representation holds:*

$$L(s)\widehat{a}_r(s)\mu(s) = \frac{P_L(s)}{\widehat{\varrho}_L(s)}\,,$$

where $P_L(s)$ is an integral function in s.

8.4 Representing the $\mathcal{H}_2$-norm in Terms of the System Function

1. In this section on the basis of (8.21)–(8.25), we construct expressions for the value $\|\mathcal{S}\|_2^2$ for the standard sampled-data system, using the representation of the PTM $w(s,t)$ in terms of the system function $\theta(\zeta)$ defined by (8.51). From (8.51), we have

$$\underline{w}'(s,t) = w'(-s,t) = \xi'(-s)\widehat{\theta}'(-s)\psi'(-s,t) + \eta'(-s,t)\,.$$

Multiplying this with the function (8.51), we receive

$$\begin{aligned} w'(-s,t)w(s,t) &= \eta'(-s,t)\eta(s,t) + \xi'(-s)\widehat{\theta}'(-s)\psi'(-s,t)\psi(s,t)\widehat{\theta}(s)\xi(s) \\ &\quad + \xi'(-s)\widehat{\theta}'(-s)\psi'(-s,t)\eta(s,t) + \eta'(-s,t)\psi(s,t)\widehat{\theta}(s)\xi(s)\,. \end{aligned}$$

Substituting this into (8.21), we obtain

$$\tilde{w}(s) = \frac{1}{T}\int_0^T w'(-s,t)w(s,t)\,\mathrm{d}t = g_1(s) - g_2(s) - g_3(s) + g_4(s)\,,$$

where

$$\begin{aligned} g_1(s) &= \xi'(-s)\widehat{\theta}'(-s)\frac{1}{T}\int_0^T \psi'(-s,t)\psi(s,t)\,\mathrm{d}t\,\widehat{\theta}(s)\xi(s)\,, \\ g_2(s) &= -\xi'(-s)\widehat{\theta}'(-s)\frac{1}{T}\int_0^T \psi'(-s,t)\eta(s,t)\,\mathrm{d}t\,, \qquad (8.70) \\ g_3(s) &= g_2'(-s) = -\frac{1}{T}\int_0^T \eta'(-s,t)\psi(s,t)\,\mathrm{d}t\,\widehat{\theta}(s)\xi(s)\,, \end{aligned}$$

and

$$g_4(s) = \frac{1}{T}\int_0^T \eta'(-s,t)\eta(s,t)\,\mathrm{d}t\,.$$

2. Next, we calculate Matrices (8.70). First of all with regard to (8.52), we find

$$\begin{aligned}\widehat{A}_L(s) &\triangleq \frac{1}{T}\int_0^T \psi'(-s,t)\psi(s,t)\,\mathrm{d}t \\ &= \widehat{a}_r'(-s)\frac{1}{T}\int_0^T \varphi_{L'\mu}(T,-s,t)\varphi_{L\mu}(T,s,t)\,\mathrm{d}t\ \widehat{a}_r(s)\,.\end{aligned} \tag{8.71}$$

Since

$$\begin{aligned}\varphi_{L\mu}(T,s,t) &= \frac{1}{T}\sum_{k=-\infty}^{\infty} L(s+kj\omega)\mu(s+kj\omega)\mathrm{e}^{kj\omega t}\,, \\ \varphi_{L'\mu}(T,-s,t) &= \frac{1}{T}\sum_{k=-\infty}^{\infty} L'(-s+kj\omega)\mu(-s+kj\omega)\mathrm{e}^{kj\omega t}\,,\end{aligned} \tag{8.72}$$

after substituting (8.72) into (8.71) and integration, we receive

$$\widehat{A}_L(s) = \widehat{a}_r'(-s)\frac{1}{T}\widehat{D}_{\underline{L}'L\mu\underline{\mu}}(T,s,0)\widehat{a}_r(s)\,, \tag{8.73}$$

where

$$\widehat{D}_{\underline{L}'L\mu\underline{\mu}}(T,s,0) = \frac{1}{T}\sum_{k=-\infty}^{\infty} L'(-s-kj\omega)L(s+kj\omega)\mu(s+kj\omega)\mu(-s-kj\omega)\,.$$

Using (8.73) in (8.70), we obtain

$$g_1(s) = \xi'(-s)\widehat{\theta}'(-s)\widehat{A}_L(s)\widehat{\theta}(s)\xi(s)\,.$$

3. To calculate the matrices $g_2(s)$ and $g_3(s)$, we denote

$$Q(s) \triangleq -\frac{1}{T}\int_0^T \eta'(-s,t)\psi(s,t)\,\mathrm{d}t\,. \tag{8.74}$$

Then,

$$Q'(-s) \triangleq -\frac{1}{T}\int_0^T \psi'(-s,t)\eta(s,t)\,\mathrm{d}t\,. \tag{8.75}$$

Using (8.52), we find

$$\begin{aligned}-\psi'(-s,t)\eta(s,t) = \widehat{a}_r'(-s)&\varphi_{L'\mu}(T,-s,t)\varphi_{L\mu}(T,s,t)\widehat{\beta}_{0r}(s)\widehat{a}_l(s)M(s) \\ &+ \widehat{a}_r'(-s)\varphi_{L'\mu}(T,-s,t)K(s)\,.\end{aligned}$$

Substituting this into (8.75) and taking account of (8.72) after integration, we find

$$Q'(-s) = \widehat{a}'_r(-s)\frac{1}{T}\widehat{D}_{\underline{L}'L\mu\underline{\mu}}(T,s,0)\widehat{\beta}_{0r}(s)\widehat{a}_l(s)M(s) + \frac{1}{T}\widehat{a}'_r(-s)L'(-s)\mu(-s)K(s)$$

and

$$Q(s) = M'(-s)\widehat{a}'_l(-s)\widehat{\beta}'_{0r}(-s)\frac{1}{T}\widehat{D}_{\underline{L}'L\mu\underline{\mu}}(T,s,0)\widehat{a}_r(s) + \frac{1}{T}K'(-s)L(s)\mu(s)\widehat{a}_r(s)$$

considering the identity

$$\widehat{\underline{D}}'_{\underline{L}'L\mu\underline{\mu}}(T,-s,0) = \widehat{D}_{\underline{L}'L\mu\underline{\mu}}(T,s,0)\,.$$

4. Using the above relations and (8.22)–(8.24), we obtain

$$\|\mathcal{S}\|_2^2 = \frac{1}{2\pi \mathrm{j}}\int_{-\mathrm{j}\infty}^{\mathrm{j}\infty} \operatorname{trace} \tilde{w}(s)\,\mathrm{d}s = J_1 + J_2\,, \tag{8.76}$$

where

$$\begin{aligned} J_1 = \frac{1}{2\pi \mathrm{j}}\int_{-\mathrm{j}\infty}^{\mathrm{j}\infty} \operatorname{trace}\Big[&\xi'(-s)\widehat{\theta}'(-s)\widehat{A}_L(s)\widehat{\theta}(s)\xi(s) \\ &- \xi'(-s)\widehat{\theta}'(-s)Q'(-s) - Q(s)\widehat{\theta}(s)\xi(s)\Big]\,\mathrm{d}s\,, \end{aligned} \tag{8.77}$$

$$J_2 = \frac{1}{2\pi \mathrm{j}}\int_{-\mathrm{j}\infty}^{\mathrm{j}\infty} \operatorname{trace} g_4(s)\,\mathrm{d}s\,.$$

Under the given assumptions, these integrals converge absolutely, *i.e.* all the integrands as $|s| \to \infty$ tend to zero as $|s|^{-2}$.

5. Since for a given initial controller, the value J_2 is a constant, we have to consider only (8.77). With regard to (8.5) from (8.77), we obtain

$$\begin{aligned} J_1 = \frac{1}{2\pi \mathrm{j}}\int_{-\mathrm{j}\infty}^{\mathrm{j}\infty} \operatorname{trace}\Big[&\widehat{\theta}'(-s)\widehat{A}_L(s)\widehat{\theta}(s)\xi(s)\xi'(-s) \\ &- \widehat{\theta}'(-s)Q'(-s)\xi'(-s) - \xi(s)Q(s)\widehat{\theta}(s)\Big]\,\mathrm{d}s\,, \end{aligned}$$

where the integral on the right-hand side converges absolutely. From (8.52), (8.74) and (8.75), we have

$$\xi(s)Q(s) = \widehat{a}_l(s)M(s)M'(-s)\widehat{a}'_l(-s)\widehat{\beta}'_{0r}(-s)\frac{1}{T}\widehat{D}_{\underline{L'}L\mu\underline{\mu}}(T,s,0)\widehat{a}_r(s)$$
$$+\frac{1}{T}\widehat{a}_l(s)M(s)K'(-s)L(s)\mu(s)\widehat{a}_r(s)\,,$$
$$Q'(-s)\xi'(-s) = \widehat{a}'_r(-s)\frac{1}{T}\widehat{D}_{\underline{L'}L\mu\underline{\mu}}(T,s,0)\widehat{\beta}_{0r}(s)\widehat{a}_l(s)M(s)M'(-s)\widehat{a}'_l(-s)$$
$$+\frac{1}{T}\widehat{a}'_r(-s)L'(-s)K(s)M'(-s)\mu(-s)\widehat{a}'_l(-s)\,.$$

Substitute this into the equation before and pass to an integral with finite integration limits. Then using (8.12), we obtain the functional

$$J_1 = \tag{8.78}$$
$$\frac{T}{2\pi \mathrm{j}}\int_{-\mathrm{j}\omega/2}^{\mathrm{j}\omega/2}\operatorname{trace}\left[\widehat{\theta}'(-s)\widehat{A}_L(s)\widehat{\theta}(s)\widehat{A}_M(s)-\widehat{\theta}'(-s)\widehat{C}'(-s)-\widehat{C}(s)\widehat{\theta}(s)\right]\mathrm{d}s\,,$$

where $\omega = 2\pi/T$, the matrix $\widehat{A}_L(s)$ is given by (8.73),

$$\begin{aligned}\widehat{A}_M(s) &= \frac{1}{T}\sum_{k=-\infty}^{\infty}\xi(s+k\mathrm{j}\omega)\xi'(-s-k\mathrm{j}\omega)\\ &= \widehat{a}_l(s)\frac{1}{T}\sum_{k=-\infty}^{\infty}M(s+k\mathrm{j}\omega)M'(-s-k\mathrm{j}\omega)\;\widehat{a}_l(-s)\\ &= \widehat{a}_l(s)\widehat{D}_{M\underline{M'}}(T,s,0)\widehat{a}'_l(-s)\end{aligned}\tag{8.79}$$

and

$$\begin{aligned}\widehat{C}(s) &= \frac{1}{T}\sum_{k=-\infty}^{\infty}\xi(s+k\mathrm{j}\omega)Q(s+k\mathrm{j}\omega)\\ &= \widehat{A}_M(s)\widehat{\beta}'_{0r}(-s)\frac{1}{T}\widehat{D}_{\underline{L'}L\mu\underline{\mu}}(T,s,0)\widehat{a}_r(s)\\ &\quad+\widehat{a}_l(s)\frac{1}{T}\widehat{D}_{M\underline{K'}L\mu}(T,s,0)\widehat{a}_r(s)\,,\\ \widehat{C}'(-s) &= \widehat{a}'_r(-s)\frac{1}{T}\widehat{D}_{\underline{L'}L\mu\underline{\mu}}(T,s,0)\widehat{\beta}_{0r}(s)\widehat{A}_M(s)\\ &\quad+\widehat{a}'_r(-s)\frac{1}{T}\widehat{D}_{\underline{L'}K\underline{M'}\underline{\mu}}(T,s,0)\widehat{a}'_l(-s)\,.\end{aligned}\tag{8.80}$$

6. Let us note several useful properties of the matrices $\widehat{A}_L(s)$, $\widehat{A}_M(s)$ and $\widehat{C}(s)$ appearing in the functional (8.78). We shall assume that (8.66) holds, because this is true in almost all applied problems.

Theorem 8.16. *The matrices* $\widehat{A}_L(s)$ *(8.73),* $\widehat{A}_M(s)$ *(8.79) and* $\widehat{C}(s)$ *(8.80) are rational periodic and admit the representations*

$$\begin{aligned} \widehat{C}(s) &= \frac{\widehat{B}_C(s)}{\widehat{\varrho}_L(s)\widehat{\varrho}_L(-s)\widehat{\varrho}_M(s)\widehat{\varrho}_M(-s)} = \frac{\widehat{B}_C(s)}{\widehat{\varrho}(s)\widehat{\varrho}(-s)}\,, \\ \widehat{A}_L(s) &= \frac{\widehat{B}_L(s)}{\widehat{\varrho}_L(s)\widehat{\varrho}_L(-s)}\,, \qquad \widehat{A}_M(s) = \frac{\widehat{B}_M(s)}{\widehat{\varrho}_M(s)\widehat{\varrho}_M(-s)}\,, \end{aligned} \tag{8.81}$$

where the numerators are finite sums of the forms

$$\begin{aligned} \widehat{B}_L(s) &= \sum_{k=-\alpha}^{\alpha} l_k \mathrm{e}^{-ksT}\,, \quad l_k = l'_{-k}\,, \\ \widehat{B}_M(s) &= \sum_{k=-\beta}^{\beta} m_k \mathrm{e}^{-ksT}\,, \quad m_k = m'_{-k}\,, \\ \widehat{B}_C(s) &= \sum_{k=-\gamma}^{\delta} c_k \mathrm{e}^{-ksT}\,. \end{aligned} \tag{8.82}$$

Herein, α, β, γ *and* δ *are non-negative integers and* l_k, m_k *and* c_k *are constant real matrices.*

Proof. Let us prove the claim for $\widehat{A}_L(s)$. First of all, the matrix $\widehat{A}_L(s)$ is rational periodical, as follows from the general properties given in Chapter 6.

Due to Corollary 8.15, the following irreducible representations exist:

$$L(s)\mu(s)\widehat{a}_r(s) = \frac{P_L(s)}{\widehat{\varrho}_L(s)}\,, \quad \widehat{a}'_r(-s)L'(-s)\mu(-s) = \frac{P'_L(-s)}{\widehat{\varrho}_L(-s)}\,,$$

where the matrix $P_L(s)$ is an integral function of s. Using (8.73), we can write

$$\begin{aligned} T\widehat{A}_L(s) = \frac{1}{T}\sum_{k=-\infty}^{\infty} &\left[\widehat{a}'_r(-s-kj\omega)L'(-s-kj\omega)\mu(-s-kj\omega)\right] \\ &\cdot \left[L(s+kj\omega)\mu(s+kj\omega)\widehat{a}_r(s+kj\omega)\right]. \end{aligned} \tag{8.83}$$

Each summand on the right-hand side can have poles only at the roots of the product $\widehat{\varrho}_1(s) \stackrel{\triangle}{=} \widehat{\varrho}_L(s)\widehat{\varrho}_L(-s)$. Under the given assumptions, the series (8.83) converges absolutely and uniformly in any restricted part of the complex plain containing no roots of the function $\widehat{\varrho}_1(s)$. Hence the sum of the series (8.83) can have poles only at the roots of the function $\widehat{\varrho}_1(s)$. Therefore, the matrix $\widehat{B}_L(s) = T\widehat{A}(s)\widehat{\rho}_1(s)$ has no poles. Moreover, since $\widehat{A}'_L(-s) = \widehat{A}_L(s)$, we have $\widehat{B}'_L(-s) = \widehat{B}_L(s)$, and the first relation in (8.82) is proven. The remaining formulae in (8.82) are proved in a similar way using (8.66). ∎

Corollary 8.17. *When the original system is stabilisable, then the matrices (8.81) do not possess poles on the imaginary axis.* ■

Corollary 8.18. *If the standard sampled-data system is modal controllable, then the matrices (8.81) are free of poles, i.e., they are integral functions of the argument s.* ■

7. Passing in the integral (8.78) to the variable $\zeta = \mathrm{e}^{-sT}$, we obtain

$$J_1 = \frac{1}{2\pi\mathrm{j}} \oint_{\Gamma} \operatorname{trace}\left[\hat{\theta}(\zeta)A_L(\zeta)\theta(\zeta)A_M(\zeta) - \hat{\theta}(\zeta)\hat{C}(\zeta) - C(\zeta)\theta(\zeta)\right]\frac{\mathrm{d}\zeta}{\zeta} \tag{8.84}$$

with the notation

$$\hat{F}(\zeta) \stackrel{\triangle}{=} F'(\zeta^{-1}) .$$

Perform integration along the unit circle Γ in positive direction (anti-clockwise). The matrices $A_L(\zeta)$, $A_M(\zeta)$, $C(\zeta)$ and $\hat{C}(\zeta)$ appearing in (8.84) admit the representations

$$\begin{aligned} A_L(\zeta) &= \frac{1}{T}\hat{a}_r(\zeta)D_{\underline{L}'L\mu\underline{\mu}}(T,\zeta,0)a_r(\zeta)\,, \\ A_M(\zeta) &= a_l(\zeta)D_{M\underline{M}'}(T,\zeta,0)\hat{a}_l(\zeta) \end{aligned} \tag{8.85}$$

and

$$\begin{aligned} C(\zeta) &= A_M(\zeta)\hat{\beta}_{0r}(\zeta)\frac{1}{T}D_{\underline{L}'L\mu\underline{\mu}}(T,\zeta,0)a_r(\zeta) + \frac{1}{T}a_l(\zeta)D_{M\underline{K}'L\mu}(T,\zeta,0)a_r(\zeta)\,, \\ \hat{C}(\zeta) &= \hat{a}_r(\zeta)\frac{1}{T}D_{\underline{L}'L\mu\underline{\mu}}(T,\zeta,0)\beta_{0r}(\zeta)A_M(\zeta) + \frac{1}{T}\hat{a}_r(\zeta)D_{\underline{L}'K\underline{M}'\mu}(T,\zeta,0)\hat{a}_l(\zeta)\,. \end{aligned} \tag{8.86}$$

Per construction,

$$\hat{A}_L(\zeta) = A_L(\zeta), \quad \hat{A}_M(\zeta) = A_M(\zeta)\,.$$

8. A rational matrix $F(\zeta)$ having no poles except for $\zeta = 0$ will be called a *quasi-polynomial matrix*. Any quasi-polynomial matrix $F(\zeta)$ can be written in the form

$$F(\zeta) = \sum_{k=-\mu}^{\nu} F_k\zeta^k\,,$$

where μ and ν are nonnegative integers and the F_k are constant matrices.

Substituting $\mathrm{e}^{-sT} = \zeta$ in (8.81) and (8.82), we obtain

$$\begin{aligned} A_L(\zeta) &= \frac{B_L(\zeta)}{\varrho_L(\zeta)\varrho_L(\zeta^{-1})}\,, \quad A_M(\zeta) = \frac{B_M(\zeta)}{\varrho_M(\zeta)\varrho_M(\zeta^{-1})}\,, \\ C(\zeta) &= \frac{B_C(\zeta)}{\varrho_L(\zeta)\varrho_L(\zeta^{-1})\varrho_M(\zeta)\varrho_M(\zeta^{-1})}\,, \end{aligned} \tag{8.87}$$

where the numerators are the quasi-polynomial matrices

$$\begin{aligned} B_L(\zeta) &= \sum_{k=-\alpha}^{\alpha} l_k\zeta^k\,, \quad l_k = l'_{-k} \\ B_M(\zeta) &= \sum_{k=-\beta}^{\beta} m_k\zeta^k\,, \quad m_k = m'_{-k} \\ B_C(\zeta) &= \sum_{k=-\gamma}^{\delta} c_k\zeta^k\,. \end{aligned} \tag{8.88}$$

Here, α, β, γ, and δ are nonnegative integers and l_k, m_k, and c_k are constant real matrices. Per construction,

$$\hat{B}_L(\zeta) = B_L(\zeta), \quad \hat{B}_M(\zeta) = B_M(\zeta)\,. \tag{8.89}$$

Remark 8.19. If the standard sampled-data system is modal controllable, then the matrices (8.87) are quasi-polynomial matrices.

Remark 8.20. Hereinafter, the matrix $B_M(\zeta)$ will be called a *quasi-polynomial matrix of type 1*, and the matrix $B_L(\zeta)$ a *quasi-polynomial matrix of type 2.*

8.5 Wiener-Hopf Method

1. In this section, we consider a method for solution of the $\mathcal{H}_2$-optimisation problem for the standard sampled-data system based on minimisation of the integral in (8.84). Such an approach was previously applied for solving $\mathcal{H}_2$-problems for continuous-time and discrete-time LTI systems and there, it was called Wiener-Hopf method [196, 47, 5, 80].

2. The following theorem provides a substantiation for applying the Wiener-Hopf method to sampled-data systems.

Theorem 8.21. *Suppose the standard sampled-data system be stabilisable and let us have the IMFDs (8.36) and (8.42). Furthermore, let $(\alpha_{0r}(\zeta), \beta_{0r}(\zeta))$ be any right initial controller, which has a dual left controller and the stable rational matrix $\theta^o(\zeta)$ ensures the minimal value $J_{1\,\min}$ of the integral (8.84). Then the transfer function of the optimal controller $w_d^o(\zeta)$, for which the standard sampled-data system is stable and the functional (8.29) approaches the minimal value, has the form*

$$w_d^o(\zeta) = V_{2o}(\zeta)V_{1o}^{-1}(\zeta)\,, \tag{8.90}$$

where

$$\begin{aligned} V_{1o}(\zeta) &= \alpha_{0r}(\zeta) - \zeta b_r(\zeta)\theta^o(\zeta)\,, \\ V_{2o}(\zeta) &= \beta_{0r}(\zeta) - a_r(\zeta)\theta^o(\zeta)\,. \end{aligned} \tag{8.91}$$

If in addition, we have IMFDs

$$\theta^o(\zeta) = D_l^{-1}(\zeta)M_l(\zeta) = M_r(\zeta)D_r^{-1}(\zeta)\,,$$

then the characteristic polynomial $\Delta^o(\zeta)$ of the optimal system satisfies the relation

$$\Delta^o(\zeta) \approx \varrho(\zeta)\det D_l(\zeta) \approx \varrho(\zeta)\det D_r(\zeta)\,.$$

Proof. Since the matrix $\theta^o(\zeta)$ is stable, the rational matrices (8.91) are also stable. Then using the Bezout identity and the ILMFD (8.36), we have

$$\begin{aligned} &a_l(\zeta)V_{1o}(\zeta) - \zeta b_l(\zeta)V_{2o}(\zeta) \\ &\qquad = a_l(\zeta)\alpha_{0r}(\zeta) - \zeta b_l(\zeta)\beta_{0r}(\zeta) - \zeta\left[a_l(\zeta)b_r(\zeta) - b_l(\zeta)a_r(\zeta)\right]\theta^o(\zeta) = I_n\,. \end{aligned}$$

Hence due to Theorem 5.58, $w_d^o(\zeta)$ is the transfer function of a stabilising controller. Using the ILMFDs (8.36) and (8.48), we find Matrix (8.49) as

$$\begin{aligned} R_N^o(\zeta) &= w_d^o(\zeta)\left[I_n - D_{N\mu}(T,\zeta,0)w_d^o(\zeta)\right]^{-1} \\ &= \beta_{0r}(\zeta)a_l(\zeta) - a_r(\zeta)\theta^o(\zeta)a_l(\zeta)\,. \end{aligned}$$

For $\zeta = \mathrm{e}^{-sT}$, we have

$$\begin{aligned} \widehat{R}^o_N(s) &= \widehat{w}^o_d(s)\left[I_n - \widehat{D}_{N\mu}(T,s,0)\widehat{w}^o_d(s)\right]^{-1} \\ &= \widehat{\beta}_{0r}(s)\widehat{a}_l(s) - \widehat{a}_r(s)\widehat{\theta}^o(s)\widehat{a}_l(s)\,. \end{aligned}$$

After substituting this equation into (8.29) and some transformations, the integral (8.29) can be reduced to the form

$$\|S\|_2^2 = J_{1o} + J_2\,,$$

where J_{1o} is given by (8.84) for $\theta(\zeta) = \theta^o(\zeta)$ and the value J_2 is constant. Per construction, the value $\|S\|_2^2$ is minimal. Therefore, Formula (8.90) gives the transfer matrix of an optimal stabilising controller. ∎

8.6 Algorithm for Realisation of Wiener-Hopf Method

1. According to the aforesaid, we shall consider the problem of minimising the functional (8.84) over the set of stable rational matrices, where the matrices $A_L(\zeta)$, $A_M(\zeta)$ and $C(\zeta)$ satisfy Conditions (8.87), (8.88) and (8.89). Moreover, if the stabilisability conditions hold for the system, then the matrices $A_L(\zeta)$, $A_M(\zeta)$, and $C(\zeta)$ do not possess poles on the integration path.

2. The following proposition presents a theoretical basis for the application of the Wiener-Hopf method to the solution of the $\mathcal{H}_2$-problem.

Lemma 8.22. *Let us have a functional of the form*

$$J_w = \frac{1}{2\pi \mathrm{j}} \oint \operatorname{trace} \left[\hat{\Gamma}(\zeta)\hat{\Psi}(\zeta)\hat{\Pi}(\zeta)\Pi(\zeta)\Psi(\zeta)\Gamma(\zeta) - \hat{\Psi}(\zeta)\hat{C}(\zeta) - C(\zeta)\Psi(\zeta) \right] \frac{\mathrm{d}\zeta}{\zeta}, \tag{8.92}$$

where the integration is performed along the unit circle in positive direction and $\Gamma(\zeta)$, $\Pi(\zeta)$ and $C(\zeta)$ are rational matrices having no poles on the integration path. Furthermore, let the matrices $\Pi(\zeta)$, $\Gamma(\zeta)$ be invertible and stable together with their inverses. Then, there exists a stable matrix $\Psi^o(\zeta)$ that minimises the functional (8.92). The matrix $\Psi^o(\zeta)$ can be constructed using the following algorithm:

a) Construct the rational matrix

$$R(\zeta) = \hat{\Pi}^{-1}(\zeta)\hat{C}(\zeta)\hat{\Gamma}^{-1}(\zeta). \tag{8.93}$$

b) Perform the separation

$$R(\zeta) = R_+(\zeta) + R_-(\zeta), \tag{8.94}$$

where the rational matrix $R_-(\zeta)$ is strictly proper and its poles incorporate all unstable poles of $R(\zeta)$. Such a separation will be called principal separation.

c) The optimal matrix $\Psi^o(\zeta)$ is determined by the formula

$$\Psi^o(\zeta) = \Pi^{-1}(\zeta)R_+(\zeta)\Gamma^{-1}(\zeta). \tag{8.95}$$

Proof. Using (8.5), we have

$$\begin{aligned} \operatorname{trace}[C(\zeta)\Psi(\zeta)] &= \operatorname{trace}[C(\zeta)\Psi(\zeta)\Gamma(\zeta)\Gamma^{-1}(\zeta)] \\ &= \operatorname{trace}[\Gamma^{-1}(\zeta)C(\zeta)\Psi(\zeta)\Gamma(\zeta)]. \end{aligned}$$

Therefore, the functional (8.92) can be represented in the form

$$\begin{aligned} J_w = \frac{1}{2\pi \mathrm{j}} \oint \operatorname{trace} &\left[\hat{\Gamma}(\zeta)\hat{\Psi}(\zeta)\hat{\Pi}(\zeta)\Pi(\zeta)\Psi(\zeta)\Gamma(\zeta) \right. \\ &\left. - \hat{\Gamma}(\zeta)\hat{\Psi}(\zeta)\hat{\Phi}(\zeta) - \Phi(\zeta)\Psi(\zeta)\Gamma(z) \right] \frac{\mathrm{d}\zeta}{\zeta}, \end{aligned} \tag{8.96}$$

where

$$\Phi(\zeta) = \Gamma^{-1}(\zeta)C(\zeta).$$

The identity

$$\begin{aligned} &\hat{\Gamma}(\zeta)\hat{\Psi}(\zeta)\hat{\Pi}(\zeta)\Pi(\zeta)\Psi(\zeta)\Gamma(\zeta) - \hat{\Gamma}(\zeta)\hat{\Psi}(\zeta)\hat{\Phi}(\zeta) - \Phi(\zeta)\Psi(\zeta)\Gamma(\zeta) \\ &\quad = [\Pi(\zeta)\Psi(\zeta)\Gamma(\zeta) - R(\zeta)]^{\wedge} [\Pi(\zeta)\Psi(\zeta)\Gamma(\zeta) - R(\zeta)] - \hat{R}(\zeta)R(\zeta) \end{aligned}$$

can easily be verified. Using the separation (8.94), the last relation can be transformed into

$$\begin{aligned}\hat{\Gamma}(\zeta)\hat{\Psi}(\zeta)\hat{\Pi}(\zeta)\Pi(\zeta)\Psi(\zeta)\Gamma(\zeta) - \hat{\Gamma}(\zeta)\hat{\Psi}(\zeta)\hat{\Phi}(\zeta) - \Phi(\zeta)\Psi(\zeta)\Gamma(\zeta) \\ = [\Pi(\zeta)\Psi(\zeta)\Gamma(\zeta) - R_+(\zeta)]^{\frown}[\Pi(\zeta)\Psi(\zeta)\Gamma(\zeta) - R_+(\zeta)] \\ + \hat{R}_-(\zeta)R_-(\zeta) - \hat{R}_-(\zeta)[\Pi(\zeta)\Psi(\zeta)\Gamma(\zeta) - R_+(\zeta)] \\ - [\Pi(\zeta)\Psi(\zeta)\Gamma(\zeta) - R_+(\zeta)]^{\frown}R_-(\zeta) - \hat{R}(\zeta)R(\zeta)\,.\end{aligned}$$

Hence the functional (8.96) can be written in the form

$$J_w = J_{w1} + J_{w2} + J_{w3} + J_{w4}\,,$$

where

$$\begin{aligned} J_{w1} &= \frac{1}{2\pi \mathrm{j}} \oint \operatorname{trace}\,[\Pi(\zeta)\Psi(\zeta)\Gamma(\zeta) - R_+(\zeta)]^{\frown}[\Pi(\zeta)\Psi(\zeta)\Gamma(\zeta) - R_+(\zeta)]\,\frac{\mathrm{d}\zeta}{\zeta}\,, \\ J_{w2} &= -\frac{1}{2\pi \mathrm{j}} \oint \operatorname{trace}\hat{R}_-(\zeta)\,[\Pi(\zeta)\Psi(\zeta)\Gamma(\zeta) - R_+(\zeta)]\,\frac{\mathrm{d}\zeta}{\zeta}\,, \\ J_{w3} &= -\frac{1}{2\pi \mathrm{j}} \oint \operatorname{trace}\,[\Pi(\zeta)\Psi(\zeta)\Gamma(\zeta) - R_+(\zeta)]^{\frown}R_-(\zeta)\frac{\mathrm{d}\zeta}{\zeta}\,, \\ J_{w4} &= \frac{1}{2\pi \mathrm{j}} \oint \operatorname{trace}\left[\hat{R}_-(\zeta)R_-(\zeta) - \hat{R}(\zeta)R(\zeta)\right]\frac{\mathrm{d}\zeta}{\zeta}\,. \end{aligned} \tag{8.97}$$

The integral J_{w4} is independent of $\Psi(\zeta)$. As for the scalar case [146], it can be shown that $J_{w2} = J_{w3} = 0$. The integral J_{w1} is nonnegative, its minimal value $J_{w1} = 0$ can be reached for (8.95). ■

Corollary 8.23. *The minimal value of the integral (8.92) is*

$$J_{w\,\min} = J_{w4}\,.$$

3. Using Lemma 8.22, we can formulate a proposition deriving a solution to the $\mathcal{H}_2$-optimisation problem for the standard sampled-data system.

Theorem 8.24. *Let the quasi-polynomial matrices $B_L(\zeta)$ and $B_M(\zeta)$ in (8.88) admit the factorisations*

$$B_L(\zeta) = \hat{\lambda}_L(\zeta)\lambda_L(\zeta), \quad B_M(\zeta) = \lambda_M(\zeta)\hat{\lambda}_M(\zeta)\,, \tag{8.98}$$

where $\lambda_L(\zeta)$ and $\lambda_M(\zeta)$ are invertible real stable polynomial matrices. Let also Condition (8.66) hold. Then the optimal matrix $\theta^o(\zeta)$ can be found using the following algorithm:

a) Construct the matrix

$$R(\zeta) = \frac{\hat{\lambda}_L^{-1}(\zeta)B_C'(\zeta^{-1})\hat{\lambda}_M^{-1}(\zeta)}{\varrho_L(\zeta)\varrho_M(\zeta)} = \frac{\hat{\lambda}_L^{-1}(\zeta)\hat{B}_C(\zeta)\hat{\lambda}_M^{-1}(\zeta)}{\varrho(\zeta)}\,. \tag{8.99}$$

b) Perform the principal separation

$$R(\zeta) = R_+(\zeta) + R_-(\zeta)\,,$$

where

$$R_+(\zeta) = \frac{\tilde{R}_+(\zeta)}{\varrho_L(\zeta)\varrho_M(\zeta)} = \frac{\tilde{R}_+(\zeta)}{\varrho(\zeta)} \tag{8.100}$$

with a polynomial matrix $\tilde{R}_+(\zeta)$.

c) The optimal system function $\theta^o(\zeta)$ is given by the formula

$$\theta^o(\zeta) = \lambda_L^{-1}(\zeta)\tilde{R}_+(\zeta)\lambda_M^{-1}(\zeta)\,. \tag{8.101}$$

Proof. Let the factorisations (8.98) hold. Since for the stabilisability of the system the polynomials $\varrho_L(\zeta)$ and $\varrho_M(\zeta)$ must be stable, the following factorisations hold:

$$A_L(\zeta) = \hat{\Pi}(\zeta)\Pi(\zeta)\,, \quad A_M(\zeta) = \Gamma(\zeta)\hat{\Gamma}(\zeta)\,, \tag{8.102}$$

where

$$\Pi(\zeta) = \frac{\lambda_L(\zeta)}{\varrho_L(\zeta)}\,, \quad \Gamma(\zeta) = \frac{\lambda_M(\zeta)}{\varrho_M(\zeta)} \tag{8.103}$$

are rational matrices, which are stable together with their inverses. From (8.103) we also have

$$\hat{\Pi}(\zeta) = \frac{\hat{\lambda}_L(\zeta)}{\varrho_L(\zeta^{-1})}\,, \quad \hat{\Gamma}(\zeta) = \frac{\hat{\lambda}_M(\zeta)}{\varrho_M(\zeta^{-1})}\,. \tag{8.104}$$

Regarding (8.102)–(8.104), the integral (8.84) can be represented in the form (8.92) with

$$C(\zeta) = \frac{B_C(\zeta)}{\varrho_L(\zeta)\varrho_L(\zeta^{-1})\varrho_M(\zeta)\varrho_M(\zeta^{-1})} = \frac{B_C(\zeta)}{\varrho(\zeta)\varrho(\zeta^{-1})}\,,$$

$$\hat{C}(\zeta) = \frac{B'_C(\zeta^{-1})}{\varrho_L(\zeta)\varrho_L(\zeta^{-1})\varrho_M(\zeta)\varrho_M(\zeta^{-1})} = \frac{\hat{B}_C(\zeta)}{\varrho(\zeta)\varrho(\zeta^{-1})}\,.$$

Then the matrix $R(\zeta)$ in (8.93) appears to be equal to (8.99).

The matrices $\hat{\lambda}_L^{-1}(\zeta)$ and $\hat{\lambda}_M^{-1}(\zeta)$ can have only unstable poles, because the polynomial matrices $\lambda_L(\zeta)$ and $\lambda_M(\zeta)$ are stable. The matrix $B'_C(\zeta^{-1})$ can have unstable pole only at $\zeta = 0$. Therefore, the set of stable poles of Matrix (8.99) belongs to the set of roots of the polynomial $\varrho(\zeta) \approx \varrho_L(\zeta)\varrho_M(\zeta)$. Hence the matrix $R_+(\zeta)$ in the principal separation (8.99) has the form (8.100), where $\tilde{R}_+(\zeta)$ is a polynomial matrix. Equation (8.101) can be derived from (8.100), (8.103), and (8.95). ∎

Corollary 8.25. *If the system is modal controllable, then $R_+(\zeta)$ is a polynomial matrix.* ∎

Corollary 8.26. *The characteristic polynomial of a modal controllable closed-loop system* $\Delta_d(\zeta) \approx \det D_l(\zeta) \approx \det D_r(\zeta)$ *is a divisor of the polynomial* $\det \lambda_L(\zeta) \det \lambda_M(\zeta)$. *Hereby, if the right-hand side of (8.101) is an irreducible DMFD (this is often the case in applications), then*

$$\Delta_d(\zeta) \approx \det \lambda_L(\zeta) \det \lambda_M(\zeta)\,.$$

∎

4. From (8.76) and (8.97), it follows that the minimal value of $\|\mathcal{S}\|_2^2$ is

$$\|\mathcal{S}\|_2^2 = \frac{1}{2\pi \mathrm{j}} \oint \operatorname{trace}\left[\hat{R}_-(\zeta)R_-(\zeta) - \hat{R}(\zeta)R(\zeta)\right] \frac{\mathrm{d}\zeta}{\zeta} + \frac{1}{2\pi \mathrm{j}} \int_{-\mathrm{j}\infty}^{\mathrm{j}\infty} \operatorname{trace} g_4(s)\, \mathrm{d}s\,.$$

8.7 Modified Optimisation Algorithm

1. The method for solving the $\mathcal{H}_2$-problems described in Section 8.5 requires for given IMFDs (8.36) and (8.42) of the plant, that the basic controller $[\alpha_{0r}(\zeta), \beta_{0r}(\zeta)]$ has to be previously found. This causes some numerical difficulties. In the present section, we describe a modified optimisation procedure, which does not need the basic controller. This method will be called the *modified Wiener-Hopf method.*

2. Let $F(\zeta)$ be a rational matrix and $F_+(\zeta)$, $F_-(\zeta)$ be the results of the principal separation (8.94). Then for the matrix $F_-(\zeta)$, we shall use the notation

$$F_-(\zeta) = \langle F(\zeta) \rangle_-\,.$$

Obviously,

$$\langle F_1(\zeta) + F_2(\zeta) \rangle_- = \langle F_1(\zeta) \rangle_- + \langle F_2(\zeta) \rangle_-\,.$$

3. Consider Matrix (8.93) in detail. Using the above relations as well as (8.86) and (8.102), we find

$$R(\zeta) = \hat{\Pi}^{-1}(\zeta)\hat{C}(\zeta)\hat{\Gamma}^{-1}(\zeta) = R_1(\zeta) + R_2(\zeta)\,, \tag{8.105}$$

where

$$R_1(\zeta) = \frac{1}{T}\hat{\Pi}^{-1}(\zeta)\hat{a}_r(\zeta)D_{\underline{L}'L\mu\underline{\mu}}(T,\zeta,0)\beta_{0r}(\zeta)A_M(\zeta)\hat{\Gamma}^{-1}(\zeta)\,, \tag{8.106}$$

$$R_2(\zeta) = \frac{1}{T}\hat{\Pi}^{-1}(\zeta)\hat{a}_r(\zeta)D_{\underline{L}'K\underline{M}'\underline{\mu}}(T,\zeta,0)\hat{a}_l(\zeta)\hat{\Gamma}^{-1}(\zeta)\,. \tag{8.107}$$

Since

$$\hat{a}_r(\zeta)\frac{1}{T}D_{\underline{L}'L\mu\underline{\mu}}(T,\zeta,0)a_r(\zeta) = A_L(\zeta)\,, \tag{8.108}$$

Matrix (8.106) can be written in the form

$$R_1(\zeta) = \hat{\Pi}^{-1}(\zeta)A_L(\zeta)a_r^{-1}(\zeta)\beta_{0r}(\zeta)A_M(\zeta)\hat{\Gamma}^{-1}(\zeta)\,. \tag{8.109}$$

With respect to (8.102), we obtain

$$R_1(\zeta) = \Pi(\zeta)a_r^{-1}(\zeta)\beta_{0r}(\zeta)\Gamma(\zeta)\,. \tag{8.110}$$

On the basis of (8.105)–(8.110), the following lemma can be proved.

Lemma 8.27. *In the principal separation*

$$R(\zeta) = R_+(\zeta) + \langle R(\zeta)\rangle_-\,, \tag{8.111}$$

the matrix $\langle R(\zeta)\rangle_-$ is independent of the choice of the basic controller.

Proof. If we choose another right initial controller with a matrix $\beta^*_{0r}(\zeta)$, we obtain a new matrix $R^*(\zeta)$ of the form

$$R^*(\zeta) = R^*_1(\zeta) + R_2(\zeta)$$

with

$$R^*_1(\zeta) = \Pi(\zeta)a_r^{-1}(\zeta)\beta^*_{0r}(\zeta)\Gamma(\zeta)\,, \tag{8.112}$$

where the matrix $R_2(\zeta)$ is the same as in (8.105). Therefore, to prove the lemma, it is sufficient to show

$$\langle R_1(\zeta)\rangle_- = \langle R^*_1(\zeta)\rangle_-\,. \tag{8.113}$$

But from (4.39), it follows that

$$\beta^*_{0r}(\zeta) = \beta_{0r}(\zeta) - a_r(\zeta)Q(\zeta) \tag{8.114}$$

with a polynomial matrix $Q(\zeta)$. Substituting this formula into (8.112), we find

$$R^*_1(\zeta) = R_1(\zeta) - \Pi(\zeta)Q(\zeta)\Gamma(\zeta)\,,$$

where the second term on the right-hand side is a stable matrix, because the matrices $\Pi(\zeta)$ and $\Gamma(\zeta)$ are stable. Therefore, (8.113) holds. ∎

4.

Lemma 8.28. *The transfer matrix of an optimal controllers $w^o_d(\zeta)$ is independent of the choice of the initial controller.*

Proof. Using (8.105)–(8.111), we have

$$R_+(\zeta) = R(\zeta) - \langle R(\zeta)\rangle_- = \Pi(\zeta)a_r^{-1}(\zeta)\beta_{0r}(\zeta)\Gamma(\zeta) + \Lambda(\zeta)\,, \tag{8.115}$$

where

$$\Lambda(\zeta) = \frac{1}{T}\hat{\Pi}^{-1}(\zeta)\hat{a}_r(\zeta)D_{\underline{L}'K\underline{M}'\mu}(T,\zeta,0)\hat{a}_l(\zeta)\hat{\Gamma}^{-1}(\zeta) - \langle R(\zeta)\rangle_- \,. \tag{8.116}$$

Using Lemma 8.27 and (8.116), we find that the matrix $\Lambda(\zeta)$ does not depend on the choice of initial controller. Hence only the first term on the right-hand side of (8.115) depends on the initial controller. From (8.115) and (8.95), we find the optimal system function

$$\theta^o(\zeta) = a_r^{-1}(\zeta)\beta_{0r}(\zeta) + \Pi^{-1}(\zeta)\Lambda(\zeta)\Gamma^{-1}(\zeta)\,.$$

Using (8.91), we obtain the optimal matrices $V_{1o}(\zeta)$ and $V_{2o}(\zeta)$:

$$V_{2o}(\zeta) = \beta_{0r}(\zeta) - a_r(\zeta)\theta^o(\zeta) = -a_r(\zeta)\Pi^{-1}(\zeta)\Lambda(\zeta)\Gamma^{-1}(\zeta) \tag{8.117}$$

and

$$\begin{aligned} V_{1o}(\zeta) &= \alpha_{0r}(\zeta) - \zeta b_r(\zeta)\theta^o(\zeta) \\ &= \alpha_{0r}(\zeta) - \zeta b_r(\zeta)a_r^{-1}(\zeta)\beta_{0r}(\zeta) - \zeta b_r(\zeta)\Pi^{-1}(\zeta)\Lambda(\zeta)\Gamma^{-1}(\zeta)\,. \end{aligned}$$

The Bezout identity guarantees

$$\begin{aligned} \alpha_{0r}(\zeta) - \zeta b_r(\zeta)a_r^{-1}(\zeta)\beta_{0r}(\zeta) &= \alpha_{0r}(\zeta) - \zeta a_l^{-1}(\zeta)b_l(\zeta)\beta_{0r}(\zeta) \\ &= a_l^{-1}(\zeta)\left[a_l(\zeta)\alpha_{0r}(\zeta) - \zeta b_l(\zeta)\beta_{0r}(\zeta)\right] = a_l^{-1}(\zeta)\,. \end{aligned}$$

Together these relations yield

$$V_{1o}(\zeta) = a_l^{-1}(\zeta) - \zeta b_r(\zeta)\Pi^{-1}(\zeta)\Lambda(\zeta)\Gamma^{-1}(\zeta)\,. \tag{8.118}$$

The matrices (8.117) and (8.118) are independent of the matrix $\beta_{0r}(\zeta)$. Then using (8.90), we can find an expression for the transfer matrix of the optimal controller that is independent of $\beta_{0r}(\zeta)$. ∎

5. As follows from Lemma 8.28, we can find an expression for the optimal matrix $w_d^o(\zeta)$, when we manage to find an expression for $\langle R(\zeta)\rangle_-$ that is independent of β_{0r}. We will show a possibility for deriving such an expression. Assume that the matrix

$$\gamma_l(\zeta) \triangleq b_l(\zeta)b_l'(\zeta^{-1}) = b_l(\zeta)\hat{b}_l(\zeta)$$

is invertible and the poles of the matrix

$$\delta_l(\zeta) \triangleq \zeta^{-1}b_l'(\zeta^{-1})\left[b_l(\zeta)b_l'(\zeta^{-1})\right]^{-1} \tag{8.119}$$

coincide neither with eigenvalues of the matrix $a_r(\zeta)$ nor with poles of the matrices $\Pi(\zeta)$ and $\Gamma(\zeta)$. Then, Equation (8.110) can be written in the form

$$R_1(\zeta) = \Pi(\zeta)a_r^{-1}(\zeta)\beta_{0r}(\zeta)b_l(\zeta)\left[\zeta^{-1}b_l'(\zeta^{-1})\right]\left[b_l(\zeta)b_l'(\zeta^{-1})\right]^{-1}\Gamma(\zeta)\,. \tag{8.120}$$

Due to the inverse Bezout identity (4.32), we have

$$-\zeta\beta_{0r}(\zeta)b_l(\zeta) + a_r(\zeta)\alpha_{0l}(\zeta) = I_m \,.$$

Thus,

$$\zeta a_r^{-1}(\zeta)\beta_{0r}(\zeta)b_l(\zeta) = \alpha_{0l}(\zeta) - a_r^{-1}(\zeta) \,.$$

Hence Equation (8.120) yields

$$R_1(\zeta) = R_{11}(\zeta) + R_{12}(\zeta) \,,$$

where

$$\begin{aligned} R_{11}(\zeta) &= -\Pi(\zeta)a_r^{-1}(\zeta)\zeta^{-1}b_l'(\zeta^{-1})\left[b_l(\zeta)b_l'(\zeta^{-1})\right]^{-1}\Gamma(\zeta) \\ &= -\Pi(\zeta)a_r^{-1}(\zeta)\delta_l(\zeta)\Gamma(\zeta) \,, \end{aligned} \tag{8.121}$$

$$\begin{aligned} R_{12}(\zeta) &= \Pi(\zeta)\alpha_{0l}(\zeta)\zeta^{-1}b_l'(\zeta^{-1})\left[b_l(\zeta)b_l'(\zeta^{-1})\right]^{-1}\Gamma(\zeta) \\ &= \Pi(\zeta)\alpha_{0l}(\zeta)\delta_l(\zeta)\Gamma(\zeta) \,. \end{aligned} \tag{8.122}$$

Consider the separation

$$R_{11}(\zeta) = R_{11}^{+}(\zeta) + \langle R_{11}(\zeta)\rangle_{-}^{a} + \langle R_{11}(\zeta)\rangle^{\delta} \,, \tag{8.123}$$

where $\langle R_{11}(\zeta)\rangle_{-}^{a}$ is a strictly proper function, whose poles are the unstable eigenvalues of the matrix $a_r(\zeta)$; $\langle R_{11}\rangle^{\delta}$ is a strictly proper function, whose poles are the poles of Matrix (8.119); and $R_{11}^{+}(\zeta)$ is a rational function, whose poles are the stable eigenvalues of the matrix $a_r(\zeta)$ as well as the poles of the matrices $\Pi(\zeta)$ and $\Gamma(\zeta)$. Similarly to (8.123) for (8.122), we find

$$R_{12}(\zeta) = R_{12}^{+}(\zeta) + \langle R_{12}(\zeta)\rangle^{\delta} \,, \tag{8.124}$$

where $\langle R_{12}(\zeta)\rangle^{\delta}$ is a strictly proper function, whose poles are the poles of Matrix (8.119), and $R_{12}^{+}(\zeta)$ is a stable rational matrix. Summing up Equations (8.123) and (8.124), we obtain

$$\begin{aligned} R_1(\zeta) &= R_{11}(\zeta) + R_{12}(\zeta) \\ &= R_{11}^{+}(\zeta) + R_{12}^{+}(\zeta) + \langle R_{11}(\zeta)\rangle_{-}^{a} + \langle R_{11}(\zeta)\rangle^{\delta} + \langle R_{12}(\zeta)\rangle^{\delta} \,. \end{aligned}$$

But

$$\langle R_{11}(\zeta)\rangle^{\delta} + \langle R_{12}(\zeta)\rangle^{\delta} = \langle R_{11}(\zeta) + R_{12}(\zeta)\rangle^{\delta} = O_{mn} \,,$$

because from (8.110), it follows that under the given assumptions, the matrix $R_1(\zeta)$ has no poles that are simultaneously poles of the matrix $\delta_l(\zeta)$. Then,

$$\langle R_1(\zeta)\rangle_{-} = \langle R_{11}(\zeta)\rangle_{-}^{a} \,,$$

where the matrix on the right-hand side is independent of the choice of the initial controller. Using the last relation and (8.105), we obtain

$$\langle R(\zeta)\rangle_{-} = \langle R_1(\zeta) + R_2(\zeta)\rangle_{-} = \langle R_{11}(\zeta)\rangle_{-}^{a} + \langle R_2(\zeta)\rangle_{-} \,.$$

Per construction, this matrix is also independent of the choice of the initial controller. Substituting the last equation into (8.116) and using (8.117), (8.118) and (8.90), an expression can be derived for the optimal transfer matrix $w_d^o(\zeta)$ that is independent of the choice of the initial controller.

6. A similar approach to the modified optimisation method can be proposed for the case, when the matrix $\hat{b}_r(\zeta)b_r(\zeta)$ is invertible, where $b_r(\zeta)$ is the polynomial matrix appearing in the IRMFD (8.42). In this case, (8.110) can be written in the form

$$R_1(\zeta) = \Pi(\zeta)\left[\hat{b}_r(\zeta)b_r(\zeta)\right]^{-1}\zeta^{-1}\hat{b}_r(\zeta)\zeta b_r(\zeta)a_r^{-1}(\zeta)\beta_{0r}(\zeta)\Gamma(\zeta)\,.$$

Due to the inverse Bezout identity, we have

$$\begin{aligned} a_r^{-1}(\zeta)\beta_{0r}(\zeta) &= \beta_{0l}(\zeta)a_l^{-1}(\zeta)\,,\\ \alpha_{0r}(\zeta)a_l(\zeta) - \zeta b_r(\zeta)\beta_{0l}(\zeta) &= I_n\,,\\ \zeta b_r(\zeta)\beta_{0l}(\zeta)a_l^{-1}(\zeta) &= \alpha_{0r}(\zeta) - a_l^{-1}(\zeta)\,. \end{aligned}$$

Using these relations, we obtain

$$\begin{aligned} R_1(\zeta) = \Pi(\zeta)&\left[\hat{b}_r(\zeta)b_r(\zeta)\right]^{-1}\zeta^{-1}\hat{b}_r(\zeta)\alpha_{0r}(\zeta)\Gamma(\zeta) - \\ &- \Pi(\zeta)\left[\hat{b}_r(\zeta)b_r(\zeta)\right]^{-1}\zeta^{-1}\hat{b}_r(\zeta)a_l^{-1}(\zeta)\Gamma(\zeta)\,. \end{aligned}$$

Assuming that no pole of the matrix

$$\delta_r(\zeta) \triangleq \zeta^{-1}\left[\hat{b}_r(\zeta)b_r(\zeta)\right]^{-1}\hat{b}_r(\zeta)$$

coincides with an eigenvalue of the matrices $a_l(\zeta)$, $\Pi(\zeta)$ or $\Gamma(\zeta)$, we find that the further procedure of constructing the optimal controller is similar to that described above.

8.8 Transformation to Forward Model

1. To make the reading easier, we will use some additional terminology and notation.

The optimisation algorithms described above make it possible to find the optimal system matrix $\theta^o(\zeta)$. Using the ILMFD

$$\theta^o(\zeta) = D_l^{-1}(\zeta)M_l(\zeta) = M_r(\zeta)D_r^{-1}(\zeta)$$

and Formulae (8.41), (8.44) and (8.46), we are able to construct the transfer matrix of the optimal controller $w_{db}^{o}(\zeta)$, which will be called the *backward transfer matrix*. Using the ILMFD

$$w_{db}^{o}(\zeta) = \alpha_l^{-1}(\zeta)\beta_l(\zeta)\,,$$

the matrix

$$Q_b(\zeta, \alpha_l, \beta_l) = \begin{bmatrix} I_\chi - \zeta e^{AT} & O_{\chi n} & -\zeta e^{AT}\mu(A)B_2 \\ -C_2 & I_n & O_{nm} \\ O_{m\chi} & -\beta_l(\zeta) & \alpha_l(\zeta) \end{bmatrix} \tag{8.125}$$

can be constructed. This matrix is called the *backward characteristic matrix*. As shown above, the characteristic polynomial of the backward model $\Delta_b(\zeta)$ is determined by the relation

$$\Delta_b(\zeta) = \det Q_b(\zeta, \alpha_l, \beta_l) = \varrho_b(\zeta)\Delta_{db}(\zeta)\,, \tag{8.126}$$

where $\varrho_b(\zeta)$ and $\Delta_{db}(\zeta)$ are polynomials. Especially, we know

$$\Delta_{db}(\zeta) = \det \begin{bmatrix} a_l(\zeta) & -\zeta b_l(\zeta) \\ -\beta_l(\zeta) & \alpha_l(\zeta) \end{bmatrix}, \tag{8.127}$$

where the matrices $a_l(\zeta)$ and $b_l(\zeta)$ are given by the ILMFD

$$w_N(\zeta) = \zeta C_2 \left(I_\chi - \zeta e^{AT}\right)^{-1} e^{AT}\mu(A)B_2 = \zeta a_l^{-1}(\zeta)b_l(\zeta)\,. \tag{8.128}$$

Moreover, the polynomial $\varrho_b(\zeta)$ appearing in (8.126) is determined by the relation

$$\varrho_b(\zeta) = \frac{\det(I_\chi - \zeta e^{AT})}{\det a_l(\zeta)} \tag{8.129}$$

and is independent of the choice of the controller. Further, the value

$$\delta_b \stackrel{\triangle}{=} \operatorname{ord} Q_b(\zeta, \alpha_l, \beta_l) = \deg\det Q_b(\zeta, \alpha_l, \beta_l) = \deg \varrho_b(\zeta) + \deg \Delta_{db}(\zeta)$$

will be called the *order of the optimal backward model*. As shown above,

$$\deg\det D_l(\zeta) = \deg \Delta_{db}(\zeta)\,.$$

2. For applied calculations and simulation, it is often convenient to use the forward system model instead of the backward model. In the present section we consider the realisation of such a transformation and investigate some general properties of the forward model.

Hereinafter, the matrix

$$w_{df}^{o}(z) = w_{db}^{o}(z^{-1})$$

will be called the *forward transfer function* of the optimal controller. Using the ILMFD

$$w_{df}^{o}(z) = \alpha_f^{-1}(z)b_f(z)\,,$$

a controllable forward model of the optimal discrete controller is found:

$$\alpha_f(q)\psi_k = \beta_f(q)y_k\,.$$

Together with (7.16), this equation determines a discrete forward model of the optimal system

$$\begin{aligned} qv_k &= \mathrm{e}^{AT}v_k + \mathrm{e}^{AT}\mu(A)B_2\psi_k + g_k \\ y_k &= C_2 v_k \\ \alpha_f(q)\psi_k &= \beta_f(q)y_k\,. \end{aligned}$$

These difference equations are associated with the matrix

$$Q_f(z,\alpha_f,\beta_f) = \begin{bmatrix} zI_\chi - \mathrm{e}^{AT} & O_{\chi n} & -\mathrm{e}^{AT}\mu(A)B_2 \\ -C_2 & I_n & O_{nm} \\ O_{m\chi} & -\beta_f(z) & \alpha_f(z) \end{bmatrix} \tag{8.130}$$

that will be called the *forward characteristic matrix* of the optimal system. Below, we formulate some propositions determining some properties of the characteristic matrices (8.125) and (8.130).

3. Similarly to (8.126)–(8.129), it can be shown that the polynomial

$$\Delta_f(z) \triangleq \det Q_f(z,\alpha_f,\beta_f)\,,$$

which is called the *characteristic polynomial of the forward model*, satisfies the relation

$$\Delta_f(z) = \varrho_f(z)\Delta_{df}(z)\,, \tag{8.131}$$

where $\varrho_f(z)$ and $\Delta_{df}(z)$ are polynomials. Moreover,

$$\Delta_{df}(z) = \det \begin{bmatrix} a_f(z) & -b_f(z) \\ -\beta_f(z) & \alpha_f(z) \end{bmatrix}, \tag{8.132}$$

where the matrices $a_f(z)$ and $b_f(z)$ are determined by the ILMFD

$$w_f(z) \triangleq C_2\left(zI_\chi - \mathrm{e}^{AT}\right)^{-1}\mathrm{e}^{AT}\mu(A)B_2 = a_f^{-1}(z)b_f(z)\,. \tag{8.133}$$

The polynomial $\varrho_f(z)$ appearing in (8.131) satisfies the equation

$$\varrho_f(z) = \frac{\det(zI_\chi - \mathrm{e}^{AT})}{\det a_f(z)}\,. \tag{8.134}$$

The value

$$\delta_f \triangleq \operatorname{ord} Q_f(z,\alpha_f,\beta_f) = \deg\det Q_f(z,\alpha_f,\beta_f) = \deg \varrho_f(z) + \deg \Delta_{df}(z)$$

will be called the *order of the optimal forward model.*

4. A connection between the polynomials (8.129) and (8.134) is determined by the following lemma.

Lemma 8.29. *The following equation holds:*

$$\delta \triangleq \deg \varrho_b(\zeta) = \deg \varrho_f(z)\,. \tag{8.135}$$

Moreover,

$$\varrho_f(z) \approx z^\delta \varrho_b(z^{-1})\,, \quad \varrho_b(\zeta) \approx \zeta^\delta \varrho_f(\zeta^{-1})\,. \tag{8.136}$$

Proof. Since Matrix (8.133) is strictly proper, there exists a minimal standard realisation in the form

$$w_f(z) = C_\nu(zI_q - U)^{-1}B_\nu\,, \quad q \le \chi\,.$$

Then for the ILMFD on the right-hand side of (8.133), we have

$$\det a_f(z) \approx \det(zI_q - U)\,. \tag{8.137}$$

Hence

$$\varrho_f(z) \approx \frac{\det(zI_\chi - \mathrm{e}^{AT})}{\det(zI_q - U)}\,. \tag{8.138}$$

From (8.138), it follows that the matrix U is nonsingular (this is a consequence of the non-singularity of the matrix e^{AT}). Comparing (8.128) with (8.133), we find

$$w_N(\zeta) = \zeta C_\nu(I_q - \zeta U)^{-1}B_\nu\,,$$

where the PMD $(I_q - \zeta U, \zeta B_\nu, C_\nu)$ is irreducible due to Lemma 5.35. Thus for the ILMFD (8.128), we obtain

$$\det a_l(\zeta) \approx \det(I_q - \zeta U)\,, \tag{8.139}$$

where

$$\operatorname{ord} a_l(\zeta) = \deg\det(I_q - \zeta U) = q\,,$$

because the matrix U is nonsingular. From (8.129) and (8.139), we obtain

$$\varrho_b(\zeta) \approx \frac{\det(I_\chi - \zeta\mathrm{e}^{AT})}{\det(I_q - \zeta U)}$$

and comparing this with (8.138) results in (8.135) and (8.136). ∎

5. The following propositions determine a property of the matrices

$$Q_{1b}(\zeta) = \begin{bmatrix} a_l(\zeta) & -\zeta b_l(\zeta) \\ -\beta_l(\zeta) & \alpha_l(\zeta) \end{bmatrix}, \quad Q_{1f}(z) = \begin{bmatrix} a_f(z) & -b_f(z) \\ -\beta_f(z) & \alpha_f(z) \end{bmatrix}.$$

Denote

$$\gamma_f = \deg\det\alpha_f(z)\,, \quad \gamma_b = \deg\det\alpha_l(\zeta)\,.$$

Lemma 8.30. *The following equation holds:*

$$\operatorname{ord} Q_{1f}(z) = q + \gamma_f \,. \tag{8.140}$$

Proof. Applying Formula (4.12) to $Q_{1f}(z)$, we obtain

$$\det Q_{1f}(z) = \det a_f(z) \det \alpha_f(z) \det \left[I_m - w^o_{df}(z) w_f(z) \right] . \tag{8.141}$$

Since the optimal controller $(\alpha_l(\zeta), \beta_l(\zeta))$ is stabilising, it is causal, *i.e.*, the backward transfer function $w^o_{db}(\zeta)$ is analytical at the point $\zeta = 0$. Hence the forward transfer matrix of the controller $w^o_{df}(z)$ is at least proper. Therefore, the matrix $w_f(z)$ is strictly proper. Thus, the product $w^o_{df}(z)w_f(z)$ is a strictly proper matrix. Moreover, the rational fraction $\det \left[I_m - w^o_{df}(z) w_f(z) \right]$ is proper, because of

$$\lim_{z \to \infty} \det \left[I_m - w^o_{df}(z) w_f(z) \right] = 1 \,.$$

Thus, it follows that there exists a representation of the form

$$\det \left[I_m - w^o_{df}(z) w_f(z) \right] = \frac{z^\lambda + b_1 z^{\lambda - 1} + \ldots}{z^\lambda + a_1 z^{\lambda - 1} + \ldots} \,,$$

where λ a nonnegative integer. Substituting this into (8.141), we find

$$\det Q_{1f}(z) = \det a_f(z) \det \alpha_f(z) \cdot \frac{z^\lambda + b_1 z^{\lambda - 1} + \ldots}{z^\lambda + a_1 z^{\lambda - 1} + \ldots} \,.$$

Since the right-hand side is a polynomial, we obtain

$$\deg \det Q_{1f}(z) = \deg \det a_f(z) + \deg \det \alpha_f(z) \,.$$

With regard to (8.137), this equation is equivalent to (8.140). ∎

Lemma 8.31. *Let us have unimodular matrices $m(z)$ and $\mu(z)$, such that the matrices*

$$\tilde a_f(z) = m(z) a_f(z) \,, \quad \tilde \alpha_f(z) = \mu(z) \alpha_f(z)$$

are row reduced. Then the matrix

$$\tilde Q_{1f}(z) = \operatorname{diag}\{m(z), \mu(z)\}\, Q_{1f}(z) \tag{8.142}$$

is row reduced.

Proof. Rewrite (8.142) in the form

$$\tilde Q_{1f}(z) = \begin{bmatrix} m(z) a_f(z) & -m(z) b_f(z) \\ -\mu(z) \beta_f(z) & \mu(z) \alpha_f(z) \end{bmatrix} = \begin{bmatrix} \tilde a_f(z) & -\tilde b_f(z) \\ -\tilde \beta_f(z) & \tilde \alpha_f(z) \end{bmatrix} , \tag{8.143}$$

where the pairs $(\tilde{a}_f(z), \tilde{b}_f(z))$ and $(\tilde{\alpha}_f(z), \tilde{\beta}_f(z))$ are irreducible. Since the matrices $\tilde{a}_f(z)$ and $\tilde{\alpha}_f(z)$ are row reduced, we receive a representation of the form (1.21)

$$\begin{aligned}\tilde{a}_f(z) &= \operatorname{diag}\{z^{a_1}, \dots, z^{a_n}\} A_0 + \tilde{a}_{1f}(z), \quad a_1 + \dots + a_n = q,\\ \tilde{\alpha}_f(z) &= \operatorname{diag}\{z^{\alpha_1}, \dots, z^{\alpha_n}\} B_0 + \tilde{\alpha}_{1f}(z), \quad \alpha_1 + \dots + \alpha_n = \gamma_f,\end{aligned}$$

where $\det A_0 \neq 0$ and $\det B_0 \neq 0$. Hereby, the degree of the i-th row of the matrix $\tilde{a}_{1f}(z)$ is less than a_i, and the degree of the i-th row of the matrix $\tilde{\alpha}_{1f}(z)$ is less than α_i. Moreover,

$$\begin{aligned}\tilde{a}_f^{-1}(z)\tilde{b}_f(z) &= a_f^{-1}(z)b_f(z) = w_f(z),\\ \tilde{\alpha}_f^{-1}(z)\tilde{\beta}_f(z) &= \alpha_f^{-1}(z)\beta_f(z) = w_{df}^{o}(z).\end{aligned}$$

Since the matrix $w_f(z)$ is strictly proper and the matrix $w_{df}^{o}(z)$ is at least proper, the degree of the i-th row of the matrix $\tilde{b}_f(z)$ is less than $\alpha_i - 1$, and the degree of the i-th row of $\tilde{\beta}_f(z)$ is not more than α_i. Therefore, Matrix (8.143) can be represented in the form (1.21)

$$\tilde{Q}_{1f}(z) = \operatorname{diag}\{z^{a_1}, \dots, z^{a_n}, z^{\alpha_1}, \dots, z^{\alpha_m}\} D_0 + \tilde{Q}_{2f}(z), \tag{8.144}$$

where D_0 is the constant matrix

$$D_0 = \begin{bmatrix} A_0 & O_{nm} \\ C_0 & B_0 \end{bmatrix}.$$

Since $\det D_0 \neq 0$, Matrix (8.142) is row reduced. ∎

Lemma 8.32. *The polynomials $\Delta_{db}(\zeta)$ and $\Delta_{df}(z)$ given by (8.127) and (8.132) are connected by*

$$\Delta_{db}(\zeta) \approx \zeta^{q+\gamma_f} \Delta_{df}(\zeta^{-1}), \tag{8.145}$$

i.e., $\Delta_{db}(\zeta)$ is equivalent to the reciprocal polynomial for $\Delta_{df}(z)$.

Proof. Since Matrix (8.143) is row reduced and has the form (8.144), the matrix

$$\tilde{Q}_{1b}(\zeta) = \operatorname{diag}\{\zeta^{a_1}, \dots, \zeta^{a_n}, \zeta^{\alpha_1}, \dots, \zeta^{\alpha_m}\} \tilde{Q}_{1f}(\zeta^{-1}) \tag{8.146}$$

defines a backward eigenoperator associated with the operator $\tilde{Q}_{1f}(z)$. Then the following formula stems from Corollary 5.39:

$$\det \tilde{Q}_{1b}(\zeta) \approx \zeta^{q+\gamma_f} \det \tilde{Q}_{1f}(\zeta^{-1}). \tag{8.147}$$

Per construction,

$$\tilde{\Delta}_{df}(z) \stackrel{\triangle}{=} \det \tilde{Q}_{1f}(z) \approx \det Q_{1f}(z) = \Delta_{df}(z)\,, \tag{8.148}$$

because the matrices $Q_{1f}(z)$ and $\tilde{Q}_{1f}(z)$ are left-equivalent. Let us show the relation

$$\tilde{\Delta}_{db}(\zeta) \stackrel{\triangle}{=} \det \tilde{Q}_{1b}(\zeta) \approx \det Q_{1b}(\zeta) = \Delta_{db}(\zeta)\,. \tag{8.149}$$

Notice that Matrix (8.146) can be represented in the form

$$\begin{aligned}\tilde{Q}_{1b}(\zeta) &= \begin{bmatrix} \operatorname{diag}\{\zeta^{a_1},\ldots,\zeta^{a_n}\}\,\tilde{a}_f(\zeta^{-1}) & -\operatorname{diag}\{\zeta^{a_1},\ldots,\zeta^{a_n}\}\,\tilde{b}_f(\zeta^{-1}) \\ -\operatorname{diag}\{\zeta^{\alpha_1},\ldots,\zeta^{\alpha_m}\}\,\tilde{\beta}_f(\zeta^{-1}) & \operatorname{diag}\{\zeta^{\alpha_1},\ldots,\zeta^{\alpha_m}\}\,\tilde{\alpha}_f(\zeta^{-1}) \end{bmatrix} \\ &= \begin{bmatrix} a_{1l}(\zeta) & -\zeta b_{1l}(\zeta) \\ -\beta_{1l}(\zeta) & \alpha_{1l}(\zeta) \end{bmatrix},\end{aligned} \tag{8.150}$$

where the pairs $(a_{1l}(\zeta), \zeta b_{1l}(\zeta))$ and $(\alpha_{1l}(\zeta), \beta_{1l}(\zeta))$ are irreducible due to Lemma 5.34. We have

$$\zeta a_{1l}^{-1}(\zeta) b_{1l}(\zeta) = \tilde{a}_f^{-1}(\zeta^{-1})\tilde{b}_f(\zeta^{-1}) = w_f(\zeta^{-1}) = w_b(\zeta)$$

and the left-hand side is an ILMFD. On the other hand, the right-hand side of (8.128) is also an ILMFD and the following equations hold:

$$a_{1l}(\zeta) = \phi(\zeta) a_l(\zeta), \quad b_{1l}(\zeta) = \phi(\zeta) b_l(\zeta)\,, \tag{8.151}$$

where $\phi(\zeta)$ is a unimodular matrix. In a similar way, it can be shown that

$$\alpha_{1l}(\zeta) = \psi(\zeta)\alpha_l(\zeta), \quad \beta_{1l}(\zeta) = \psi(\zeta)\beta_l(\zeta) \tag{8.152}$$

with a unimodular matrix $\psi(\zeta)$. Substituting (8.151) and (8.152) into (8.150), we find

$$\tilde{Q}_{1l}(\zeta) = \operatorname{diag}\{\phi(\zeta), \psi(\zeta)\}\, Q_{1b}(\zeta)\,,$$

i.e., the matrices $Q_{1b}(\zeta)$ and $\tilde{Q}_{1b}(\zeta)$ are left-equivalent. Therefore, Relations (8.149) hold. Then Relation (8.145) directly follows from (8.147)–(8.149). ∎

On the basis of Lemmata 8.29–8.32, the following theorem will be proved.

Theorem 8.33. *Let $\Delta_f(z)$ and $\Delta_b(\zeta)$ be the forward and backward characteristic polynomials of the optimal system. Then,*

$$\begin{aligned}\delta_f &= \deg \Delta_f(z) = \delta + q + \gamma_f\,, \\ \delta_b &= \deg \Delta_b(\zeta) = \delta + \deg\det D_l(\zeta)\,,\end{aligned}$$

where the number δ is determined by (8.135) and the number δ_0 of zero roots of the polynomial $\Delta_f(z)$ is

$$\delta_0 = q + \gamma_f - \deg\det D_l(\zeta)\,.$$

In this case, the polynomials $\Delta_f(z)$ and $\Delta_b(\zeta)$ are related by

$$\Delta_b(\zeta) = \zeta^{\delta_f} \Delta_f(\zeta^{-1})\,,$$

i.e., $\Delta_b(\zeta)$ is equivalent to the reciprocal polynomial for $\Delta_f(z)$.

Proof. The proof is left as exercise for the reader. ∎

9

$\mathcal{H}_2$ Optimisation of a Single-loop Multivariable SD System

9.1 Single-loop Multivariable SD System

1. The aforesaid approach for solving the $\mathcal{H}_2$-problem is fairly general and can be applied to any sampled-data system that can be represented in the standard form. Nevertheless, as will be shown in this chapter, the specific structure of a system and algebraic properties of the transfer matrices of the continuous-time blocks can play an important role for solving the $\mathcal{H}_2$-problem. In this case, we can find possibilities for some additional cancellations, extracting matrix divisors and so on. Moreover, this makes it possible to investigate important additional properties of the optimal solutions.

2. In this chapter, the above ideas are exemplarily illustrated by the single-loop system shown in Fig. 9.1, where $F(s)$, $Q(s)$ and $G(s)$ are rational matrices

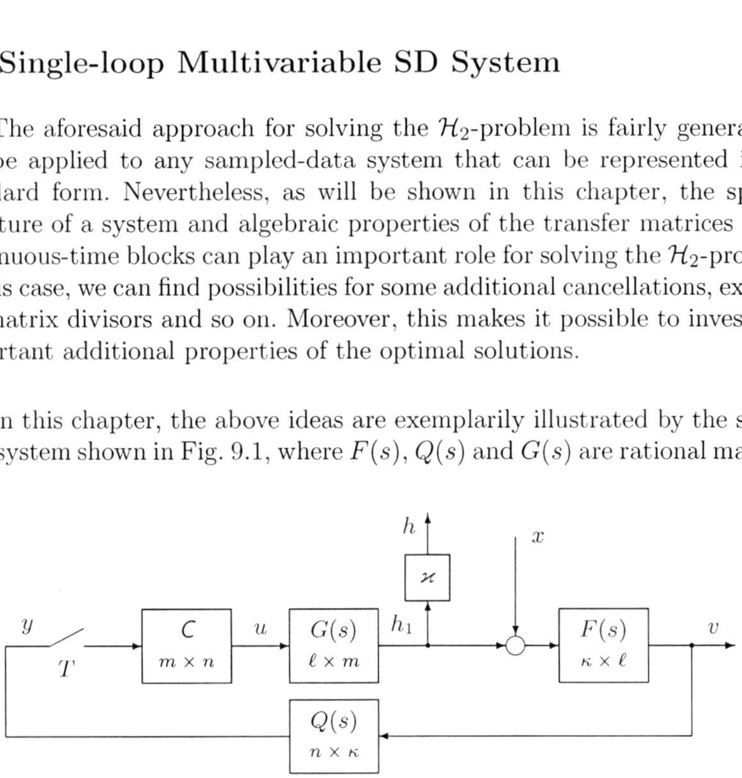

Fig. 9.1. Single-loop sampled-data system

of compatible dimensions and $\varkappa$ is a constant. The vector

$$z(t) = \begin{bmatrix} \varkappa h_1(t) \\ v(t) \end{bmatrix} = \begin{bmatrix} h(t) \\ v(t) \end{bmatrix} \tag{9.1}$$

will be taken as the output of the system. If the system is internally stable and $x(t)$ is a stationary centered vector, the covariance matrix of the quasi-stationary output is given by

$$K_z(t_1, t_2) = \mathsf{E}\left[z(t_1)z'(t_2)\right].$$

Since

$$z(t_1)z'(t_2) = \begin{bmatrix} \varkappa^2 h_1(t_1)h_1'(t_2) & \varkappa h_1(t_1)v'(t_2) \\ \varkappa v(t_1)h_1'(t_2) & v(t_1)v'(t_2) \end{bmatrix},$$

we have

$$\mathrm{trace}[z(t_1)z'(t_2)] = \mathrm{trace}[v(t_1)v'(t_2)] + \varkappa^2\,\mathrm{trace}[h_1(t_1)h_1'(t_2)].$$

For $\Phi_x(s) = I_\ell$, we obtain the square of the $\mathcal{H}_2$-norm of the system $\mathcal{S}$ in Fig. 9.1 as

$$\|\mathcal{S}\|_2^2 = \bar{r}_z = \frac{1}{T}\int_0^T \mathrm{trace}\left[K_z(t,t)\right]\,\mathrm{d}t = \varkappa^2 \bar{d}_{h_1} + \bar{d}_v\,, \tag{9.2}$$

where $\bar{d}_{h_1}$ and $\bar{d}_v$ are the mean variances of the corresponding output vectors. To solve the $\mathcal{H}_2$-problem, it is required to find a stabilising discrete controller, such that the right-hand side of (9.2) reaches the minimum.

9.2 General Properties

1. Using the general methods described in Section 7.3, we construct the PTM $w(s,t)$ of the single-loop system from the input x to the output (9.1). Let us show that such a construction can easily be done directly on the basis of the block-diagram shown in Fig. 9.1 without transformation to the standard form. In the given case, we realise

$$w(s,t) = \overset{\ell}{\begin{bmatrix} w_{hx}(s,t) \\ w_{vx}(s,t) \end{bmatrix}} \begin{matrix} \ell \\ \kappa \end{matrix}\,, \tag{9.3}$$

where $w_{hx}(s,t)$ and $w_{vx}(s,t)$ are the PTMs of the system from the input x to the outputs h and v, respectively.

2. To find the PTM $w_{vx}(s,t)$, we assume according to the previously exposed approach

$$x(t) = \mathrm{e}^{st}I_\ell\,, \quad v(t) = w_{vx}(s,t)\mathrm{e}^{st}, \quad w_{vx}(s,t) = w_{vx}(s,t+T)$$

and

$$y(t) = w_{yx}(s,t)\mathrm{e}^{st}, \quad w_{yx}(s,t) = w_{yx}(s,t+T)\,, \tag{9.4}$$

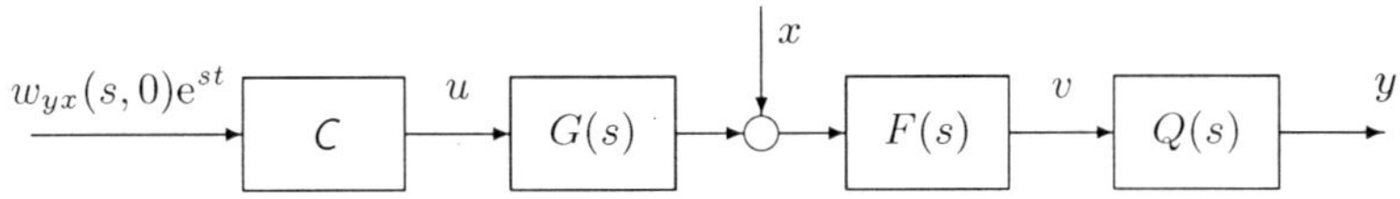

Fig. 9.2. Open-loop sampled-data system

where $w_{yx}(s,t)$ is the PTM from the input x to the output y. The matrix $w_{yx}(s,t)$ is assumed to be continuous in t. For our purpose, it suffices to assume that the matrix $Q(s)F(s)G(s)$ is strictly proper.

Consider the open-loop system shown in Fig. 9.2. The exp.per. output $y(t)$ is expressed by

$$y(t) = \varphi_{QFG\mu}(T,s,t)\widehat{w}_d(s)w_{yx}(s,0)e^{st} + Q(s)F(s)e^{st}.$$

Comparing the formulae for $y(t)$ here and in (9.4), we obtain

$$w_{yx}(s,t) = \varphi_{QFG\mu}(T,s,t)\widehat{w}_d(s)w_{yx}(s,0) + Q(s)F(s).$$

Hence for $t = 0$, we have

$$w_{yx}(s,0) = \left[I_n - \widehat{D}_{QFG\mu}(T,s,0)\widehat{w}_d(s)\right]^{-1} Q(s)F(s).$$

Returning to Fig. 9.1 and using the last equation, we immediately get

$$w_{vx}(s,t) = \varphi_{FG\mu}(T,s,t)\widehat{R}_{QFG}(s)Q(s)F(s), +F(s) \tag{9.5}$$

where

$$\widehat{R}_{QFG}(s) = \widehat{w}_d(s)\left[I_n - \widehat{D}_{QFG\mu}(T,s,0)\widehat{w}_d(s)\right]^{-1}.$$

Comparing (9.5) and (7.30), we find that in the given case

$$\begin{aligned} K(p) &= F(p), & L(p) &= F(p)G(p), \\ M(p) &= Q(p)F(p), & N(p) &= Q(p)F(p)G(p), \end{aligned} \tag{9.6}$$

i.e., Matrix (7.2) has the form

$$w_v(p) = \begin{bmatrix} F(p) & F(p)G(p) \\ Q(p)F(p) & Q(p)F(p)G(p) \end{bmatrix}.$$

The matrix $F(p)G(p)$ is assumed to be at least proper and the remaining blocks should be strictly proper. In a similar way, it can be shown that

$$w_{hx}(s,t) = \varkappa\varphi_{G\mu}(T,s,t)\widehat{R}_{QFG}(s)Q(s)F(s) \tag{9.7}$$

and the corresponding matrix $w_h(p)$ (7.2) is equal to

$$w_h(p) = \begin{bmatrix} O_{\ell\ell} & \varkappa G(p) \\ Q(p)F(p) & Q(p)F(p)G(p) \end{bmatrix},$$

where the matrix $G(p)$ is assumed to be at least proper. Combining (9.5) and (9.7), we find that the PTM (9.3) has the form

$$w(s,t) = \varphi_{L\mu}(T,s,t)\widehat{R}_N(s)M(s) + K(s) \tag{9.8}$$

with

$$\begin{gathered} K(s) = \overset{\ell}{\begin{bmatrix} O_{\ell\ell} \\ F(s) \end{bmatrix}} \begin{matrix} \ell \\ \kappa \end{matrix}, \qquad L(s) = \overset{m}{\begin{bmatrix} \varkappa G(s) \\ F(s)G(s) \end{bmatrix}} \begin{matrix} \ell \\ \kappa \end{matrix}, \\ M(s) = \overset{\ell}{Q(s)F(s)}\ n\,, \qquad N(s) = \overset{m}{Q(s)F(s)G(s)}\ n\,. \end{gathered} \tag{9.9}$$

Under the given assumptions, the matrix $L(s)$ is at least proper and the remaining matrices in (9.9) are strictly proper. The matrix $w(p)$ associated with the PTM (9.8) has the form

$$w(p) = \begin{bmatrix} K(p) & L(p) \\ M(p) & N(p) \end{bmatrix} = \left[\begin{array}{c:c} \overset{\ell}{O_{\ell\ell}} & \overset{m}{\varkappa G(p)} \\ F(p) & F(p)G(p) \\ \hdashline Q(p)F(p) & Q(p)F(p)G(p) \end{array}\right] \begin{matrix} \ell \\ \kappa \\ \\ n\,. \end{matrix} \tag{9.10}$$

3. Hereinafter the standard form (2.21) of a rational matrix $R(s)$ is written as

$$R(s) = \frac{N_R(s)}{d_R(s)}\,. \tag{9.11}$$

The further exposition is based on the following three assumptions I–III, which usually hold in applications.

I The matrices

$$Q(s) = \frac{N_Q(s)}{d_Q(s)}\,, \quad F(s) = \frac{N_F(s)}{d_F(s)}\,, \quad G(s) = \frac{N_G(s)}{d_G(s)} \tag{9.12}$$

are normal.

II The fraction

$$N(s) = Q(s)F(s)G(s) = \frac{N_Q(s)N_F(s)N_G(s)}{d_Q(s)d_F(s)d_G(s)} \tag{9.13}$$

is irreducible.

III The poles of the matrix $N(s)$ should satisfy the strict conditions for non-pathological behavior (6.124) and (6.125). Moreover, it is assumed that the number of inputs and outputs of any continuous-time block does not exceed the McMillan-degree of its transfer function.

These assumptions are introduced for the sake of simplicity of the solution. They are satisfied for the vast majority of applied problems.

4. Let us formulate a number of propositions following from the above assumptions.

Lemma 9.1. *The following subordination relations hold:*

$$Q(s)F(s) \underset{l}{\prec} N(s), \quad F(s)G(s) \underset{r}{\prec} N(s), \quad G(s) \underset{r}{\prec} F(s)G(s). \tag{9.14}$$

Proof. The proof follows immediately from Theorem 3.14. ∎

Lemma 9.2. *All matrices (9.6) are normal.*

Proof. The claim follows immediately from Theorem 3.8. ∎

Lemma 9.3. *All matrices (9.9) are normal.*

Proof. Obviously, it suffices to prove the claim for $L(s)$. But

$$L(s) = L_1(s)G(s), \tag{9.15}$$

where

$$L_1(s) = \begin{bmatrix} \varkappa I_\ell \\ F(s) \end{bmatrix} = \frac{\begin{bmatrix} \varkappa d_F(s) I_\ell \\ N_F(s) \end{bmatrix}}{d_F(s)} = \frac{N_{L_1}(s)}{d_F(s)}.$$

Let us show that this matrix is normal. Indeed, since the matrix $F(s)$ is normal, all second-order minors of the matrix $N_F(s)$ are divisible by $d_F(s)$. Obviously, the same is true for all second-order minors of the matrix $N_{L_1}(s)$. Thus, both factors on the right-hand side of (9.15) are normal matrices and its product is irreducible, because the product $F(s)G(s)$ is irreducible. Therefore, the matrix $L(s)$ is normal. ∎

Corollary 9.4. *The matrix $F(s)G(s)$ dominates in the matrix $L(s)$.* ∎

Lemma 9.5. *Matrix (9.10) is normal. Moreover, the matrix $N(s)$ dominates in $w(s)$.*

Proof. Matrix (9.10) can be written in the form

$$w(p) = \operatorname{diag}\{I_\ell, I_\kappa, Q(p)\} \begin{bmatrix} O_{\ell\ell} & \varkappa I_\ell \\ F(p) & F(p) \\ F(p) & F(p) \end{bmatrix} \operatorname{diag}\{I_\ell, G(p)\}. \tag{9.16}$$

Each factor on the right-hand side of (9.16) is a normal matrix. This statement is proved similarly to the proof of Lemma 9.3. Moreover, Matrix (9.13) is irreducible, such that the product on the right-hand side of (9.16) is irreducible. Hence Matrix (9.16) is normal. It remains to prove that the matrix $N(s) = Q(s)F(s)G(s)$ dominates in Matrix (9.10).

Denote

$$\delta_Q \stackrel{\triangle}{=} \deg d_Q(s), \quad \delta_F \stackrel{\triangle}{=} \deg d_F(s), \quad \delta_G \stackrel{\triangle}{=} \deg d_G(s).$$

Then by virtue of Theorem 3.13,

$$\operatorname{Mdeg} N(p) = \delta_Q + \delta_F + \delta_G \stackrel{\triangle}{=} \chi .$$

On the other hand, using similar considerations, we find for Matrix (9.16)

$$\operatorname{Mdeg} w(p) = \delta_Q + \delta_F + \delta_G = \chi .$$

Hence

$$\operatorname{Mdeg} w(p) = \operatorname{Mdeg} N(p) .$$

This equation means that the matrix $N(p)$ dominates in Matrix (9.10). ■

5. Let a minimal standard realisation of the matrix $N(p)$ have the form

$$N(p) = C_2(pI_\chi - A)^{-1}B_2 , \tag{9.17}$$

where the matrix A is cyclic. Then, as follows from Theorem 2.67, the minimal standard realisation of the matrix $w(p)$ can be written in the form

$$w(p) = \begin{bmatrix} C_1(pI_\chi - A)^{-1}B_1 & C_1(pI_\chi - A)^{-1}B_2 + D_L \\ C_2(pI_\chi - A)^{-1}B_1 & C_2(pI_\chi - A)^{-1}B_2 \end{bmatrix}, \tag{9.18}$$

or equivalently,

$$w(p) = \begin{bmatrix} C_1 \\ C_2 \end{bmatrix} (pI_\chi - A)^{-1} \begin{bmatrix} B_1 & B_2 \end{bmatrix} + \begin{bmatrix} O_{\ell+\kappa,\ell} & D_L \\ O_{n\ell} & O_{nm} \end{bmatrix}. \tag{9.19}$$

Equation (9.19) is associated with the state equations

$$\begin{aligned} \frac{\mathrm{d}v}{\mathrm{d}t} &= Av + B_1x + B_2u \\ z &= C_1v + D_Lu , \quad y = C_2v . \end{aligned}$$

9.3 Stabilisation

1.

Theorem 9.6. *Let Assumptions I–III on page 350 hold. Then the single-loop system shown in Fig. 9.1 is modal controllable (hence, is stabilisable).*

Proof. Under the given assumptions, the matrix

$$D_{N\mu}(T,\zeta,0) = \widehat{D}_{QFG\mu}(T,s,0)\,|_{\mathrm{e}^{-sT}=\zeta} = \zeta C_2\left(I_\chi - \zeta\mathrm{e}^{AT}\right)^{-1}\mathrm{e}^{AT}\mu(A)B_2$$

is normal. Thus, in the IMFDs

$$D_{N\mu}(T,\zeta,0) = \zeta a_l^{-1}(\zeta)b_l(\zeta) = \zeta b_r(\zeta)a_r^{-1}(\zeta)\,, \tag{9.20}$$

the matrices $a_l(\zeta)$, $a_r(\zeta)$ are simple and we have

$$\det a_l(\zeta) \approx \det a_r(\zeta) \approx \det\left(I_\chi - \zeta\mathrm{e}^{AT}\right) \approx \Delta_Q(\zeta)\Delta_F(\zeta)\Delta_G(\zeta)\,. \tag{9.21}$$

In (9.21) and below, $\Delta_R(\zeta)$ denotes the discretisation of the polynomial $d_R(s)$ given by (9.11). Thus, in the given case, we have

$$\varrho(\zeta) = \frac{\det(I_\chi - \zeta\mathrm{e}^{AT})}{\det a_l(\zeta)} = \text{const.} \neq 0\,, \tag{9.22}$$

whence the claim follows. ∎

2.

Remark 9.7. In the general case, the assumption on irreducibility of the right-hand side of (9.13) is essential. If the right-hand side of (9.13) is reducible by an unstable factor, then the system in Fig. 9.1 is not stabilisable despite the fact that all other assumptions of Theorem 9.6 hold.

Example 9.8. Let us have instead of (9.12), the irreducible representations

$$Q(p) = \frac{N_Q(p)}{d_Q(p)}\,, \quad F(p) = \frac{N_{1F}(p)}{p\,d_{1F}(p)}\,, \quad G(p) = \frac{p\,N_{1G}(p)}{d_G(p)}\,, \tag{9.23}$$

where $d_{1F}(0) \neq 0$, $d_Q(0) \neq 0$, $d_G(0) \neq 0$ and the matrix

$$\tilde{N}(p) = \frac{N_Q(p)N_{1F}(p)N_{1G}(p)}{d_Q(p)d_{1F}(p)d_G(p)} \tag{9.24}$$

is irreducible. Then, Matrix (9.10) takes the form

$$\tilde{w}(p) = \left[\begin{array}{c|c} O_{\ell\ell} & \dfrac{p\,N_{1G}(p)}{d_G(p)} \\ \dfrac{N_{1F}(p)}{p\,d_{1F}(p)} & \dfrac{N_{1F}(p)N_{1G}(p)}{d_{1F}(p)d_G(p)} \\ \hline \dfrac{N_Q(p)N_{1F}(p)}{p\,d_Q(p)d_{1F}(p)} & \dfrac{N_Q(p)N_{1F}(p)N_{1G}(p)}{d_Q(p)d_{1F}(p)d_G(p)} \end{array}\right] = \begin{bmatrix} \tilde{K}(p) & \tilde{L}(p) \\ \tilde{M}(p) & \tilde{N}(p) \end{bmatrix}.$$

In this case, the matrix $\tilde{N}(p)$ is not dominant in the matrix $\tilde{w}(p)$, because it is analytical for $p = 0$, while some elements of the matrix $\tilde{w}(p)$ have poles at $p = 0$. Then the matrix A in the minimal standard realisation (9.19) will have the eigenvalue zero. At the same time, the matrix $\tilde{A}$ in the minimal representation

$$\tilde{N}(p) = \tilde{C}_2(pI_\chi - \tilde{A})^{-1}\tilde{B}_2$$

has no eigenvalue zero, because the right-hand side of (9.24) is analytical for $p = 0$. Therefore, in the given case, (9.22) is not a stable polynomial. Hence the single-loop system with (9.23) is not stabilisable. □

9.4 Wiener-Hopf Method

1. Using the results of Chapter 8, we find that in this case, the $\mathcal{H}_2$-problem reduces to the minimisation of a functional of the form (8.84)

$$J_1 = \frac{1}{2\pi \mathrm{j}} \oint \operatorname{trace}\left[\hat{\theta}(\zeta)A_L(\zeta)\theta(\zeta)A_M(\zeta) - \hat{\theta}(\zeta)\hat{C}(\zeta) - C(\zeta)\theta(\zeta)\right]\frac{\mathrm{d}\zeta}{\zeta}$$

over the set of stable rational matrices. The matrices $A_L(\zeta)$, $A_M(\zeta)$, $C(\zeta)$ and $\hat{C}(\zeta)$ can be calculated using Formulae (8.85) and (8.86). Then referring to (9.9) and (8.86), we find

$$\begin{aligned} A_L(\zeta) &= \frac{1}{T}\,\hat{a}_r(\zeta)D_{\underline{L}'L\mu\underline{\mu}}(T,\zeta,0)a_r(\zeta) \\ &= \frac{1}{T}\,\hat{a}_r(\zeta)\left[\varkappa^2 D_{\underline{G}'G\mu\underline{\mu}}(T,\zeta,0) + D_{\underline{G}'\underline{F}'FG\mu\underline{\mu}}(T,\zeta,0)\right]a_r(\zeta)\,, \\ A_M(\zeta) &= a_l(\zeta)D_{M\underline{M}'}(T,\zeta,0)\hat{a}_l(\zeta) = a_l(\zeta)D_{QF\underline{F}'\underline{Q}'}(T,\zeta,0)\hat{a}_l(\zeta)\,. \end{aligned} \tag{9.25}$$

Applying (9.9) and (8.78), we obtain

$$
\begin{aligned}
C(\zeta) &= A_M(\zeta)\hat{\beta}_{0r}(\zeta)\frac{1}{T}\,D_{\underline{L}'L\mu\underline{\mu}}(T,\zeta,0)a_r(\zeta) + \frac{1}{T}\,a_l(\zeta)D_{M\underline{K}'L\mu}(T,\zeta,0)a_r(\zeta) \\
&= A_M(\zeta)\hat{\beta}_{0r}(\zeta)\frac{1}{T}\left[\varkappa^2 D_{\underline{G}'G\mu\underline{\mu}}(T,\zeta,0) + D_{\underline{G}'\underline{F}'FG\mu\underline{\mu}}(T,\zeta,0)\right] a_r(\zeta) + \\
&\quad + \frac{1}{T}\,a_l(\zeta)D_{QF\underline{F}'FG\mu}(T,\zeta,0)a_r(\zeta)\,, \\
\hat{C}(\zeta) &= \hat{a}_r(\zeta)\frac{1}{T}\left[\varkappa^2 D_{\underline{G}'G\mu\underline{\mu}}(T,\zeta,0) + D_{\underline{G}'\underline{F}'FG\mu\underline{\mu}}(T,\zeta,0)\right] \beta_{0r}(\zeta)A_M(\zeta) \\
&\quad + \frac{1}{T}\,\hat{a}_r(\zeta)D_{\underline{G}'\underline{F}'F\underline{F}'\underline{Q}'\mu}(T,\zeta,0)\hat{a}_l(\zeta)\,,
\end{aligned}
\tag{9.26}
$$

where the matrices $a_l(\zeta)$ and $a_r(\zeta)$ are determined by the IMFDs (9.20).

2. Since under the given assumptions, the single-loop system is modal controllable, all matrices in (9.25) and (9.26) are quasi-polynomials and can have poles only at $\zeta = 0$. Assume the following factorisations

$$
A_L(\zeta) = \hat{\Pi}(\zeta)\Pi(\zeta), \quad A_M(\zeta) = \hat{\Gamma}(\zeta)\Gamma(\zeta)\,, \tag{9.27}
$$

where the polynomial matrices $\Pi(\zeta)$ and $\Gamma(\zeta)$ are stable. Then, there exists an optimal controller, which can be found using the algorithm described in Chapter 8:

a) Calculate the matrix

$$
R(\zeta) = \hat{\Pi}^{-1}(\zeta)\hat{C}(\zeta)\hat{\Gamma}^{-1}(\zeta)\,.
$$

b) Perform the principal separation

$$
R(\zeta) = R_+(\zeta) + \langle R(\zeta)\rangle_-\,, \tag{9.28}
$$

where $R_+(\zeta)$ is a polynomial matrix and the matrix $\langle R(\zeta)\rangle_-$ is strictly proper.

c) The optimal system function is given by the formula

$$
\theta^o(\zeta) = \Pi^{-1}(\zeta)R_+(\zeta)\Gamma^{-1}(\zeta)\,.
$$

d) The transfer matrix of the optimal controller $w_d^o(\zeta)$ is given by (8.90) and (8.91).

9.5 Factorisation of Quasi-polynomials of Type 1

1. One of the fundamental steps in the Wiener-Hopf-method requires to factorise the quasi-polynomial (9.25) according to (9.27). In the present section, we investigate special features of the factorisation of a quasi-polynomial $A_M(\zeta)$ of type 1 determined by the given assumptions.

2. Let us formulate some auxiliary propositions. Suppose

$$M(s) = Q(s)F(s) = \frac{N_M(s)}{d_M(s)}$$

with

$$\begin{aligned} &d_M(s) = d_Q(s)d_F(s) = (s - m_1)^{\mu_1} \cdots (s - m_\rho)^{\mu_\rho}\,, \\ &\deg d_M(s) = \mu_1 + \ldots + \mu_\rho = \deg d_Q(s) + \deg d_F(s) = \delta_Q + \delta_F \stackrel{\triangle}{=} \gamma\,. \end{aligned} \tag{9.29}$$

Let us have a corresponding minimal standard realisation similar to (9.17)

$$M(s) = C_M(sI_\gamma - A_M)^{-1}B_M\,,$$

where I_γ is the identity matrix of compatible dimension. As follows from the subordination relations (9.14) and (9.18),

$$M(s) = C_2(sI_\chi - A)^{-1}B_1\,, \tag{9.30}$$

where B_1 and C_2 are constant matrices. For $0 < t < T$, let us have as before

$$D_M(T,\zeta,t) = C_M\left(I_\gamma - \zeta e^{A_M T}\right)^{-1} e^{A_M t}B_M\,. \tag{9.31}$$

Matrix (9.31) is normal for all t. Let us have an ILMFD

$$C_M\left(I_\gamma - \zeta e^{A_M T}\right)^{-1} = a_M^{-1}(\zeta)b_M(\zeta)\,. \tag{9.32}$$

Then under the given assumptions, the formulae

$$D_M(T,\zeta,t) = a_M^{-1}(\zeta)b_M(\zeta,t), \quad b_M(\zeta,t) = b_M(\zeta)e^{A_M t}B_M \tag{9.33}$$

determine an ILMFD of Matrix (9.31). Moreover, the matrix $a_M(\zeta)$ is simple and

$$\det a_M(\zeta) \approx \Delta_Q(\zeta)\Delta_F(\zeta)\,. \tag{9.34}$$

On the other hand using (9.30), we have

$$D_M(T,\zeta,t) = C_2\left(I_\chi - \zeta e^{AT}\right)^{-1} e^{At}B_1\,. \tag{9.35}$$

Consider the ILMFD

$$C_2\left(I_\chi - \zeta e^{AT}\right)^{-1} = a_l^{-1}(\zeta)\tilde{b}_l(\zeta)\,. \tag{9.36}$$

The set of matrices $a_l(\zeta)$ satisfying (9.36) coincides with the set of matrices $a_l(\zeta)$ for the ILMFD (9.20), which satisfy Condition (9.21). This fact follows from the minimality of the PMD $\left(I_\chi - \zeta e^{AT}, \zeta e^{At}B_2, C_2\right)$. From (9.35) and (9.36), we obtain an LMFD for the matrix $D_M(T,\zeta,t)$

$$D_M(T,\zeta,t) = a_l^{-1}\left[\tilde{b}_l(\zeta)\mathrm{e}^{At}B_1\right] = a_l^{-1}(\zeta)\tilde{b}_M(\zeta,t)\,.$$

Since Equation (9.33) is an ILMFD, we obtain

$$a_l(\zeta) = a_1(\zeta)a_M(\zeta)\,, \tag{9.37}$$

where $a_1(\zeta)$ is a polynomial matrix. Here the matrix $a_1(\zeta)$ is simple and with respect to (9.34) and (9.35), we find

$$\det a_1(\zeta) = \frac{\det a_l(\zeta)}{\det a_M(\zeta)} \approx \Delta_G(\zeta)\,. \tag{9.38}$$

3. Consider the sum of the series

$$D_{M\underline{M}'}(T,\zeta,0) = \left[\frac{1}{T}\sum_{k=-\infty}^{\infty} M(s+k\mathrm{j}\omega)M'(-s-k\mathrm{j}\omega)\right]_{|\,\mathrm{e}^{-sT}=\zeta}$$

and the matrix

$$P_M(\zeta) = a_M(\zeta)D_{M\underline{M}'}(T,\zeta,0)\hat{a}_M(\zeta)\,. \tag{9.39}$$

Lemma 9.9. *Matrix (9.39) is a symmetric quasi-polynomial matrix of the form (8.88).*

Proof. Since

$$\begin{aligned}\widehat{D}_M(T,s,t) &= \frac{1}{T}\sum_{k=-\infty}^{\infty} M(s+k\mathrm{j}\omega)\mathrm{e}^{(s+k\mathrm{j}\omega)t}\,,\\ \widehat{D}_{M'}(T,-s,t) &= \frac{1}{T}\sum_{k=-\infty}^{\infty} M'(-s+k\mathrm{j}\omega)\mathrm{e}^{(-s+k\mathrm{j}\omega)t}\,,\end{aligned} \tag{9.40}$$

we have the equality

$$\widehat{D}_{M\underline{M}'}(T,s,0) = \int_0^T \widehat{D}_M(T,s,t)\widehat{D}_{M'}(T,-s,t)\,\mathrm{d}t\,. \tag{9.41}$$

This result can be proved by substituting (9.40) into (9.41) and integrating term-wise. Substituting ζ for e^{-sT} in (9.41), we find

$$D_{M\underline{M}'}(T,\zeta,0) = \int_0^T D_M(T,\zeta,t)D_{M'}(T,\zeta^{-1},t)\,\mathrm{d}t\,. \tag{9.42}$$

As follows from (9.33) for $0 < t < T$,

$$\begin{aligned}D_{M'}(T,\zeta^{-1},t) &= D_M'(T,\zeta^{-1},t) = b_M'(\zeta^{-1},t)\left[a_M'(\zeta^{-1})\right]^{-1}\\ &= \hat{b}_M(\zeta,t)\hat{a}_M^{-1}(\zeta)\,.\end{aligned}$$

Then substituting this and (9.33) into (9.42), we receive

$$D_{M\underline{M}'}(T,\zeta,0) = a_M^{-1}(\zeta) \int_0^T b_M(\zeta,t)\hat{b}_M(\zeta,t)\,\mathrm{d}t\,\hat{a}_M^{-1}(\zeta)\,. \tag{9.43}$$

Hence with account for (9.39),

$$P_M(\zeta) = \int_0^T b_M(\zeta,t)\hat{b}_M(\zeta,t)\,\mathrm{d}t\,. \tag{9.44}$$

Obviously, the right-hand side of (9.44) is a quasi-polynomial matrix and

$$\hat{P}_M(\zeta) = P_M'(\zeta^{-1}) = P_M(\zeta)\,,$$

i.e., the quasi-polynomial (9.39) is symmetric. ∎

4. The symmetric quasi-polynomial $P(\zeta) = \hat{P}(\zeta)$ of dimension $n \times n$ will be called *nonnegative (positive) on the unit circle*, if for any vector $x \in \mathbb{C}_{1n}$ and $|\zeta| = 1$, we have

$$xP(\zeta)\bar{x}' \geq 0, \quad (xP(\zeta)\bar{x}' > 0)\,,$$

where the overbar denotes the complex conjugate value [133].

Lemma 9.10. *The quasi-polynomial (9.44) is nonnegative on the unit circle.*

Proof. Since we have $\zeta^{-1} = \bar{\zeta}$ on the unit circle, Equation (9.44) yields

$$xP_M(\zeta)\bar{x}' = \int_0^T [xb_M(\zeta,t)]\,[\overline{xb_M(\zeta,t)}\,]'\,\mathrm{d}t = \int_0^T |xb_M(\zeta,t)|^2\,\mathrm{d}t \geq 0\,,$$

where $|\cdot|$ denotes the absolute value of the complex row vector. ∎

Corollary 9.11. *Since under the given assumptions, the matrix $b_M(\zeta,t)$ is continuous with respect to t, the quasi-polynomial $P_M(\zeta)$ is nonnegative on the unit circle, if and only if there exists a constant nonzero row x_0 such that*

$$x_0 b_M(\zeta,t) = O_{1n}\,.$$

If such row does not exist, then the quasi-polynomial matrix $P_M(\zeta)$ is positive on the unit circle. ∎

Remark 9.12. In applied problems, the quasi-polynomial matrix $P_M(\zeta)$ is usually positive on the unit circle.

5.

Lemma 9.13. *Let under the given assumptions the matrix* $a_M(\zeta)$ *in the ILMFD (9.32) be row reduced. Then,*

$$P_M(\zeta) = \sum_{k=-\lambda}^{\lambda} m_k \zeta^k, \quad m_k = m'_{-k} \tag{9.45}$$

with

$$0 \le \lambda \le \rho_M - 1, \tag{9.46}$$

where

$$\rho_M \triangleq \deg a_M(\zeta) \le \deg \det a_M(\zeta) = \gamma. \tag{9.47}$$

Proof. Since the matrix $D_M(T, \zeta, t)$ is strictly proper for $0 < t < T$, due to Corollary 2.23 for the ILMFD (9.33), we have

$$\deg b_M(\zeta, t) < \deg a_M(\zeta), \quad (0 < t < T).$$

At the same time, since the matrix $a_M(\zeta)$ is row reduced, Equation (9.47) follows from (9.29). Then using (9.44), we obtain (9.45) and (9.46). ∎

6. Denote

$$q_M(\zeta) \triangleq \det P_M(\zeta).$$

Lemma 9.14. *Let the matrix* $M(s)$ *be normal and the product*

$$\tilde{M}(s) \triangleq M(s)M'(-s) = \frac{N_M(s)N'_M(-s)}{d_M(s)d_M(-s)} \tag{9.48}$$

be irreducible. Let also the roots of the polynomial

$$g_M(s) \triangleq d_M(s)d_M(-s)$$

satisfy the conditions for non-pathological behavior (6.106). Then,

$$q_M(\zeta) = \sum_{k=-\xi}^{\xi} q_k \zeta^k, \quad q_k = q'_{-k}, \tag{9.49}$$

where the q_k *are real constants and*

$$0 \le \xi \le \gamma - n, \tag{9.50}$$

where n *is the dimension of the vector* y *in Fig. 9.1.*

Proof. a) From (9.31) after transposition and substitution of ζ^{-1} for ζ, we obtain

$$D_{M'}(T,\zeta^{-1},t) = B'_M \mathrm{e}^{A'_M t}\left(I_\gamma - \zeta^{-1}\mathrm{e}^{A'_M T}\right)^{-1} C'_M .$$

Substituting this and (9.31) into (9.42), we find

$$D_{M\underline{M}'}(T,\zeta,0) = C_M \left(I_\gamma - \zeta \mathrm{e}^{A_M T}\right)^{-1} J \left(I_\gamma - \zeta^{-1}\mathrm{e}^{A'_M T}\right)^{-1} C'_M , \quad (9.51)$$

where

$$J = \int_0^T \mathrm{e}^{A_M t} B_M B'_M \mathrm{e}^{A'_M t}\,\mathrm{d}t .$$

From (9.51), it follows that the matrix $D_{M\underline{M}'}(T,\zeta,0)$ is strictly proper and can be written in the form

$$D_{M\underline{M}'}(T,\zeta,0) = \frac{\zeta K(\zeta)}{\Delta_M(\zeta)\Delta_{\underline{M}}(\zeta)} , \quad (9.52)$$

where

$$\begin{aligned} \Delta_M(\zeta) &= \left(\zeta - \mathrm{e}^{-m_1 T}\right)^{\mu_1} \cdots \left(\zeta - \mathrm{e}^{-m_\rho T}\right)^{\mu_\rho} , \\ \Delta_{\underline{M}}(\zeta) &= \left(\zeta - \mathrm{e}^{m_1 T}\right)^{\mu_1} \cdots \left(\zeta - \mathrm{e}^{m_\rho T}\right)^{\mu_\rho} , \end{aligned} \quad (9.53)$$

and $K(\zeta)$ is a polynomial matrix.

b) Let us show that Matrix (9.52) is normal. Indeed, using (9.48), we can write

$$D_{M\underline{M}'}(T,\zeta,0) = D_{\tilde{M}}(T,\zeta,0) .$$

Since the matrix $M(s)$ is normal, the matrix $\underline{M}(s) = M(-s)$ is normal. Hence $\tilde{M}(s)$ is also normal as a product of irreducible normal matrices. Moreover, since the poles of $\tilde{M}(s)$ satisfy Condition (6.106), the normality of Matrix (9.52) follows from Corollary 6.25.

c) Since Matrix (9.52) is strictly proper and normal, using (2.92) and (2.93), we have

$$f_M(\zeta) \triangleq \det D_{M\underline{M}'}(T,\zeta,0) = \frac{\zeta^n u_M(\zeta)}{\Delta_M(\zeta)\Delta_{\underline{M}}(\zeta)} , \quad (9.54)$$

where $u_M(\zeta)$ is a polynomial, such that $\deg u_M(\zeta) \le 2\gamma - 2n$. From (9.53), we find

$$\Delta_{\underline{M}}(\zeta) = (-1)^\gamma \mathrm{e}^{T\sum_{i=1}^{\rho} m_i\mu_i}\Delta_M(\zeta^{-1})\zeta^\gamma . \quad (9.55)$$

Substituting (9.55) into (9.54), we obtain

$$f_M(\zeta) = \frac{\psi_M(\zeta)}{\Delta_M(\zeta)\Delta_M(\zeta^{-1})} , \quad (9.56)$$

where $\psi_M(\zeta)$ is a quasi-polynomial of the form

$$\psi_M(\zeta) = (-1)^\gamma \mathrm{e}^{-T\sum_{i=1}^{\rho} m_i\mu_i}\zeta^{(n-\gamma)} u_M(\zeta) . \quad (9.57)$$

Substituting ζ^{-1} for ζ in (9.56) and using the fact that $f_M(\zeta) = f_M(\zeta^{-1})$, we obtain

$$\psi_M(\zeta) = \psi_M(\zeta^{-1}), \tag{9.58}$$

i.e., the quasi-polynomial (9.57) is symmetric. Moreover, the product $\zeta^{\gamma-n}\psi(\zeta)$ is a polynomial due to (9.57).
Calculating the determinants of both sides on (9.43), we find

$$f_M(\zeta) = \frac{q_M(\zeta)}{\det a_M(\zeta) \det a_M(\zeta^{-1})}\,.$$

Since

$$\det a_M(\zeta) = \lambda \Delta_M(\zeta)\,, \qquad \det a_M(\zeta^{-1}) = \lambda \Delta_M(\zeta^{-1})$$

with $\lambda = \text{const.} \neq 0$, a comparison of (9.56) with (9.58) yields

$$\psi(\zeta) = \nu q_M(\zeta)$$

with $\nu = \text{const.} \neq 0$. Therefore, the product $\zeta^{\gamma-n} q_M(\zeta)$ is a polynomial as (9.49) and (9.50) claim.

∎

7. Next, we provide some important properties of the quasi-polynomial matrix $A_M(\zeta)$ in (9.25).

Theorem 9.15. *Let Assumptions I-III on page 350 hold and let us have the ILMFD (9.36)*

$$C_2\left(I_\chi - \zeta e^{AT}\right)^{-1} = a_l^{-1}(\zeta)\tilde{b}_l(\zeta)\,. \tag{9.59}$$

Then, the matrices $a_l(\zeta)$ in (9.59) and $a_M(\zeta)$ in the ILMFD (9.32) can be chosen in such a way that the following equality holds:

$$A_M(\zeta) = a_l(\zeta) D_{M\underline{M}'}(T,\zeta,0)\hat{a}_l(\zeta) = \sum_{k=-\eta}^{\eta} a_k \zeta^k\,, \quad a_k = a'_{-k}\,, \tag{9.60}$$

where

$$0 \leq \eta \leq \chi - 1 \tag{9.61}$$

and

$$\chi = \deg d_N(s) = \deg d_Q(s) + \deg d_F(s) + \deg d_G(s)\,.$$

Proof. As was proved above, the sets of matrices $a_l(\zeta)$ from (9.59) and (9.20) coincide. Taking into account (9.37) and (9.43), we rewrite the matrix $A_M(\zeta)$ in the form

$$A_M(\zeta) = a_1(\zeta) P_M(\zeta) \hat{a}_1(\zeta)\,, \tag{9.62}$$

where $P_M(\zeta)$ is the quasi-polynomial (9.39). Using (9.44) from (9.62), we obtain

$$A_M(\zeta) = \int_0^T [a_1(\zeta)b_M(\zeta,t)]\,[a_1(\zeta)b_M(\zeta,t)]^{\frown}\,\mathrm{d}t\,. \tag{9.63}$$

Let the matrix $a_M(\zeta)$ be row reduced. Then we have

$$\deg b_M(\zeta,t) < \deg a_M(\zeta) \leq \deg\det a_M(\zeta) = \gamma\,. \tag{9.64}$$

Moreover, if we have the ILMFD (9.59), any pair $(\xi(\zeta)a_l(\zeta), \xi(\zeta)\tilde{b}_l(\zeta))$ with any unimodular matrix $\xi(\zeta)$ is also an ILMFD for Matrix (9.59). As a special case, the matrix $\xi(\zeta)$ can be chosen in such a way, that the matrix $a_1(\zeta)$ in (9.37) is row reduced. Then with respect to (9.38), we obtain

$$\deg a_1(\zeta) \leq \deg\det a_1(\zeta) = \deg \Delta_G(\zeta) = \delta_G\,.$$

If this and (9.64) hold, we have

$$\deg\,[a_1(\zeta)b_M(\zeta,t)] \leq \delta_M + \delta_G - 1 = \chi - 1\,.$$

From this and (9.63), the validity of (9.60) and (9.61) follows. ∎

Theorem 9.16. *Denote*

$$r_M(\zeta) \stackrel{\triangle}{=} \det A_M(\zeta)\,.$$

Under Assumptions I-III on page 350, we have

$$r_M(\zeta) = \sum_{k=-\sigma}^{\sigma} r_k\zeta^k\,, \quad r_k = r'_{-k} \tag{9.65}$$

with

$$0 \leq \sigma \leq \chi - n\,. \tag{9.66}$$

Proof. From (9.62), we have

$$r_M(\zeta) = \det a_1(\zeta)\det a_1(\zeta^{-1})\det P_M(\zeta) = \det a_1(\zeta)\det a_1(\zeta^{-1})q_M(\zeta)\,. \tag{9.67}$$

Then with regard to (9.49) and (9.50),

$$r_M(\zeta) = \det a_1(\zeta)\det a_1(\zeta^{-1})\sum_{k=-\xi}^{\xi} q_k\zeta^k\,,$$

which is equivalent to (9.65) and (9.66), because $\deg\det a_1(\zeta) = \delta_G$, $0 \leq \xi \leq \gamma - n$ and $\delta_G + \gamma = \chi$. ∎

Corollary 9.17. *As follows from (9.67), the set of zeros of the function $r_M(\zeta)$ includes the set of roots of the polynomial $\Delta_G(\zeta)$ as well as those of the quasi-polynomial $\Delta_G(\zeta^{-1})$.* ∎

8. Using the above auxiliary relations, we can formulate an important proposition about the factorisation of quasi-polynomials of type 1.

Theorem 9.18. *Let Assumptions I-III on page 350 and the propositions of Lemma 9.14 hold. Let also the quasi-polynomial $A_M(\zeta)$ be positive on the unit circle. Then, there exists a factorisation*

$$A_M(\zeta) = \Gamma(\zeta)\hat{\Gamma}(\zeta) = \Gamma(\zeta)\Gamma'(\zeta^{-1}), \tag{9.68}$$

where $\Gamma(\zeta)$ is a stable real $n \times n$ polynomial matrix. Under these conditions, there exists a factorisation

$$r_M(\zeta) = \det A_M(\zeta) = r_M^+(\zeta) r_M^+(\zeta^{-1}), \tag{9.69}$$

where $r_M^+(\zeta)$ is a real stable polynomial with $\deg r_M^+(\zeta) \leq \chi - n$. Moreover,

$$\det \Gamma(\zeta) \approx r_M^+(\zeta)$$

and the matrices $a_l(\zeta)$, $a_M(\zeta)$ in the ILMFDs (9.32), (9.59) can be chosen in such a way that

$$\deg \Gamma(\zeta) \leq \chi - 1\,.$$

Proof. With respect to the above results, the proof is a direct corollary of the general theorem about factorisation given in [133]. ∎

Remark 9.19. As follows from Corollary 9.17, if the polynomial $d_G(s)$ has roots on the imaginary axis, then the quasi-polynomial $r_M(\zeta)$ has roots on the unit circle. In this case, the factorisations (9.68) and (9.69) are impossible.

Remark 9.20. Let the polynomial

$$d_G(s) = (s - g_1)^{\lambda_1} \cdots (s - g_\sigma)^{\lambda_\sigma}$$

be free of roots on the imaginary axis. Let also be

$$\operatorname{Re} g_i < 0, \quad (i = 1, \ldots, \kappa); \operatorname{Re} g_i > 0, \quad (i = \kappa + 1, \ldots, \sigma)\,.$$

Then the polynomial $r_M^+(\zeta)$ can be represented in the form

$$r_M^+(\zeta) = \Delta_G^+(\zeta) r_{1M}^+(\zeta)\,,$$

where $\Delta_G^+(\zeta)$, $r_{1M}^+(\zeta)$ are stable polynomials and

$$\Delta_G^+(\zeta) = \left(\zeta - \mathrm{e}^{-g_1 T}\right)^{\lambda_1} \cdots \left(\zeta - \mathrm{e}^{-g_\kappa T}\right)^{\lambda_\kappa} \left(\zeta - \mathrm{e}^{g_{\kappa+1} T}\right)^{\lambda_{\kappa+1}} \cdots \left(\zeta - \mathrm{e}^{g_\sigma T}\right)^{\lambda_\sigma},$$

i.e., the numbers $\mathrm{e}^{-g_i T}$, $(i = 1, \ldots, \kappa)$ and $\mathrm{e}^{g_i T}$, $(i = \kappa + 1, \ldots, \sigma)$ are found among the roots of the polynomial $r_M^+(\zeta)$.

9.6 Factorisation of Quasi-polynomials of Type 2

1. Let the matrix $L(s)$ be at least proper and have the standard form

$$L(s) = \frac{N_L(s)}{d_L(s)}\,, \quad \operatorname{Mdeg} L(s) = \beta\,,$$

where

$$d_L(s) = (s-\lambda_1)^{\eta_1}\cdots(s-\lambda_m)^{\eta_m}\,, \quad \eta_1+\ldots+\eta_m = \beta$$

with the minimal standard representation

$$L(s) = C_L(sI_\beta - A_L)^{-1}B_L + D_L\,.$$

Let also

$$\mu(s) = \int_0^T \mathrm{e}^{-s\tau} m(\tau)\,\mathrm{d}\tau$$

be the transfer function of the forming element. Then for $0 < t < T$ from (6.90), (6.100) and (6.103), we have

$$\begin{aligned} \widehat{D}_{L\mu}(T,s,t) &= \frac{1}{T}\sum_{k=-\infty}^{\infty} L(s+kj\omega)\mu(s+kj\omega)\mathrm{e}^{(s+kj\omega)t} \\ &= C_L h_\mu^*(A_L,t)B_L + C_L\mu(A_L)\mathrm{e}^{A_L t}\left(I_\beta - \mathrm{e}^{-sT}\mathrm{e}^{A_L T}\right)^{-1}B_L + D_L m(t)\,, \end{aligned} \tag{9.70}$$

where

$$h_\mu^*(A_L,t) = -\int_t^T \mathrm{e}^{A_L(t-\tau)}m(\tau)\,\mathrm{d}\tau\,.$$

Replacing $\mathrm{e}^{-sT} = \zeta$ in (9.70), we find the rational matrix

$$\begin{aligned} D_{L\mu}(T,\zeta,t) &= \widehat{D}_{L\mu}(T,s,t)\,|_{\mathrm{e}^{-sT}=\zeta} \\ &= C_L\mu(A_L)\mathrm{e}^{A_L t}w_L(\zeta) + D_L(t)\,, \end{aligned} \tag{9.71}$$

where

$$w_L(\zeta) = \left(I_\beta - \zeta\mathrm{e}^{A_L T}\right)^{-1}B_L\,, \tag{9.72}$$

and

$$D_L(t) = C_L h_\mu^*(A_L,t)B_L + D_L m(t) \tag{9.73}$$

is a matrix independent of ζ. Let us have an IRMFD

$$w_L(\zeta) = \left(I_\beta - \zeta\mathrm{e}^{A_L T}\right)^{-1}B_L = b_L(\zeta)a_L^{-1}(\zeta)\,. \tag{9.74}$$

Then using (9.71), we find an RMFD

$$D_{L\mu}(T,\zeta,t) = b_L(\zeta,t)a_L^{-1}(\zeta)\,, \tag{9.75}$$

where the matrix

$$b_L(\zeta, t) = C_L e^{A_L t} \mu(A_L) b_L(\zeta) + D_L(t) a_L(\zeta)$$

is a polynomial in ζ for all t. When Assumptions I-III on page 350 hold, then the matrix $a_L(\zeta)$ is simple and we have

$$\det a_L(\zeta) \approx \Delta_F(\zeta)\Delta_G(\zeta)\,.$$

2. Consider the sum of the series

$$\widehat{D}_{\underline{L}' L\mu\underline{\mu}}(T, s, 0) = \frac{1}{T} \sum_{k=-\infty}^{\infty} L'(-s - kj\omega) L(s + kj\omega) \mu(s + kj\omega) \mu(-s - kj\omega)$$

and the rational matrices

$$D_{\underline{L}' L\mu\underline{\mu}}(T, \zeta, 0) = \widehat{D}_{\underline{L}' L\mu\underline{\mu}}(T, s, 0) \,|_{e^{-sT}=\zeta} \tag{9.76}$$

and

$$P_L(\zeta) = a'_L(\zeta^{-1}) D_{\underline{L}' L\mu\underline{\mu}}(T, \zeta, 0) a_L(\zeta)\,. \tag{9.77}$$

Let us formulate a number of propositions determining some properties of the matrices (9.76) and (9.77) required below.

Lemma 9.21. *Matrix (9.77) is a symmetric quasi-polynomial.*

Proof. Since

$$\widehat{D}_{L'\mu}(T, -s, t) = \frac{1}{T} \sum_{k=-\infty}^{\infty} L'(-s + kj\omega) \mu(-s + kj\omega) e^{(-s+kj\omega)t}$$

regarding (9.70) after integration, we find

$$\begin{aligned}
&\int_0^T \widehat{D}_{L'\mu}(T, -s, t) \widehat{D}_{L\mu}(T, s, t)\, dt \\
&\quad = \frac{1}{T} \sum_{k=-\infty}^{\infty} L'(-s - kj\omega) L(s + kj\omega) \mu(-s - kj\omega) \mu(s + kj\omega) \\
&\quad = \widehat{D}_{\underline{L}' L\mu\underline{\mu}}(T, s, 0)\,.
\end{aligned}$$

Substituting ζ for e^{-sT}, we find

$$D_{\underline{L}' L\mu\underline{\mu}}(T, \zeta, 0) = \int_0^T D_{L'\mu}(T, \zeta^{-1}, t) D_{L\mu}(T, \zeta, t)\, dt\,. \tag{9.78}$$

Nevertheless from (9.75), we have

$$
\begin{aligned}
D_{L'\mu}(T,\zeta^{-1},t) &= D'_{L\mu}(T,\zeta^{-1},t) \\
&= \left[a'_L(\zeta^{-1})\right]^{-1} b'_L(\zeta^{-1},t) = \hat{a}_L^{-1}(\zeta)\hat{b}_L(\zeta,t)\,.
\end{aligned} \tag{9.79}
$$

Using (9.75) and (9.79) in (9.78), we find

$$
D_{\underline{L}'L\mu\underline{\mu}}(T,\zeta,0) = \hat{a}_L^{-1}(\zeta)\int_0^T \hat{b}_L(\zeta,t)b_L(\zeta,t)\,\mathrm{d}t\,a_L^{-1}(\zeta)\,.
$$

Hence

$$
P_L(\zeta) = \int_0^T \hat{b}_L(\zeta,t)b_L(\zeta,t)\,\mathrm{d}t \tag{9.80}
$$

is a symmetric quasi-polynomial. ∎

Lemma 9.22. *The quasi-polynomial $P_L(\zeta)$ is nonnegative on the unit circle.*

Proof. The proof is similar to that given for Lemma 9.10. ∎

Lemma 9.23. *Let Assumptions I-III on page 350 hold and the matrix $a_L(\zeta)$ from the IRMFD (9.74) be column reduced. Then,*

$$
P_L(\zeta) = \sum_{k=-\nu}^{\nu} \ell_k\zeta^k\,, \quad \ell_k = \ell'_{-k}
$$

with

$$
0 \leq \nu \leq \rho_L\,,
$$

where

$$
\rho_L \triangleq \deg a_L(\zeta) \leq \deg\det a_L(\zeta) = \beta\,.
$$

Proof. Under the given assumptions, the matrix $w_L(\zeta)$ is normal. Hence

$$
\deg\det a_L(\zeta) = \deg \Delta_L(\zeta) = \beta = \delta_F + \delta_G
$$

and since the matrix $a_L(\zeta)$ is column reduced, we have

$$
\deg a_L(\zeta) \leq \beta\,.
$$

Moreover, since Matrix (9.71) is at least proper, due to Corollary 2.23, we obtain

$$
\deg b_L(\zeta,t) \leq \deg a_L(\zeta) \leq \beta\,.
$$

The claim of the lemma follows from (9.80) and the last relations. ∎

3. Introduce the following additional notations

$$q_L(\zeta) \stackrel{\triangle}{=} \det P_L(\zeta)$$

and

$$\tilde{\mu}(s) \stackrel{\triangle}{=} \mu(s)\mu(-s)\,.$$

Lemma 9.24. *Let Assumptions I-III on page 350 hold and the product*

$$\tilde{L}(s) = L'(-s)L(s) = \frac{N'_L(s)N_L(s)}{d_L(s)d_L(-s)} \tag{9.81}$$

be irreducible. Let also the roots $\tilde{\lambda}_1, \dots, \tilde{\lambda}_\rho$ *of the polynomial*

$$\tilde{d}_L(s) \stackrel{\triangle}{=} d_L(s)d_L(-s)$$

satisfy the conditions for non-pathological behavior (6.106) and moreover,

$$\tilde{\mu}(\tilde{\lambda}_i) \neq 0, \quad (i = 1, \dots, \rho)\,. \tag{9.82}$$

Then,

$$q_L(\zeta) = \sum_{k=-\tilde{\nu}}^{\tilde{\nu}} \tilde{q}_k \zeta^k\,,$$

where the $\tilde{q}_k = \tilde{q}'_{-k}$ *are real constants and*

$$0 \leq \tilde{\nu} \leq \beta\,.$$

Proof. a) First of all, we show that the rational matrix $D_{\underline{L}'L\mu\underline{\mu}}(T,\zeta,0)$ is at least proper. With this aim in view, recall that (9.71) yields

$$D_{L'\mu}(T,\zeta^{-1},t) = w'_L(\zeta^{-1})\mu'(A_L)\mathrm{e}^{A'_L t}C'_L + D'_L(t)\,.$$

Using (9.72) and (9.73), it can be easily established that this matrix is at least proper. Then the product

$$D_{L'\mu}(T,\zeta^{-1},t)D_{L\mu}(T,\zeta,t)$$

is also at least proper. Further from (9.78), it follows that the matrix $D_{\underline{L}'L\mu\underline{\mu}}(T,\zeta,0)$ is at least proper. Then, if

$$d_L(s) = (s-\lambda_1)^{\eta_1}\cdots(s-\lambda_m)^{\eta_m}\,, \quad \eta_1 + \ldots + \eta_M = \beta\,,$$

then

$$D_{\underline{L}'L\mu\underline{\mu}}(T,\zeta,0) = \frac{\mathcal{L}(\zeta)}{\Delta_L(\zeta)\Delta_{\underline{L}}(\zeta)} \tag{9.83}$$

with

$$\begin{aligned} \Delta_L(\zeta) &= \left(\zeta - \mathrm{e}^{-\lambda_1 T}\right)^{\eta_1}\cdots\left(\zeta - \mathrm{e}^{-\lambda_m T}\right)^{\eta_m}\,, \\ \Delta_{\underline{L}}(\zeta) &= \left(\zeta - \mathrm{e}^{\lambda_1 T}\right)^{\eta_1}\cdots\left(\zeta - \mathrm{e}^{\lambda_m T}\right)^{\eta_m}\,, \end{aligned} \tag{9.84}$$

where $\mathcal{L}(\zeta)$ is a polynomial matrix, such that $\deg\mathcal{L}(\zeta) \leq 2\beta$.

b) Let us show that Matrix (9.83) is normal. For this purpose using (9.81), we write

$$\begin{aligned}\widehat{D}_{\underline{L}'L\mu\underline{\mu}}(T,s,0) &= D_{\tilde{L}\mu\underline{\mu}}(T,s,0)\\ &= \frac{1}{T}\sum_{k=-\infty}^{\infty}\tilde{L}(s+kj\omega)\mu(s+kj\omega)\mu(-s-kj\omega)\,.\end{aligned} \tag{9.85}$$

Using the fact that

$$\mu(-s) = \int_0^T e^{s\tau} m(\tau)\,d\tau$$

from (9.85), we can derive

$$\begin{aligned}\widehat{D}_{\underline{L}'L\mu\underline{\mu}}(T,s,0) &= \frac{1}{T}\sum_{k=-\infty}^{\infty}\tilde{L}(s+kj\omega)\mu(s+kj\omega)\int_0^T e^{(s+kj\omega)\tau}m(\tau)\,d\tau\\ &= \int_0^T \left[\frac{1}{T}\sum_{k=-\infty}^{\infty}\tilde{L}(s+kj\omega)\mu(s+kj\omega)e^{(s+kj\omega)\tau}\right]m(\tau)\,d\tau\\ &= \int_0^T \widehat{D}_{\tilde{L}\mu}(T,s,\tau)m(\tau)\,d\tau\,.\end{aligned} \tag{9.86}$$

Under the given assumptions, the matrix $L'(-s)$ is normal. Hence Matrix (9.81) is also normal as a product of irreducible normal matrices. Therefore, the minimal standard representation of the matrix $\tilde{L}(s)$ can be written in the form

$$\tilde{L}(s) = \tilde{C}_L\left(sI_{2\beta} - \tilde{A}_L\right)^{-1}\tilde{B}_L + \tilde{D}_L\,.$$

Then similarly to (9.70)–(9.73) for $0 < t < T$, we obtain

$$\widehat{D}_{\tilde{L}\mu}(T,s,t) = \tilde{C}_L e^{\tilde{A}_L t}\mu(\tilde{A}_L)\left(sI_{2\beta} - e^{-sT}e^{\tilde{A}_L T}\right)^{-1}\tilde{B}_L + \tilde{D}_L(t)\,, \tag{9.87}$$

where

$$\tilde{D}_L(t) = \tilde{C}_2 h_\mu^*(\tilde{A}_L,t)\tilde{B}_L + \tilde{D}_L m(t)\,.$$

Using (9.87) in (9.86), after integration and substitution $e^{-sT} = \zeta$, we obtain

$$\begin{aligned}D_{\underline{L}'L\mu\underline{\mu}}(T,\zeta,0) = \tilde{C}_L\mu(-\tilde{A}_L)\mu(\tilde{A}_L)\left(I_{2\beta} - \zeta e^{\tilde{A}_L T}\right)^{-1}\tilde{B}_L\\ + \int_0^T \tilde{D}_L(t)m(t)\,dt\,,\end{aligned}$$

where

$$\mu(-\tilde{A}_L) = \int_0^T e^{\tilde{A}_L t} m(t)\, dt .$$

Under the given assumptions, the matrices $\tilde{A}_L$ and $e^{\tilde{A}_L T}$ are cyclic, the pair $(e^{\tilde{A}_L T}, \tilde{B}_L)$ is controllable and the pair $[e^{\tilde{A}_L T}, \tilde{C}_L]$ is observable. Moreover, the matrix

$$\tilde{\mu}(\tilde{A}_L) = \mu(-\tilde{A}_L)\mu(\tilde{A}_L)$$

is commutative with the matrix $e^{\tilde{A}_L}$ and due to (9.82), it is nonsingular. Therefore, Matrix (9.83) is normal.

c) With respect to the normality of Matrix (9.83), calculating the determinants on both sides of (9.83), we find

$$f_L(\zeta) \stackrel{\triangle}{=} \det D_{\underline{L}' L \mu \underline{\mu}}(T, \zeta, 0) = \frac{u_L(\zeta)}{\Delta_L(\zeta)\Delta_{\underline{L}}(\zeta)}, \tag{9.88}$$

where $\Delta_L(\zeta)$, $\Delta_{\underline{L}}(\zeta)$ are the polynomials (9.84), and $u_L(\zeta)$ is a polynomial with $\deg u_L(\zeta) \leq 2\beta$. Per construction, we have $f_L(\zeta) = f_L(\zeta^{-1})$. Therefore, similarly to (9.54) from (9.88), we obtain

$$f_L(\zeta) = \frac{\psi_L(\zeta)}{\Delta_L(\zeta)\Delta_L(\zeta^{-1})}, \tag{9.89}$$

where

$$\psi_L(\zeta) = (-1)^\beta e^{-T\sum_{i=1}^n \lambda_i \eta_i} \zeta^{-\beta} u_L(\zeta)$$

is a symmetric quasi-polynomial. Moreover, since the product $\psi_L(\zeta)\zeta^\beta$ is a polynomial, we receive

$$\psi_L(\zeta) = \sum_{k=-\tilde{\nu}}^{\tilde{\nu}} \psi_k \zeta^k, \quad \psi_k = \psi'_{-k},$$

where

$$0 \leq \tilde{\nu} \leq \beta .$$

Using (9.89), (9.77) and the relations

$$\det a_L(\zeta) \approx \Delta_L(\zeta), \quad \det a'_L(\zeta^{-1}) = \det a_L(\zeta^{-1}),$$

we obtain the equality

$$q_L(\zeta) = \det a_L(\zeta) \det a'_L(\zeta^{-1}) f_L(\zeta) = k_L \psi_L(\zeta), \quad k_L = \text{const.}$$

This completes the proof. ∎

4. Using the above auxiliary results under Assumptions I-III on page 350, we consider some properties of the quasi-polynomial $A_L(\zeta)$.

Using a minimal standard representation (9.18), introduce the matrix

$$\tilde{w}_L(\zeta) = \left(I_\chi - \zeta e^{AT}\right)^{-1} B_2 \tag{9.90}$$

and an arbitrary IRMFD

$$\tilde{w}_L(\zeta) = \tilde{b}_L(\zeta) a_r^{-1}(\zeta)\,. \tag{9.91}$$

Since Matrix (9.90) is normal, the matrix $a_r(\zeta)$ is simple and

$$\det a_r(\zeta) \approx \det\left(I_\chi - \zeta e^{AT}\right) \approx \Delta_Q(\zeta)\Delta_F(\zeta)\Delta_G(\zeta)\,.$$

From (9.18), it follows that the representation

$$L(s) = C_1(sI_\chi - A)^{-1}B_2 + D_L$$

exists. Hence together with (9.71), we have

$$D_{L\mu}(T,\zeta,t) = C_1 h_\mu^*(A,t)B_2 + C_1\mu(A)e^{At}\left(I_\chi - \zeta e^{AT}\right)^{-1} B_2 + m(t)D_L\,.$$

Thus with account for (9.91), we obtain the RMFD

$$D_{L\mu}(T,\zeta,t) = \tilde{b}_L(\zeta,t)a_r^{-1}(\zeta)\,, \tag{9.92}$$

where

$$\tilde{b}_L(\zeta,t) = C_1 h_\mu^*(A,t)B_2 a_r(\zeta) + C_1\mu(A)e^{At}\tilde{b}_L(\zeta) + m(t)D_L a_r(\zeta)\,.$$

Simultaneously with the RMFD (9.92), we have the IRMFD (9.75), therefore

$$a_r(\zeta) = a_L(\zeta)a_2(\zeta)\,. \tag{9.93}$$

Moreover, the polynomial matrix $a_2(\zeta)$ is simple and

$$\det a_2(\zeta) = \frac{\det a_r(\zeta)}{\det a_L(\zeta)} \approx \Delta_Q(\zeta)\,. \tag{9.94}$$

Theorem 9.25. *The set of matrices $a_r(\zeta)$ in the IRMFD (9.91) coincides with the set of matrices $a_r(\zeta)$ in the IRMFD (9.20). Moreover, the matrices $a_r(\zeta)$ in (9.91) and $a_L(\zeta)$ in the IRMFD (9.74) can be chosen in such a way, that the following representation holds:*

$$A_L(\zeta) = \frac{1}{T}\hat{a}_r(\zeta)D_{\underline{L}'L\mu\underline{\mu}}(T,\zeta,0)a_r(\zeta) = \sum_{k=-\nu}^{\nu} a_k\zeta^k\,, \quad a_k = a'_{-k}\,, \tag{9.95}$$

where

$$0 \leq \nu \leq \chi\,. \tag{9.96}$$

Proof. The coincidence of the sets of matrices $a_r(\zeta)$ in (9.80) and (9.91) stems from the minimality of the PMD

$$\left(I_\chi - \zeta e^{AT}, e^{At}\mu(A)B_2, C_2\right).$$

Using (9.79), (9.80) and (9.93), the matrix $A_L(\zeta)$ can be written in the form

$$A_L(\zeta) = \hat{a}_2(\zeta)P_L(\zeta)a_2(\zeta), \tag{9.97}$$

where $P_L(\zeta)$ is the quasi-polynomial matrix (9.77). Using (9.80) from (9.97), we find

$$A_L(\zeta) = \frac{1}{T}\int_0^T \left[\widehat{b_L(\zeta,t)a_2(\zeta)}\right]\left[b_L(\zeta,t)a_2(\zeta)\right] \mathrm{d}t\,. \tag{9.98}$$

Let the matrix $a_L(\zeta)$ be column reduced. Then as before, we have

$$\deg b_L(\zeta,t) \le \deg a_L(\zeta) \le \deg \det a_l(\zeta) = \beta\,.$$

Moreover, if we have the IRMFD (9.91), then any pair $[a_r(\zeta)\phi(\zeta), \tilde{b}_L(\zeta)\phi(\zeta)]$, where $\phi(\zeta)$ is any unimodular matrix, determines an IRMFD for Matrix (9.91). In particular, the matrix $\phi(\zeta)$ can be chosen in such a way that the matrix $a_2(\zeta)$ in (9.93) becomes column reduced. In this case, we receive

$$\deg a_2(\zeta) \le \deg \det a_2(\zeta) = \deg \Delta_Q(\zeta) = \delta_q\,.$$

The last two estimates yield

$$\deg\left[b_L(\zeta,t)a_2(\zeta)\right] \le \beta + \delta_q = \chi\,.$$

Equations (9.95) and (9.96) follow from (9.98) and the last estimate. ∎

Theorem 9.26. *Denote*

$$r_L(\zeta) \overset{\triangle}{=} \det A_L(\zeta)\,.$$

Then under Assumptions I-III on page 350 and the conditions of Lemma 9.23, we have

$$r_L(\zeta) = \sum_{k=-\kappa}^{\kappa} \tilde{r}_k\zeta^k\,, \quad \tilde{r}_k = \tilde{r}'_{-k}\,,$$

where

$$0 \le \kappa \le \chi\,.$$

Proof. From (9.97), we have

$$r_L(\zeta) = \delta \det a_2(\zeta)\det a_2(\zeta^{-1})q_L(\zeta), \quad \delta = \text{const.} \ne 0 \tag{9.99}$$

that is equivalent to the claim, because $\deg\det a_2(\zeta) = \delta_Q$ and $\delta_Q + \beta = \chi$. ∎

Corollary 9.27. *From (9.99), it follows that the set of roots of the function $r_L(\zeta)$ includes the set of roots of the polynomial $\Delta_Q(\zeta)$ and the set of roots of the quasi-polynomial $\Delta_Q(\zeta^{-1})$.* ∎

5. Using the above results, we prove a theorem about factorisation of quasi-polynomials of type 2.

Theorem 9.28. *Let Assumptions I-III on page 350 and the conditions of Lemmata 9.23 and 9.24 hold. Let also the quasi-polynomial $A_L(\zeta)$ be positive on the unit circle. Then, there exists a factorisation*

$$A_L(\zeta) = \hat{\Pi}(\zeta)\Pi(\zeta) = \Pi'(\zeta^{-1})\Pi(\zeta)\,, \tag{9.100}$$

where $\Pi(\zeta)$ is a stable polynomial matrix. Under the same conditions, the following factorisation is possible:

$$r_L(\zeta) = \det A_L(\zeta) = r_L^+(\zeta) r_L^+(\zeta^{-1})\,, \tag{9.101}$$

where $r_L^+(\zeta)$ is a real stable polynomial with $\deg r_L^+(\zeta) \le \chi$. Moreover,

$$\det \Pi(\zeta) \approx r_L^+(\zeta)$$

and the matrices $a_r(\zeta)$, $a_L(\zeta)$ in the IRMFDs (9.91), (9.74) can be chosen, such that

$$\deg \Pi(\zeta) \le \chi\,.$$

Proof. As for Theorem 9.16, the proof is a direct corollary of the theorem about factorisation from [133] with account for our auxiliary results. ∎

Remark 9.29. From Corollary 9.27, it follows that, when the polynomial $d_Q(s)$ has roots on the imaginary axis, then the quasi-polynomial (9.99) has roots on the unit circle and the factorisations (9.100) and (9.101) are impossible.

Remark 9.30. Let the polynomial

$$d_Q(s) = (s - q_1)^{\delta_1} \cdots (s - q_\lambda)^{\delta_\lambda}\,, \quad \delta_1 + \ldots, \delta_\lambda = \delta_Q$$

be free of roots on the imaginary axis. Let also be

$$\operatorname{Re} q_i < 0\,,\ (i = 1, \ldots, m); \quad \operatorname{Re} q_i > 0\,,\ (i = m+1, \ldots, \lambda)\,.$$

Then the polynomial $r_L^+(\zeta)$ can be represented in the form

$$r_L^+(\zeta) = d_Q^+(\zeta) r_{1L}^+(\zeta)\,,$$

where $d_Q^+(\zeta)$ and $r_{1L}^+(\zeta)$ are stable polynomials and

$$d_Q^+(\zeta) = \left(\zeta - e^{-q_1 T}\right)^{\delta_1} \cdots \left(\zeta - e^{-q_m T}\right)^{\delta_m} \left(\zeta - e^{q_{m+1} T}\right)^{\delta_{m+1}} \cdots \left(\zeta - e^{q_\lambda T}\right)^{\delta_\lambda}\,,$$

i.e., the numbers $e^{-q_i T}$, $(i = 1, \ldots, m)$ and $e^{q_i T}$, $(i = m+1, \ldots, \lambda)$ are found among the roots of the polynomial $r_L^+(\zeta)$.

9.7 Characteristic Properties of Solution for Single-loop System

1. Using the above results, we can formulate some characteristic properties of the solution to the $\mathcal{H}_2$-problem for the single-loop system shown in Fig. 9.1. We assume that Assumptions I-III on page 350 and the conditions of Lemmata 9.14, 9.24 hold.

2. Let the quasi-polynomials $A_M(\zeta)$ and $A_L(\zeta)$ be positive on the unit circle. Then, there exist factorisations (9.68) and (9.100). Moreover, the optimal system matrix $\theta^o(\zeta)$ has the form

$$\theta^o(\zeta) = \Pi^{-1}(\zeta) R_+(\zeta) \Gamma^{-1}(\zeta),$$

where $R_+(\zeta)$ is a polynomial matrix and the relations $\deg\det \Pi(\zeta) \leq \chi$ and $\deg\det \Gamma(\zeta) \leq \chi - n$ hold. Due to Lemma 2.8, there exists an LMFD

$$R_+(\zeta)\Gamma^{-1}(\zeta) = \Gamma_1^{-1}(\zeta) R_{1+}(\zeta)$$

with $\deg\det \Gamma_1(\zeta) = \deg\det \Gamma(\zeta) \leq \chi - n$. From the last two equations, we obtain the LMFD

$$\theta^o(\zeta) = \left[\Gamma_1(\zeta)\Pi(\zeta)\right]^{-1} R_{1+}(\zeta),$$

where $\deg\det[\Gamma_1(\zeta)\Pi(\zeta)] \leq 2\chi - n$.

On the other hand, let us have an ILMFD

$$\theta^o(\zeta) = D_l^{-1}(\zeta) M_l(\zeta).$$

Then the function

$$\frac{\det \Gamma_1(\zeta) \det \Pi(\zeta)}{\det D_l(\zeta)} = \frac{\det \Gamma(\zeta) \det \Pi(\zeta)}{\det D_l(\zeta)}$$

is a polynomial. Since the system under consideration is modal controllable, due to the properties of the system function, the polynomial $\det D_l(\zeta)$ is equivalent to the characteristic polynomial of the optimal system $\Delta^o(\zeta)$. Then we obtain

$$\deg \Delta^o(\zeta) = \deg\det D_l(\zeta) \leq 2\chi - n.$$

3. Let $g_1, \dots, g_\kappa$ be the stable and $g_{\kappa+1}, \dots, g_\sigma$ the unstable poles of the matrix $G(s)$; and $q_1, \dots, q_m$; $q_{m+1}, \dots, q_\lambda$ be the corresponding sequences of poles of the matrix $Q(s)$. Then the characteristic polynomial has in the general case its roots at the points $\zeta_1 = \mathrm{e}^{-g_1 T}, \dots, \zeta_\kappa = \mathrm{e}^{-g_\kappa T}$; $\zeta_{\kappa+1} = \mathrm{e}^{g_{\kappa+1} T}, \dots, \zeta_\sigma = \mathrm{e}^{g_\sigma T}$; and $\tilde{\zeta}_1 = \mathrm{e}^{-q_1 T}, \dots, \tilde{\zeta}_m = \mathrm{e}^{-q_m T}$; $\tilde{\zeta}_{m+1} = \mathrm{e}^{q_{m+1} T}, \dots, \tilde{\zeta}_\lambda = \mathrm{e}^{q_\lambda T}$.

4. The single-loop system shown in Fig. 9.1 will be called *critical*, if at least one of the matrices $Q(s)$, $F(s)$ or $G(s)$ has poles on the imaginary axis. These poles will also be called *critical*. The following important conclusions stem from the above reasoning.

a) The presence of critical poles of the matrix $F(s)$ does not change the $\mathcal{H}_2$-optimisation procedure.
b) If any of the matrices $Q(s)$ or $G(s)$ has a critical pole, then the corresponding factorisations (9.100) or (9.68) appear to be impossible, because at least one of the polynomials $\det a_2(\zeta)$ or $\det a_1(\zeta)$ has roots on the unit circle. In this case, formal following the Wiener-Hopf procedure leads to a controller that does not stabilise.
c) As follows from the aforesaid, for solving the $\mathcal{H}_2$-optimisation problems for sampled-data systems with critical continuous-time elements, it is necessary to take into account some special features of the system structure, as well as the placement of the critical elements with respect to the system input and output.

9.8 Simplified Method for Elementary System

1. In principle for the $\mathcal{H}_2$-optimisation of the single-loop structure, we can use the modified Wiener-Hopf-method described in Section 8.6. However in some special cases, a simplified optimisation procedure can be used that does not need the inversion of the matrices $b_l(\zeta)b_l'(\zeta^{-1})$ or $b_r'(\zeta^{-1})b_r(\zeta)$. In this section, such a possibility is illustrated by the example shown in Fig. 9.3, where $F(s) \in \mathbb{R}_{nm}(s)$. Hereinafter, such a system will be called *elementary*.

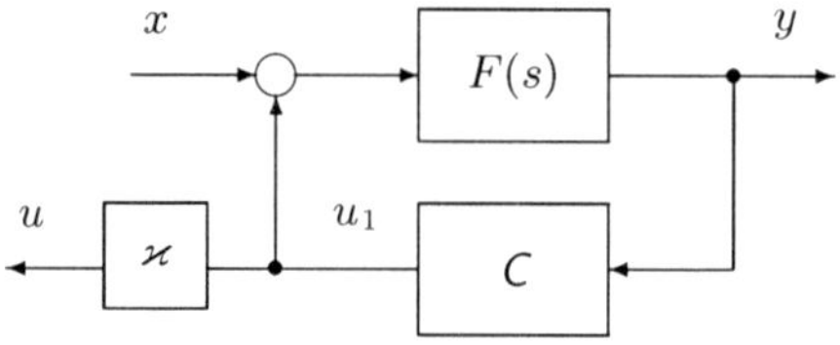

Fig. 9.3. Simplified sampled-data system

The elementary system is a special case of the single-loop system in Fig. 9.1, when $Q(s) = I_n$ and $G(s) = I_m$. In this case from (9.9), we have

$$K(s) = \begin{bmatrix} O_{mm} \\ F(s) \end{bmatrix}, \quad L(s) = \begin{bmatrix} \varkappa I_m \\ F(s) \end{bmatrix}, \qquad M(s) = F(s), \quad N(s) = F(s). \tag{9.102}$$

It is assumed that the matrix

$$F(s) = \frac{N_F(s)}{d_F(s)}$$

with

$$d_F(s) = (s - f_1)^{\nu_1} \cdots (s - f_\psi)^{\nu_\psi}, \quad \nu_1 + \ldots + \nu_\psi = \chi \tag{9.103}$$

is strictly proper and normal. Moreover, the fractions

$$F(s)F'(-s) = \frac{N_F(s)N_F'(-s)}{d_F(s)d_F(-s)}, \quad F'(-s)F(s) = \frac{N_F'(-s)N_F(s)}{d_F(s)d_F(-s)}$$

are assumed to be irreducible. If

$$g(s) \triangleq d_F(s)d_F(-s) = (s - g_1)^{\kappa_1} \cdots (s - g_\rho)^{\kappa_\rho},$$

we shall assume that

$$\tilde{\mu}(gi) = \mu(g_i)\mu(-g_i) \neq 0, \quad (i = 1, \ldots, \rho)$$

and the set of numbers g_i satisfy Conditions (6.106).

2. To solve the $\mathcal{H}_2$-optimisation problem, we apply the general relations of Chapter 8. Hereby, Equation (9.102) leads to a number of serious simplifications.

Using (9.102), we have

$$L'(-s)L(s) = \varkappa^2 I_m + F'(-s)F(s).$$

Then,

$$\widehat{D}_{\underline{L}'L\mu\underline{\mu}}(T,s,0) = \varkappa^2 \frac{1}{T} \sum_{k=-\infty}^{\infty} \mu(s + kj\omega)\mu(-s - kj\omega) + \widehat{D}_{\underline{F}'F\mu\underline{\mu}}(T,s,0).$$

This series can be easily summarised. Indeed, using (6.36) and (6.39), we obtain

$$\begin{aligned}\frac{1}{T}\sum_{k=-\infty}^{\infty} \mu(s + kj\omega)\mu(-s - kj\omega) &= \frac{1}{T}\sum_{k=-\infty}^{\infty} \mu(s + kj\omega)\int_0^T e^{(s+kj\omega)t} m(t)\, dt \\ &= \int_0^T \left[\frac{1}{T}\sum_{k=-\infty}^{\infty} \mu(s + kj\omega)e^{(s+kj\omega)t}\right] m(t)\, dt = \int_0^T m^2(t)\, dt \triangleq m_2.\end{aligned}$$

Hence

$$\widehat{D}_{\underline{L}'L\mu\underline{\mu}}(T,s,0) = \varkappa^2 m_2 + \widehat{D}_{\underline{F}'F\mu\underline{\mu}}(T,s,0).$$

Let us have a minimal standard realisation

$$F(s) = C(sI_\chi - A)^{-1}B$$

and IMFDs

$$C\left(I_\chi - \zeta e^{AT}\right)^{-1} = a_l^{-1}(\zeta)b_l(\zeta),$$
$$\left(I_\chi - \zeta e^{AT}\right)^{-1}B = b_r(\zeta)a_r^{-1}(\zeta),$$

where

$$\det a_l(\zeta) \approx \det a_r(\zeta) \approx \left(\zeta - e^{-f_1 T}\right)^{\nu_1} \cdots \left(\zeta - e^{-f_\psi T}\right)^{\nu_\psi} = \Delta_F(\zeta).$$

In general, the determinant of the matrix

$$\begin{aligned} A_L(\zeta) &= \hat{a}_r(\zeta)\frac{1}{T}D_{\underline{L}'L\mu\underline{\mu}}(T,\zeta,0)a_r(\zeta) \\ &= \frac{1}{T}\varkappa^2 m_2\,\hat{a}_r(\zeta)a_r(\zeta) + \hat{a}_r(\zeta)\frac{1}{T}D_{\underline{F}'F\mu\underline{\mu}}(T,\zeta,0)a_r(\zeta) \end{aligned} \tag{9.104}$$

will not vanish at the roots of the function $\det a_r(\zeta)\det\hat{a}_r(\zeta)$, because these roots are cancelled in the second summand.

Similarly, the determinant of the quasi-polynomial

$$A_M(\zeta) = a_l(\zeta)D_{F\underline{F}'}(T,\zeta,0)\hat{a}_l(\zeta) \tag{9.105}$$

is not zero at these points due to cancellations on the right-hand side.

3.

Theorem 9.31. *Let the above formulated assumptions hold in this section. Let the quasi-polynomials (9.104) and (9.105) be positive on the unit circle, so that there exist factorisations (9.68) and (9.100). Let also the set of eigenvalues of the matrices $\Pi(\zeta)$ and $\Gamma(\zeta)$ does not include the numbers $\zeta_i^\pm = e^{\pm f_i T}$, where f_i are the roots of the polynomial (9.103). Then the following propositions hold:*

a) The matrix

$$R_2(\zeta) = \frac{1}{T}\hat{\Pi}^{-1}(\zeta)\hat{a}_r(\zeta)D_{\underline{F}'F\underline{F}'\underline{\mu}}(T,\zeta,0)\hat{a}_l(\zeta)\hat{\Gamma}^{-1}(\zeta) \tag{9.106}$$

admits a unique separation

$$R_2(\zeta) = R_{21}(\zeta) + R_{22}(\zeta), \tag{9.107}$$

where $R_{22}(\zeta)$ is a strictly proper rational matrix having only unstable poles; it is analytical at the points $\zeta_i^- = e^{-f_i T}$. Moreover, $R_{21}(\zeta)$ is a rational matrix having its poles at the points ζ_i^-.

b) The transfer function of the optimal controller $w_d^o(\zeta)$ *is given by*

$$w_d^o(\zeta) = V_{2o}(\zeta)V_{1o}^{-1}(\zeta)\,, \tag{9.108}$$

where

$$\begin{aligned} V_{1o}(\zeta) &= a_l^{-1}(\zeta) - b_r(\zeta)\Pi^{-1}(\zeta)R_{21}(\zeta)\Gamma^{-1}(\zeta)\,, \\ V_{2o}(\zeta) &= -a_r(\zeta)\Pi^{-1}(\zeta)R_{21}(\zeta)\Gamma^{-1}(\zeta)\,. \end{aligned} \tag{9.109}$$

c) The matrices (9.109) are stable and analytical at the points ζ_i^-, *and the set of their poles is included in the set of poles of the matrices* $\Pi^{-1}(\zeta)$ *and* $\Gamma^{-1}(\zeta)$.

d) The characteristic polynomial of the optimal system $\Delta^o(\zeta)$ *is a divisor of the polynomial* $\det \Pi(\zeta) \det \Gamma(\zeta)$.

Proof. Applying (8.105)–(8.110) to the case under consideration, we have

$$R(\zeta) = \hat{\Pi}^{-1}(\zeta)\hat{C}(\zeta)\hat{\Gamma}^{-1}(\zeta) = R_1(\zeta) + R_2(\zeta)\,, \tag{9.110}$$

where

$$R_1(\zeta) = \Pi(\zeta)a_r^{-1}(\zeta)\beta_{0r}(\zeta)\Gamma(\zeta) \tag{9.111}$$

and the matrix $R_2(\zeta)$ is given by (9.106). Under the given assumptions owing to Remark 8.19, the matrix $\hat{C}(\zeta)$ is a quasi-polynomial. Therefore, the matrix $R(\zeta)$ can have unstable poles only at the point $\zeta = 0$ and at the poles of the matrices $\hat{\Pi}^{-1}(\zeta)$ and $\hat{\Gamma}^{-1}(\zeta)$. Hence under the given assumptions, Matrix (9.110) is analytical at the points $\zeta_i^- = \mathrm{e}^{-f_i T}$. Simultaneously, all nonzero poles of the matrix

$$\hat{a}_r(\zeta)D_{\underline{F}'F\underline{F}'\underline{\mu}}(T,\zeta,0)\hat{a}_l(\zeta)$$

belong to the set of the numbers ζ_i^-, because the remaining poles are cancelled against the factors $\hat{a}_r(\zeta)$ and $\hat{a}_l(\zeta)$. Then it follows immediately that Matrix (9.106) admits a unique separation (9.107). Using (9.111) and (9.107) from (9.110), we obtain

$$R(\zeta) = \left[\Pi(\zeta)a_r^{-1}(\zeta)\beta_{0\eta}(\zeta)\Gamma(\zeta) + R_{21}(\zeta)\right] + R_{22}(\zeta)\,. \tag{9.112}$$

Per construction, $R_{22}(\zeta)$ is a strictly proper rational matrix, whose poles include all poles of the matrix $R(\zeta)$, which are all unstable. Also per construction, the expression in the square brackets can have poles at the points ζ_i^-. But under the given assumptions, the matrix $R(\zeta)$ is analytical at these points. Hence the matrix in the square brackets in (9.112) is a polynomial. Then the right-hand side of (9.112) coincides with the principal separation (9.28) and from (8.115), we obtain

$$\begin{aligned} R_+(\zeta) &= \Pi(\zeta)a_r^{-1}(\zeta)\beta_{0r}(\zeta)\Gamma(\zeta) + R_{21}(\zeta)\,, \\ R_-(\zeta) &= \langle R(\zeta)\rangle_- = R_{22}(\zeta)\,, \quad R_{21}(\zeta) = \Lambda(\zeta)\,. \end{aligned}$$

Using (8.95), we find the optimal system matrix

$$\theta^o(\zeta) = a_r^{-1}(\zeta)\beta_{0r}(\zeta) + \Pi^{-1}(\zeta)R_{21}(\zeta)\Gamma^{-1}(\zeta)\,,$$

which is stable and analytical at the points ζ_i^-. Therefore, the matrices (9.109) calculated by (8.117)–(8.118) are stable and analytical at the points ζ_i^-. The remaining claims of the theorem follow from the constructions of Section 8.7. ∎

Example 9.32. Consider the simple SISO system shown in Fig. 9.4 with

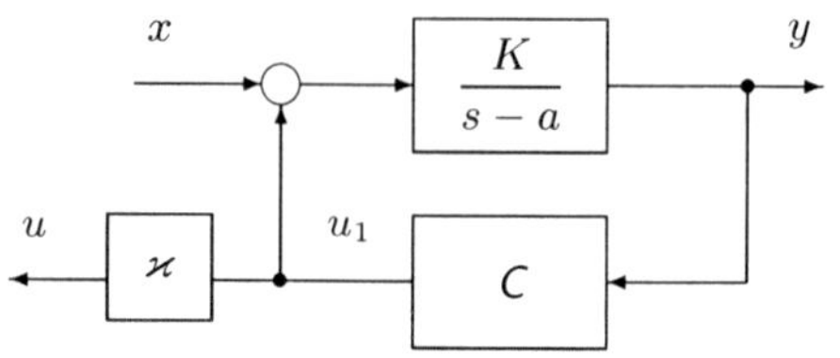

Fig. 9.4. Example of elementary sampled-data system

$$F(s) = \frac{K}{s-a}\,,$$

where K and a are constants. Moreover, assume that $x(t)$ is unit white noise. It is required to find the transfer function of a discrete controller $w_d^o(\zeta)$, which stabilises the closed-loop system and minimises the value

$$\|S\|_2^2 = \varkappa^2 \bar{d}_{u_1} + \bar{d}_y\,.$$

In the given case from (6.72) and (6.86), it follows that

$$\begin{aligned}
\widehat{D}_F(T,s,t) &= \frac{K e^{at}}{1 - e^{aT}e^{-sT}}\,, \quad 0 < t < T\,,\\
\widehat{D}_{F\mu}(T,s,t) &= \frac{K\mu(a)e^{at}}{e^{sT}e^{-aT} - 1} + K\int_0^t e^{a(t-\tau)}m(\tau)\,d\tau\,, \quad 0 \le t \le T\,.
\end{aligned} \tag{9.113}$$

Moreover,

$$D_{F\mu}(T,\zeta,0) = \frac{\zeta K\mu(a)e^{aT}}{1-\zeta e^{aT}}\,.$$

Hence we can take

$$\begin{aligned}
a_l(\zeta) &= 1 - \zeta e^{aT}, \quad & b_l(\zeta) &= \zeta K\mu(a)e^{aT},\\
a_r(\zeta) &= a_l(\zeta), & b_r(\zeta) &= b_l(\zeta)\,.
\end{aligned} \tag{9.114}$$

As follows from (9.51), in this case

$$D_{F\underline{F}'}(T,\zeta,0) = \frac{\gamma}{(1-\zeta \mathrm{e}^{aT})(1-\zeta^{-1}\mathrm{e}^{aT})}\,, \tag{9.115}$$

where $\gamma > 0$ is a known constant. Moreover, using (9.113), (9.78) and (9.104), it can be easily shown that

$$\begin{aligned} D_{\underline{L}'L\mu\underline{\mu}}(T,\zeta,0) &= \varkappa^2 \bar{m} + D_{\underline{F}'F\mu\underline{\mu}}(T,\zeta,0) \\ &= \frac{q(1-\zeta\nu)(1-\zeta^{-1}\nu)}{(1-\zeta \mathrm{e}^{aT})(1-\zeta^{-1}\mathrm{e}^{aT})}\,, \end{aligned} \tag{9.116}$$

where $q > 0$ and ν are constants, such that $|\nu| < 1$.

As follows from (9.113) and (9.78),

$$\hat{a}_l(\zeta) = \hat{a}_r(\zeta) = 1 - \zeta^{-1}\mathrm{e}^{aT} = \frac{\zeta - \mathrm{e}^{aT}}{\zeta}\,. \tag{9.117}$$

Then using (8.85), (9.115) and (9.116), we find

$$A_M(\zeta) = \gamma, \quad A_L(\zeta) = K_1(1-\zeta\nu)(1-\zeta^{-1}\nu)\,,$$

where $K_1 > 0$ is a constant. Thus, we obtain that in the factorisations (9.68), (9.100), we can take

$$\Gamma(\zeta) = \eta_1\,, \quad \Pi(\zeta) = \eta_2(1-\zeta\nu)\,, \tag{9.118}$$

where η_1 and η_2 are real constants.

For further calculations, we notice that in the given case, Formulae (6.92) and (6.93) yield

$$D_{F\underline{F}'F\mu}(T,\zeta,0) = \frac{\zeta(\ell_2 + \ell_1\zeta + \ell_0\zeta^2)}{(1-\zeta \mathrm{e}^{aT})^2(1-\zeta \mathrm{e}^{-aT})} \tag{9.119}$$

with constants ℓ_0, ℓ_1, and ℓ_2. Since

$$D_{\underline{F}'F\underline{F}'\underline{\mu}}(T,\zeta,0) = D'_{F\underline{F}'F\mu}(T,\zeta^{-1},0)\,,$$

from (9.119), we find

$$D_{\underline{F}'F\underline{F}'\underline{\mu}}(T,\zeta,0) = \frac{\ell_0 + \ell_1\zeta + \ell_2\zeta^2}{(\zeta - \mathrm{e}^{aT})^2(\zeta - \mathrm{e}^{-aT})}\,.$$

Hence using (9.117), we obtain

$$\frac{1}{T}\,\hat{a}_r(\zeta) D_{\underline{F}'F\underline{F}'\underline{\mu}}(T,\zeta,0)\hat{a}_l(\zeta) = -\frac{\mathrm{e}^{aT}}{T}\,\frac{\ell_0 + \ell_1\zeta + \ell_2\zeta^2}{\zeta^2(1-\zeta \mathrm{e}^{aT})}\,. \tag{9.120}$$

Owing to

$$\hat{\Pi}(\zeta) = m_2(1 - \zeta^{-1}\nu) = \frac{m_2(\zeta - \nu)}{\zeta}, \qquad \hat{\Gamma}(\zeta) = m_1$$

and using (9.120), we find the function (9.106) in the form

$$R_2(\zeta) = \frac{n_0 + n_1\zeta + n_2\zeta^2}{\zeta(\zeta - \nu)(1 - \zeta e^{aT})},$$

where n_0, n_1, n_2 are known constants. Performing the separation (9.107) with $|\nu| < 1$, we obtain

$$R_{21}(\zeta) = \frac{\lambda}{1 - \zeta e^{aT}},$$

where λ is a known constant. From this form and (9.118), we find

$$\Pi^{-1}(\zeta)R_{21}(\zeta)\Gamma^{-1}(\zeta) = \frac{\lambda_1}{(1 - \zeta e^{aT})(1 - \zeta\nu)} \tag{9.121}$$

with a known constant λ_1. Taking into account (9.114), we obtain the function $V_{1o}(\zeta)$ in (9.109):

$$V_{1o}(\zeta) = \frac{\zeta\lambda_2}{(1 - \zeta e^{aT})(1 - \zeta\nu)} + \frac{1}{1 - \zeta e^{aT}}$$

with a known constant λ_2. Due to Theorem 9.31, the function $V_{1o}(\zeta)$ is analytical at the point $\zeta = e^{-aT}$. Hence

$$1 + e^{-aT}\lambda_2 - e^{-aT}\nu = 0\,.$$

From the last two equations, we receive

$$V_{1o}(\zeta) = \frac{1}{1 - \zeta\nu}\,.$$

Furthermore using (9.121) and (9.109), we obtain

$$V_{2o}(\zeta) = \frac{\lambda_3}{1 - \zeta\nu}, \qquad \lambda_3 = \text{const.}$$

Therefore, Formula (9.108) yields

$$w_d^o(\zeta) = \lambda_3 = \text{const.}$$

and the characteristic polynomial of the closed-loop appears as

$$\Delta^o(\zeta) \approx 1 - \zeta\nu\,.$$

□

10

$\mathcal{L}_2$-Design of SD Systems for $0 < t < \infty$

10.1 Problem Statement

1. Let the input of the standard sampled-data system for $t \geq 0$ be acted upon by a vector input signal $x(t)$ of dimension $\ell \times 1$, and let $z(t)$ be the $r \times 1$ output vector under zero initial energy. Then the system performance can be evaluated by the value

$$\tilde{J} = \int_0^\infty z'(t)z(t)\,\mathrm{d}t = \sum_{i=1}^{r} \int_0^\infty z_i^2(t)\,\mathrm{d}t\,, \tag{10.1}$$

where $z_i(t)$, $(i = 1, \dots, r)$ are the components of the output vector $z(t)$. It is assumed that the conditions for the convergence of the integral (10.1) hold. It is known [206] that the value

$$\|z\|_{\mathcal{L}_2} = +\sqrt{\tilde{J}}$$

determines the $\mathcal{L}_2$-norm of the output signal $z(t)$. Thus, the following optimisation problem is formulated.

> **$\mathcal{L}_2$-problem.** Given the matrix $w(p)$ in (7.2), the input vector $x(t)$, the sampling period T and the form of the control impulse $m(t)$. Find a stabilising controller (8.35) that ensures the internal stability of the standard sampled-data system and the minimal value of $\|z(t)\|_{\mathcal{L}_2}$.

2. It should be noted that the general problem formulated above include, for different choice of the vector $z(t)$, many important applied problems, also including the tracking problem. Indeed, let us consider the block-diagram shown in Fig. 10.1, where the dotted box denotes the initial standard system that will be called *nominal*. Moreover in Fig. 10.1, $Q(p)$ denotes the transfer matrix of an ideal transition. To evaluate the tracking performance, it is natural to use the value

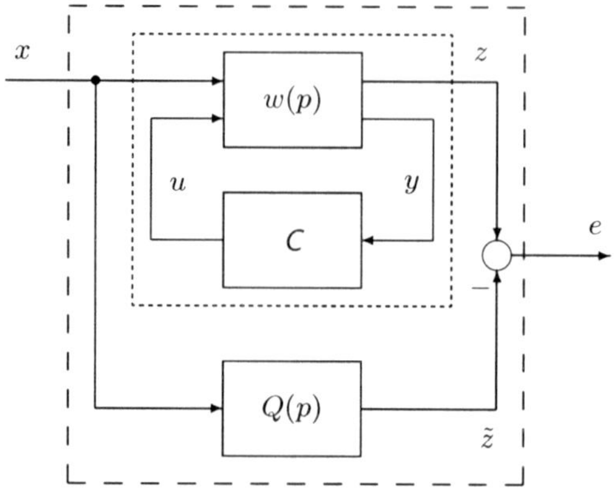

Fig. 10.1. Tracking control loop

$$\tilde{J}_e = \int_0^\infty e'(t)e(t)\,\mathrm{d}t = \int_0^\infty [z(t) - \tilde{z}(t)]'\,[z(t) - \tilde{z}(t)]\,\mathrm{d}t\,. \tag{10.2}$$

If the PTM of the nominal system $w(s,t)$ has the form (7.30)

$$w(s,t) = \varphi_L(T,s,t)\widehat{R}_N(s)M(s) + K(s)\,, \tag{10.3}$$

then the tracking error $e(t)$ can be considered as a transformed result of the input signal $x(t)$ by a new standard sampled-data system with the PTM

$$w_e(s,t) = w(s,t) - Q(s) = \varphi_L(T,s,t)\widehat{R}_N(s)M(s) + K(s) - Q(s)\,. \tag{10.4}$$

This system is fenced by a dashed line in Fig. 10.1. The standard sampled-data system with the PTM (10.4) is associated with a continuous-time LTI plant having the transfer matrix

$$w_e(p) = \begin{bmatrix} K(p) - Q(p) & L(p) \\ M(p) & N(p) \end{bmatrix}.$$

Then, the integral (10.2) coincides with (10.1) for the new standard sampled-data system.

3. Under some restrictions formulated below and using Parseval's formula [181], the integral (10.1) can be transformed into

$$\tilde{J} = \frac{1}{2\pi\mathrm{j}} \int_{-\mathrm{j}\infty}^{\mathrm{j}\infty} Z'(-s)Z(s)\,\mathrm{d}s\,, \tag{10.5}$$

where $Z(s)$ is the Laplace transform of the output $z(t)$. Thus, the $\mathcal{L}_2$-problem formulated above can be considered as a problem of choosing a stabilising controller which minimises the integral (10.5). This problem will be considered in the present chapter.

10.2 Pseudo-rational Laplace Transforms

1. According to the above statement of the problem, to consider the integral (10.5), we have to find the Laplace transform of the output $z(t)$ for the standard sampled-data system under zero initial energy and investigate its properties as a function of s. In this section, we describe some properties of a class of transforms used below.

2. Henceforth, we denote by Λ_γ the set of functions (matrices) $f(t)$ that are zero for $t < 0$, have bounded variation for $t \geq 0$ and satisfy the estimation

$$|f(t)| < de^{\gamma t}, \quad t > 0,$$

where $d > 0$ and γ are constants. It is known [39] that for any function $f(t) \in \Lambda_\gamma$ for $\operatorname{Re} s > \gamma$, there exists the Laplace transform

$$F(s) = \int_0^\infty f(t)\mathrm{e}^{-st}\,\mathrm{d}t \tag{10.6}$$

and for any $t \geq 0$ the following inversion formula holds:

$$\tilde{f}(t) = \frac{1}{2\pi\mathrm{j}} \lim_{a\to\infty} \int_{c-\mathrm{j}a}^{c+\mathrm{j}a} F(s)\mathrm{e}^{st}\,\mathrm{d}s\,, \quad c > \gamma\,,$$

where

$$\tilde{f}(t) = \frac{f(t-0) + f(t+0)}{2}\,.$$

As follows from the general properties of the Laplace transformation [39], under the given assumptions in any half-plane $\operatorname{Re} s \geq \gamma_1 > \gamma$, we have

$$\lim_{s\to\infty} |F(s)| = 0$$

for s increasing to infinity along any contour. Then for $f(t) \in \Lambda_\gamma$ and $s = x + \mathrm{j}y$, $x \geq \gamma_1 > \gamma$, the following estimation holds [22] :

$$|F(x + k\mathrm{j}y)| \leq \frac{c}{|k|}\,, \quad c = \text{const.} \tag{10.7}$$

Hereinafter, the elements $f(t)$ of the set Λ_γ will be called *originals* and denoted by small letters, while the corresponding Laplace transforms (10.6) will be called *images* and denoted by capital letters.

3. For any original $f(t) \in \Lambda_\gamma$ for $\operatorname{Re} s \geq \gamma_1 > \gamma$, the following series converges:

$$\varphi_f(T, s, t) = \sum_{k=-\infty}^{\infty} f(t + kT)\mathrm{e}^{-s(t+kT)}\,, \quad -\infty < t < \infty\,. \tag{10.8}$$

According to [148], $\varphi_f(T,s,t)$ is the displaced pulse frequency response (DPFR). The function $\varphi_f(T,s,t)$ is periodic in t, therefore, it can be associated with a Fourier series of the form (6.15)

$$\varphi_F(T,s,t) = \frac{1}{T}\sum_{k=-\infty}^{\infty} F(s+kj\omega)e^{kj\omega t}\,, \quad \omega = \frac{2\pi}{T}\,. \tag{10.9}$$

Hereinafter, we shall assume that the function $\varphi_f(T,s,t)$ is of bounded variation over the interval $0 \le t \le T$. This holds as a rule in applications. Then, we have the equality

$$\varphi_f(T,s,t) = \varphi_F(T,s,t)\,,$$

that should be understood in the sense that

$$\varphi_f(T,s,t_0) = \varphi_F(T,s,t_0)$$

for any t_0, where the function $\varphi_f(T,s,t)$ is continuous and

$$\varphi_F(T,s,t_0) = \frac{\varphi_f(T,s,t_0-0)+\varphi_f(T,s,t_0+0)}{2}\,,$$

if $\varphi_f(T,s,t)$ at $t=t_0$ has a break of the first kind (finite break).

4. Together with the DPFR (10.8), we consider the *discrete Laplace transform* (DLT) $\overset{\frown}{D}_f(T,s,t)$ of the function $f(t)$:

$$\overset{\frown}{D}_f(T,s,t) = \sum_{k=-\infty}^{\infty} f(t+kT)e^{-ksT} = \varphi_f(T,s,t)e^{st} \tag{10.10}$$

and the associated series (6.67)

$$\overset{\frown}{D}_F(T,s,t) = \frac{1}{T}\sum_{k=-\infty}^{\infty} F(s+kj\omega)e^{(s+kj\omega)t} = \varphi_F(T,s,t)e^{st}\,, \tag{10.11}$$

which will be called the *discrete Laplace transform* of the image $F(s)$.

Let us have a strictly proper rational matrix

$$F(s) = \frac{N_F(s)}{d_F(s)}\,,$$

where $N_F(s)$ is a polynomial matrix and

$$d_F(s) = (s-f_1)^{\nu_1}\cdots(s-f_\psi)^{\nu_\psi}\,, \quad \nu_1+\ldots+\nu_\psi = \chi\,.$$

Then as follows from (6.70)–(6.72), the function (matrix)

$$D_F(T,\zeta,t) = \overset{\frown}{D}_F(T,s,t)\,|_{e^{-sT}=\zeta}$$

is rational in ζ for all t, and for $0 < t < T$ it can be represented in the form

$$D_F(T,\zeta,t) = \frac{\sum_{k=0}^{m} d_k(t)\zeta^k}{\Delta_F(\zeta)}, \tag{10.12}$$

where

$$\Delta_F(\zeta) = \left(\zeta - \mathrm{e}^{-f_1 T}\right)^{\nu_1} \cdots \left(\zeta - \mathrm{e}^{-f_\psi T}\right)^{\nu_\psi}$$

is the discretisation of the polynomial $d_F(s)$, and $d_k(t)$ are functions of bounded variation on the interval $0 < t < T$.

5. Below it will be shown that some images $F(s)$, which are not rational functions, may possess a DLT of the form (10.12).

Henceforth, the image $F(s)$ will be called *pseudo-rational,* if its DLT for $0 < t < T$ can be represented in the form (10.12). The set of all pseudo-rational images are determined by the following Lemma.

Lemma 10.1. *A necessary and sufficient condition for an image $F(s)$ to be pseudo-rational is, that it can be represented as*

$$F(s) = \frac{\sum_{k=0}^{m} \mathrm{e}^{-ksT} \int_0^T d_k(t)\mathrm{e}^{-st}\,\mathrm{d}t}{\widehat{\Delta}_F(s)}, \tag{10.13}$$

where

$$\widehat{\Delta}_F(s) = \Delta_F(\zeta)\,|_{\zeta=\mathrm{e}^{-sT}} = \left(\mathrm{e}^{-sT} - \mathrm{e}^{-f_1 T}\right)^{\nu_1} \cdots \left(\mathrm{e}^{-sT} - \mathrm{e}^{-f_\psi T}\right)^{\nu_\psi}.$$

Proof. Necessity: Let the matrix $D_F(T,\zeta,t)$ have the form (10.12). Then we have

$$\widehat{D}_F(T,s,t) = D(T,\zeta,t)\,|_{\zeta=\mathrm{e}^{-sT}} = \frac{\sum_{k=0}^{m} \mathrm{e}^{-ksT} d_k(t)}{\widehat{\Delta}_F(s)}. \tag{10.14}$$

Moreover for the image $F(s)$, we have [148]

$$F(s) = \int_0^T \widehat{D}_F(T,s,t)\mathrm{e}^{-st}\,\mathrm{d}t. \tag{10.15}$$

With respect to (10.14), this yields (10.13).

Sufficiency: Denote

$$D_k(s) \triangleq \int_0^T d_k(t)\mathrm{e}^{-st}\,\mathrm{d}t.$$

Then, (6.36)–(6.39) yield

$$\frac{1}{T}\sum_{n=-\infty}^{\infty} D_k(s + nj\omega)\mathrm{e}^{(s+nj\omega)t} = d_k(t), \quad 0 < t < T.$$

Hence using (10.13) and (10.11) for $0 < t < T$, we obtain

$$\begin{aligned} D_F(T,s,t) &= \frac{\sum_{k=0}^m e^{-ksT} \frac{1}{T} \sum_{n=-\infty}^{\infty} D_k(s+nj\omega)e^{(s+nj\omega)t}}{\widehat{\Delta}_F(s)} \\ &= \frac{\sum_{k=0}^m e^{-ksT} d_k(t)}{\widehat{\Delta}_F(s)}, \end{aligned}$$

which proves the claim. ■

Lemma 10.2. *Let the images $F(s)$ and $G(s)$ be pseudo-rational. Then, the product $H(s) = F(s)G(s)$ is pseudo-rational.*

Proof. Together with (10.12), let us have

$$D_G(T,s,t) = \frac{\sum_{i=0}^r e^{-isT} b_i(t)}{\Delta_G(\zeta)}, \quad 0 < t < T, \tag{10.16}$$

where

$$\widehat{\Delta}_G(s) = \left(e^{-sT} - e^{-g_1 T}\right)^{\lambda_1} \cdots \left(e^{-sT} - e^{-g_\psi T}\right)^{\lambda_\psi}.$$

We will show that

$$\begin{aligned} \widehat{D}_{FG}(T,s,t) &= \int_0^T \widehat{D}_F(T,s,t-\tau)\widehat{D}_G(T,s,\tau)\,d\tau \\ &= \int_0^T \widehat{D}_F(T,s,\tau)\widehat{D}_G(T,s,t-\tau)\,d\tau. \end{aligned} \tag{10.17}$$

Indeed, we have

$$\begin{aligned} \widehat{D}_F(T,s,t-\tau) &= \frac{1}{T} \sum_{k=-\infty}^{\infty} F(s+kj\omega)e^{(s+kj\omega)(t-\tau)} \\ \widehat{D}_G(T,s,\tau) &= \frac{1}{T} \sum_{k=-\infty}^{\infty} G(s+kj\omega)e^{(s+kj\omega)\tau}. \end{aligned}$$

Substituting this into the middle part of (10.17) after integration, we prove the first equality in (10.17). The second one is proved in a similar way.

Using known properties of the DLT and (10.14), we have

$$\begin{aligned} \widehat{D}_F(T,s,t-\tau) &= \widehat{\mathcal{D}}_F(T,s,t-\tau) \\ &= \frac{\sum_{k=0}^m e^{-ksT} d_k(t-\tau)}{\widehat{\Delta}_F(s)}, \qquad 0 < t-\tau < T, \\ \widehat{D}_F(T,s,t-\tau) &= \widehat{\mathcal{D}}_F(T,s,t-\tau+T)e^{-sT} \\ &= \frac{\sum_{k=0}^m e^{-(k+1)sT} d_k(t-\tau+T)}{\widehat{\Delta}_F(s)}, \qquad -T < t-\tau < 0. \end{aligned}$$

Substituting these forms and (10.16) into (10.17), we obtain an expression of the form

$$\widehat{D}_{FG}(T,s,t) = \frac{\sum_{k=0}^{q} \mathrm{e}^{-ksT} h_k(t)}{\widehat{\Delta}_F(s)\widehat{\Delta}_G(s)}, \quad 0 < t < T, \tag{10.18}$$

where $h_k(t)$ are known functions. This expression has the form (10.14), so that the image $H(s) = F(s)G(s)$ is pseudo-rational. ∎

Corollary 10.3. *Using estimates similar to (10.7), it can be shown that in the given case, the function $\widehat{D}_{FG}(T,s,t)$ is continuous for all t. Then from (10.18) for $t = 0$ and $\mathrm{e}^{-sT} = \zeta$, we obtain*

$$D_{FG}(T,\zeta,0) = \frac{N_{FG}(\zeta)}{\Delta_F(\zeta)\Delta_G(\zeta)},$$

where $N_{FG}(\zeta)$ is a polynomial matrix. ∎

6. We note that many important originals encountered in applications have pseudo-rational images, including finite length originals $f(t)$. Indeed, suppose $f(t) = 0$ outside the interval $0 \le t \le a$. Without loss of generality, we can assume $a = nT$, where $n \ge 1$ is an integer. Moreover, if

$$f(t) = f_i(t), \quad iT < t < (i+1)T, \ (i = 0, 1, \ldots, n-1)$$

from (10.10) and (10.11), we have

$$\widehat{D}_F(T,s,t) = \widehat{\mathcal{D}}_F(T,s,t) = \sum_{i=0}^{n-1} f_i(t+iT)\mathrm{e}^{-isT}, \quad 0 < t < T,$$

whence it immediately follows that the Laplace transform $F(s)$ of a finite length impulse is a pseudo-rational function, because the function $D_F(T,\zeta,t)$ is a polynomial in ζ.

10.3 Laplace Transforms of Standard SD System Output

1. Hereinafter we assume that the input $x(t)$ of the system under consideration is in Λ_γ and its image $X(s)$ is pseudo-rational, *i.e.*,

$$\widehat{D}_X(T,s,t) = \frac{\sum_{k=0}^{\ell} \mathrm{e}^{-ksT} x_k(t)}{\widehat{\Delta}_X(s)}, \quad 0 < t < T, \tag{10.19}$$

where

$$\widehat{\Delta}_X(s) = \left(\mathrm{e}^{-sT} - \mathrm{e}^{-x_1 T}\right)^{q_1} \cdots \left(\mathrm{e}^{-sT} - \mathrm{e}^{-x_\eta T}\right)^{q_\eta}. \tag{10.20}$$

The polynomial $\Delta_X(\zeta)$ is assumed to be stable, *i.e.*, free of roots in the closed unit disk. As a special case, the vector (10.19) can be a polynomial in $\zeta = \mathrm{e}^{-sT}$. The continuous-time parts of the standard sampled-data system will be described in the form of the state equations

$$\begin{aligned}\frac{\mathrm{d}v}{\mathrm{d}t} &= Av + B_1 x + B_2 u \\ z &= C_1 v + D_L u\,, \qquad y = C_2 v\,.\end{aligned} \tag{10.21}$$

The equations of the digital controller have the form (7.4), (7.6) and (7.7):

$$\begin{aligned}\xi_k &= y(kT)\,, \quad (k = 0, \pm 1, \dots) \\ \alpha_0\psi_k + \ldots + \alpha_q\psi_{k-q} &= \beta_0\xi_k + \ldots \beta_q\xi_{k-q} \\ u(t) &= m(t - kT)\psi_k\,, \quad kT < t < (k+1)T\,.\end{aligned} \tag{10.22}$$

As follows from [148] for $x(t) \in \Lambda_\gamma$, all solutions $z(t)$ of the system (10.21), (10.22) belong to the set Λ_β, where β is a sufficiently large number. In particular, if $\gamma < 0$ and the system (10.21)– (10.22) is internally stable, then we can take $\beta < 0$.

2. The following theorem gives a general expression for the Laplace transform of the output $Z(s)$ under zero initial energy.

Theorem 10.4. *For* $\operatorname{Re} s > \beta$ *with sufficiently large* β*, there exists the Laplace transform of the solution* $z(t)$ *of the system (10.21)–(10.22) under zero initial energy. This image* $Z(s)$ *has the form*

$$Z(s) = L(s)\mu(s)\widehat{R}_N(s)\widehat{D}_{MX}(T,s,0) + K(s)X(s)\,, \tag{10.23}$$

where

$$\begin{aligned}K(s) &= C_1(sI_\chi - A)^{-1}B_1\,, \qquad L(s) = C_1(sI_\chi - A)^{-1}B_2 + D_L\,, \\ M(s) &= C_2(sI_\chi - A)^{-1}B_1\,, \qquad N(s) = C_2(sI_\chi - A)^{-1}B_2\end{aligned} \tag{10.24}$$

and

$$\widehat{D}_{MX}(T,s,0) = \frac{1}{T}\sum_{k=-\infty}^{\infty} M(s + k\mathrm{j}\omega)X(s + k\mathrm{j}\omega)\,.$$

Moreover, as before in (10.23), we have

$$\widehat{R}_N(s) = \widehat{w}_d(s)\left[I_n - \widehat{D}_{N\mu}(T,s,0)\widehat{w}_d(s)\right]^{-1}\,, \tag{10.25}$$

where

$$\widehat{D}_{N\mu}(T,s,0) = \frac{1}{T}\sum_{k=-\infty}^{\infty} N(s+kj\omega)\mu(s+kj\omega)$$

and

$$\mu(s) = \int_0^T e^{-s\tau} m(\tau)\, d\tau\,. \tag{10.26}$$

The matrix $\widehat{w}_d(s)$ *in (10.25) is determined by the relation*

$$\widehat{w}_d(s) = \widehat{\alpha}^{-1}(s)\widehat{\beta}(s)\,, \tag{10.27}$$

where

$$\widehat{\alpha}(s) = \alpha_0 + \alpha_1 e^{-sT} + \ldots + \alpha_q e^{-qsT}\,,$$

$$\widehat{\beta}(s) = \beta_0 + \beta_1 e^{-sT} + \ldots + \beta_q e^{-qsT}\,.$$

Proof. a) Taking the Laplace images for the first equation in (10.21) under zero initial conditions, we obtain

$$sV(s) = AV(s) + B_2U(s) + B_1X(s)\,,$$

whence it follows

$$V(s) = w_2(s)U(s) + w_1(s)X(s) \tag{10.28}$$

with the notation

$$w_2(s) = (sI_\chi - A)^{-1}B_2\,, \quad w_1(s) = (sI_\chi - A)^{-1}B_1\,. \tag{10.29}$$

b) The image $U(s)$ appearing in Equation (10.28) is determined by the relations

$$\begin{aligned} U(s) &= \int_0^\infty e^{-st}u(t)\, dt = \sum_{k=0}^{\infty}\int_{kT}^{(k+1)T} e^{-st}m(t-kT)\, dt\, \psi_k \\ &= \int_0^T e^{-st}m(t)\, dt \sum_{k=0}^{\infty}\psi_k e^{-ksT} = \mu(s)\widehat{\psi}^*(s)\,, \end{aligned} \tag{10.30}$$

where

$$\widehat{\psi}^*(s) = \sum_{k=0}^{\infty}\psi_k e^{-ksT} \tag{10.31}$$

is the discrete Laplace transform for the vector sequence $\{\psi_k\} \triangleq \{\psi(kT)\}$, $(k = 0, 1, \ldots)$. From (10.28) and (10.30), we have

$$V(s) = w_2(s)\mu(s)\widehat{\psi}^*(s) + w_1(s)X(s)\,. \tag{10.32}$$

c) Let us find an expression for the vector $\overset{\frown}{\psi}^*(s)$ appearing in (10.31). With this aim in view, we notice that (10.32) for $k = 0, \pm 1, \dots$ yields

$$V(s + kj\omega) = w_2(s + kj\omega)\mu(s + kj\omega)\overset{\frown}{\psi}^*(s) + w_1(s + kj\omega)X(s + kj\omega)\,.$$

Then, we find

$$\begin{aligned}\overset{\frown}{D}_V(T, s, t) &= \frac{1}{T}\sum_{k=-\infty}^{\infty} V(s + kj\omega)\mathrm{e}^{(s+kj\omega)t} \\ &= \overset{\frown}{D}_{w_2\mu}(T, s, t)\overset{\frown}{\psi}^*(s) + \overset{\frown}{D}_{w_1X}(T, s, t)\,.\end{aligned} \tag{10.33}$$

Under the taken assumptions, the components of the vectors $V(s + kj\omega)$ vanish as $|k|^{-2}$ when k increases. Therefore, the right side of (10.33) is continuous in t and we can substitute $t = 0$ in (10.33):

$$\overset{\frown}{D}_V(T, s, 0) = \overset{\frown}{D}_{w_2\mu}(T, s, 0)\overset{\frown}{\psi}^*(s) + \overset{\frown}{D}_{w_1X}(T, s, 0)\,. \tag{10.34}$$

Moreover, from (10.8) and (10.9) for $t = 0$, we have

$$\begin{aligned}\overset{\frown}{D}_V(T, s, 0) &= \frac{1}{T}\sum_{k=-\infty}^{\infty} V(s + kj\omega) = \varphi_V(T, s, 0) = \varphi_v(T, s, 0) \\ &= \sum_{k=0}^{\infty} v(kT)\mathrm{e}^{-ksT} = \overset{\frown}{v}^*(s)\,,\end{aligned} \tag{10.35}$$

where $\overset{\frown}{v}^*(s)$ is the discrete Laplace transform of the sequence $\{v_k\} \triangleq \{v(kT)\}$, $(k = 0, 1, \dots)$. With regard to (10.35), Relation (10.34) can be written in the form

$$\overset{\frown}{v}^*(s) = \overset{\frown}{D}_{w_2\mu}(T, s, 0)\overset{\frown}{\psi}^*(s) + \overset{\frown}{D}_{w_1X}(T, s, 0)\,. \tag{10.36}$$

Further, taking the ζ-transform of the second equation in (10.22) under zero initial energy and substituting e^{-sT} for ζ, we obtain

$$\overset{\frown}{\alpha}(s)\overset{\frown}{\psi}^*(s) = \overset{\frown}{\beta}(s)\overset{\frown}{y}^*(s)\,, \tag{10.37}$$

where $\overset{\frown}{y}^*(s)$ is the discrete Laplace transform of the vector sequence $\{y_k\} \triangleq \{y(kT)\}$, $(k = 0, 1, \dots)$. From (10.37) as to (10.27), we have

$$\overset{\frown}{\psi}^*(s) = \overset{\frown}{w}_d(s)\overset{\frown}{y}^*(s)\,. \tag{10.38}$$

Substituting (10.38) into (10.36), we find

$$\overset{\frown}{v}^*(s) = \overset{\frown}{D}_{w_2\mu}(T, s, 0)\overset{\frown}{w}_d(s)\overset{\frown}{y}^*(s) + \overset{\frown}{D}_{w_1X}(T, s, 0)\,.$$

Multiplying from left by the matrix C_2 and using the fact that

$$C_2\overset{\frown}{v^*}(s) = \overset{\frown}{y^*}(s) \tag{10.39}$$

from (10.24), we find

$$\overset{\frown}{y^*}(s) = \overset{\frown}{D}_{N\mu}(T,s,0)\overset{\frown}{w}_d(s)\overset{\frown}{y^*}(s) + \overset{\frown}{D}_{MX}(T,s,0)\,.$$

Then,

$$\overset{\frown}{y^*}(s) = \left[I_n - \overset{\frown}{D}_{N\mu}(T,s,0)\overset{\frown}{w}_d(s)\right]^{-1}\overset{\frown}{D}_{MX}(T,s,0)\,,$$

and with regard to (10.38), we find

$$\begin{aligned}\overset{\frown}{\psi^*}(s) = \overset{\frown}{w}_d(s)\overset{\frown}{y^*}(s) &= \overset{\frown}{w}_d(s)\left[I_n - \overset{\frown}{D}_{N\mu}(T,s,0)\overset{\frown}{w}_d(s)\right]^{-1}\overset{\frown}{D}_{MX}(T,s,0) \\ &= \overset{\frown}{R}_N(s)\overset{\frown}{D}_{MX}(T,s,0)\,.\end{aligned} \tag{10.40}$$

d) Substituting (10.40) into (10.32), we have

$$V(s) = w_2(s)\mu(s)\overset{\frown}{R}_N(s)\overset{\frown}{D}_{MX}(T,s,0) + w_1(s)X(s)\,. \tag{10.41}$$

Multiplying this equation from the left by C_1, we have

$$C_1V(s) = C_1(sI_\chi - A)^{-1}B_2\mu(s)\overset{\frown}{R}_N(s)\overset{\frown}{D}_{MX}(T,s,0) + K(s)X(s)\,. \tag{10.42}$$

But, (10.21) yields

$$Z(s) = C_1V(s) + D_LU(s)\,. \tag{10.43}$$

From (10.30) and (10.40), it follows that

$$U(s) = \mu(s)\overset{\frown}{\psi^*}(s) = \mu(s)\overset{\frown}{R}_N(s)D_{MX}(T,s,0)\,.$$

Finally, substituting (10.42) into (10.43), we obtain

$$Z(s) = \left[C_1(sI_\chi - A)^{-1}B_2 + D_L\right]\mu(s)\overset{\frown}{R}_N(s)\overset{\frown}{D}_{MX}(T,s,0) + K(s)X(s)\,.$$

With respect to (10.24), this is equivalent to (10.23). ∎

3. Equations (10.21) are associated with a realisation of the matrix $w(p)$ in (7.8). It is easily seen that the image $Z(s)$ is independent of the specific form of this realisation. This is caused by the fact that Formula (10.23) is completely determined by Matrices (10.24), that are independent of the realisation of the matrix $w(p)$. Therefore, without loss of generality, we will assume that the pair $(A, \begin{bmatrix} B_1 & B_2 \end{bmatrix})$ in (10.21) is controllable and the pair $\left[A, \begin{bmatrix} C_1 \\ C_2 \end{bmatrix}\right]$ is observable.

4. Equation (10.23) can be derived in an alternative way. As follows from [148] under zero initial energy, the connection between the input $x(t)$ and output $z(t)$ of the standard sampled-data system is determined by a linear periodic operator with the PTM $w(s,t)$ defined by Formula (10.3). Moreover, the discrete Laplace transform of the output $\mathcal{D}_Z(T,s,t)$ is given by the formula [148]

$$\begin{aligned}\mathcal{D}_Z(T,s,t) &= \frac{1}{T}\sum_{k=-\infty}^{\infty} Z(s+kj\omega)\mathrm{e}^{(s+kj\omega)t} \\ &= \frac{1}{T}\sum_{k=-\infty}^{\infty} w(s+kj\omega,t)X(s+kj\omega)\mathrm{e}^{(s+kj\omega)t}\,.\end{aligned} \tag{10.44}$$

The image-vector $Z(s)$ can be found from (10.15) as

$$Z(s) = \int_0^T \mathcal{D}_Z(T,s,t)\mathrm{e}^{-st}\,\mathrm{d}t\,.$$

Substituting here the right side of (10.44) and the expression for $w(s,t)$ from (10.3) after some transformations, we obtain a result equivalent to (10.23).

10.4 Investigation of Poles of the Image $Z(s)$

1. From (10.23) it immediately follows that the image $Z(s)$ is a meromorphic function in the variable s, *i.e.*, all its singular points are poles. For an effective application of the Laplace transformation theory to the investigation of sampled-data systems, it is important to investigate the properties of these poles. From the first glance, the image (10.23) must have its poles at the poles of the matrices $L(s)$ and $K(s)$, and at the poles of the matrix $\overset{\frown}{\mathcal{D}}_{MX}(T,s,0)$ determined by $M(s)$. This feature makes the application of (10.23) to the solution of practical problems very difficult. Nevertheless, a more detailed investigation shows that, generally speaking, the image $Z(s)$ can be free of poles caused by the poles of the matrices $K(s)$, $L(s)$, and $M(s)$. The present section deals with this question in detail.

2.

Theorem 10.5. *Denote by $\mathcal{P}_X$ the set of roots of the equation*

$$\overset{\frown}{\Delta}_X(s) = \left(\mathrm{e}^{-sT} - \mathrm{e}^{-x_1T}\right)^{q_1}\cdots\left(\mathrm{e}^{-sT} - \mathrm{e}^{-x_\eta T}\right)^{q_\eta} = 0$$

and by $\mathcal{P}_\Delta$ the set of all roots of the equation

$$\overset{\frown}{\Delta}(s) = \det\overset{\frown}{Q}(s,\overset{\frown}{\alpha},\overset{\frown}{\beta}) = 0\,,$$

where $\widehat{Q}(s,\widehat{\alpha},\widehat{\beta})$ *is Matrix (7.38):*

$$\widehat{Q}(s,\widehat{\alpha},\widehat{\beta}) = \begin{bmatrix} I_\chi - \mathrm{e}^{-sT}\mathrm{e}^{AT} & O_{\chi n} & -\mathrm{e}^{-sT}\mathrm{e}^{AT}\mu(A)B_2 \\ -C_2 & I_n & O_{nm} \\ O_{m\chi} & -\widehat{\beta}(s) & \widehat{\alpha}(s) \end{bmatrix}.$$

Then the set of all poles of the image (10.23) belongs to $\mathcal{P}_X \cup \mathcal{P}_\Delta$.

Proof. a) From (6.75) and (6.76), it follows that

$$\begin{aligned} &\frac{1}{T}\sum_{k=-\infty}^{\infty} \left[(s+kj\omega)I_\chi - A\right]^{-1}\mathrm{e}^{(s+kj\omega)(t-\tau)} \\ &\quad = \begin{cases} \left(\mathrm{e}^{sT}\mathrm{e}^{-AT} - I_\chi\right)^{-1}\mathrm{e}^{A(t-\tau)} + \mathrm{e}^{A(t-\tau)}, & 0 < t-\tau < T \\ \left(\mathrm{e}^{sT}\mathrm{e}^{-AT} - I_\chi\right)^{-1}\mathrm{e}^{A(t-\tau)}, & -T < t-\tau < 0. \end{cases} \end{aligned} \tag{10.45}$$

b) Let us have a matrix

$$f(t) \triangleq \begin{cases} O_{nm}, & t < 0 \\ C\mathrm{e}^{At}B, & t > 0, \end{cases}$$

where A, B, and C are constant matrices of compatible dimensions. Then for satisfactory great $\operatorname{Re} s$, we have

$$F(s) = \int_0^\infty \mathrm{e}^{-st}f(t)\,\mathrm{d}t = C(sI_\chi - A)^{-1}B.$$

Moreover, using (10.45), we obtain

$$\begin{aligned} &\widehat{D}_F(T,s,t-\tau) \\ &\quad = \begin{cases} C(\mathrm{e}^{sT}\mathrm{e}^{-AT} - I_\chi)^{-1}\mathrm{e}^{A(t-\tau)}B + C\mathrm{e}^{A(t-\tau)}B, & 0 < t-\tau < T \\ C(\mathrm{e}^{sT}\mathrm{e}^{-AT} - I_\chi)^{-1}\mathrm{e}^{A(t-\tau)}B & -T < t-\tau < 0. \end{cases} \end{aligned} \tag{10.46}$$

Let $\widehat{D}_X(T,s,t)$ be the DLT of an image $X(s)$. Then using (10.17), we receive

$$\widehat{D}_{FX}(T,s,t) = \int_0^T \widehat{D}_F(T,s,t-\tau)\widehat{D}_X(T,s,\tau)\,\mathrm{d}\tau.$$

Substituting here (10.46), we find

$$\begin{aligned} \widehat{D}_{FX}(T,s,t) &= C\left(\mathrm{e}^{sT}\mathrm{e}^{-AT} - I_\chi\right)^{-1}\int_0^T \mathrm{e}^{A(t-\tau)}B\widehat{D}_X(T,s,\tau)\,\mathrm{d}\tau \\ &\quad + \int_0^t C\mathrm{e}^{A(t-\tau)}B\widehat{D}_X(T,s,\tau)\,\mathrm{d}\tau, \quad 0 < t < T. \end{aligned} \tag{10.47}$$

As a special case for $C = I$, Equation (10.47) yields

$$\widehat{D}_{FX}(T,s,t) = \left(\mathrm{e}^{sT}\mathrm{e}^{-AT} - I_\chi\right)^{-1} \int_0^T \mathrm{e}^{A(t-\tau)} B\widehat{D}_X(T,s,\tau)\,\mathrm{d}\tau + \int_0^t \mathrm{e}^{A(t-\tau)} B\widehat{D}_X(T,s,\tau)\,\mathrm{d}\tau\,, \quad 0 < t < T\,. \tag{10.48}$$

c) Let the image $X(s)$ be pseudo-rational. Then for all t, the matrix $\widehat{D}_{FX}(T,s,t)$ defined by (10.47) is continuous in t. For $t = 0$ from (10.48), we obtain

$$\widehat{D}_{FX}(T,s,0) = \left(\mathrm{e}^{sT}\mathrm{e}^{-AT} - I_\chi\right)^{-1} \int_0^T \mathrm{e}^{-A\tau} B\widehat{D}_X(T,s,\tau)\,\mathrm{d}\tau\,. \tag{10.49}$$

Regarding (10.49) from (10.48), we find

$$\widehat{D}_{FX}(T,s,t) = \mathrm{e}^{At}\widehat{D}_{FX}(T,s,0) + \int_0^t \mathrm{e}^{A(t-\tau)} B\widehat{D}_X(T,s,\tau)\,\mathrm{d}\tau\,.$$

d) Now proceed to the proof of the theorem. From (10.29), (6.89) and (6.100), it follows for $t = 0$ that

$$\widehat{D}_{w_2\mu}(T,s,0) = \mu(A)\left(\mathrm{e}^{sT}\mathrm{e}^{-AT} - I_\chi\right)^{-1} B_2\,.$$

Therefore, with respect to (10.49), Equation (10.36) can be written in the form

$$\widehat{v}^*(s) = \left(\mathrm{e}^{sT}\mathrm{e}^{-AT} - I_\chi\right)^{-1} \mu(A)B_2\widehat{\psi}^*(s) + \left(\mathrm{e}^{sT}\mathrm{e}^{-AT} - I_\chi\right)^{-1} \int_0^T \mathrm{e}^{-A\tau} B_1\widehat{D}_X(T,s,\tau)\,\mathrm{d}\tau\,. \tag{10.50}$$

Hence

$$\left(I_\chi - \mathrm{e}^{-sT}\mathrm{e}^{AT}\right)\widehat{v}^*(s) = \mathrm{e}^{-sT}\mathrm{e}^{AT}\mu(A)B_2\widehat{\psi}^*(s) + \mathrm{e}^{-sT}\mathrm{e}^{AT} \int_0^T \mathrm{e}^{-A\tau} B_1\widehat{D}_X(T,s,\tau)\,\mathrm{d}\tau\,.$$

Combining this with (10.39) and (10.37), we obtain the system of equations

$$\begin{aligned}
\left(I_\chi - \mathrm{e}^{-sT}\mathrm{e}^{AT}\right)\widehat{v}^*(s) - \mathrm{e}^{-sT}\mathrm{e}^{AT}\mu(A)B_2\widehat{\psi}^*(s) &= \mathrm{e}^{-sT}\mathrm{e}^{AT} \int_0^T \mathrm{e}^{-A\tau} B_1\widehat{D}_X(T,s,\tau)\,\mathrm{d}\tau\,,\\
-C_2\widehat{v}^*(s) + \widehat{y}^*(s) &= O_{n1}\,,\\
-\widehat{\beta}(s)\widehat{y}^*(s) + \widehat{\alpha}(s)\widehat{\psi}^*(s) &= O_{m1}\,.
\end{aligned} \tag{10.51}$$

Introduce the $(\chi + n + m) \times 1$ vector

$$\widehat{G^*}(s) = \left[\left[\widehat{v^*}(s)\right]' \left[\widehat{y^*}(s)\right]' \left[\widehat{\psi^*}(s)\right]' \right]' .$$

Then the system of equations (10.51) can be written in the form

$$\widehat{Q}(s, \widehat{\alpha}, \widehat{\beta})\widehat{G^*}(s) = \widehat{D}(s) ,$$

where $\widehat{D}(s)$ is given by

$$\widehat{D}(s) = \left[\widehat{D_1}'(s) \;\; O_{1n} \;\; O_{1m} \right]' ,$$

where

$$\widehat{D}_1(s) = \mathrm{e}^{-sT}\mathrm{e}^{AT} \int_0^T \mathrm{e}^{-A\tau} B_1 \widehat{D}_X(T, s, \tau)\, \mathrm{d}\tau .$$

Thus, substituting herein (10.19), we find

$$\widehat{D}_1(s) = \frac{\mathrm{e}^{-sT}\mathrm{e}^{AT} \sum_{k=0}^{\ell} \mathrm{e}^{-ksT} \int_0^T \mathrm{e}^{-A\tau} B_1 x_k(\tau)\, \mathrm{d}\tau}{\widehat{\Delta}_X(s)} .$$

Since the numerator of this expression is an integral function of s, it follows that the set of poles of the vector $\widehat{D}_1(s)$ belongs to the set $\mathcal{P}_X$. Obviously, the same is true for the vector $\widehat{D}(s)$. Then with account for

$$\widehat{G^*}(s) = \widehat{Q}^{-1}(s, \widehat{\alpha}, \widehat{\beta})\widehat{D}(s) ,$$

we find that the set of all poles of the vectors $\widehat{v^*}(s)$ and $\widehat{y^*}(s)$, $\widehat{\psi^*}(s)$ belongs to the union of the sets $\mathcal{P}_X$ and $\mathcal{P}_\Delta$.

e) Substituting the relation

$$\widehat{D}_{w_2\mu}(T, s, t) = \int_0^t \mathrm{e}^{A(t-\tau)} m(\tau)\, \mathrm{d}\tau\, B_2 + \left(\mathrm{e}^{sT}\mathrm{e}^{-AT} - I_\chi\right)^{-1} \mu(A)\mathrm{e}^{At} B_2$$

and Equation (10.48) with $F(s) = w_2(s)$ into (10.33), we find for $0 < t < T$

$$\begin{aligned}
&\widehat{D}_V(T, s, t) \\
&\quad = \left[\int_0^t \mathrm{e}^{A(t-\tau)} m(\tau)\, \mathrm{d}\tau\, B_2 + \left(\mathrm{e}^{sT}\mathrm{e}^{-AT} - I_\chi\right)^{-1} \mu(A)\mathrm{e}^{At} B_2 \right] \widehat{\psi^*}(s) \\
&\qquad + \left(\mathrm{e}^{sT}\mathrm{e}^{-AT} - I_\chi\right)^{-1} \int_0^T \mathrm{e}^{A(t-\tau)} B_1 \widehat{D}_X(T, s, \tau)\, \mathrm{d}\tau \\
&\qquad + \int_0^t \mathrm{e}^{A(t-\tau)} B_1 \widehat{D}_X(T, s, \tau)\, \mathrm{d}\tau .
\end{aligned}$$

After rearrangement, we get

$$\begin{aligned}
&\widehat{D}_V(T,s,t) \\
&\quad = \mathrm{e}^{At}\left[\left(\mathrm{e}^{sT}\mathrm{e}^{-AT} - I_\chi\right)^{-1}\mu(A)B_2\widehat{\psi}^*(s) + \int_0^T \mathrm{e}^{-A\tau}B_1\widehat{D}_X(T,s,\tau)\,\mathrm{d}\tau\right] \\
&\qquad + \int_0^t \mathrm{e}^{A(t-\tau)}m(\tau)\,\mathrm{d}\tau\, B_2\widehat{\psi}^*(s) + \int_0^t \mathrm{e}^{A(t-\tau)}B_1\widehat{D}_X(T,s,\tau)\,\mathrm{d}\tau\,.
\end{aligned}$$

According to (10.50), the expression in the square brackets equals $\widehat{v}^*(s)$. Therefore,

$$\begin{aligned}
\widehat{D}_V(T,s,t) = \mathrm{e}^{At}\widehat{v}^*(s) &+ \int_0^t \mathrm{e}^{A(t-\tau)}m(\tau)\,\mathrm{d}\tau\, B_2\widehat{\psi}^*(s) \\
&+ \int_0^t \mathrm{e}^{A(t-\tau)}B_1\widehat{D}_X(T,s,\tau)\,\mathrm{d}\tau\,.
\end{aligned}$$

Using the relation

$$V(s) = \int_0^T \widehat{D}_V(T,s,t)\mathrm{e}^{-st}\,\mathrm{d}t\,,$$

we obtain $V(s)$ in the form

$$V(s) = G_1(s)\widehat{v}^*(s) + G_2(s)\widehat{\psi}^*(s) + G_3(s)\,,$$

where

$$\begin{aligned}
G_1(s) &= \int_0^T \mathrm{e}^{At}\mathrm{e}^{-st}\,\mathrm{d}t = (A - sI_\chi)^{-1}\left(\mathrm{e}^{-sT}\mathrm{e}^{AT} - I_\chi\right)^{-1}\,, \\
G_2(s) &= \int_0^T \mathrm{e}^{-st}\int_0^T \mathrm{e}^{A(t-\tau)}m(\tau)\,\mathrm{d}\tau\,\mathrm{d}t\,, \\
G_3(s) &= \int_0^T \mathrm{e}^{-st}\int_0^t \mathrm{e}^{A(t-\tau)}B_1\widehat{D}_X(T,s,\tau)\,\mathrm{d}\tau\,\mathrm{d}t\,.
\end{aligned}$$

Obviously, the matrices $G_1(s)$ and $G_2(s)$ are integral functions of s and the poles of the matrix $G_3(s)$ belong to the set $\mathcal{P}_X$. As was proved above, the poles of the vectors $\widehat{v}^*(s)$ and $\widehat{\psi}^*(s)$ belong to the union of $\mathcal{P}_X$ and $\mathcal{P}_\Delta$. Hence the poles of the image $V(s)$ belong to $\mathcal{P}_X \cup \mathcal{P}_\Delta$.

f) To conclude the proof, we notice that (10.30) and (10.43) yield

$$Z(s) = C_1V(s) + D_L\mu(s)\widehat{\psi}^*(s)\,.$$

Due to (10.26), $\mu(s)$ is an integral function so that the set of all poles of the vector $Z(s)$ belongs to the set $\mathcal{P}_X \cup \mathcal{P}_\Delta$. ∎

Corollary 10.6. *Under the assumptions of Theorem 10.5, the image* $Z(s)$ *admits a representation of the form*

$$Z(s) = \frac{P_Z(s)}{\overset{\frown}{\Delta}(s)\overset{\frown}{\Delta}_X(s)},$$

where $P_Z(s)$ *is an integral function in* s *and* $\overset{\frown}{\Delta}_X(s)$ *is given by (10.20). According to (7.93),*

$$\Delta(\zeta) = \varrho(\zeta)\Delta_d(\zeta), \tag{10.52}$$

where the polynomial $\varrho(\zeta)$ *is independent of the choice of the discrete controller.* ∎

10.5 Representing the Output Image in Terms of the System Function

1. When not stated otherwise in this section for simplicity, we assume that the standard sampled-data system is modal controllable. Then we have $\varrho(\zeta) =$ const. $\neq 0$ and from (10.52), it follows that

$$\Delta(\zeta) \approx \Delta_d(\zeta).$$

Moreover, we stick to all definitions and notation of Section 8.3.

2. In analogy with Section 8.3, the image (10.23) can be represented in terms of the system function (system matrix) $\theta(\zeta)$. With this aim in view, we substitute (8.50) in (10.23):

$$\overset{\frown}{R}_N(s) = \overset{\frown}{\beta}_{0r}(s)\overset{\frown}{a}_l(s) - \overset{\frown}{a}_r(s)\overset{\frown}{\theta}(s)\overset{\frown}{a}_l(s). \tag{10.53}$$

As a result, we obtain

$$Z(s) = p(s)\overset{\frown}{\theta}(s)\overset{\frown}{q}(s) + r(s), \tag{10.54}$$

where

$$\begin{aligned}
p(s) &= -L(s)\mu(s)\overset{\frown}{a}_r(s), \\
\overset{\frown}{q}(s) &= \overset{\frown}{a}_l(s)\overset{\frown}{D}_{MX}(T,s,0), \\
r(s) &= L(s)\mu(s)\overset{\frown}{\beta}_{0r}(s)\overset{\frown}{a}_l(s)\overset{\frown}{D}_{MX}(T,s,0) + K(s)X(s), \\
&= L(s)\mu(s)\overset{\frown}{\beta}_{0r}(s)\overset{\frown}{q}(s) + K(s)X(s).
\end{aligned} \tag{10.55}$$

Hereinafter, we shall call Equation (10.54) a *representation of the image* $Z(s)$ *in terms of the system function.* The matrix $p(s)$ and the vectors $\overset{\frown}{q}(s)$ and

$r(s)$ will be called the *coefficients* of this representation. As follows from (10.55) under the given assumptions, the coefficients (10.55) are meromorphic functions of the argument s. In the present section, we investigate the set of poles of the matrices (10.55).

3.

Theorem 10.7. *Let the standard sampled-data system be modal controllable. Then the matrix $p(s)$ is an integral function of s and the set of all poles of the vectors $\widehat{q}(s)$ and $r(s)$ belongs to the set $\mathcal{P}_X$.*

Proof. The claim regarding the matrix $p(s)$ follows immediately from Corollary 8.15.

Then we consider the vector $\widehat{q}(s)$ in (10.55). From (10.17), it follows

$$\widehat{D}_{MX}(T,s,t) = \int_0^T \widehat{D}_M(T,s,\tau)\widehat{D}_X(T,s,t-\tau)\,\mathrm{d}\tau\,.$$

For $t = 0$, we have

$$\widehat{D}_{MX}(T,s,0) = \int_0^T \widehat{D}_M(T,s,\tau)\widehat{D}_X(T,s,-\tau)\,\mathrm{d}\tau\,.$$

Hence using (10.55), we obtain

$$\widehat{q}(s) = \widehat{a}_l(s)\widehat{D}_{MX}(T,s,0) = \int_0^T \left[\widehat{a}_l(s)\widehat{D}_M(T,s,\tau)\right]\widehat{D}_X(T,s,-\tau)\,\mathrm{d}\tau\,. \tag{10.56}$$

Due to Corollary 8.14, the matrix in the square brackets is an integral function in s. Therefore, with respect to (10.19) and (6.71), we obtain the claim of the theorem regarding the vector $\widehat{q}(s)$.

To prove the claim about the vector $r(s)$, we notice that for $\theta(z) = O_{mn}$, we have

$$Z(s) = r(s)$$

and the further proof is performed similarly to the proof of Theorem 8.4 using Theorem 10.5. ∎

Remark 10.8. From (10.56) it follows that

$$\widehat{q}(s) = \frac{\widehat{N}_q(s)}{\widehat{\Delta}_X(s)}\,, \tag{10.57}$$

where $N_q(\zeta)$ is a polynomial vector and the function $\widehat{\Delta}_X(s)$ is given by (10.20). Hence

$$q(\zeta) = \frac{N_q(\zeta)}{\Delta_X(\zeta)}\,. \tag{10.58}$$

Moreover, we have

$$r(s) = \frac{P_r(s)}{\widehat{\Delta}_X(s)}, \tag{10.59}$$

where the numerator is an integral function in s.

10.6 Representing the $\mathcal{L}_2$-norm in Terms of the System Function

1. Using (10.54), we obtain

$$\underline{Z}'(s) = \widehat{\underline{q}}'(s)\widehat{\underline{\theta}}'(s)\underline{p}'(s) + \underline{r}'(s), \tag{10.60}$$

where, as before, we use the notation

$$\underline{f}(s) \stackrel{\triangle}{=} f(-s)$$

and the following relations hold:

$$\begin{aligned}
\underline{p}'(s) &= -\widehat{\underline{a}}'_r(s)\underline{L}'(s)\underline{\mu}(s), \\
\widehat{\underline{q}}'(s) &= \widehat{\underline{D}}'_{MX}(T,s,0)\widehat{\underline{a}}'_l(s) = \widehat{D}_{\underline{X}'\underline{M}'}(T,s,0)\widehat{\underline{a}}'_l(s), \\
\underline{r}'(s) &= \widehat{D}_{\underline{X}'\underline{M}'}(T,s,0)\widehat{\underline{a}}'_l(s)\widehat{\underline{\beta}}'_{0r}(s)\underline{L}'(s)\underline{\mu}(s) + \underline{X}'(s)\underline{K}'(s), \\
&= \widehat{\underline{q}}'(s)\widehat{\underline{\beta}}'_{0r}(s)\underline{L}'(s)\underline{\mu}(s) + \underline{X}'(s)\underline{K}'(s).
\end{aligned} \tag{10.61}$$

Multiplying (10.54) and (10.60), we obtain

$$\begin{aligned}
\underline{Z}'(s)Z(s) &= Z'(-s)Z(s) \\
&= \widehat{\underline{q}}'(s)\widehat{\underline{\theta}}'(s)\underline{p}'(s)p(s)\widehat{\theta}(s)\widehat{q}(s) + \underline{r}'(s)r(s) + \\
&\quad + \widehat{\underline{q}}'(s)\widehat{\underline{\theta}}'(s)\underline{p}'(s)r(s) + \underline{r}'(s)p(s)\widehat{\theta}(s)\widehat{q}(s).
\end{aligned}$$

Substituting this expression into (10.5) yields

$$\tilde{J} = \tilde{J}_1 + \tilde{J}_2, \tag{10.62}$$

where

$$\begin{aligned}
\tilde{J}_1 &= \frac{1}{2\pi \mathrm{j}} \int_{-\mathrm{j}\infty}^{\mathrm{j}\infty} \Big[\widehat{\underline{q}}'(s)\widehat{\underline{\theta}}'(s)\underline{p}'(s)p(s)\widehat{\theta}(s)\widehat{q}(s) + \\
&\quad + \widehat{\underline{q}}'(s)\widehat{\underline{\theta}}'(s)\underline{p}'(s)r(s) + \underline{r}'(s)p(s)\widehat{\theta}(s)\widehat{q}(s) \Big] \, \mathrm{d}s, \\
\tilde{J}_2 &= \frac{1}{2\pi \mathrm{j}} \int_{-\mathrm{j}\infty}^{\mathrm{j}\infty} \underline{r}'(s)r(s)\, \mathrm{d}s.
\end{aligned} \tag{10.63}$$

As follows from the above relations, the integral $\tilde{J}_2$ is independent of the system function, *i.e.*, it is independent of the choice of the controller.

2. Let us transform the integral (10.63). With this purpose, we pass to finite integration limits in (10.63):

$$\begin{aligned}\tilde{J}_1 = \frac{T}{2\pi \mathrm{j}} \int_{-\mathrm{j}\omega/2}^{\mathrm{j}\omega/2} \Big[& \underline{\widehat{q}}'(s)\underline{\widehat{\theta}}'(s)\widehat{A}_{L1}(s)\widehat{\theta}(s)\widehat{q}(s) \\ & -\underline{\widehat{q}}'(s)\underline{\widehat{\theta}}'(s)\underline{\widehat{B}}'(s) - \widehat{B}(s)\widehat{\theta}(s)\widehat{q}(s)\Big] \, \mathrm{d}s\,,\end{aligned} \tag{10.64}$$

where

$$\begin{aligned}\widehat{A}_{L1}(s) &= \frac{1}{T}\sum_{k=-\infty}^{\infty} \underline{p}'(s+k\mathrm{j}\omega)p(s+k\mathrm{j}\omega)\,, \\ \widehat{B}(s) &= -\frac{1}{T}\sum_{k=-\infty}^{\infty} \underline{r}'(s+k\mathrm{j}\omega)p(s+k\mathrm{j}\omega)\,, \\ \underline{\widehat{B}}'(s) &= -\frac{1}{T}\sum_{k=-\infty}^{\infty} \underline{p}'(s+k\mathrm{j}\omega)r(s+k\mathrm{j}\omega)\,.\end{aligned} \tag{10.65}$$

Let us find expanded expressions for the matrices (10.65). From Chapter 8, it follows that

$$\begin{aligned}\widehat{A}_{L1}(s) &= \frac{1}{T}\underline{\widehat{a}}'_r(s)\sum_{k=-\infty}^{\infty} \underline{L}'(s+k\mathrm{j}\omega)L(s+k\mathrm{j}\omega)\underline{\mu}(s+k\mathrm{j}\omega)\mu(s+k\mathrm{j}\omega)\widehat{a}_r(s) \\ &= \underline{\widehat{a}}'_r(s)\widehat{D}_{\underline{L}'L\underline{\mu}\mu}(T,s,0)\widehat{a}_r(s) = T\widehat{A}_L(s)\,,\end{aligned}$$

where the matrix $\widehat{A}_L(s)$ is defined in (8.73). To calculate the matrix $\widehat{B}(s)$, we note that, due to (10.55) and (10.61),

$$-\underline{r}'(s)p(s) = \underline{\widehat{q}}'(s)\underline{\widehat{\beta}}'_{0r}(s)\underline{L}'(s)L(s)\mu(s)\underline{\mu}(s)\widehat{a}_r(s) + \underline{X}'(s)\underline{K}'(s)L(s)\mu(s)\widehat{a}_r(s)\,.$$

Therefore, for any integer k

$$\begin{aligned}&-\underline{r}'(s+k\mathrm{j}\omega)p(s+k\mathrm{j}\omega) \\ &\quad = \underline{\widehat{q}}'(s)\underline{\widehat{\beta}}'_{0r}(s)\underline{L}'(s+k\mathrm{j}\omega)L(s+k\mathrm{j}\omega)\mu(s+k\mathrm{j}\omega)\underline{\mu}(s+k\mathrm{j}\omega)\widehat{a}_r(s) \\ &\qquad + \underline{X}'(s+k\mathrm{j}\omega)\underline{K}'(s+k\mathrm{j}\omega)L(s+k\mathrm{j}\omega)\mu(s+k\mathrm{j}\omega)\widehat{a}_r(s)\,.\end{aligned} \tag{10.66}$$

Substituting (10.66) into (10.65), we find

$$\overset{\frown}{B}(s) = \overset{\frown}{\underline{q}}'(s)\overset{\frown}{\underline{\beta}}'_{0r}(s)\overset{\frown}{D}_{\underline{L}'L\mu\underline{\mu}}(T,s,0)\overset{\frown}{a}_r(s) + \overset{\frown}{D}_{\underline{X}'\underline{K}'L\mu}(T,s,0)\overset{\frown}{a}_r(s)\,.$$

Placing $-s$ for s and transposing, we obtain

$$\overset{\frown}{\underline{B}}'(s) = \overset{\frown}{\underline{a}}'_r(s)\overset{\frown}{D}_{\underline{L}'L\mu\underline{\mu}}(T,s,0)\overset{\frown}{\beta}_{0r}(s)\overset{\frown}{q}(s) + \overset{\frown}{\underline{a}}'_r(s)\overset{\frown}{D}_{\underline{L}'KX\underline{\mu}}(T,s,0)\,.$$

As shown in Chapter 8, the matrix $\overset{\frown}{A}_L(s)$ is an integral function in s. Moreover, since the matrix $p(s)$ is integral, then as follows from (10.59), the poles of each product (10.66) belong to the set of roots of the function $\overset{\frown}{\underline{\Delta}}_X(s)$. Thus for the matrix $\tilde{B}(s)$, the series (10.65) converges uniformly in any closed area that is free of roots of the function $\overset{\frown}{\underline{\Delta}}_X(s)$. Therefore, the matrix $\overset{\frown}{B}(s)$ may possess poles only among the roots of the function $\overset{\frown}{\underline{\Delta}}_X(s)$. Hence there exists the following representation:

$$\overset{\frown}{B}(s) = \frac{\overset{\frown}{N}_{B1}(s)}{\overset{\frown}{\underline{\Delta}}_X(s)}\,,$$

where $N_{B1}(\zeta)$ is a quasi-polynomial vector. From (10.52), it follows that

$$\overset{\frown}{\underline{B}}'(s) = \frac{\overset{\frown}{N}_B(s)}{\overset{\frown}{\Delta}_X(s)}\,, \tag{10.67}$$

where $N_B(\zeta) = N'_{B1}(\zeta^{-1})$ is a quasi-polynomial vector.

3. Placing the variable $\zeta = \mathrm{e}^{-sT}$ for s in (10.64), we obtain

$$\tilde{J}_1 = \frac{1}{2\pi\mathrm{j}}\oint\left[\hat{q}(\zeta)\hat{\theta}(\zeta)A_{L1}(\zeta)\theta(\zeta)q(\zeta) - \hat{q}(\zeta)\hat{\theta}(\zeta)\hat{B}(\zeta) - B(\zeta)\theta(\zeta)q(\zeta)\right]\frac{\mathrm{d}\zeta}{\zeta} \tag{10.68}$$

where, as before, we used the notation

$$\hat{f}(\zeta) \triangleq f'(\zeta^{-1})$$

and the integration is performed along the unit circle in positive direction. The matrices appearing in (10.68) are given by

$$\begin{aligned}
A_{L1}(\zeta) &= TA_L(\zeta) = \hat{a}_r(\zeta)D_{\underline{L}'L\mu\underline{\mu}}(T,\zeta,0)a_r(\zeta)\,,\\
B(\zeta) &= \hat{q}(\zeta)\hat{\beta}_{0r}(\zeta)D_{\underline{L}'L\mu\underline{\mu}}(T,\zeta,0)a_r(\zeta) + D_{\underline{X}'\underline{K}'L\mu}(T,\zeta,0)a_r(\zeta)\,, \\
\hat{B}(\zeta) &= \hat{a}_r(\zeta)D_{\underline{L}'L\mu\underline{\mu}}(T,\zeta,0)\beta_{0r}(\zeta)q(\zeta) + \hat{a}_r(\zeta)D_{\underline{L}'KX\underline{\mu}}(T,\zeta,0)\,.
\end{aligned} \tag{10.69}$$

In Chapter 8, it was shown that the matrix $A_L(\zeta)$ is a symmetric quasi-polynomial. Moreover from (10.67), it follows that $\hat{B}(\zeta)$ is a rational matrix, which can have poles at the roots of the polynomial $\Delta_X(\zeta)$ and at the point $\zeta = 0$. Therefore, there exists a representation

$$\hat{B}(\zeta) = \frac{\tilde{N}_B(\zeta)}{\zeta^\delta \Delta_X(\zeta)}, \tag{10.70}$$

where $\tilde{N}_B(\zeta)$ is a polynomial vector and $\delta \geq 0$ is an integer.

10.7 Wiener-Hopf Method

1. Since the second summand in (10.62) is independent of the choice of the system matrix, the problem of minimising the functional (10.62) is equivalent to the problem of minimising the integral (10.68) over the set of stable rational matrices $\theta(\zeta)$. This fact is substantiated by the following theorem.

Theorem 10.9. *Let $\theta^o(\zeta)$ be a stable rational matrix minimising the integral (10.68). Then the transfer function of the optimal controller $w_d^o(\zeta)$ ensuring the stability of the standard sampled-data system and minimising the functional (10.5) is given by*

$$w_d^o(\zeta) = V_{2o}(\zeta) V_{1o}^{-1}(\zeta),$$

where

$$V_{1o}(\zeta) = \alpha_{0r}(\zeta) - \zeta\, b_r(\zeta)\theta^o(\zeta),$$

$$V_{2o}(\zeta) = \beta_{0r}(\zeta) - a_r(\zeta)\theta^o(\zeta),$$

and the polynomial matrices $a_r(\zeta)$, $b_r(\zeta)$ are determined by the IRMFD (8.42) and moreover, $[\alpha_{0r}, \beta_{0r}]$ is a right initial controller solving Equation (4.37). Furthermore, if we have both the IMFDs

$$\theta^o(\zeta) = D_l^{-1}(\zeta) M_l(\zeta) = M_r(\zeta) D_r^{-1}(\zeta)$$

and the system is modal controllable, then the characteristic polynomial of the optimal system $\Delta^o(\zeta)$ satisfies the relation

$$\Delta^o(\zeta) \approx \Delta_d(\zeta),$$

where

$$\Delta_d(\zeta) \approx \det D_l(\zeta) \approx \det D_r(\zeta).$$

Proof. The proof is similar to that of Theorem 8.21. ∎

2. We recognise that the form of the functional (10.68) coincides with that of (8.92). Nevertheless, a direct application of the Wiener-Hopf method in the form given in Section 8.6 is impossible, because the matrix $\Gamma(\zeta) = q(\zeta)$ in (10.68) is not invertible. This means that the functional (10.68) is singular. Therefore, the minimisation needs special approaches. Below we describe such an approach based on an idea of [124]. With this aim in view, we derive a number of auxiliary transformations.

a) Consider the rational vector (10.58)

$$q(\zeta) = a_l(\zeta) D_{MX}(T, \zeta, 0) = \frac{N_q(\zeta)}{\Delta_X(\zeta)}\,. \tag{10.71}$$

As follows from Remark 3.4, there exists a unimodular matrix $R(\zeta)$, such that

$$R(\zeta) N_q(\zeta) = \ell(\zeta)\mathbf{1}_n\,, \quad \mathbf{1}_n = \begin{bmatrix} 1 \\ 0 \\ \vdots \\ 0 \end{bmatrix} n\,,$$

where $\ell(\zeta)$ is a greatest common divisor of the elements of the column $N_q(\zeta)$. Thus with (10.71), we have

$$q(\zeta) = \frac{\ell(\zeta)}{\Delta_X(\zeta)} Y(\zeta)\mathbf{1}_n\,, \quad \hat{q}(\zeta) = \frac{\ell(\zeta^{-1})}{\Delta_X(\zeta^{-1})}\mathbf{1}_n' \hat{Y}(\zeta) \tag{10.72}$$

with a unimodular matrix

$$Y(\zeta) \stackrel{\triangle}{=} R^{-1}(\zeta)\,.$$

b) Substituting (10.72) into (10.68), we obtain

$$\begin{aligned} \tilde{J}_1 = \frac{1}{2\pi \mathrm{j}} \oint \Bigg[& \mathbf{1}_n' \hat{Y}(\zeta)\hat{\theta}(\zeta) \frac{\ell(\zeta^{-1}) A_{L1}(\zeta)\ell(\zeta)}{\Delta_X(\zeta^{-1})\Delta_X(\zeta)} \theta(\zeta) Y(\zeta)\mathbf{1}_n \\ & - \mathbf{1}_n' \hat{Y}(\zeta)\hat{\theta}(\zeta)\hat{B}(\zeta)\frac{\ell(\zeta^{-1})}{\Delta_X(\zeta^{-1})} - B(\zeta)\theta(\zeta)Y(\zeta)\mathbf{1}_n \frac{\ell(\zeta)}{\Delta_X(\zeta)} \Bigg] \frac{\mathrm{d}\zeta}{\zeta} \end{aligned} \tag{10.73}$$

Introduce the matrix

$$\theta_1(\zeta) \stackrel{\triangle}{=} \theta(\zeta) Y(\zeta)\,. \tag{10.74}$$

Then with account for (10.69), the functional (10.73) takes the form

$$\begin{aligned} \tilde{J}_1 = \frac{1}{2\pi \mathrm{j}} \oint \Bigg[& \mathbf{1}_n' \hat{\theta}_1(\zeta) \frac{T\ell(\zeta^{-1}) A_L(\zeta)\ell(\zeta)}{\Delta_X(\zeta^{-1})\Delta_X(\zeta)} \theta_1(\zeta)\mathbf{1}_n \\ & - \mathbf{1}_n' \hat{\theta}_1(\zeta)\hat{B}(\zeta)\frac{\ell(\zeta^{-1})}{\Delta_X(\zeta^{-1})} - B(\zeta)\theta_1(\zeta)\mathbf{1}_n \frac{\ell(\zeta)}{\Delta_X(\zeta)} \Bigg] \frac{\mathrm{d}\zeta}{\zeta}\,. \end{aligned} \tag{10.75}$$

Since the matrix $Y(\zeta)$ is unimodular, the problems of minimising the functionals (10.73) and (10.75) are equivalent in the sense that, if a stable matrix $\theta_1^o(\zeta)$ minimises the integral (10.75), then the matrix

$$\theta^o(\zeta) = \theta_1^o(\zeta)R(\zeta) \tag{10.76}$$

is stable and minimises the integral (10.73). The converse proposition is also valid.

c) Let us have

$$\theta_1(\zeta) = \begin{bmatrix} \overset{1}{\Omega_1(\zeta)} & \overset{n-1}{\Omega_2(\zeta)} \end{bmatrix} m\,,$$

where $\Omega_1(\zeta)$ denotes the first column of the matrix $\theta_1(\zeta)$. Then obviously,

$$\theta_1(\zeta)\mathbf{1}_n = \Omega_1(\zeta)$$

and the integral (10.75) can be written in a form depending only on the column $\Omega_1(\zeta)$:

$$\begin{aligned}\tilde{J}_1 = \frac{1}{2\pi \mathrm{j}} \oint \Bigg[& \hat{\Omega}_1(\zeta)\frac{T\ell(\zeta^{-1})A_L(\zeta)\ell(\zeta)}{\Delta_X(\zeta^{-1})\Delta_X(\zeta)}\Omega_1(\zeta) \\ & - \hat{\Omega}_1(\zeta)\hat{B}(\zeta)\frac{\ell(\zeta^{-1})}{\Delta_X(\zeta^{-1})} - B(\zeta)\Omega_1(\zeta)\frac{\ell(\zeta)}{\Delta_X(\zeta)}\Bigg]\frac{\mathrm{d}\zeta}{\zeta}\,.\end{aligned} \tag{10.77}$$

Assume that there exists a stable rational column $\Omega_1^o(\zeta)$ minimising the functional (10.77). Then the stable rational matrix

$$\theta_1^o(\zeta) = \begin{bmatrix} \Omega_1^o(\zeta) & \Omega_2(\zeta) \end{bmatrix}$$

with any stable rational $m \times (n-1)$ matrix $\Omega_2(\zeta)$ minimises the integral (10.75), because this integral depends only on the first column of $\theta_1(\zeta)$. Furthermore, using (10.74), we find that the stable rational matrix

$$\theta^o(\zeta) = \theta_1^o(\zeta)R(\zeta) = \begin{bmatrix} \Omega_1^o(\zeta) & \Omega_2(\zeta) \end{bmatrix} R(\zeta) \tag{10.78}$$

minimises the functional (10.68) for any stable $m \times (n-1)$ matrices $\Omega_2(\zeta)$. Therefore, if $n > 1$ and there exists an optimal column $\Omega_1(\zeta)$, then there exists also a set of optimal system functions depending on the stable matrix parameter $\Omega_2(\zeta)$. Since each optimal system function (10.76) is associated by Theorem 10.9 with an optimal stabilising controller, we conclude for the $\mathcal{L}_2$-problem, that the existence of one optimal controller means that there exists a set of optimal stabilising controllers depending on the choice of the stable matrix $\Omega_2(\zeta)$.
This result principally differs from the situation for the $\mathcal{H}_2$-problem and it is originated by the singularity of the functional (10.68).

3. Consider the minimisation of the functional (10.77) on the basis of the Wiener-Hopf method. Suppose that there exists a factorisation

$$A_L(\zeta) = \hat{\Pi}(\zeta)\Pi(\zeta) \tag{10.79}$$

with a stable invertible $m \times m$ polynomial matrix $\Pi(\zeta)$ and

$$\ell(\zeta)\ell(\zeta^{-1}) = \ell^+(\zeta)\ell^+(\zeta^{-1}) \tag{10.80}$$

with a stable polynomial $\ell^+(\zeta)$. Since the polynomial $\Delta_X(\zeta)$ is stable, there exists a factorisation

$$T\frac{\ell(\zeta^{-1})A_L(\zeta)\ell(\zeta)}{\Delta_X(\zeta^{-1})\Delta_X(\zeta)} = \hat{\Pi}_1(\zeta)\Pi_1(\zeta)\,, \tag{10.81}$$

where the rational matrix

$$\Pi_1(\zeta) = \frac{\sqrt{T}\,\Pi(\zeta)\ell^+(\zeta)}{\Delta_X(\zeta)} \tag{10.82}$$

is stable together with its inverse. Further, according to the Wiener-Hopf minimisation algorithm, we find the matrix

$$R(\zeta) = \hat{\Pi}_1^{-1}(\zeta)\hat{B}(\zeta)\frac{\ell(\zeta^{-1})}{\Delta_X(\zeta^{-1})}\,. \tag{10.83}$$

Using (10.70) and (10.82), we can write (10.83) in the form

$$R(\zeta) = \frac{1}{\sqrt{T}}\,\frac{\hat{\Pi}^{-1}(\zeta)\tilde{N}_B(\zeta)}{\zeta^{\delta}\Delta_X(\zeta)}\,\frac{\ell(\zeta^{-1})}{\ell^+(\zeta^{-1})}\,.$$

The next stage of the minimisation requires to perform the principal separation

$$R(\zeta) = R_+(\zeta) + R_-(\zeta) = R_+(\zeta) + \langle R(\zeta)\rangle_-\,, \tag{10.84}$$

where $R_-(\zeta)$ is a strictly proper rational matrix incorporating all unstable poles of the matrix $R(\zeta)$ and finally, $R_+(\zeta)$ is a stable rational matrix. The general form of the matrix $R_+(\zeta)$ is given in the following Theorem.

Theorem 10.10. *The matrix $R_+(\zeta)$ admits a representation*

$$R_+(\zeta) = \frac{N_R^+(\zeta)}{\Delta_X(\zeta)} \tag{10.85}$$

with a polynomial matrix(-column) $N_R^+(\zeta)$.

Proof. a) First of all, we show that the function

$$\varrho_\ell(\zeta) \stackrel{\triangle}{=} \frac{\ell(\zeta^{-1})}{\ell^+(\zeta^{-1})} \tag{10.86}$$

possesses only unstable poles. Let

$$\ell(\zeta) = (\zeta - \ell_1)^{b_1} \cdots (\zeta - \ell_m)^{b_m}, \quad b_1 + \ldots + b_m = g$$

and

$$\ell(\zeta^{-1}) = \zeta^{-g}(1 - \zeta\ell_1)^{b_1} \cdots (1 - \zeta\ell_m)^{b_m}. \tag{10.87}$$

Assume

$$|\ell_i| > 1, \ (i = 1, \ldots, \kappa); \quad |\ell_i| < 1, \ (i = \kappa + 1, \ldots, m).$$

Then, we can take

$$\ell^+(\zeta) = (\zeta - \ell_1)^{b_1} \cdots (\zeta - \ell_\kappa)^{b_\kappa} (1 - \zeta\ell_{\kappa+1})^{b_{\kappa+1}} \cdots (1 - \zeta\ell_m)^{b_m},$$

whence it follows

$$\ell^+(\zeta^{-1}) = \zeta^{-g}(1 - \zeta\ell_1)^{b_1} \cdots (1 - \zeta\ell_\kappa)^{b_\kappa} (\zeta - \ell_{\kappa+1})^{b_{\kappa+1}} \cdots (\zeta - \ell_m)^{b_m}.$$

Hence using (10.87), we obtain

$$\varrho_\ell(\zeta) = \frac{(1 - \zeta\ell_{\kappa+1})^{b_{\kappa+1}} \cdots (1 - \zeta\ell_m)^{b_m}}{(\zeta - \ell_{\kappa+1})^{b_{\kappa+1}} \cdots (\zeta - \ell_m)^{b_m}},$$

i.e., the function (10.86) has only unstable poles.

b) Let us show that the matrix $\hat{\Pi}^{-1}(\zeta)$ possesses only unstable poles. Indeed per construction, the matrix $\Pi^{-1}(\zeta)$ has only stable poles. Hence the matrix $\Pi^{-1}(\zeta^{-1})$ can have only unstable poles. Therefore, the matrix $\hat{\Pi}(\zeta) = \left[\Pi^{-1}(\zeta^{-1})\right]'$ can also have only unstable poles.

c) From a) and b), it follows that the stable poles of the matrix $R(\zeta)$ belong to the set of roots of the polynomial $\Delta_X(\zeta)$, whence it follows (10.85). ∎

4. Using (10.85) and (10.82), we find the optimal column

$$\Omega_1^o(\zeta) = \Pi_1^{-1}(\zeta)R_+(\zeta) = \frac{1}{\sqrt{T}} \frac{\Pi^{-1}(\zeta)N_R^+(\zeta)}{\ell^+(\zeta)}. \tag{10.88}$$

In particular, when we have $\ell^+(\zeta) = \text{const.} \neq 0$, which is true almost always in applied problems, then we obtain

$$\Omega_1^o(\zeta) = \Pi^{-1}(\zeta)\tilde{N}_R^+(\zeta),$$

where $\tilde{N}_R^+(\zeta)$ is a polynomial vector.

10.8 General Properties of Optimal Systems

1. Let us have (10.88) and let $\Omega_2(\zeta)$ be a stable rational $m \times (n-1)$ matrix. Then, the expression

$$\theta_1^o(\zeta) = \left[\frac{1}{\sqrt{T}} \frac{\Pi^{-1}(\zeta) N_R^+(\zeta)}{\ell^+(\zeta)} \quad \Omega_2(\zeta) \right]$$

determines the set of all optimal matrices $\theta_1^o(\zeta)$. Thus with respect to (10.78), we find that the set of all optimal system matrices $\theta^o(\zeta)$ is determined by

$$\theta^o(\zeta) = \left[\frac{1}{\sqrt{T}} \frac{\Pi^{-1}(\zeta) N_R^+(\zeta)}{\ell^+(\zeta)} \quad \Omega_2(\zeta) \right] R(\zeta)\,.$$

Placing e^{-sT} for ζ, we obtain

$$\widehat{\theta}^o(s) = \left[\frac{1}{\sqrt{T}} \frac{\widehat{\Pi}^{-1}(s) \widehat{N}_R^+(s)}{\widehat{\ell}^+(s)} \quad \widehat{\Omega}_2(s) \right] \widehat{R}(s)\,.$$

Substituting this into (10.54), we find the image of the optimal transient process

$$Z^o(s) = p(s) \left[\frac{1}{\sqrt{T}} \frac{\widehat{\Pi}^{-1}(s) \widehat{N}_R^+(s)}{\widehat{\ell}^+(s)} \quad \widehat{\Omega}_2(s) \right] \widehat{R}(s) \widehat{q}(s) + r(s)\,.$$

However from (10.72) for $\zeta = \mathrm{e}^{-sT}$, we have

$$\widehat{R}(s) \widehat{q}(s) = \frac{\widehat{\ell}(s)}{\widehat{\Delta}_X(s)} \mathbf{1}_n\,.$$

With regard to the last two relations, we obtain

$$Z^o(s) = \frac{1}{\sqrt{T}} \frac{\widehat{\ell}(s)}{\widehat{\Delta}_X(s) \widehat{\ell}^+(s)} \, p(s) \widehat{\Pi}^{-1}(s) \widehat{N}_R^+(s) + r(s)\,.$$

This expression is independent of the choice of the matrix $\Omega_2(\zeta)$. So we arrive at the important conclusion that the optimal transient under zero initial energy is independent of the choice of the matrix $\Omega_2(\zeta)$. [1]

[1] The authors' attention to this fact was attracted by Dr. K. Polyakov.

2. Let us have an ILMFD

$$\theta_1^o(\zeta) = D_l^{-1}(\zeta)\tilde{M}_l(\zeta)\,. \tag{10.89}$$

Then by Lemma 2.16, the expression

$$\theta^o(\zeta) = D_l^{-1}\left[\tilde{M}_l(\zeta)R(\zeta)\right]$$

is an ILMFD for the matrix $\theta^o(\zeta)$. Thus for the characteristic polynomial of the optimal system $\Delta^o(\zeta)$, we have

$$\Delta^o(\zeta) \approx \det D_l(\zeta)\,. \tag{10.90}$$

The following theorem states an important property of the polynomial $\Delta^o(\zeta)$.

Theorem 10.11. *Let us have an ILMFD for the optimal column (10.88):*

$$\Omega_1^o(\zeta) = a_\Omega^{-1}(\zeta)b_\Omega(\zeta)\,. \tag{10.91}$$

Then independently of the choice of the matrix $\Omega_2(\zeta)$, the function

$$\varrho_\Omega(\zeta) = \frac{\Delta^o(\zeta)}{\det a_\Omega(\zeta)}$$

is a polynomial.

Proof. Let us have ILMFDs (10.89) and (10.91). Then,

$$D_l(\zeta)\theta_1^o(\zeta) = D_l(\zeta)\left[\,\Omega_1^o(\zeta)\;\Omega_2(\zeta)\right] = \left[\,D_l(\zeta)\Omega_1^o(\zeta)\;D_l(\zeta)\Omega_2(\zeta)\right]$$

is a polynomial matrix. Hence the matrix $D_l(\zeta)\Omega_1^o(\zeta)$ is also a polynomial and the polynomial matrix $D_l(\zeta)$ is a cancelling polynomial for the column $\Omega_1^o(\zeta)$. Since the right-hand side of (10.91) is an ILMFD, we obtain

$$D_l(\zeta) = a_0(\zeta)a_\Omega(\zeta)\,,$$

where $a_0(\zeta)$ a polynomial matrix and the function

$$\varrho_1(\zeta) = \frac{\det D_l(\zeta)}{\det a_\Omega(\zeta)}$$

is a polynomial. Then from (10.90), it follows immediately that $\varrho_\Omega(\zeta)$ is a polynomial. ∎

The claim of Theorem 10.11 is an important applied result. It shows that despite of a wide choice of optimal system functions (and, consequently, optimal controllers), there are some limitations on the attainable degree of stability. These limitations are imposed by the eigenvalues of the matrix $a_\Omega(\zeta)$ and they are independent of the choice of the matrix $\Omega_2(\zeta)$.

10.9 Modified Optimisation Algorithm

1. For solving the $\mathcal{L}_2$-problem as well as the $\mathcal{H}_2$-problem, we can apply a modified method that does not use a basic controller. The following lemma gives a substantiation for this method.

Lemma 10.12. *In the principal separation (10.84), the matrix* $R_-(\zeta) = \langle R(\zeta) \rangle_-$ *is independent of the choice of the initial basic controller.*

Proof. From (10.69) and (10.72), we have

$$\hat{B}(\zeta) = \hat{a}_r(\zeta) D_{\underline{L}'L\mu\underline{\mu}}(T,\zeta,0)\beta_{0r}(\zeta)Y(\zeta)\mathbf{1}_n \,\frac{\ell(\zeta)}{\Delta_X(\zeta)} + \hat{a}_r(\zeta)D_{\underline{L}'KX\underline{\mu}}(T,\zeta,0)\,,$$

whence

$$\begin{aligned}\hat{B}(\zeta)\frac{\ell(\zeta^{-1})}{\Delta_X(\zeta^{-1})} &= \hat{a}_r(\zeta)\,\frac{T\ell(\zeta^{-1})A_L(\zeta)\ell(\zeta)}{\Delta_X(\zeta^{-1})\Delta_X(\zeta)}\,\beta_{0r}(\zeta)Y(\zeta)\mathbf{1}_n \\ &\quad + \hat{a}_r(\zeta)D_{\underline{L}'KX\underline{\mu}}(T,\zeta,0)\frac{\ell(\zeta^{-1})}{\Delta_X(\zeta^{-1})}\,.\end{aligned}$$

Using (10.81), this can be written in the form

$$\begin{aligned}\hat{B}(\zeta)\frac{\ell(\zeta^{-1})}{\Delta_X(\zeta^{-1})} &= \hat{\Pi}_1(\zeta)\Pi_1(\zeta)a_r^{-1}(\zeta)\beta_{0r}(\zeta)Y(\zeta)\mathbf{1}_n \\ &\quad + \hat{a}_r(\zeta)D_{\underline{L}'KX\underline{\mu}}(T,\zeta,0)\frac{\ell(\zeta^{-1})}{\Delta_X(\zeta^{-1})}\,.\end{aligned}$$

Using (10.83), we obtain

$$\begin{aligned}R(\zeta) &= \hat{\Pi}_1^{-1}(\zeta)\hat{B}(\zeta)\frac{\ell(\zeta^{-1})}{\Delta_X(\zeta^{-1})} \\ &= \Pi_1(\zeta)a_r^{-1}(\zeta)\beta_{0r}(\zeta)Y(\zeta)\mathbf{1}_n + \hat{\Pi}_1^{-1}(\zeta)\hat{a}_r(\zeta)D_{\underline{L}'KX\underline{\mu}}(T,\zeta,0)\frac{\ell(\zeta^{-1})}{\Delta_X(\zeta^{-1})}\,.\end{aligned}$$

If we choose instead of $\beta_{0r}(\zeta)$ another matrix (8.114)

$$\beta_{0r}^*(\zeta) = \beta_{0r}(\zeta) - a_r(\zeta)Q(\zeta)\,,$$

where $Q(\zeta)$ is a stable rational matrix, then with the new matrix, we get

$$R^*(\zeta) = R(\zeta) - \Pi_1(\zeta)Q(\zeta)Y(\zeta)\,\mathbf{1}_n\,.$$

The second term on the right-hand side is a stable rational matrix, because the matrix $\Pi_1(\zeta)$ is stable. Therefore,

$$\langle R^*(\zeta)\rangle_- = \langle R(\zeta)\rangle_-\,. \qquad \blacksquare$$

2. With account for Theorem 10.20, we can extend the modified method given in Section 8.6 onto the problem at hand. This makes it possible to find the optimal vector $\Omega_1^o(\zeta)$ without calculating an initial basic controller.

10.10 Single-loop Control System

1. As an example, in the present section, we consider the $\mathcal{L}_2$-optimisation problem for the system shown in Fig. 10.2. In Fig. 10.2, we use the same

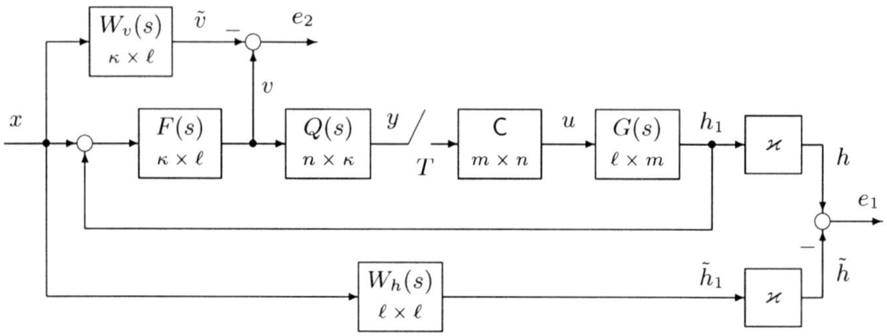

Fig. 10.2. Single-loop control system

notation as in Fig. 9.1. In addition, $W_v(s)$ and $W_h(s)$ are the transfer functions of the ideal LTI convolution operators

$$\tilde{v}(t) = \int_{-\infty}^{t} g_v(t-\tau)x(\tau)\,\mathrm{d}\tau$$

$$\tilde{h}_1(t) = \int_{-\infty}^{t} g_h(t-\tau)x(\tau)\,\mathrm{d}\tau\,,$$

where $g_v(t)$ and $g_h(t)$ are the corresponding impulse responses.

Hereinafter, we assume that the matrices

$$W_v(s) = \int_0^{\infty} g_v(t)\mathrm{e}^{-st}\,\mathrm{d}t, \quad W_h(s) = \int_0^{\infty} g_h(t)\mathrm{e}^{-st}\,\mathrm{d}t$$

are pseudo-rational and all their poles are stable. Hence for $0 < t < T$, we have

$$D_{W_v}(T,\zeta,t) = \frac{N_v(\zeta,t)}{\Delta_v(\zeta)}\,, \quad D_{W_h}(T,\zeta,t) = \frac{N_h(\zeta,t)}{\Delta_h(\zeta)}\,, \tag{10.92}$$

where $N_v(\zeta,t)$ and $N_h(\zeta,t)$ are polynomial matrices in ζ, while $\Delta_v(\zeta)$ and $\Delta_h(\zeta)$ are stable scalar polynomials having no roots inside the closed unit disk.

2. The input of the system is the vector $x(t)$ with image $X(s)$, which should be stable and pseudo-rational. The error under zero initial energy

$$e(t) = \begin{bmatrix} e_1(t) \\ e_2(t) \end{bmatrix} = \begin{bmatrix} \varkappa\left[h_1(t) - \tilde{h}_1(t)\right] \\ v(t) - \tilde{v}(t) \end{bmatrix} \tag{10.93}$$

is chosen as output vector. The system performance will be evaluated by

$$\begin{aligned} J_e = \int_0^\infty e'(t)e(t)\,\mathrm{d}t &= \int_0^\infty [v(t) - \tilde{v}(t)]'\,[v(t) - \tilde{v}(t)]\,\mathrm{d}t \\ &+ \varkappa^2 \int_0^\infty \left[h_1(t) - \tilde{h}_1(t)\right]'\left[h_1(t) - \tilde{h}_1(t)\right]\mathrm{d}t\,, \end{aligned}$$

where J_e is assumed to be finite.

3. Under the given assumptions, there exist the images

$$\tilde{V}(s) = W_v(s)X(s)\,, \quad \tilde{H}(s) = \varkappa W_h(s)X(s)$$

as stable pseudo-rational vectors. Moreover, we have $\tilde{v}(t) \in \Lambda_\gamma$ and $\tilde{h}(t) \in \Lambda_\gamma$ for some $\gamma < 0$.

4. For $W_v(s) = O_{\kappa\ell}$ and $W_h(s) = O_{\ell\ell}$, the system shown in Fig. 10.2 is transformed into the single-loop system shown in Fig. 9.1 with the PTM

$$w(s,t) = \varphi_{L\mu}(T,s,t)\widehat{R}_N(s)M(s) + K(s)\,,$$

where

$$\begin{aligned} K(s) &= \begin{bmatrix} O_{\ell\ell} \\ F(s) \end{bmatrix}, & L(s) &= \begin{bmatrix} \varkappa G(s) \\ F(s)G(s) \end{bmatrix}, \\ M(s) &= Q(s)F(s), & N(s) &= Q(s)F(s)G(s)\,. \end{aligned} \tag{10.94}$$

Assume that the matrix $L(s)$ is at least proper and the remaining matrices in (10.94) are strictly proper.

5. Under the given assumptions, for the output of the nominal system

$$z(t) = \begin{bmatrix} h(t) \\ v(t) \end{bmatrix} \tag{10.95}$$

by Theorem 10.4, there exists the Laplace transform

$$Z(s) = L(s)\mu(s)\widehat{R}_N(s)\widehat{D}_{MX}(T,s,0) + K(s)X(s)\,, \tag{10.96}$$

which converges in the half-plane $\operatorname{Re} s > \beta$, where β is a sufficiently large number. Using (10.94), we rewrite (10.96) in the form

$$H(s) = \varkappa G(s)\mu(s)\overset{\frown}{R}_{QFG}(s)\overset{\frown}{D}_{QFX}(T,s,0)\,,$$

$$V(s) = F(s)G(s)\mu(s)\overset{\frown}{R}_{QFG}(s)\overset{\frown}{D}_{QFX}(T,s,0) + F(s)X(s)\,.$$

In particular, if the system is internally stable, then we can take $\beta < 0$ and the vector (10.95) has a finite $\mathcal{L}_2$-norm. Then for the error vector (10.93), we have $\mathrm{e}(t) \in \Lambda_\lambda$, where $\lambda < 0$. For $\operatorname{Re} s > \lambda$, there exists the image of the error vector

$$E(s) = \begin{bmatrix} H(s) \\ V(s) \end{bmatrix} - \begin{bmatrix} \tilde{H}(s) \\ \tilde{V}(s) \end{bmatrix},$$

that can be represented using the above relations in the form

$$\begin{aligned} E(s) &= L(s)\mu(s)\overset{\frown}{R}_N(s)\overset{\frown}{D}_{MX}(T,s,0) + K(s)X(s) - W_e(s)X(s) \\ &= Z(s) - W_e(s)X(s)\,, \end{aligned} \tag{10.97}$$

where

$$W_e(s) = \begin{bmatrix} \varkappa W_h(s) \\ W_v(s) \end{bmatrix}.$$

The right-hand sides of (10.97) and (10.96) differ only by the last term. Thus, for application of the Wiener-Hopf method in the given case, we can use the above general relations, changing the matrix $K(s)$ for

$$K_e(s) = K(s) - W_e(s) = \begin{bmatrix} -\varkappa W_h(s) \\ F(s) - W_v(s) \end{bmatrix}. \tag{10.98}$$

The matrix $W_e(s)$ is not rational in the general case, but this fact does not affect the optimisation procedure, because under the given assumptions, all matrices in the cost functional are rational.

10.11 Wiener-Hopf Method for Single-loop Tracking System

1. Using (10.53), (10.94), and (10.98), the image of the error (10.97) can be written in the form

$$Z(s) = p(s)\overset{\frown}{\theta}(s)\overset{\frown}{q}(s) + r_e(s)\,, \tag{10.99}$$

where

$$p(s) = -L(s)\mu(s)\widehat{a}_r(s) = \begin{bmatrix} -\varkappa G(s)\mu(s)\widehat{a}_r(s) \\ -F(s)G(s)\mu(s)\widehat{a}_r(s) \end{bmatrix},$$

$$\widehat{q}(s) = \widehat{a}_l(s)\widehat{D}_{MX}(T,s,0) = \widehat{a}_l(s)\widehat{D}_{QFX}(T,s,0)\,, \tag{10.100}$$

$$\begin{aligned} r_e(s) &= L(s)\mu(s)\widehat{\beta}_{0r}(s)\widehat{a}_l(s)\widehat{D}_{QFX}(T,s,0) + K_e(s)X(s) \\ &= \begin{bmatrix} \varkappa G(s)\mu(s)\widehat{\beta}_{0r}(s)\widehat{a}_l(s)\widehat{D}_{QFX}(T,s,0) - \varkappa W_h(s)X(s) \\ F(s)G(s)\mu(s)\widehat{\beta}_{0r}(s)\widehat{a}_l(s)\widehat{D}_{QFX}(T,s,0) + F(s)X(s) - W_v(s)X(s) \end{bmatrix}. \end{aligned}$$

Let the nominal system be modal controllable. Then by Theorem 10.7, the matrix $p(s)$ is an integral function in s and the matrices $\widehat{q}(s)$ and $r(s)$ admit representations (10.57) and (10.59):

$$\widehat{q}(s) = \frac{\widehat{N}_q(s)}{\widehat{\Delta}_X(s)}, \qquad r(s) = \frac{P_r(s)}{\widehat{\Delta}_X(s)}.$$

Hence for the vector $R_e(s)$, we have

$$r_e(s) = \frac{P_e(s)}{\widehat{\Delta}_h(s)\widehat{\Delta}_v(s)\widehat{\Delta}_X(s)}, \tag{10.101}$$

where $\Delta_h(\zeta)$ and $\Delta_v(\zeta)$ are the denominators of the functions in (10.92) and $P_e(s)$ is an integral function of the argument s.

2. With account for (10.99) and (10.100) in the given case, the functional (10.64) takes the form

$$\begin{aligned} \tilde{J}_1 = \frac{T}{2\pi\mathrm{j}} \int_{-\mathrm{j}\omega/2}^{\mathrm{j}\omega/2} \Big[&\widehat{\underline{q}}'(s)\widehat{\underline{\theta}}'(s)\widehat{A}_{L1}(s)\widehat{\theta}(s)\widehat{q}(s) - \widehat{\underline{q}}'(s)\widehat{\underline{\theta}}'(s)\widehat{B}'_e(s) \\ &- \widehat{B}_e(s)\widehat{\theta}(s)\widehat{q}(s)\Big]\,\mathrm{d}s\,, \end{aligned} \tag{10.102}$$

where

$$\begin{aligned} \widehat{A}_{L1}(s) &= \frac{1}{T}\sum_{k=-\infty}^{\infty} \underline{p}'(s+k\mathrm{j}\omega)p(s+k\mathrm{j}\omega) = \widehat{\underline{a}}'_r(s)\widehat{D}_{\underline{L}'L\mu\underline{\mu}}(T,s,0)\widehat{a}_r(s) \\ &= \widehat{\underline{a}}'_r(s)\left[\varkappa^2\widehat{D}_{\underline{G}'G\mu\underline{\mu}}(T,s,0) + \widehat{D}_{\underline{G}'\underline{F}'FG\mu\underline{\mu}}(T,s,0)\right]\widehat{a}_r(s) \end{aligned} \tag{10.103}$$

and

$$
\begin{aligned}
\overset{\frown}{B}_e(s) &= -\frac{1}{T}\sum_{k=-\infty}^{\infty} \underline{p}'(s+kj\omega)r_e(s+kj\omega) \\
&= \overset{\frown}{\underline{a}'}_r(s)\overset{\frown}{D}_{\underline{L}'L\mu\underline{\mu}}(T,s,0)\overset{\frown}{\beta}_{0r}(s)\overset{\frown}{a}_l(s)\overset{\frown}{D}_{MX}(T,s,0) \\
&\quad + \overset{\frown}{\underline{a}'}_r(s)\overset{\frown}{D}_{\underline{L}'K_eX\underline{\mu}}(T,s,0) \\
&= \overset{\frown}{\underline{a}'}_r(s)\left[\varkappa^2\overset{\frown}{D}_{\underline{G}'G\mu\underline{\mu}}(T,s,0)\right. \\
&\qquad\qquad \left. + \overset{\frown}{D}_{\underline{G}'\underline{F}'FG\mu\underline{\mu}}(T,s,0)\right]\overset{\frown}{\beta}_{0r}(s)\overset{\frown}{a}_l(s)\overset{\frown}{D}_{MX}(T,s,0) \\
&\quad + \varkappa^2\overset{\frown}{D}_{\underline{G}'W_hX\underline{\mu}}(T,s,0) + \overset{\frown}{\underline{a}'}_r(s)\overset{\frown}{D}_{\underline{G}'\underline{F}'FX\underline{\mu}}(T,s,0) \\
&\quad - \overset{\frown}{\underline{a}'}_r(s)\overset{\frown}{D}_{\underline{G}'\underline{F}'W_vX\underline{\mu}}(T,s,0)\,.
\end{aligned}
$$

As was proved before, the rational periodic matrix (10.103) has no poles and with respect to (10.101), the matrix $\overset{\frown}{\underline{B}'}_e(s)$ admits a representation of the form

$$
\overset{\frown}{\underline{B}'}_e(s) = \frac{\overset{\frown}{Q}_e(s)}{\overset{\frown}{\Delta}_h(s)\overset{\frown}{\Delta}_v(s)\overset{\frown}{\Delta}_X(s)}\,,
$$

where $\overset{\frown}{Q}_e(s)$ is an integral rational periodic function.

3. Using the variable $\zeta = \mathrm{e}^{-sT}$, we can transform the functional (10.102) into the form (10.68)

$$
\tilde{J}_1 = \frac{1}{2\pi\mathrm{j}}\oint\left[\hat{q}(\zeta)\hat{\theta}(\zeta)A_{L1}(\zeta)\theta(\zeta)q(\zeta) - \hat{q}(\zeta)\hat{\theta}(\zeta)\hat{B}_e(\zeta) - B_e(\zeta)\theta(\zeta)q(\zeta)\right]\frac{\mathrm{d}\zeta}{\zeta}\,,
$$

where

$$
A_{L1}(\zeta) = \hat{a}_r(\zeta)\left[\varkappa^2 D_{\underline{G}'G\mu\underline{\mu}}(T,\zeta,0) + D_{\underline{G}'\underline{F}'FG\mu\underline{\mu}}(T,\zeta,0)\right]a_r(\zeta) = TA_L(\zeta)\,,
$$

$$
q(\zeta) = a_l(\zeta)D_{QFX}(T,\zeta,0) = \frac{N_q(\zeta)}{\Delta_X(\zeta)}\,,
$$

$$
\hat{B}_e(\zeta) = \frac{Q_e(\zeta)}{\Delta_h(\zeta)\Delta_v(\zeta)\Delta_X(\zeta)}\,,
$$

and the vector $Q_e(\zeta)$ is a quasi-polynomial. Per construction, these rational matrices have no poles at the unit circle. Let there exist the factorisation

$$
A_L(\zeta) = \hat{\Pi}(\zeta)\Pi(\zeta)\,, \tag{10.104}
$$

where $\Pi(\zeta)$ is a stable invertible polynomial matrix and similarly to (10.72),

$$q(\zeta) = \frac{\ell(\zeta)}{\Delta_X(\zeta)} Y(\zeta) \mathbf{1}_n \,,$$

where $Y(\zeta)$ is a unimodular matrix and $\ell(\zeta)$ is a scalar polynomial that admits a factorisation of the form (10.80). Then according to the general procedure given in Section 10.7, we can construct the set of optimal controllers. The matrix $R_+(\zeta)$, which is found as a result of the principal separation, admits the representation

$$R_+(\zeta) = \frac{N_e^+(\zeta)}{\Delta_h(\zeta)\Delta_v(\zeta)\Delta_X(\zeta)}$$

with a polynomial matrix $N_e^+(\zeta)$. The optimal column $\Omega_1^o(\zeta)$ can be written, in analogy with (10.88), in the form

$$\Omega_1^o(\zeta) = \frac{1}{\sqrt{T}} \frac{\Pi^{-1}(\zeta) N_e^+(\zeta)}{\Delta_h(\zeta)\Delta_v(\zeta)\ell^+(\zeta)} \,. \tag{10.105}$$

4. Let the conditions of Lemma 9.24 hold. Let us note some special features of the optimisation procedure for this case.

a) As follows from Section 9.6, if the polynomial $d_Q(s)$ has no roots on the imaginary axis, then for the factorisation (10.104) we have

$$\det \Pi(\zeta) = \Delta_Q^+(\zeta)\tilde{\Delta}_Q^+(\zeta)\lambda(\zeta) \,, \tag{10.106}$$

where all factors are stable polynomials and

$$\deg \det \Pi(\zeta) = \deg d_Q(\zeta) + \deg d_F(\zeta) + \deg d_G(\zeta) = \chi \,.$$

Then, if the polynomial $d_Q(s)$ is factored into stable and unstable cofactors

$$d_Q(s) = d_Q^+(s) d_Q^-(s) \,,$$

then $\Delta_Q^+(\zeta)$ is the discretisation of the polynomial $d_Q^+(s)$. The polynomial $\tilde{\Delta}_Q^+(\zeta)$ in (10.106) is constructed as follows. Let $\Delta_Q^-(\zeta)$ be the discretisation of the polynomial $d_Q^-(s)$. Then,

$$\tilde{\Delta}_Q^+(\zeta) = \Delta_Q^-(\zeta^{-1})\zeta^\delta \,,$$

where δ is the least nonnegative integer transforming the right-hand side into a polynomial.

b) If under the same conditions, the polynomial $d_Q(s)$ has roots on the imaginary axis, the factorisation (10.104) is impossible and a formal application of the Wiener-Hopf method provides a controller that does not stabilise.

c) Let $\ell(\zeta) = \text{const.} \neq 0$. Then using (10.105) and (10.106), we obtain

$$\Omega_1^o(\zeta) = \frac{N_\Omega(\zeta)}{\Delta_h(\zeta)\Delta_v(\zeta)\Delta_Q^+(\zeta)\tilde{\Delta}_Q^+(\zeta)\lambda(\zeta)} \,,$$

where $N_\Omega(\zeta)$ is a polynomial vector. Let the right-hand side be irreducible. Then from Remark 3.4, it follows that $\Omega_1^o(\zeta)$ is a normal matrix and for the ILMFD (10.91), we have

$$\det a_\Omega(\zeta) \approx \Delta_h(\zeta)\Delta_v(\zeta)\Delta_Q^+(\zeta)\tilde{\Delta}_Q^+(\zeta)\lambda(\zeta)\,.$$

Moreover, an optimal controller can be chosen in such a way that the characteristic polynomial of the optimal system $\tilde{\Delta}^o(\zeta)$ has the form

$$\tilde{\Delta}^o(\zeta) \approx \det a_\Omega(\zeta)\,.$$

For any choice of an optimal controller, the characteristic polynomial of the optimal system $\Delta^o(\zeta)$ is divisible by the polynomial $\tilde{\Delta}^o(\zeta)$. Therefore, all roots of the product $\Delta_h(\zeta)\Delta_v(\zeta)\Delta_Q^+(\zeta)\tilde{\Delta}_Q^+(\zeta)\lambda(\zeta)$, which are independent of the choice of optimal controller, are always among the roots of the polynomial $\Delta^o(\zeta)$.

10.12 $\mathcal{L}_2$ Redesign of Continuous-time LTI Systems under Persistent Excitation

1. In the preceding sections of this chapter, it was essential that all poles of the image of the input signal are located in the left half-plane. This condition is equivalent to vanishing of all input signals for $t \to \infty$. Nevertheless, many applied problems are connected with situations, where the system is acted upon by persistent signals including constant excitations like step signals

$$x(t) = \begin{cases} 0 & \text{for } t < 0 \\ x_0 & \text{for } t \geq 0 \end{cases}$$

with a constant vector x_0.

In principle in many cases, the approach described above can be extended onto the case of non-vanishing input signals.

Such a possibility is illustrated in this section by an example of the $\mathcal{L}_2$-redesign problem for the standard continuous-time system.

2. The two compared systems are shown in Fig. 10.3: a given continuous-time LTI system I, which will be called the *reference system*, and the standard sampled-data system II. As before, $w(s)$ in Fig. 10.3 is a rational matrix

$$w(s) = \begin{array}{c} \begin{matrix} \;\;\ell & \;\;\;m \end{matrix} \\ \left[\begin{matrix} K(s) & L(s) \\ M(s) & N(s) \end{matrix}\right] \end{array} \begin{matrix} r \\ n \end{matrix}\,. \tag{10.107}$$

It is assumed that $L(s)$ is at least proper, while the other elements of the matrix $w(s)$ are strictly proper. Moreover, we assume that the standard system

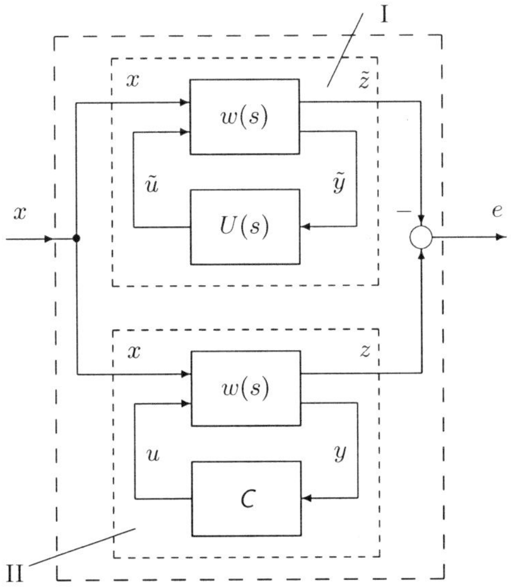

Fig. 10.3. Structure for redesign

is internally stable and modal controllable and the forming element is a zero-order hold, *i.e.*

$$\mu(s) = \mu_0(s) = \frac{1 - \mathrm{e}^{-sT}}{s}. \tag{10.108}$$

Suppose that the reference system I is stable and the rational matrix in the feedback $U(s)$ is analytical at the point $s = 0$.

3. With account for (10.108) under zero initial energy, the image of the output of the standard sampled-data system has the form

$$\begin{aligned} Z(s) = L(s)\mu_0(s)\widehat{w}_d(s)\left[I_n - \widehat{D}_{N\mu_0}(T,s,0)\widehat{w}_d(s)\right]^{-1}\widehat{D}_{MX}(T,s,0) \\ + K(s)X(s). \end{aligned} \tag{10.109}$$

Under similar assumptions, the image of the output of the reference system $\tilde{Z}(s)$ is

$$\tilde{Z}(s) = w_c(s)X(s), \tag{10.110}$$

where the transfer matrix of the reference system

$$w_c(s) = L(s)U(s)\left[I_n - N(s)U(s)\right]^{-1}M(s) + K(s) \tag{10.111}$$

is assumed to be strictly proper.

4. Let the inputs of both systems be acted upon by a step signal

$$x(t) = x_0 \, \mathbb{1}(t), \quad X(s) = \frac{x_0}{s}, \quad D_X(T,s,t) = \frac{x_0}{1 - \mathrm{e}^{-sT}} \quad (0 < t < T), \tag{10.112}$$

where x_0 is a constant vector and

$$\mathbb{1}(t) = \begin{cases} 0, & t < 0 \\ 1, & t > 0 \end{cases}$$

is the unit step. Then after vanishing of the transient processes, the constant output $\tilde{z}_\infty$ in the reference system has the form

$$\tilde{z}_\infty = \lim_{s \to 0} s w_c(s) X(s) = \lim_{s \to 0} w_c(s) x_0 = w_c(0) x_0 , \tag{10.113}$$

where we used the fact that the image $X(s)$ has the form (10.112). Similarly, if the standard sampled-data system is internally stable and (10.108) and (10.112) hold, then there exists the limit

$$z_\infty = \lim_{t \to \infty} z(t) = \text{const.}, \tag{10.114}$$

which can be found by the formula

$$z_\infty = \lim_{s \to 0} s Z(s) .$$

As follows from (10.113) and (10.114) under (10.112), the output signals of both systems have in the general case infinite $\mathcal{L}_2$-norms. Nevertheless under the condition

$$\tilde{z}_\infty = z_\infty , \tag{10.115}$$

the difference

$$e(t) = z(t) - \tilde{z}(t)$$

has a finite $\mathcal{L}_2$-norm, *i.e.* the following integral converges:

$$\tilde{J} = \int_0^\infty e'(t) e(t) \, \mathrm{d}t = \int_0^\infty \left[z(t) - \tilde{z}(t) \right]' \left[z(t) - \tilde{z}(t) \right] \mathrm{d}t .$$

Using the Parseval formula, we can write this integral in the form

$$\begin{aligned} \tilde{J} &= \frac{1}{2\pi \mathrm{j}} \int_{-\mathrm{j}\infty}^{\mathrm{j}\infty} E'(-s) E(s) \, \mathrm{d}s \\ &= \frac{1}{2\pi \mathrm{j}} \int_{-\mathrm{j}\infty}^{\mathrm{j}\infty} \left[Z(-s) - \tilde{Z}(-s) \right]' \left[Z(s) - \tilde{Z}(s) \right] \mathrm{d}s , \end{aligned} \tag{10.116}$$

where $Z(s)$ and $\tilde{Z}(s)$ are the images (10.109) and (10.110). Then the following optimisation problem is quite logical.

$\mathcal{L}_2$-redesign problem. Let a reference system I and a sampling period T be given, and suppose (10.108). Find the transfer function (-matrix) $w_d(\zeta)$ of a discrete-time controller such that the standard sampled-data system II is internally stable, satisfies Condition (10.115), and the integral (10.116) reaches its minimum.

Henceforth this problem will be called the problem of $\mathcal{L}_2$-redesign of the reference system. The general solution described below is based on the general approach by Wiener and Hopf. Firstly, the set of all stabilising controllers ensuring Condition (10.115) is constructed. Then the problem is reduced to the minimisation of a quadratic functional. In this case, there will arise some special features that are not encountered in the $\mathcal{H}_2$-problem.

5. According to the above statement of the problem first of all, we must construct the set of stabilising controllers for the standard sampled-data system that guarantee Condition (10.115). One such possibility is given by the following lemma.

Lemma 10.13. *Let the poles of the matrix $w(s)$ satisfy Conditions (6.106) for non-pathological behavior, the reference system be asymptotically stable and the standard sampled-data system be internally stable. Let also the rational matrix $U(s)$ be analytical at $s = 0$. Then for Condition (10.115) to be valid, it is sufficient that*

$$\overset{\frown}{w}_d(0) = w_d(1) = U(0)\,. \tag{10.117}$$

Proof. a) From (10.108) in the vicinity of the point $s = 0$, we have

$$\mu_0(s) = T + \dots\,. \tag{10.118}$$

Hereinafter, the dots denote the sum of terms that vanish as $s \to 0$. Moreover, we have

$$\mu_0(kj\omega) = 0\,, \quad (k = \pm 1, \pm 2, \dots\,). \tag{10.119}$$

b) Consider the series

$$\overset{\frown}{D}_{N\mu_0}(T, s, 0) = \frac{1}{T}\sum_{k=-\infty}^{\infty} N(s + kj\omega)\mu(s + kj\omega)\,.$$

Using (10.118) and (10.119), it can be easily shown that in the vicinity of the point $s = 0$

$$\overset{\frown}{D}_{N\mu_0}(T, s, 0) = N(s) + \dots\,.$$

c) Since $X(s) = s^{-1}x_0$ similarly to this equation, we obtain

$$sD_{MX}(T, s, 0) = \frac{s}{T}\sum_{k=-\infty}^{\infty}\frac{M(s + kj\omega)x_0}{s + kj\omega} = \frac{1}{T}M(s)x_0 + \dots\,.$$

Using Relations (10.109) and (10.118), it can be shown that

$$sZ(s) = L(s)\widehat{w}_d(s)\left[I_n - N(s)\widehat{w}_d(s)\right]^{-1} M(s)x_0 + K(s)x_0 + \dots \,. \tag{10.120}$$

At the same time, Equation (10.110) yields

$$s\tilde{Z}(s) = L(s)U(s)\left[I_n - N(s)U(s)\right]^{-1} M(s)x_0 + K(s)x_0 \,.$$

Since the reference system is asymptotically stable, the right-hand side tends to the finite value $\tilde{z}_\infty$ (10.113) as $s \to 0$. Comparing this with (10.120), we find that under Condition (10.117), the right-hand side of (10.120) tends to the finite value $z_\infty = \tilde{z}_\infty$ as $s \to 0$. ■

6. Hereinafter, we assume that the conditions of Lemma 10.13 hold.

Lemma 10.14. *Let the conditions of Lemma 10.13 hold and the matrix*

$$\widehat{a}_r(0) - U(0)\widehat{b}_r(0) = a_r(1) - U(0)b_r(1) \tag{10.121}$$

be nonsingular. Assume

$$\theta_0 \stackrel{\triangle}{=} \left[a_r(1) - U(0)b_r(1)\right]^{-1}\left[\beta_{0r}(1) - U(0)\alpha_{0r}(1)\right] \,. \tag{10.122}$$

Then the set of all system matrices ensuring the internal stability of the standard sampled-data system and guaranteeing (10.115) has the form

$$\theta(\zeta) = \theta_0 + (1-\zeta)\theta_\rho(\zeta)\,, \tag{10.123}$$

where θ_ρ is any stable rational matrix.

Proof. As follows from (8.45), the set of transfer functions of all stabilising controllers are given by

$$\begin{aligned}\widehat{w}_d(s) &= w_d(\zeta)\,\big|_{\zeta = \mathrm{e}^{-sT}} \\ &= \left[\widehat{\beta}_{0r}(s) - \widehat{a}_r(s)\widehat{\theta}(s)\right]\left[\widehat{\alpha}_{0r}(s) - \mathrm{e}^{-sT}\widehat{b}_r(s)\widehat{\theta}(s)\right]^{-1} .\end{aligned}$$

For $s \to 0$ with regard to (10.117), we obtain

$$U(0) = \left[\widehat{\beta}_{0r}(0) - \widehat{a}_r(0)\widehat{\theta}(0)\right]\left[\widehat{\alpha}_{0r}(0) - \widehat{b}_r(0)\widehat{\theta}(0)\right]^{-1} .$$

For a nonsingular matrix (10.121), we receive

$$\widehat{\theta}(0) = \theta_0\,,$$

where θ_0 is given by (10.122). This condition is equivalent to

$$\theta(1) = \theta_0\,,$$

whence it follows (10.123). ■

Corollary 10.15. *From (10.123) for* $\zeta = \mathrm{e}^{-sT}$*, we obtain*

$$\widehat{\theta}(s) = \theta_0 + \left(1 - \mathrm{e}^{-sT}\right) \widehat{\theta}_\rho(s)\,. \tag{10.124}$$

■

7. Using (10.109)–(10.111), the image of the error can be written in the form

$$\begin{aligned} E(s) &= Z(s) - \tilde{Z}(s) \\ &= L(s)\mu_0(s)\widehat{w}_d(s)\left[I_n - \widehat{D}_{N\mu}(T,s,0)\widehat{w}_d(s)\right]^{-1}\widehat{D}_{MX}(T,s,0) \\ &\quad - L(s)U(s)\left[I_n - N(s)U(s)\right]^{-1} M(s)X(s)\,. \end{aligned} \tag{10.125}$$

Under Condition (10.117), the right-hand side of this relation is analytical for $s = 0$. Let us find a representation of the error image $E(s)$ in terms of the new system matrix $\theta_\rho(\zeta)$. For this purpose, we use Equation (10.124). Then from (10.53) and (10.124), we have

$$\begin{aligned} \widehat{R}_N(s) &= \widehat{w}_d(s)\left[I_n - D_{N\mu_0}(T,s,0)\widehat{w}_d(s)\right]^{-1} \\ &= \widehat{\beta}_{0r}(s)\widehat{a}_l(s) - \widehat{a}_r(s)\widehat{\theta}(s)\widehat{a}_l(s) \\ &= \widehat{\beta}_{0r}(s)\widehat{a}_l(s) - \widehat{a}_r(s)\theta_0\widehat{a}_l(s) - \left(1 - \mathrm{e}^{-sT}\right)\widehat{a}_r(s)\widehat{\theta}_\rho(s)\widehat{a}_l(s)\,. \end{aligned}$$

From these relations and (10.125), we find

$$E(s) = p_0(s)\widehat{\theta}_\rho(s)\widehat{q}_0(s) + r_0(s)\,,$$

where

$$\begin{aligned} p_0(s) &= -L(s)\mu_0(s)\widehat{a}_r(s)\,, \\ \widehat{q}_0(s) &= \left(1 - \mathrm{e}^{-sT}\right)\widehat{a}_l(s)\widehat{D}_{MX}(T,s,0)\,, \\ r_0(s) &= L(s)\mu_0(s)\left[\widehat{\beta}_{0r}(s)\widehat{a}_l(s) - \widehat{a}_r(s)\theta_0\widehat{a}_l(s)\right]\widehat{D}_{MX}(T,s,0) \\ &\quad - L(s)U(s)\left[I_n - N(s)U(s)\right]^{-1} M(s)X(s)\,. \end{aligned} \tag{10.126}$$

Let us prove some properties of the matrices (10.126), which will be important. In the present section it is always assumed that the conditions of Lemma 10.13 hold.

Lemma 10.16. *The matrices* $p_0(s)$ *and*

$$p_1(s) = L(s)\widehat{a}_r(s) \tag{10.127}$$

are integral functions of s.

Proof. The claim about the matrix $p_0(s)$ was proved above. Further, we have

$$L(s)\widehat{a}_r(s) = -\frac{p_0(s)}{\mu_0(s)}\,. \tag{10.128}$$

The left-hand side of (10.128) can have poles only at poles of the matrix $L(s)$ and due to (10.108), the right-hand side can have poles only at the points $s_k = kj\omega$, $(k = \pm 1, \pm 2, \dots)$. But by assumptions, the matrix $L(s)$ is analytical at the points s_k. Therefore, the left-hand side of (10.128) has no poles, *i.e.* is an integral function of s. ∎

Lemma 10.17. *The vector $\widehat{q}_0(s)$ is an integral function of s.*

Proof. Let us transform the expression for the vector $\widehat{q}_0(s)$ using the relation

$$\widehat{\mathcal{D}}_{MX}(T,s,0) = \int_0^T \widehat{D}_M(T,s,\tau)\widehat{D}_X(T,s,-\tau)\,\mathrm{d}\tau\,. \tag{10.129}$$

From (10.112) and (6.71) for $0 < \tau < T$, we have

$$\widehat{D}_X(T,s,-\tau) = \widehat{D}_X(T,s,T-\tau)\mathrm{e}^{-sT} = \frac{\mathrm{e}^{-sT}}{1-\mathrm{e}^{-sT}}\,x_0\,.$$

Substituting this result into (10.129), we find

$$\widehat{D}_{MX}(T,s,0) = \frac{\mathrm{e}^{-sT}}{1-\mathrm{e}^{-sT}}\int_0^T \widehat{D}_M(T,s,\tau)\,\mathrm{d}\tau\,x_0\,.$$

Finally using (10.126), we obtain

$$\widehat{q}_0(s) = \widehat{q}_1(s)x_0, \qquad \widehat{q}_1(s) = \mathrm{e}^{-sT}\int_0^T \widehat{a}_l(s)\widehat{D}_M(T,s,\tau)\,\mathrm{d}\tau\,,$$

whence it immediately follows that the vector $\widehat{q}_1(s)$ has no poles, because we have already proved that the matrix $\widehat{a}_l(s)\widehat{D}_M(T,s,\tau)$ has no poles. ∎

Lemma 10.18. *If the conditions of Lemma 10.13 hold, the vector $r_0(s)$ has no poles on the imaginary axis.*

Proof. Consider the vector

$$Z_1(s) = L(s)\mu_0(s)\left[\widehat{\beta}_{0r}(s)\widehat{a}_l(s) - \widehat{a}_r(s)\theta_0\widehat{a}_l(s)\right]\widehat{D}_{MX}(T,s,0) + K(s)X(s)\,. \tag{10.130}$$

This vector is the image of the output for $\theta(\zeta) = \theta_0$. Since the standard sampled-data system is modal controllable, due to Theorem 10.5, the following representation holds:

$$Z_1(s) = \frac{D(s)x_0}{1-\mathrm{e}^{-sT}}\,, \tag{10.131}$$

where $D(s)$ is an integral function. Then the image (10.130) can have single pure imaginary poles only at the points $s_k = kj\omega = 2k\pi j/T$, $(k = 0, \pm 1, \dots)$. On the other hand, with account for (10.108), (10.126) and (10.127), from (10.130), we obtain

$$Z_1(s) = \frac{Z_2(s)}{s} x_0\,, \tag{10.132}$$

where

$$Z_2(s) = L(s)\widehat{\beta}_{0r}(s)\widehat{q}_1(s) - p_1(s)\theta_0\widehat{q}_1(s) + K(s)\,.$$

The matrix $Z_2(s)$ is analytical at the points $s_k = kj\omega$, $(k = 0, \pm 1, \dots)$, because under the given assumptions the right-hand side has no poles due to (6.106), Lemmata 10.16 and 10.17. Moreover, the matrix $Z_2(s)$ has no poles at $s = 0$. Indeed, if we assume the converse from (10.132), we find that the image (10.130) has a pole at $s = 0$ with a multiplicity greater than one, but this contradicts (10.131). Hence the matrix $Z_2(s)$ is an integral function of s. Therefore, with respect to (10.132), the vector $r_0(s)$ can be written in the form

$$r_0(s) = [Z_2(s) - w_c(s)]\,\frac{1}{s}\,x_0\,, \tag{10.133}$$

where $w_c(s)$ is the transfer function of the reference system (10.111). Its poles are located in the left half-plane due to the assumption on its stability. From (10.133), we obtain that the vector $r_0(s)$ can have a simple pole at the point $s = 0$. But, due to the choice $\theta(\zeta) = \theta_0$, we have

$$\lim_{s \to 0} s r_0(s) = z_\infty - \tilde{z}_\infty = [Z_2(0) - w_c(0)]\,x_0 = O_{\kappa 1}\,,$$

so that the vector $r_0(s)$ is analytical at $s = 0$. Hence this vector is analytical on the whole imaginary axis. ∎

Corollary 10.19. *Let us have the standard form*

$$w_c(s) = \frac{N_c(s)}{d_c(s)}\,. \tag{10.134}$$

Then the vector $r_0(s)$ can be represented in the form

$$r_0(s) = \frac{P_c(s)}{d_c(s)}\,, \tag{10.135}$$

where the vector $P_c(s)$ is an integral function of s.

Proof. From the proof of Lemma 10.18, it follows that the matrix $Z_2(s)$ is an integral function of s. Moreover, the right-hand side of (10.133) is analytical at the point $s = 0$. Therefore, substituting (10.134) into (10.133), we obtain the claim of the corollary. ∎

8. Using (10.126) and repeating the derivations of Section 10.6, we obtain that under the given assumptions the $\mathcal{L}_2$-optimal redesign problem reduces to minimising the functional

$$\tilde{J}_1 = \frac{T}{2\pi\mathrm{j}} \int_{-\mathrm{j}\omega/2}^{\mathrm{j}\omega/2} \Big[\widehat{\underline{q}'}_0(s) \widehat{\underline{\theta}'}_\rho(s) \widehat{A}_{L1}(s) \widehat{\theta}_\rho(s) \widehat{q}_0(s) \\ - \widehat{\underline{q}'}_0(s) \widehat{\underline{\theta}'}_\rho(s) \widehat{\underline{B}'}_0(s) - \widehat{B}_0(s) \widehat{\theta}(s) \widehat{q}_0(s) \Big] \,\mathrm{d}s \,, \tag{10.136}$$

where

$$\widehat{A}_{L1}(s) = \widehat{\underline{a}'}_r(s) \widehat{D}_{\underline{L'} L \mu_0 \underline{\mu_0}}(T, s, 0) \widehat{a}_r(s) \,. \tag{10.137}$$

Moreover,

$$\widehat{B}_0(s) = -\frac{1}{T} \sum_{k=-\infty}^{\infty} \underline{r}'_0(s + k\mathrm{j}\omega) p_0(s + k\mathrm{j}\omega) \,, \\ \widehat{\underline{B}'}_0(s) = -\frac{1}{T} \sum_{k=-\infty}^{\infty} \underline{p}'_0(s + k\mathrm{j}\omega) r_0(s + k\mathrm{j}\omega) \tag{10.138}$$

and after transformations, we obtain

$$\widehat{\underline{B}'}_0(s) = \widehat{\underline{a}'_r}(s) \widehat{D}_{\underline{L'} L \mu_0 \underline{\mu_0}}(T, s, 0) \left[\widehat{\beta}_{0r}(s) \widehat{a}_l(s) - \widehat{a}_r(s) \theta_0 \widehat{a}_l(s) \right] \widehat{D}_{MX}(T, s, 0) \\ - \widehat{\underline{a}'}_r(s) \widehat{D}_{\underline{L'} w_c X \underline{\mu}_0}(T, s, 0) + \widehat{\underline{a}'}_r(s) \widehat{D}_{\underline{L'} K X \underline{\mu_0}}(T, s, 0) \,. \tag{10.139}$$

The matrices (10.137)–(10.139) are rational periodic. Therefore, Matrix (10.137) is an integral function. Moreover, since the matrix $p_0(s)$ is an integral function and (10.135) holds, we have

$$\widehat{\underline{B}'}_0(s) = \frac{\widehat{D}_0(s)}{\widehat{\Delta}_c(s)} \,, \tag{10.140}$$

where the polynomial $\widehat{\Delta}_c(s)$ is the discretisation of the polynomial $d_c(s)$ and $\widehat{D}_0(s)$ is an integral rational periodic function.

9. Using the new variable $\zeta = \mathrm{e}^{-sT}$ in (10.136), we obtain the functional

$$\tilde{J}_1 = \frac{1}{2\pi\mathrm{j}} \oint \Big[\hat{q}_0(\zeta) \hat{\theta}_\rho(\zeta) A_{L1}(\zeta) \theta_\rho(\zeta) q_0(\zeta) \\ - \hat{q}_0(\zeta) \hat{\theta}_\rho(\zeta) \hat{B}_0(\zeta) - B_0(\zeta) \theta(\zeta) q_0(\zeta) \Big] \frac{\mathrm{d}\zeta}{\zeta} \,, \tag{10.141}$$

that should be minimised over the set of stable rational matrices $\theta_\rho(\zeta)$. Similarly to (10.72), we obtain

$$G(\zeta)q_0(\zeta) = \ell_0(\zeta)\mathbf{1}_n\,,$$

where $\ell_0(\zeta)$ is a scalar polynomial and $G(\zeta)$ is a unimodular matrix. Then

$$\theta_2(\zeta) \triangleq \theta_\rho(\zeta)G^{-1}(\zeta) = \overset{\quad 1 \qquad\quad n-1}{\left[\,\phi_1(\zeta)\;\phi_2(\zeta)\,\right]}\, m\,,$$

so that the functional (10.141) can be written in a form similar to (10.77):

$$\tilde{J}_1 = \frac{1}{2\pi \mathrm{j}}\oint_\Gamma \Big[\hat{\phi}_1(\zeta)\ell_0(\zeta^{-1})A_{L1}(\zeta)\ell_0(\zeta)\phi_1(\zeta) \\ - \hat{\phi}_1(\zeta)\ell_0(\zeta^{-1})\hat{B}_0(\zeta) - B_0(\zeta)\phi_1(\zeta)\ell_0(\zeta)\Big]\frac{\mathrm{d}\zeta}{\zeta}\,.$$

If the factorisations (10.79) and (10.80) hold, the optimisation problem is solvable, because the matrix $B_0(\zeta)$ has no poles on the unit circle. When we have

$$\ell_0(\zeta)\ell_0(\zeta^{-1}) = \ell_0^+(\zeta)\ell_0^+(\zeta^{-1})$$

with a stable polynomial $\ell_0^+(\zeta)$ and (10.79), then according to the general Wiener-Hopf method, we construct the matrix

$$R_0(\zeta) = \hat{\Pi}_1^{-1}(\zeta)\hat{B}_0(\zeta)\frac{\ell_0(\zeta^{-1})}{\ell_0^+(\zeta^{-1})}\,.$$

From (10.140) for $\mathrm{e}^{-sT} = \zeta$, we receive

$$\hat{B}_0(\zeta) = \frac{D_0(\zeta)}{\Delta_c(\zeta)}\,.$$

From the last two equations, we conclude that the set of stable poles of the matrix $R_0(\zeta)$ belongs to the set of roots of the polynomial $\Delta_c(\zeta)$. Therefore, as a result of the principal separation (10.84), we obtain

$$R_{0+}(\zeta) = \frac{N_0(\zeta)}{\Delta_c(\zeta)}$$

with a polynomial matrix $N_0(\zeta)$. The optimal vector $\phi_1^o(\zeta)$ has the form

$$\phi_1^o(\zeta) = \frac{1}{\sqrt{T}}\frac{\Pi^{-1}(\zeta)N_0(\zeta)}{\ell_0^+(\zeta)\Delta_c(\zeta)}\,. \tag{10.142}$$

The further procedure for constructing the set of optimal controllers is the same as in Section 10.7.

10. Similarly to Section 10.7, it can be found that the characteristic polynomial of the optimal system $\Delta^o(\zeta)$ is divisible by the polynomial $\Delta_c(\zeta)$. Hence if in particular, the function (10.142) is irreducibile, then the characteristic polynomial of the optimal standard sampled-data system is divisible by the discretisation of the characteristic polynomial of the reference model $d_c(s)$, and this fact does not dependent on the choice of the controller which minimises Functional (10.141).

10.13 $\mathcal{L}_2$ Redesign of a Single-loop LTI System

1. To illustrate the general approach given in Section 10.12, we consider in the present section the $\mathcal{L}_2$ redesign problem for a single-loop continuous-time system shown in Fig. 10.4. Here the matrices $F(s)$, $Q(s)$, $G(s)$ are the same

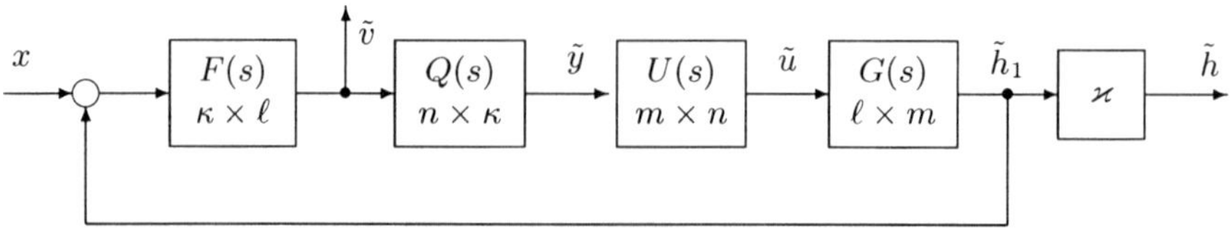

Fig. 10.4. Single-loop reference system

as in Fig. 9.1, and $U(s)$ is a given rational matrix of compatible dimensions. The output vector of the reference system has the form

$$\tilde{z}(t) = \begin{bmatrix} \varkappa \tilde{h}_1(t) \\ \tilde{v}(t) \end{bmatrix}. \tag{10.143}$$

Suppose the reference system in Fig. 10.4 is asymptotically stable, the input signal has the form (10.112) and the matrix $U(s)$ is analytical at the point $s = 0$. Then the $\mathcal{L}_2$-redesign problem can be formulated as follows.

For the sampled-data system in Fig. 9.1, let us have the sampling period T, the transfer function of the forming element (10.108) and the input signal (10.112). Let $\tilde{z}(t)$ denote the output vector (10.143) of the LTI system under zero initial energy and the vector $z(t)$ is the output of the sampled-data system under similar assumptions. It is required to find the transfer function of the discrete controller $w_d(\zeta)$ satisfying the following conditions:

a) The following equality holds:

$$\tilde{z}_\infty = \lim_{t\to\infty} \tilde{z}(t) = \lim_{t\to\infty} z(t) = z_\infty\,.$$

b) The sampled-data system is internally stable.
c) The integral

$$\begin{aligned}\tilde{J} &= \int_0^\infty [z(t) - \tilde{z}(t)]' \, [z(t) - \tilde{z}(t)] \, \mathrm{d}t \\ &= \int_0^\infty [v(t) - \tilde{v}(t)]' \, [v(t) - \tilde{v}(t)] \, \mathrm{d}t \\ &\quad + \varkappa^2 \int_0^\infty \left[h_1(t) - \tilde{h}_1(t)\right]' \left[h_1(t) - \tilde{h}_1(t)\right] \mathrm{d}t\end{aligned}$$

takes the minimal value.

2. Let us show that the problem formulated above can be reduced to the general scheme considered in Section 10.12. With this aim in view, we show that the system shown in Fig. 10.4 can be presented in form of a reference system I from Fig. 10.3. Notice that the transfer matrix of the LTI system $w_c(s)$ from the input x to the output $\tilde{z}$ can be represented in the form (10.111). Indeed, using the standard structural transformations, it is easy to find the transfer matrices $w_{\tilde{h}x}(s)$ and $w_{\tilde{v}x}(s)$ from the input $x(t)$ to the outputs $\tilde{h}(t)$ and $\tilde{v}(t)$:

$$w_{\tilde{h}x}(s) = \varkappa G(s)U(s)Q(s)F(s)\left[I_\ell - G(s)U(s)Q(s)F(s)\right]^{-1}, \quad (10.144)$$

$$w_{\tilde{v}x}(s) = F(s)\left[I_\ell - G(s)U(s)Q(s)F(s)\right]^{-1}. \quad (10.145)$$

Recall that for any matrices A and B of compatible dimensions, we have

$$A(I - BA)^{-1} = (I - AB)^{-1}A\,.$$

Then assuming

$$A = Q(s)F(s)\,, \quad B = G(s)U(s)\,,$$

we obtain

$$\begin{aligned} &G(s)U(s)Q(s)F(s)\left[I_\ell - G(s)U(s)Q(s)F(s)\right]^{-1} \\ &\qquad = G(s)U(s)\left[I_n - Q(s)F(s)G(s)U(s)\right]^{-1}Q(s)F(s)\,. \end{aligned} \quad (10.146)$$

Using (10.144), we obtain

$$w_{\tilde{h}x}(s) = \varkappa G(s)U(s)\left[I_n - Q(s)F(s)G(s)U(s)\right]^{-1}Q(s)F(s)\,. \quad (10.147)$$

Then we prove

$$\begin{aligned} w_{\tilde{v}x}(s) &= F(s)\left[I_\ell - G(s)U(s)Q(s)F(s)\right]^{-1} \\ &= F(s)\left\{\left[I_\ell - G(s)U(s)Q(s)F(s)\right]^{-1} - I_\ell\right\} + F(s) \\ &= F(s)G(s)U(s)Q(s)F(s)\left[I_\ell - G(s)U(s)Q(s)F(s)\right]^{-1} + F(s)\,. \end{aligned} \quad (10.148)$$

From (10.148) and (10.146), it follows

$$w_{\tilde{v}x}(s) = F(s)G(s)U(s)\left[I_n - Q(s)F(s)G(s)U(s)\right]^{-1}Q(s)F(s) + F(s)\,.$$

Using (10.147) and (10.148), we find

$$w_c(s) = \begin{bmatrix} w_{\tilde{h}x}(s) \\ w_{\tilde{v}x}(s) \end{bmatrix} = L(s)U(s)\left[I_n - N(s)U(s)\right]^{-1}M(s) + K(s)\,,$$

where

$$K(s) = \begin{bmatrix} O_{\ell\ell} \\ F(s) \end{bmatrix}, \qquad L(s) = \begin{bmatrix} \varkappa G(s) \\ F(s)G(s) \end{bmatrix}, \tag{10.149}$$
$$M(s) = Q(s)F(s), \quad N(s) = Q(s)F(s)G(s).$$

Comparing (10.149) with (9.9), we arrive to the conclusion that the problem under consideration is a special case of the general problem described in Section 10.12, whenever the elements of Matrix (10.107) have the form (10.149). Therefore, the further solution of the $\mathcal{L}_2$-redesign problem can be found using the general algorithm of Section 10.12.

3. Under some additional assumptions taking into account the special structure of the reference system, we can establish some additional important properties of the optimal system.

Theorem 10.20. *Let the conditions of the Lemma 9.24 hold, the factorisation (10.104) exist and* $\ell(\zeta) = \text{const}$. *Then there exists a set of optimal controllers. Moreover, if the ratio (10.142) is irreducible, then the optimal controller can be chosen in such a way that the characteristic polynomial of the optimal system* $\tilde{\Delta}^o(\zeta)$ *becomes*

$$\tilde{\Delta}^o(\zeta) = \Delta_c(\zeta)\lambda(\zeta),$$

where $\Delta_c(\zeta)$ *is a polynomial and* $\lambda(\zeta)$ *is a polynomial such that* $\lambda(\zeta) = \det \Pi(\zeta)$ *and*

$$\deg \lambda(\zeta) \leq \deg d_Q(s) + \deg d_F(s) + \deg d_G(s) = \chi.$$

For any choice of the optimal controller, the characteristic polynomial of the optimal system $\Delta^o(\zeta)$ *is divisible by the polynomial* $\tilde{\Delta}^o(\zeta)$. ∎

Remark 10.21. If the polynomial $d_Q(s)$ has roots on the imaginary axis, then the factorisation (10.104) is impossible.

Appendices

A

Operator Transformations of Taylor Sequences

1. Let the sequence of complex numbers

$$\{u_k\} = \{u_0, u_1, \dots\} \tag{A.1}$$

be given. This sequence is called a *Taylor sequence*, if there exist positive numbers M and ρ such that the inequalities

$$|u_i| < \frac{M}{\rho^i}, \quad (i = 0, 1, \dots) \tag{A.2}$$

are true.

2. For a Taylor sequence $\{u_k\}$ and $|\zeta| < \rho$, the series

$$u^0(\zeta) = \sum_{k=0}^{\infty} u_k \zeta^k \tag{A.3}$$

converges. The function $u^0(\zeta)$ is called ζ-transform (image) of the Taylor sequence (A.1). Relation (A.3) is symbolically written as

$$\{u_k\} \xrightarrow{\zeta} u^0(\zeta). \tag{A.4}$$

It is well known that the ζ-transform $u^0(\zeta)$ is analytical in $|\zeta| < \rho$. Thus, Relation (A.4) might be interpreted as a map from the set of Taylor sequences in the set of functions $u^0(\zeta)$ which are analytical in the point $\zeta = 0$.

3. Conversely, every function $u^0(\zeta)$, analytical in $\zeta = 0$, may be developed in an environment of the origin in its Taylor series

$$u^0(\zeta) = u_0 + u_1\zeta + \dots,$$

the coefficients of which satisfy an inequality of the form (A.2). Thus, the coefficients of this expansion always establish a Taylor sequence, for which the function $u^0(\zeta)$ proves to be the ζ-transform. Hence there exists an invertible unique map between the set of Taylor sequences and the set of function of a complex arguments that are analytical in the origin.

4. Let the ζ-transform (A.3) of the sequence (A.1) be convergent for $|\zeta| < R$. Then for $|z| > R^{-1}$, the series

$$u^*(z) = \sum_{k=0}^{\infty} u_k z^{-k} \tag{A.5}$$

converges, which is named *z-transform* of the sequence $\{u_k\}$. Relation (A.5) is denoted by

$$\{u_k\} \xrightarrow{z} u^*(z)\,. \tag{A.6}$$

Obviously, also the reverse is correct: If we have (A.6) for $|z| > R^{-1}$, then for $|\zeta| < R$ the series (A.3) converges, hence the sequence $\{u_k\}$ is a Taylor sequence. Therefore, the sequence $\{u_k\}$ possesses a z-transform and a ζ-transform, if and only if it is a Taylor sequence.

5. If we compare Formula (A.3) with (A.5), then we recognise that the ζ-transform $u^0(\zeta)$ and the z-transform $u^*(z)$ are connected by the interrelations

$$u^0(\zeta) = u^*(\zeta^{-1})\,, \quad u^*(z) = u^0(z^{-1})\,. \tag{A.7}$$

Nevertheless, it must be clear that in general, the functions $u^0(\zeta)$ and $u^*(z)$ are defined in different regions.

6. The above considerations suggest that a complex function $u^0(\zeta)$ represents a ζ-transform, if it is analytical in the origin, but the function $u^*(z)$ represent exactly then a z-transform, when it is analytical in the infinitely far point. In particular, a rational function $u^0(\zeta)$ is a ζ-transform, if it has no pole at $\zeta = 0$, while the rational function $u^*(z)$ is a z-transform, whenever it is at least proper.

7. In the control theoretical and engineering literature mainly the z-transformation was investigated, and its properties are presented in detail, e.g. in [3]. For the purposes of our book, both transformations are important. On one side, there is no need for considering the properties of the ζ-transformation in detail, because due to Relation (A.7), any formula from the theory of z-transformation can be transferred into a corresponding formula for the ζ-transformation by exchanging z against ζ^{-1}, and reverse. On the other side, however, we have to be careful, because the named transformations are defined over different regions.

8. In particular, let $\{u_k\}$ be a Taylor sequence. Then, also the displaced sequence

$$\{u_{k+\ell}\} = \{u_\ell, u_{\ell+1}, \dots\}$$

is a Taylor sequence. As known from [3], we get from (A.6)

$$\{u_{k+\ell}\} \xrightarrow{z} z^{\ell}\left[u^*(z) - \sum_{\nu=0}^{\ell-1} u_\nu z^{-\nu}\right]. \tag{A.8}$$

Substituting ζ^{-1} for z on the right side of (A.8), we obtain the corresponding formula for the ζ-transformation

$$\{u_{k+\ell}\} \xrightarrow{\zeta} \zeta^{-\ell}\left[u^0(\zeta) - \sum_{\nu=0}^{\ell-1} u_\nu \zeta^{\nu}\right], \tag{A.9}$$

where $u^0(\zeta)$ is the ζ-transform configured by (A.4).

9. Applying the ζ-transformation for solving difference equations requires to overcome certain theoretical difficulties, that arise from using the forward-shift operator (right shifting) and the backward-shift operator (left shifting). This fact was considered in [99]. The situation should be demonstrated by an example, which was already considered in [14].

Let us have the scalar difference equation

$$y_{k+1} - ay_k = u_k\,, \quad (k = 0, 1, \dots) \tag{A.10}$$

with a freely selectable initial condition $y_0 = \bar{y}$, where a is any given number. Suppose the input $\{u_k\}$ to be a Taylor sequence. Then after transition to z-transforms according to (A.8), we get

$$z\left[y^*(z) - \bar{y}\right] - ay^*(z) = u^*(z)\,, \tag{A.11}$$

which results in

$$y^*(z) = \frac{z}{z-a}\,\bar{y} + \frac{u^*(z)}{z-a}\,. \tag{A.12}$$

Particularly, $\bar{y} = 0$ implies

$$(z-a)y^*(z) = u^*(z)\,.$$

On the other side, applying the ζ-transformation on Equation (A.10), we receive with respect to (A.9)

$$\zeta^{-1}\left[y^0(\zeta) - \bar{y}\right] - ay^0(\zeta) = u^0(\zeta)$$

and thus

$$y^0(\zeta) = \frac{\bar{y}}{1-a\zeta} + \frac{\zeta}{1-a\zeta}u^0(\zeta)\,. \tag{A.13}$$

Especially for $\bar{y} = 0$, we obtain

$$(1-a\zeta)y^0(\zeta) = \zeta u^0(z)\,.$$

It is emphasised that Relation (A.13) may be derived from (A.12), if we exchange z against ζ^{-1}.

10. Assume $\zeta = \mathrm{e}^{-sT}$ in (A.3), so the function

$$\overset{\frown}{u}^{0}(s) = u^{0}(\zeta)\,|_{\zeta=\mathrm{e}^{-sT}} = \sum_{k=0}^{\infty} u_k \mathrm{e}^{-ksT} \tag{A.14}$$

is obtained, which is called *discrete Laplace transformation (DLT)* of the sequence $\{u_k\}$, [148]. Obviously, the transformation (A.14) converges, if and only if $\{u_k\}$ is a Taylor sequence. Besides, if the ζ-transform (A.3) converges in the circle $|\zeta| < \rho$, then the transform (A.14) converges in the open half-plane $\operatorname{Re} s > -\frac{1}{T} \ln R$.

B

Sums of Certain Series

1. Let the strictly proper scalar fraction

$$F(s) = \frac{m(s)}{d(s)} = \frac{m_1 s^{n-1} + \ldots + m_n}{s^n + d_1 s^{n-1} + \ldots + d_n} \tag{B.1}$$

be given and the expansion into partial fractions

$$F(s) = \sum_{i=1}^{q} \sum_{k=1}^{\nu_i} \frac{f_{ik}}{(s - s_i)^k}$$

should be valid. Then for $0 < t < T$, we obtain

$$\varphi_F(T, s, t) = \frac{1}{T} \sum_{k=-\infty}^{\infty} \frac{m(s + kj\omega)}{d(s + kj\omega)} \mathrm{e}^{kj\omega t} = \tilde{\varphi}_F(T, s, t)\,, \tag{B.2}$$

where

$$\tilde{\varphi}_F(T, s, t) = \sum_{i=1}^{q} \sum_{k=1}^{\nu_i} \frac{f_{ik}}{(k-1)!} \frac{\partial^{k-1}}{\partial s_i^{k-1}} \frac{\mathrm{e}^{(s_i - s)t}}{1 - \mathrm{e}^{(s_i - s)T}}\,.$$

Besides, if we have $m_1 \neq 0$ in (B.1), then the sum of the series $\varphi_F(T, s, t)$ possesses jumps of finite height in the points $t_n = nT$, $(n = 0, \pm 1, \ldots)$. However, when $m_1 = m_2 = \ldots = m_{\ell-1} = 0$ and $m_\ell \neq 0$ take place, then the periodic function $\varphi_F(T, s, t)$ has derivatives up to and including $(\ell - 1)$-th order, where the $(\ell - 1)$-th derivative is piecewise continuous, but the lower derivatives are continuous.

2. Multiplying (B.2) by e^{st} for $0 < t < T$, we obtain

$$\widehat{D}_F(T, s, t) = \frac{1}{T} \sum_{k=-\infty}^{\infty} \frac{m(s + kj\omega)}{d(s + kj\omega)} \mathrm{e}^{(s + kj\omega)t} = \widehat{\mathcal{D}}_F(T, s, t)\,,$$

where

$$\widehat{\mathcal{D}}_F(T,s,t) = \mathrm{e}^{st}\tilde{\varphi}_F(T,s,t) = \sum_{i=1}^{q}\sum_{k=1}^{\nu_i} \frac{f_{ik}}{(k-1)!}\frac{\partial^{k-1}}{\partial s_i^{k-1}}\frac{\mathrm{e}^{s_i t}}{1-\mathrm{e}^{(s_i-s)T}}\,.$$

If we have $m_1 \neq 0$ in (B.1), then the sum of the series $\widehat{D}_F(T,s,t)$ has jumps of finite height in the points $t_n = nT$, $(n = 0, \pm 1, \dots)$. However, if $m_1 = m_2 = \dots = m_{\ell-1} = 0$ and $m_\ell \neq 0$ are true, then the periodic function $\varphi_F(T,s,t)$ has derivatives up to and including $(\ell-1)$-th order, where the $(\ell-1)$-th derivative is piecewise continuous, but the lower derivatives are continuous.

3. Suppose

$$\mu(s) = \int_0^T \mathrm{e}^{-st} m(t)\,\mathrm{d}t\,,$$

where the function $m(t)$ is of bounded variation on the interval $0 \leq t \leq T$ and it has a finite number of jumps. Then for $0 < t < T$, we obtain

$$\widehat{D}_{F\mu}(T,s,t) = \frac{1}{T}\sum_{k=-\infty}^{\infty} F(s+kj\omega)\mu(s+kj\omega)\mathrm{e}^{(s+kj\omega)t} = \widehat{\mathcal{D}}_{F\mu}(T,s,t)\,,$$

where the function $\widehat{\mathcal{D}}_{F\mu}(T,s,t)$ is determined by each one of the equivalent formulae

$$\widehat{\mathcal{D}}_{F\mu}(T,s,t) = \sum_{i=1}^{q}\sum_{k=1}^{\nu_i} \frac{f_{ik}}{(k-1)!}\frac{\partial^{k-1}}{\partial s_i^{k-1}}\frac{\mathrm{e}^{s_i t}\mu(s_i)}{\mathrm{e}^{(s-s_i)T-1}} + h_{F\mu}(t)\,,$$

$$\widehat{\mathcal{D}}_{F\mu}(T,s,t) = \sum_{i=1}^{q}\sum_{k=1}^{\nu_i} \frac{f_{ik}}{(k-1)!}\frac{\partial^{k-1}}{\partial s_i^{k-1}}\frac{\mathrm{e}^{s_i t}}{1-\mathrm{e}^{(s_i-s)T}} + h^*_{F\mu}(t)\,.$$

The functions $h_{F\mu}(t)$ and $h^*_{F\mu}(t)$ are committed by the relations

$$h_{F\mu}(t) = \int_0^t h_F(t-\tau)m(\tau)\,\mathrm{d}\tau\,, \qquad h^*_{F\mu}(t) = -\int_t^T h_F(t-\tau)m(\tau)\,\mathrm{d}\tau\,,$$

where

$$h_F(t) = \sum_{i=1}^{q}\sum_{k=1}^{\nu_i} \frac{f_{ik}}{(k-1)!}\, t^{k-1}\mathrm{e}^{s_i t}\,.$$

Besides, the sum of the series $\mathcal{D}_{F\mu}(T,s,t)$ depends continuously on t and has a piecewise continuous derivative. For $m_1 = \dots = m_{\ell-1} = 0$, $m_\ell \neq 0$, the function $\widehat{D}_{F\mu}(T,s,t)$ possesses derivatives up to and including ℓ-th order. Hereby, the ℓ-th derivative is piecewise continuous and the lower ones are continuous.

C

DirectSDM – A Toolbox for Optimal Design of Multivariable SD Systems

C.1 Introduction

This section contains a short description of the `DirectSDM Toolbox` for MATLAB®. The toolbox is designed for solving optimisation problems for multivariable sampled-data control systems. The computational procedures used in this software are based on the frequency-domain theory of sampled-data systems developed in the present book, and on the theory of matrix polynomial equations [79, 80, 55, 56].

The `DirectSDM Toolbox` is compatible with MATLAB® 6.0 and higher and requires the `Control Toolbox`. The toolbox is not compatible with the `Polynomial Toolbox` (http://www.polyx.com), although some functions have the same names. The reader may download the `DirectSDM Toolbox` from `http://www.iat.uni-rostock.de/blampe/` .

C.2 Data Structures

For description of control system elements, the following two data structures are used:

- Polynomial and quasi-polynomial matrices;
- Real rational matrices.

Polynomial matrices are realised as objects of the class `poln`. The special variables `s`, `p`, `z`, `d`, and `q` are realised as functions, and they are used for entering polynomial matrices. For example, after the input

```
P = [ s+1     s^2+s-6
      s^3      s-12 ]
```

the MATLAB® environment creates and displays the following polynomial matrix:

```
P: polynomial matrix: 2 x 2
   s + 1     s^2 + s - 6
     s^3             s - 12
```

Moreover, the `DirectSDM Toolbox` supports operations with quasi-polynomials (by this term we mean functions having poles only at the origin) by means of the same class `poln`. For example, the input

```
P = [ z+1      z+1+z^-1
      z^2-5     1+z^-2 ]
```

creates and displays the following quasi-polynomial matrix:

```
P: quasi-polynomial matrix: 2 x 2
     z + 1      z + 1 + z^-1
   z^2 - 5          1 + z^-2
```

Real rational matrices are stored and handled as objects of standard classes of the `Control Toolbox` describing models of LTI-systems, namely, `tf` (transfer matrix), `zpk` (zero-pole-gain form), and `ss` (state-space description). The `DirectSDM Toolbox` redefines the `display` function for the classes `tf` and `zpk`. Also, some errors in the `Control Toolbox` (versions up to 5.2) have been corrected.

C.3 Operations with Polynomial Matrices

Since the synthesis procedures developed in the present book essentially exploit models in form of polynomial matrices and matrix fraction descriptions (MFD), the `DirectSDM Toolbox` supports all basic operations with polynomial and quasi-polynomial matrices.

For objects of the `poln` class, the arithmetic operations (addition, subtraction, multiplication, division) as well as concatenation, transposition and inversion (for square matrices) are overloaded. It should be noticed that all operands used in binary operations should have the same independent variable (`s`, `p`, `z`, `d`, or `q`), respectively.

Below a short list of functions for handling polynomial and quasi-polynomial matrices are given.

Basic properties of polynomial matrices:

`coef`	coefficient at term of given degree
`coldeg`	column degrees
`deg`	matrix degree
`det`	determinant (a polynomial)
`eig`	eigenvalues of square matrix (roots of the determinant)
`lcoef`	leading coefficient
`norm`	norm (Euclidean norm of coefficient matrix)
`polyval`	value for given argument value
`polyder`	derivative
`rank`	normal rank
`roots`	roots of determinant (or those of each element)
`rowdeg`	row degrees
`trace`	trace

Simple transformations:

`coladd`	column addition
`colchg`	column interchange
`colmul`	multiplication of column by polynomial
`fliplr`	flip in left/right direction
`flipud`	flip in up/down direction
`rowadd`	row addition
`rowchg`	row interchange
`rowmul`	multiplication of row by polynomial

Special forms:

`colherm`	column Hermite form
`colred`	column-reduced form
`echelonl`	left echelon form
`echelonr`	right echelon form
`ltriang`	lower triangular form
`rowherm`	row Hermite form
`rowred`	row-reduced form
`smith`	canonical Smith form
`utriang`	upper triangular form

Miscellaneous functions:

`gcld`	a greatest left common divisor
`gcrd`	a greatest right common divisor
`invuni`	inversion of unimodular matrix
`jfact`	spectral J-factorisation of Hermitian-conjugate matrix
`null`	null-space basis
`pinv`	pseudoinverse matrix
`lfact`	left spectral factorisation
`linv`	left inverse
`rfact`	right spectral factorisation
`sylv`	block Sylvester coefficient matrix

Solution of Diophantine polynomial equations:

`daxb`	equation $AX = B$
`daxbyc`	equation $AX + BY = C$
`daxybc`	equation $AX + YB = C$

C.4 Auxiliary Algorithms

The `DirectSDM Toolbox` includes a number of auxiliary functions that are necessary for realising the optimal design procedures described in the book.

Operations with MFD:

`bezout`	solution to Bezout identity
`lmfd`	left-coprime MFD
`lmfd2ss`	state-space model for left MFD
`rmfd`	right-coprime MFD
`rmfd2ss`	state-space model for right MFD
`ss2lmfd`	left MFD for state-space model
`ss2rmfd`	right MFD for state-space model
`rmfd2lmfd`	transformation from right MFD to left MFD
`lmfd2rmfd`	transformation from left MFD to right MFD

Discrete transformations:

`ztrm`	discrete Laplace transform $D_F(T, \zeta, t)$
`dtfm`	discrete transfer matrix $D_{FM}(T, \zeta, t)$ for plant with ZOH
`dtfm2`	discrete transform $D_{\underline{M}'\underline{F}'FM}(T, \zeta, 0)$ for plant with ZOH

C.5 $\mathcal{H}_2$-optimal Controller

C.5.1 Extended Single-loop System

Consider the multivariable single-loop system shown in Fig. C.1. The digital controller (in the dashed box) composed of a discrete filter with transfer matrix $C(\zeta)$ and hold circuit with transfer function $\mu(s)$ is used for stabilising a continuous-time plant. The control loop includes a plant $F(s)$, actuator $G(s)$ and dynamic negative feedback $Q(s)$. The exogenous disturbance $w(t)$ and measurement noise $m(t)$ are modeled as vector stationary stochastic processes with spectral density matrices $S_w(s) = \underline{F_w}'(s)F_w(s)$ and $S_m(s) = \underline{F_m}'(s)F_m(s)$, respectively. The signals $\xi(t)$ and $\eta(t)$ are independent unit centred white noises.

The output signal $e(t)$ denotes the stabilisation error. The controller should ensure minimal power of the error signal under restrictions imposed on the control power. The frequency-dependent weighting functions $V_e(s)$ and $V_u(s)$ are introduced in order to shape the frequency-domain properties of the system (for example, to ensure roll-off of the controller frequency response at high frequencies).

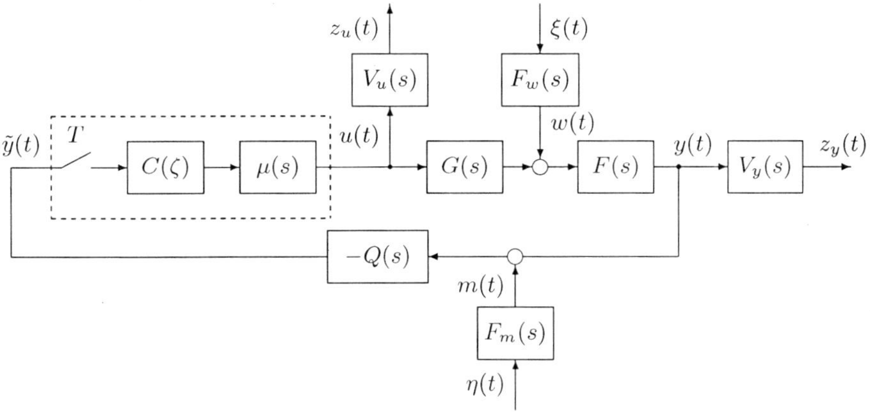

Fig. C.1. Single-loop sampled-data system

The following assumptions should hold:

1. The matrices $K(s)$, $M(s)$, and $N(s)$ are strictly proper, and $L(s)$ is at least proper.
2. The matrix $N(s)$ is irreducible in the sense of Sec. 9.2.
3. The matrices $F_w(s)$, $F_m(s)$, $V_e(s)$ and $V_u(s)$ are stable.
4. The matrices $G(s)$ and $Q(s)$ are free of poles at the imaginary axis.
5. The transfer matrices $F(s)$, $G(s)$ and $Q(s)$ are normal.
6. The sampling period T is non-pathological.

Assumption 1 is necessary for the optimisation problem to be correct (see Chapter 9). Assumptions 2, 5, and 6 ensure that the assumptions of Chapter 9 hold. In applied problems, Assumption 5 holds almost always. Assumption 3 causes no loss of generality, because for any spectral density having no poles at the imaginary axis, a stable forming filter can be derived. As was shown in Chapter 9, when Assumption 4 is violated, a formal application of the Wiener-Hopf optimisation technique leads to a non-stabilising controller.

The stabilisation quality is estimated by the average variance $\bar{v}_z$ of the output vector signal

$$z(t) = \begin{bmatrix} z_e(t) \\ z_u(t) \end{bmatrix},$$

which can be found (for centred stochastic processes) as

$$J = \bar{v}_z = \frac{1}{T}\int_0^T \mathsf{E}\left\{z^T(t)z(t)\right\}\,\mathrm{d}t = \frac{1}{T}\int_0^T \mathsf{E}\left\{\operatorname{trace} z(t)z^T(t)\right\}\,\mathrm{d}t\,, \qquad \text{(C.1)}$$

where $\mathsf{E}\{\cdot\}$ denotes the mathematical expectation.

The problem can be formulated as follows: Let the continuous-time elements, sampling period T and hold device $\mu(s)$ be given. Find the transfer matrix of a stabilising digital controller $C(\zeta)$ ensuring the minimum of the cost function (C.1).

It can be shown that this problem is a special case of the general $\mathcal{H}_2$-optimisation problem for standard sampled-data system investigated in Chapter 8. Assume

$$x(t) = \begin{bmatrix} \xi(t) \\ \eta(t) \end{bmatrix}$$

and denote by $y(t)$ the vector signal acting upon the sampling unit. Then, the operator equations of the system take the form

$$\begin{aligned} z &= K(s)x + L(s)u \\ y &= M(s)x + N(s)u\,, \end{aligned}$$

where the matrices of the associated standard system are:

$$K(s) = \begin{bmatrix} V_e(s)F(s)F_w(s) & 0 \\ 0 & 0 \end{bmatrix}, \quad L(s) = \begin{bmatrix} V_e(s)F(s)G(s) \\ V_u(s) \end{bmatrix},$$

$$M(s) = \begin{bmatrix} -Q(s)F(s)F_w(s) & -Q(s)F_m(s) \end{bmatrix}, \quad N(s) = -Q(s)F(s)G(s)\,.$$

Thus, the cost function (C.1) equals the square of the $\mathcal{H}_2$-norm of the above standard sampled-data system.

C.5.2 Function sdh2

The function **sdh2** can be used for synthesis of $\mathcal{H}_2$-optimal controllers for extended single-loop multivariable systems as described above. Consider, for example, a simplified model of course stabilisation for a "Kazbek"-type tanker [149]:

$$F(s) = \begin{bmatrix} \dfrac{0.051}{(25s+1)s} \\ \\ \dfrac{0.051}{25s+1} \end{bmatrix}, \quad G(s) = \frac{1}{s+1},$$

$$F_w(s) = 1\,, \quad F_m(s) = 0\,, \quad V_e(s) = I\,, \quad V_u(s) = 1\,, \quad T = 1\,.$$

As distinct from the problem considered in [149], the yaw angle φ and rotation rate ω_z are both measured, *i.e.*, the controller has 2 inputs and 1 output.

The system shown in Fig. C.1 must be described as a structure of MATLAB® as follows:

```
sys.F  = tf({0.051; 0.051},{[25 1 0];[25 1]});
sys.G  = tf(1, [1 1]);
sys.Fw = tf(1);
sys.Vu = tf(1);
sys.T  = 1;
```

Mandatory fields of the structure are only `sys.F`, `sys.G`, `sys.Fw`, and `sys.T`. If others are not specified, they take the following default values:

```
sys.Fm = 0;
sys.Ve = eye(n);
sys.Vu = 0;
sys.Q  = eye(n);
```

Here `n` denotes the number of outputs of the plant $F(s)$ and `eye(n)` denotes the identity matrix of the corresponding dimension.

The function call

```
[C,P] = sdh2 ( sys )
```

gives the transfer matrix of the (unique) optimal controller $C(z)$ (in the variable z!) and poles of the closed-loop system in the z-plane:

```
C: zero-pole-gain model 1 x 2

     ! 0.98247 (z-0.3679)    17.4769 (z-0.3679) !
     ! ------------------    ------------------ !
     !      (z-0.3457)           (z-0.3457)     !

Sampling time: 1

P =
    0.9627 + 0.0240i
    0.9627 - 0.0240i
    0.3679 + 0.0000i
    0.3679 - 0.0000i
```

Since all poles are inside the unit disk, the optimal closed-loop system is stable.

This example is investigated in detail in the demo script `demoh2` included in the `DirectSDM Toolbox` .

C.6 $\mathcal{L}_2$-optimal Controller

C.6.1 Extended Single-loop System

We consider the extended multivariable single-loop tracking system shown in Fig. C.2. The control loop includes a plant $F(s)$, prefilter $G(s)$ and dynamic negative feedback $Q(s)$. The input signal $x(t)$ has the Laplace transform $X(s)$. The digital controller (in the dashed box), consists of a discrete filter with transfer matrix $C(\zeta)$ and hold device with transfer function $\mu(s)$.

The transfer matrices $W_e(s)$ and $W_u(s)$ define ideal operators reflecting the requirements to the output and control transients, respectively. The frequency-dependent weighting matrices $V_y(s)$ and $V_e(s)$ can be used for shaping the

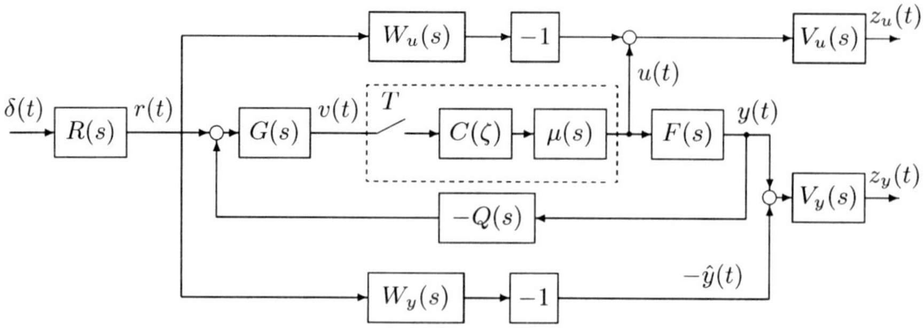

Fig. C.2. Single-loop sampled-data tracking system

frequency properties of the system and controller to be designed. Henceforth, we assume that $W_e(s)$ and $W_u(s)$ are free of unstable poles and all remaining assumptions set in the $\mathcal{H}_2$-optimisation problem hold again.

Introduce a stacked output signal

$$z(t) = \begin{bmatrix} z_e(t) \\ z_u(t) \end{bmatrix} .$$

The cost function includes the sum of weighted integral quadratic output and control errors and coincides with the square of the $\mathcal{L}_2$-norm of $z(t)$:

$$J = \|z(t)\|_2^2 = \int_0^\infty z^T(t)z(t)\,\mathrm{d}t = \int_0^\infty \left[z_e^T(t)z_e(t) + z_u^T(t)z_u(t)\right]\mathrm{d}t\,. \quad \text{(C.2)}$$

The problem is formulated as follows: Let all continuous elements of the system, hold device $\mu(s)$ and sampling interval T be given. Find a stabilising digital controller $C(\zeta)$ ensuring the minimum of the cost function (C.2).

It can be shown that the problem under consideration can be viewed as a special case of the general $\mathcal{L}_2$-optimisation problem for standard sampled-data system analysed in Chapter 10. Denote the signal acting upon the sampling unit by $y(t)$. Then the system equations in operator form appear as

$$\begin{aligned} z &= K(s)x + L(s)u \\ y &= M(s)x + N(s)u\,, \end{aligned}$$

where the matrices of the corresponding standard system have the form

$$K(s) = -\begin{bmatrix} V_e(s)W_e(s)R(s) \\ V_u(s)W_u(s)R(s) \end{bmatrix}, \quad L(s) = \begin{bmatrix} V_e(s)F(s) \\ V_u(s) \end{bmatrix},$$

$$M(s) = G(s)R(s)\,, \quad N(s) = -G(s)Q(s)F(s)\,.$$

C.6.2 Function sdl2

The function sdl2 can be used for synthesis of $\mathcal{L}_2$-optimal controllers for the extended single-loop multivariable system described above. Assume,

$$F(s) = \begin{bmatrix} \dfrac{1}{0.5s+1} \\ \dfrac{1}{(0.5s+1)s} \end{bmatrix}, \quad G(s) = 1, \quad Q(s) = I, \quad X(s) = \begin{bmatrix} \dfrac{1}{s} \\ \dfrac{1}{s} \end{bmatrix},$$

$$W_e(s) = \begin{bmatrix} 0 & 0 \\ 0 & 1 \end{bmatrix}, \quad V_e(s) = I, \quad W_u(s) = \begin{bmatrix} 0 & 0 \end{bmatrix}, \quad V_u(s) = 0, \quad T = 1.$$

The system shown in Fig. C.2 is described as a structure of MATLAB® as

```
sys.F  = tf( {1; 1}, {[0.5 1]; conv(1,[0.5 1 0])} );
sys.X  = tf( {1; 1}, {[1 0]; [1 0]} );
sys.We = tf( {0 0;0 1}, {1 1;1 1} );
sys.T  = 1;
```

Among all the fields, only sys.F, sys.R, sys.Wy, and sys.T are required. If the others are not given, they take the following default values:

```
sys.G = eye(m);
sys.Q = eye(n);
sys.Ve = eye(n);
sys.Wu = 0;
sys.Vu = 0;
```

Here n denotes the number of outputs of the plant $F(s)$, m is the dimension of the input signal $x(t)$, and eye(·) denotes the identity matrix of the corresponding dimension.

The function call

```
[C,P] = sdl2 ( sys )
```

gives the transfer matrix of an optimal controller $C(z)$ (non-unique for multivariable systems) and the poles of the optimal closed-loop system in z-plane:

```
C: zero-pole-gain model 1 x 2

!  0.6891 (z-0.1469) (z-1)      0.21306 (z-0.124) (z+3.833)  !
! -------------------------    ----------------------------  !
! (z^2  + 0.2842z + 0.869)       (z^2  + 0.2842z + 0.869)    !

Sampling time: 1

P =

         0
```

```
        0
   0.3673
  -0.2329
```

Since all poles are inside the unit disk, the closed-loop system is stable.

This example is investigated in detail in the demo script `demol2` included in the `DirectSDM Toolbox` .

D

Design of SD Systems with Guaranteed Performance

D.1 Introduction

During the projection of control systems, usually for analysis and design, nearly complete information is required about the conditions under which the system should be in function. A typical practical problem consists in the investigation of the function, when the system is disturbed by stochastic external signals. As shown in the present book, in this case for evaluating the function of a sampled-data system, the mean variance of the output could be used, and the optimisation criterion could be a weighted sum of the output variances.

For calculating the mean variance and for applying the optimisation procedure, the considered methods require the spectral density of the excitation. However, for the majority of real stochastic processes, a rough information about the spectral density is not available. For instance, there is no exact answer to the question about the spectrum of sea waves [23, 24, 122]. Therefore, it is impossible to predict, under which conditions the process will move.

The lack of rough information about the spectral density has the consequence, that the variance of the output signal cannot be calculated, thus we cannot find the optimal controller. In engineering practice, this situation is managed in the following way. For the real spectral density of the affecting disturbance, several approximations are built. Then for each approximation, the analysis or synthesis problem for the optimal system is solved. However, this way of solution never takes into account the approximation error of the spectral density. Hence the influence of this error on the performance of the system cannot be estimated under real excitations. But in practice, the situation may arise that a prescribed performance of the system must be guaranteed for any of a set of excitations. In this case, it cannot be predicted how variations in the parameters of the excitation affect the performance of the optimal system.

Hence the absence of rough information about the spectral density of the excitation leads to the following problem:

- Analyse a system under incomplete information about the external excitation.
- Design a system that guarantees an upper bound of the performance index for all excitations of a certain set.

Below, such systems are called *systems with guaranteed performance*, and the synthesis procedure is named *design for guaranteed performance.* The set of excitations, for which the performance of the system is guaranteed inside prescribed limits, is called *class of excitations.* Taking a single loop scalar systems as an instance, the present appendix considers methods for the solution of analysis and design problems for guaranteed performance. Moreover, the modelling of classes of stochastic excitations is explained which are needed for definition and solution of the named tasks.

The practical computations are realised with the MATLAB®-Toolbox *GarSD* which was particularly developed for analysis and design of sampled-data systems with guaranteed performance. The package operates together with the MATLAB®-Toolbox *DirectSD*, [130] and the Toolbox *DirectSDM* which has been presented in Appendix C.

D.2 Design for Guaranteed Performance

D.2.1 System Description

Consider the single-loop scalar sampled-data system with the structure shown in Fig. D.1. The centralised stationary stochastic excitation $g(t)$ with the

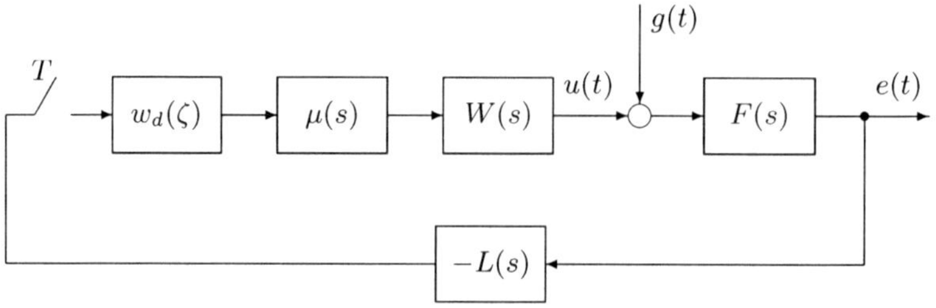

Fig. D.1. Single-loop scalar sampled-data system

spectral density $S_g(s)$ affects the continuous process with the transfer function $F(s)$. Furthermore, we have the transfer functions of the actuator $W(s)$, of the feedback $L(s)$, and of the forming element $\mu(s)$. The product $W(s)F(s)L(s)$ is assumed to be strictly proper, while $W(s)$ and $W(s)F(s)$ are at least proper. The system is controlled by the digital controller with the transfer function

$$w_d(\zeta) = \frac{\sum_{r=0}^{R} b_r \zeta^r}{\sum_{r=0}^{R} a_r \zeta^r} = \frac{B(\zeta)}{A(\zeta)}, \qquad \zeta = \mathrm{e}^{-sT}, \tag{D.1}$$

where A and B are polynomials, and $a_0 \neq 0$. The order R of the controller and the sampling period T are prescribed. The deviation $e(t)$ and the control signal $u(t)$ appear as outputs of the system. The PTFs from the input $g(t)$ to the outputs $e(t)$ and $u(t)$ become

$$w_{ge}(s,t) = F(s)\left(1 - \frac{L(s)w_d(\zeta)\varphi_{\mu WF}(T,s,t)}{1 + w_d(\zeta)\varphi_{\mu WFL}(T,s,0)}\right), \tag{D.2}$$

$$w_{gu}(s,t) = -F(s)\frac{L(s)w_d(\zeta)\varphi_{\mu W}(T,s,t)}{1 + w_d(\zeta)\varphi_{\mu WFL}(T,s,0)}, \tag{D.3}$$

where $\varphi_{\mu WF}(T,s,t)$, $\varphi_{\mu W}(T,s,t)$, $\varphi_{\mu WFL}(T,s,t)$ are the corresponding displaced pulse frequency responses (DPFR).

Let Z be the vector containing the constructive parameters, which have to be determined. The components of the vectors Z have to be committed in such a way that the transfer function of the controller $w_d(\zeta)$ is uniquely established. In this case, the PTFs (D.2), (D.3) are functions of the vector Z. These functions will be denoted by

$$w_1(s,t,Z) = w_{ge}(s,t), \quad w_2(s,t,Z) = w_{gu}(s,t).$$

Then the formulae for calculating the variance of the outputs might be written in the form

$$d_\kappa(t,Z) = \frac{1}{\pi}\int_0^\infty A_\kappa^2(\nu,t,Z)\tilde{S}_g(\nu)\,\mathrm{d}\nu, \qquad \kappa = 1,2 \tag{D.4}$$

with the magnitude of the parametric frequency response

$$A_\kappa(\nu,t,Z) = \left|w_\kappa(s,t,Z)\right|_{s=i\nu}$$

and the spectral density

$$\tilde{S}_g(\nu) \triangleq S_g(s)|_{s=i\nu}.$$

The functional

$$\bar{J}(Z) = \bar{d}_1(Z) + \rho^2 \bar{d}_2(Z) \tag{D.5}$$

is used as performance criterion, where ρ is a real weighting coefficient and $\bar{d}_1(Z)$ and $\bar{d}_2(Z)$ are the mean variances, which are determined by

$$\bar{d}_\kappa(Z) = \frac{1}{T}\int_0^T d_\kappa(t,Z)\,\mathrm{d}t, \quad \kappa = 1,2.$$

D.2.2 Problem Statement

Consider the situation, where we miss rough knowledge of the spectral density $S_g(s)$, but only the general characterisation that the excitations belong to a class M_S of stochastic disturbances.

Suppose for certain known parameters Z, that there exists an estimation $\bar{D}_\kappa(Z)$ of the mean variance for the generalised characteristics of the class M_S, such that

$$\bar{D}_\kappa(Z) \geq \bar{d}_\kappa(Z) \tag{D.6}$$

is true over the whole set M_S.

Using this estimation, in analogy to (D.5), the functional $\bar{E}(Z)$ could be written as weighted sum of the estimation $\bar{D}_\kappa(Z)$:

$$\bar{E}(Z) = \bar{D}_1(Z) + \rho^2 \bar{D}_2(Z)\,. \tag{D.7}$$

The value $\bar{E}(Z)$ is adequate to the maximal possible value of $\bar{J}(Z)$ in (D.5) for the functioning system with the parameter Z under any excitations from the class M_S. Per construction, the functional (D.7) does not depend on the concrete spectral density and its value is only determined by the vector Z of the system parameters and a general characterisation of the excitations.

Assume that we can find a vector Z_{gar}, such that the functional (D.7) takes a minimal value, *i.e.*

$$\bar{E}(Z_{gar}) = \min_{Z} \max_{M_S} \bar{J}(Z)\,. \tag{D.8}$$

The procedure for searching the vector Z_{gar} is called *design for guaranteed performance.*

Let $\bar{J}_0$ be known as the largest value of (D.5), for which the function of the system is accepted as successful. Then, if we can prove the inequality

$$\bar{E}(Z_{gar}) \leq \bar{J}_0\,,$$

then with the aid of (D.6)–(D.8), we are able to state that a successful operation of the systems with the parameter Z_{gar} for any excitation from the class M_S is guaranteed. There is no disturbance of the class M_S, for which the maximal possible value (D.5) exceeds the boundary $\bar{J}_0$.

We consider two modelling variants for the class of random excitations in problems for guaranteed performance.

In the first variant, the model involves the variance d_0 of the excitation and the totality of its N moments d_n

$$\frac{1}{\pi}\int_0^\infty \nu^{2n} S(\nu)\,\mathrm{d}\nu = d_n\,, \quad n = 0, 1, \ldots, N\,. \tag{D.9}$$

This totality is a generalised characteristic of the spectral density that is robust against variations [18, 122].

In the second variant, the class M_S is modelled by enveloping the spectral density $S_{og}(\nu)$. The construction makes sure that there exists no frequency ν, for which any element of the class M_S takes a value greater than $S_{og}(\nu)$.

D.2.3 Calculation of Performance Criterion

I. Let the class M_S of the system excitations in Fig. D.1 be given by the set d_n, $n = 0, \dots, N$ (D.9). Always suppose that the transfer function of the processes $F(s)$ and the product $F(s)L(s)$ are strictly proper. In this case, we obtain for the parametric frequency response $A_k(\nu, t, Z)$

$$\lim_{\nu \to \infty} A_\kappa(\nu, t, Z) = 0\,, \qquad \kappa = 1, 2\,,$$

i.e. the system as a whole reacts to the input $g(t)$ as a low-pass [148]. Besides, the PFR decreases not slower than $1/\nu$, when $\nu \to \infty$. Thus, the integrals (D.4) converge absolutely and the infinite limits in (D.4) might be substituted by finite values β_κ, because of

$$A_\kappa^2(\nu, t, Z) \approx 0, \quad \forall \nu > \beta_\kappa\,.$$

Moreover, the frequency domain of the spectral density is supposed to be upwards bounded

$$\tilde{S}_g(\nu) \approx 0 \qquad \forall \nu > \beta_S\,,$$

where the value β_S is known.

On basis of these suppositions, the mean variances $\bar{d}_\kappa$ of the signals $e(t)$ and $u(t)$ for known $\tilde{S}_g(\nu)$ may be calculated approximately by the formula

$$\bar{d}_\kappa(Z) = \frac{1}{\pi} \int_0^{\beta_S} \bar{A}_\kappa(\nu, Z) \tilde{S}_g(\nu)\, \mathrm{d}\nu\,,$$

where

$$\bar{A}_\kappa(\nu, Z) = \frac{1}{T} \int_0^T A_\kappa^2(\nu, t, Z)\, \mathrm{d}t, \qquad \kappa = 1, 2\,.$$

The estimates $\bar{D}_\kappa(Z)$ of the mean variances $\bar{d}_\kappa(Z)$ are calculated by

$$\bar{d}_\kappa(Z) \le \bar{D}_\kappa(Z) = \sum_{n=0}^{N} c_{n\kappa}(Z) d_n\,, \tag{D.10}$$

where $c_{n\kappa}(Z)$ are the coefficients of the polynomials

$$C_\kappa(\nu, Z) = \sum_{n=0}^{N} c_{n\kappa}(Z) \nu^{2n}\,,$$

which are determined by

$$\bar{A}_\kappa(\nu, Z) \le C_\kappa(\nu, Z) \qquad \forall \nu \in [0, \beta_S] \tag{D.11}$$

and in addition

$$C_\kappa(\nu, Z) \approx \bar{A}_\kappa(\nu, Z) \quad \forall \nu \in [0, \beta_S]\,. \tag{D.12}$$

If (D.10) is satisfied, the functional (D.7) appears in the form

$$\bar{E}(Z) = \sum_{n=0}^{N} c_{n1}(Z) d_n + \rho^2 \sum_{n=0}^{N} c_{n2}(Z) d_n \,. \tag{D.13}$$

Due to (D.11), this functional majorises the functional (D.5), and for any given Z, it constitutes an upper bound [155]. Moreover, it does not depend on the concrete spectral density, but its value is determined only by the generalised characteristic of the class M_S. The coefficients $c_{n\kappa}(Z)$ can be computed by applying known numeric procedures [121], [155].

Remark D.1. Practical computations have shown that for arbitrary excitations, the inclusion of variances higher than first order has marginal influence to the estimation of the mean variance. Therefore, in practice the calculation of two coefficients for each polynomial $C_\kappa(\nu, Z)$ is sufficient.

II. When the class M_S is given by the envelope spectral density $S_{og}(\nu)$, then an estimation of the mean variance could be found more precisely. Let for the envelope spectral density the value β_S be known, and for the system with given vector Z the value β_κ should be found. The following considerations are valid for $\beta_\kappa \le \beta_S$.

In this case, the functional (D.7) can achieve the form

$$\bar{E}(Z) = \sum_{n=0}^{N} c'_{n1}(Z) d'_{n1} + \rho^2 \sum_{n=0}^{N} c'_{n2}(Z) d'_{n2} \,,$$

where the quantities d'_{n1} and d'_{n2} are determined by integrals of the shape

$$d'_{n\kappa} = \frac{1}{\pi} \int_0^{\beta_\kappa} S_{og}(\nu) \nu^{2n} \, \mathrm{d}\nu, \quad n = 0, \ldots, N; \quad \kappa = 1, 2$$

and the coefficients $c'_{n\kappa}(Z)$ are chosen in such a way that Conditions (D.12), (D.11) are satisfied on the interval $[0, \beta_\kappa]$.

D.2.4 Minimisation of Performance Criterion Estimate for SD Systems

Consider the search process for the vector Z_{gar} that minimises (D.7) for the sampled-data system containing the digital controller with the transfer function (D.1). For this purpose, the application of genetic algorithms is suitable [131], [117].

There are two variants of using genetic algorithms for selecting a controller. Let us have to design a sampled-data system of Fig. D.1 for guaranteed performance. For the discrete transfer function $D_{WLF\mu}(T, s, t)$ of the open sampled-data system with the elements $W(s)$, $L(s)$, $F(s)$ and $\mu(s)$, the representation [148]

$$D_{\mu WLF}(T, s, 0) = \left. \frac{n(\zeta)}{d(\zeta)} \right|_{\zeta = e^{-sT}}$$

takes place, where $n(\zeta)$ and $d(\zeta)$ are polynomials.

Investigate the equation

$$A(\zeta)d(\zeta) + B(\zeta)n(\zeta) = \Delta_{des}(\zeta)\Delta^{\star}(\zeta)\,,$$

where $\Delta_{des}(\zeta)$ is a polynomial containing as roots all desired pole positions $\tilde{\zeta}_i$, and $\Delta^{\star}(\zeta)$ is a stable polynomial. This equation has to be solved for $A(\zeta)$ and $B(\zeta)$ and these polynomials establish the transfer function of a stabilising controller, which guarantees that the values $\tilde{\zeta}_i$ are among the poles of the closed-loop system.

Then the first variant consists in selecting stable poles of the closed-loop system. Hereby, it is assumed that there exists a special choice Z_{gar} of the vector Z with roots of the polynomial $\Delta^{\star}(\zeta)$, such that the corresponding transfer function of the controller minimises the functional (D.7). Besides, all roots of the polynomial $\Delta_{des}(\zeta)$ are among the poles of the closed-loop system.

The second variant consists in selecting the coefficients of the controller directly, where the order R is given. Thereto, the existence of a stabilising controller of this order has to be confirmed [14]. The elements of the vector Z are the searched coefficients a_r, $(r = 1, \dots, R)$ and b_r, $(r = 0, \dots, R)$ of the controller. The existence of a vector Z_{gar} is assumed, such that the corresponding controller guarantees in addition to the stability also the minimal value of the functionals (D.7). During the search procedure, the stability of the system must be assured.

D.3 MATLAB®-Toolbox *GarSD*

The above derived algorithm was realised in the MATLAB®-Toolbox *GarSD*, which provides the solution of analysis and design problems for sampled-data systems with guaranteed performance. The solution of the design problem needs elements of the theory of polynomial equations realised in the MATLAB®-Toolbox *DirectSD* that has to be available [130]. The package was tested on a PC under Windows XP with MATLAB® 6.5.

D.3.1 Structure

The package consists of three modules.

1. The module `SPECTRAL` contains procedures to investigate the information about the external excitation and to formulate the initial data that are needed for computations.

2. The module GarSD realises procedures for computing the functional (D.7) for the sampled-data system and for its minimisation.
3. The module GenSD realises numerical minimisation procedures by applying genetic algorithms.

For completion, the module SEAWAVE might be used. Here various evaluation methods were collected. These methods are applicable for data, which describe the effects of sea waves to a ship. The module is suitable for the solution of problems for guaranteed performance, because it contains sea-wave spectra of real measurements.

D.3.2 Setting Properties of External Excitations

The information about the external excitation is set into the structure `spectral` by the command `spt` as

```
spectral=spt(<information>);
```

The variable `information` in case of a known envelope spectral density $S_{og}(\nu)$ may have different formats:

1. Transfer function of the form filters (object `tf`)
2. Numerator and denominator polynomials of the fractional rational spectral density function
3. Vector with number values of the spectral density
4. Coefficients of the exponential functions of the spectral density.

For instance, the envelope spectral density may be given by $S_{og}(\nu) = 0.1/(\nu^4 - 2\nu^2 + 2)$. This corresponds to a form filter with the transfer function $F_{fil}(s) = 0.32/(s^2 + 0.91s + 1.41)$. It is set by one of the both commands

```
spectral=spt(0.1,[1 0 2 0 2]);
spectral=spt(tf(0.32,[1 0.91 1.41]));
```

The spectral density of the form $S_{og}(\nu) = A\nu^{-m}exp(-B\nu^{-n})$ is fed into the computer by the command

```
spectral=spt(A,B,m,n);
```

Moreover, the module `spectral` includes procedures that allow to design the envelope spectral density for a given set of excitations, or to build the set of excitations in various practical situations (for instance in case of three dimensional disturbance models or of switching between different modes in the system), or to find a rational approximation for the envelope spectral density, when several spectra are given by numerical data.

The class of excitations may be given by the totality of the variances of its excitations and their moments d_0, d_1, ... and the limit frequency β_S in the class of the spectral densities. Then the class is defined by the command

```
spectral=spt(d_i,beta_S);
```

where d_i is the vector of the excitation variances and its moments.

Moreover, the module `spectral` contains procedures for testing a set of data to turn out as variances.

D.3.3 Investigation of SD Systems

System description The Toolbox is dedicated to work with sampled-data systems with a structure as in Fig. D.1. As forming element a zero-order hold is used.

If we have to solve an analysis problem, *i.e.* when the transfer function of the controller is known, then the structure of the system is fed by the command

```
system=sys(F,W,L,wd);
```

where the variables `F, W, L, wd` are objects of the class `tf` according to the elements of the systems. If the controller is unknown, then the structure of the system is set by the command

```
system=sys(F,W,L,T);
```

where `T` is the sampling period (variable of type `double`). The structure `system` is taken inside the toolbox for the solution of analysis and design problems.

Estimation of guaranteed performance Suppose a system of Fig. D.1, where all elements are known and stored in the structures `system` and `spectral`. The toolbox contains macros, which compute in the given case an estimation of the variance (D.6) for the time instants in the interval $[0, T]$

```
[t,D]=aprsys(num,system,spectral,step);
```

where `step` is the step width of the time for estimating the time-varying variance (the value of the variable `step` must not exceed the value of the sampling period) and `num` is the identifier of the system output to be analysed. For the output $e(t)$, we take `num=12`, and `num=22` for the output $u(t)$. As a result, the vectors `t, D` of equal size will be generated, which contains the time instant of the estimated variance of the output, and the value of this estimation itself.

The value of the mean variance is estimated with the help of the command

```
[t,D]=aprsys(num,system,spectral);
```

As a result, the variable `t` takes the value of the empty matrix, and the variable `D` contains the estimation of the mean variance.

The kind of computing the estimation depends on the type of information about the excitation. Depending on the type of systems, the toolbox realises two algorithms. One of them requires only negligible computation time, but does not allow to investigate systems with multiple or nearly multiple poles.

The other one is free of these restrictions, but needs more computation time. For realising the last algorithm, some macros of the toolbox *DirectSD* were employed. The selection of the algorithm happens automatically.

Moreover, the toolbox `GarSD` contains several procedures for testing the stability of sampled-data systems, for computing the poles or the oscillation index, for construction of the PFR, for determining the transfer functions of stabilising controllers and of controllers with certain assigned poles.

Minimising performance index estimate Suppose for the system in Fig. D.1, that the transfer functions of all elements except the controller are known and stored in the structure `system`. Design a system with guaranteed performance over the class M_S, which is given by the structure `spectral`.

The solution of the minimising problem for the estimation of the performance index is provided by a genetic algorithm [117]. It is realised in the macro `regelgarsys`, which accesses to the structures `system` and `spectral`

```
[system_gar,reg_gar,D_e,D_u,E]
     =regelgarsys(type,deg,T,system,spectral,rho,num);
```

The parameters in the macro `regelgarsys` have the following meanings: The parameter `type` contains the type of the optimisation and can adopt the values `type='sta'` or `type='all'` according to the first or second variant for selecting the elements of the vectors Z. The parameter `deg` contains the order of the searched controller; for the optimisation type `'sta'`, this parameter may hold the value `deg='min'`, which is adequate to the smallest degree for stabilising controllers [148]. The parameter `T` contains the sampling period of the wanted digital controller and the parameter `rho` is the weighting coefficient in the functional (D.7). The parameter `num` commits the number of iterations of the genetic algorithm, it is usually selected between 30 and 100.

The command supplies as result: The structure `system_gar` of the system with guaranteed performance, the transfer function of the digital controller `reg_gar` as object of `tf`, `D_e, D_u` as estimates of the output variances of the systems with guaranteed performance according to the class M_S, which is given by the structure `spectral`, as well as the value `E` of the functional (D.7).

Applicative example Consider the design problem for a controller with guaranteed performance for the system in [149], where the course of the tanker “Kasbek” under the condition of continuous excitation has to be kept. The behavior of the ship and the rudder plant is described by

$$F(s) = \frac{\alpha}{s(1+\beta s)}, \qquad W(s) = 1,$$

with the values $\alpha = 0.051\,\mathrm{sec}^{-1}$, $\beta = 25\,\mathrm{sec}$. The transfer functions of the process, the actuator and the feedback are are given by

```
alpha=0.051; beta=25;
F=tf(alpha/beta,[1 1/beta 0]);
W=tf(1);  L=tf(1);
```

and this information is collected in the structure of the system with the sampling period 1 sec by the command

```
system=sys(F,W,L,1);
```

Suppose the class of excitations M_S is given by the enveloped spectral density

$$S_{og}(\nu) = \frac{0.0757}{\nu^4 - 2.489\nu^2 + 1.848} \,. \tag{D.14}$$

The structure of this class of excitations is generated by the command

```
spectral=spt(0.0757,[1 0 2.489 0 1.848]);
```

For the weighting coefficient $\rho = 0.1$, a system with guaranteed performance should be designed, where the sampling period is $T = 1\,\text{sec}$. As number of iterations for the genetic algorithm, we choose num=50. For the selection of the minimal controller order according to the first variant, the command

```
[system_gar,reg_gar,D_e,D_u,E]=
  regelgarsys('sta','min',1,system,spectral,0.1,50);
```

is used. As a result, we obtain the transfer function of the controller reg_gar in the form

$$w_{d_1}(z) = \frac{37.82z - 34.99}{z - 0.535} \,.$$

For any excitation of the class M_S, the system with this controller guarantees values of the mean variances $\bar{d}_e \leq \bar{D}_e = 0.000066$ and $\bar{d}_u \leq \bar{D}_u = 0.0105$. Besides, the value of the functional (D.7) is estimated by $\bar{E} = 0.00017$.

Applying the second variant of controller design, so for instance the controller order 1 could be chosen (the existence of a stabilising controllers of 1st order for the given system was just proven)

```
[system_gar,reg_gar,D_e,D_u,E]=
  regelgarsys('all',1,1,system,spectral,0.1,50);
```

The macro supplies the transfer function of the controller reg_gar:

$$w_{d_2}(z) = \frac{66.86z - 61.45}{z - 0.14} \,.$$

For any excitation of the class M_S, the system with this controller guarantees values of the mean variances $\bar{d}_e \leq \bar{D}_e = 0.000061$ and $\bar{d}_u \leq \bar{D}_u = 0.0173$. Besides, the value of the functional (D.7) is estimated by $\bar{E} = 0.00023$.

The property of the envelop of the spectral density obviously ensures that for any excitation of the class M_S, the value of the mean variances of the signals $u(t)$ and $e(t)$ in the systems with controllers $w_{d_1}(z)$ or $w_{d_2}(z)$, will not exceed the values of the obtained estimations.

Now, let us assume that the class M_S is given by the set of variances

$$d_0 = 0.05807, \qquad d_1 = 0.07895 \tag{D.15}$$

and the width $\beta_S = 3.04$. This new class involves the spectrum (D.14). The structure of excitations is generated by

```
spectral=spt([0.05807 0.07895],3.04);
```

Let us take the first variant for the controller design:

```
[system_gar,reg_gar,D_e,D_u,E]=
  regelgarsys('sta','min',1,system,spectral,0.1,50);
```

As a result, we obtain the transfer function of the controller `reg_gar`

$$w_{d_3}(z) = \frac{367.8z - 333.8}{z + 0.3107}\,.$$

For any excitation of the class M_S, the system with this controller guarantees values of the mean variances $\bar{d}_e \leq \bar{D}_e = 0.000056$ and $\bar{d}_u \leq \bar{D}_u = 0.0597$ as well as the value $E = 0.00065$ in (D.7).

Finally, it remains to design the controller, for instance of first order, for the totality of variances by the second variant. The existence of a stabilising controller of 1st order was proven above. The command

```
[system_gar,reg_gar,D_e,D_u,E]=
  regelgarsys('all',1,1,system,spectral,0.1,50);
```

supplies the transfer function of the controller `reg_gar`:

$$w_{d_4}(z) = \frac{1.87z - 0.24}{z + 0.83}\,.$$

For any excitation of the class M_S, the system with this controller guarantees values of the mean variances $\bar{d}_e \leq \bar{D}_e = 0.019$ and $\bar{d}_u \leq \bar{D}_u = 0.050$ as well as $E = 0.019$ in (D.7).

Now investigate the behavior of the system with the controllers $w_{d_3}(z)$ and $w_{d_4}(z)$ under the condition of various excitations of the class M_S for the given set (D.15). For certain spectral densities of the class M_S having the form

$$S(\nu) = a_1/(\nu^4 + a_2\nu^2 + a_3)\,,$$

the values of the coefficients a_1, a_2, a_3 are listed in Table D.1. In the same table, the exact values of the mean variances of the signals $e(t)$ and $u(t)$ occur for the controllers $w_{d_3}(z)$ and $w_{d_4}(z)$, respectively.

Table D.1 exemplifies that the values of the mean variances of the signals $e(t)$ and $u(t)$ for all considered excitations do not exceed the values of the calculated estimations.

Table D.1. Variances of the output of the system with guaranteed performance for various excitations from the class M_S

Spectrum	a_1	a_2	a_3	$\bar{d}_e(w_{d_3})$	$\bar{d}_u(w_{d_3})$	$\bar{d}_e(w_{d_4})$	$\bar{d}_u(w_{d_4})$
1	0.154	1.768	1.847	6.58×10^{-6}	0.043	0.0023	0.0018
2	0.265	-0.118	1.847	1.06×10^{-5}	0.048	0.0041	0.0032
3	0.225	0.672	1.847	9.23×10^{-6}	0.047	0.00349	0.0027

References

1. J. Ackermann. Entwurf durch Polvorgabe. *Regelungstechnik*, 25:173–179, 209–215, 1977.
2. J. Ackermann. *Sampled-Data Control Systems: Analysis and Synthesis, Robust System Design*. Springer-Verlag, Berlin, 1985.
3. J. Ackermann. *Abtastregelung*. Springer-Verlag, Berlin, 3 edition, 1988.
4. A.G. Alexandrov and Y.F. Orlov. Finite frequency identification of multivariable objects. In *Proc. 2nd Russian-Swedish Control Conf. (RSCC'95)*, pages 66–69, Saint Petersburg, Russia, 1995.
5. F.A. Aliev, V.B. Larin, K.I. Naumenko, and V.I. Suncev. *Optimization of linear control systems: Analytical methods and computational algorithms*. Gordon & Breach, Baffalo, 1998.
6. F.A. Aliev, V.B. Larin, K.I. Naumenko, and V.I. Suntsev. *Optimization of linear time-invariant control systems*. Naukova Dumka, Kiev, 1978. (in Russian).
7. B.D.O. Anderson. Controller design: Moving from theory to practice. *IEEE Control Systems*, 13(4):16–24, 1993.
8. B.D.O. Anderson and J.B. Moore. *Optimal Filtering*. Prentice-Hall, Englewood Cliffs, NJ, 1979.
9. M. Araki, T. Hagiwara, and Y. Ito. Frequency response of sampled-data systems II. Closed-loop considerations. In *Proc. 12th IFAC Triennial World Congr.*, volume 7, pages 293–296, Sydney, 1993.
10. M. Araki and Y. Ito. Frequency response of sampled-data systems I. Open-loop considerations. In *Proc. 12th IFAC Triennial World Congr.*, volume 7, pages 289–292, Sydney, 1993.
11. K.J. Åström. *Introduction to stochastic control theory*. Academic Press, NY, 1970.
12. K.J. Åström, P. Hagender, and J. Sternby. Zeros of sampled-data systems. *Automatica*, 20(4):31–38, 1984.
13. K.J. Åström and B. Wittenmark. *Computer controlled systems: Theory and design*. Prentice-Hall, Englewood Cliffs, NJ, 1984.
14. K.J. Åström and B. Wittenmark. *Computer Controlled Systems: Theory and Design*. Prentice-Hall, Englewood Cliffs, NJ, 3rd edition, 1997.
15. B.A. Bamieh and J.B. Pearson. A general framework for linear periodic systems with applications to $\mathcal{H}_\infty$ sampled-data control. *IEEE Trans. Autom. Contr*, AC-37(4):418–435, 1992.

16. B.A. Bamieh and J.B. Pearson. The H_2 problem for sampled-data systems. *Syst. Contr. Lett.*, 19(1):1–12, 1992.
17. B.A. Bamieh, J.B. Pearson, B.A. Francis, and A. Tannenbaum. A lifting technique for linear periodic systems with applications to sampled-data control systems. *Syst. Contr. Lett.*, 17:79–88, 1991.
18. V.A. Besekerskii and A.V. Nebylov. *Robust systems in automatic control.* Nauka, Moscow, 1983. (in Russian).
19. M.J. Blachuta. *Contributions to the theory of discrete-time control for continuous-time systems.* Habilitation thesis, Silesian Techn. University, Gliwice, Poland, 1999.
20. M.J. Blachuta. Discrete-time modeling of sampled-data control systems with direct feedthrough. *IEEE Trans. Autom. Contr*, 44(1):134–139, 1999.
21. Ch. Blanch. Sur les equation differentielles lineares a coefficients lentement variable. *Bull. technique de la Suisse romande*, 74:182–189, 1948.
22. S. Bochner. *Lectures on Fourier Integrals.* University Press, Princeton, NJ, 1959.
23. I. Boroday, V. Mohrenschildt, et al. *Behavior of ships in ocean waves.* Sudostroyenie, Leningrad, 1969.
24. I.K. Boroday and V.V. Nezetaev. *Application problems of dynamics for ships on waves.* Sudostroyenie, Leningrad, 1989.
25. G.D. Brown, M.G. Grimble, and D. Biss. A simple efficient $\mathcal{H}_\infty$ controller algorithm. In *Proc. 26th IEEE Conf. Decision Contr*, Los Angeles, 1987.
26. B.W. Bulgakov. *Schwingungen.* GITTL, Moskau, 1954. (in Russisch).
27. F.M. Callier and C.A. Desoer. *Linear system theory.* Springer-Verlag, New York, 1991.
28. M. Cantoni. Algebraic characterization of the H_∞ and H_2 norms for linear continuous-time periodic systems. In *Proc. 4th Asian Control Conference*, pages 1945–1950, Singapore, 2002.
29. S.S.L. Chang. *Synthesis of optimum control systems.* McGraw Hill, New York, Toronto, London, 1961.
30. T. Chen and B.A. Francis. *Optimal sampled-data control systems.* Springer-Verlag, Berlin, Heidelberg, New York, 1995.
31. T.A.C.M. Claasen and W.F.G. Mecklenbräuker. On stationary linear time varying systems. *IEEE Trans. Circuits and Systems*, CAS-29(2):169–184, 1982.
32. P. Colaneri. Continuous-time periodic systems in H_2 and H_∞. Part I: Theoretical aspects; Part II: State feedback control. *Kybernetika*, 36(3):211–242; 329–350, 2000.
33. R.E. Crochiere and L.R. Rabiner. *Multirate digital signal processing.* Prentice-Hall, Englewood Cliffs, NJ, 1983.
34. L. Dai. *Singular control systems.* Lecture notes in Control and Information Sciences. Springer-Verlag, New York, 1989.
35. J.A. Daletskii and M.G. Krein. *Stability of solutions of differential equations in Banach-space.* Nauka, Moscow, 1970. (in Russian).
36. R. D'Andrea. Software for modeling, analysis, and control design for multidimensional systems. In *Proc. IEEE Symp. on Computer Aided Control System Design (CACSD'99)*, pages 24–27, Kohala Coast, Island of Hawai'i, Hawai'i, USA, 1999.
37. C.E. de Souza and G.C. Goodwin. Intersample variance in discrete minimum variance control. *IEEE Trans. Autom. Contr*, AC-29:759–761, 1984.

38. B.W. Dickinson. *Systems - Analysis, Design and Computation.* Prentice Hall, Englewood Cliffs, NJ, 1991.
39. G. Doetsch. *Anleitung zum praktischen Gebrauch der Laplace Transformation und z-Transformation.* Oldenbourg, München, Wien, 1967.
40. R.C. Dorf and R.H. Bishop. *Modern control systems.* Pearson Prentice Hall, Upper Saddle River, NJ, tenth edition, 2001.
41. J.C. Doyle. Guaranteed margins for LQG regulators. *IEEE Trans. Autom. Contr*, AC-23(8):756–757, 1978.
42. J.C. Doyle, B.A. Francis, and A.R. Tannenbaum. *Feedback control theory.* Macmillan, New York, 1992.
43. S. Engell. *Lineare optimale Regelung.* Springer-Verlag, Berlin, 1988.
44. D.K. Faddeev and V.N. Faddeeva. *Numerische Methoden der linearen Algebra.* Oldenbourg, München, 1979. (with L. Bittner).
45. A. Feuer and G.C. Goodwin. Generalised sample and hold functions - frequency domain analysis of robustness, sensitivity and intersampling difficulties. *IEEE Trans. Autom. Contr*, AC-39(5):1042–1047, 1994.
46. N. Fliege. *Multiraten-Signalverarbeitung.* B.G. Teubner, Stuttgart, 1993.
47. V.N. Fomin. *Control methods for discrete multidimensional processes.* University press, Leningrad, 1985. (in Russian).
48. V.N. Fomin. *Regelungsverfahren für diskrete Mehrgrößenprozesse.* Verlag der Universität, Leningrad, 1985. (in Russisch).
49. G.F. Franklin, J.D. Powell, and A. Emami-Naeini. *Feedback Control of Dynamic Systems.* Prentice Hall, Upper Saddle River, NJ 07458, 4 edition, 2002.
50. G.F. Franklin, J.D. Powell, and H.L. Workman. *Digital control of dynamic systems.* Addison Wesley, New York, 1990.
51. F.R. Gantmacher. *The theory of matrices.* Chelsea, New York, 1959.
52. E.G. Gilbert. Controllability and observability in multivariable control systems. *SIAM J. Control*, A(1):128–151, 1963.
53. G.C. Goodwin, S.F. Graebe, and M.E. Salgado. *Control system design.* Prentice-Hall, Upper Saddle River, NJ 07458, 2001.
54. G.C. Goodwin and M. Salgado. Frequency domain sensitivity functions for continuous-time systems under sampled-data control. *Automatica*, 30(8):1263–1270, 1994.
55. M.J. Grimble. *Robust Industrial Control: Optimal Design Approach for Polynomial Systems.* International Series in Systems and Control Engineering. Prentice Hall International (UK) Ltd, Hemel Hempstead, Hertfordshire, 1994.
56. M.J. Grimble and V. Kučera, editors. *Polynomial methods for control systems design.* Springer-Verlag, London, 1996.
57. M. Günther. *Kontinuierliche und zeitdiskrete Regelungen.* B.G. Teubner, Stuttgart, 1997.
58. T. Hagiwara and M. Araki. FR-operator approach to the $\mathcal{H}_2$-analysis and synthesis of sampled-data systems. *IEEE Trans. Autom. Contr*, AC-40(8):1411–1421, 1995.
59. V. Hahn. *Direkte adaptive Regelstrategien für die diskrete Regelung von Mehrgrößensystemen.* PhD thesis, University of Bochum, 1983.
60. M.E. Halpern. Preview tracking for discrete-time SISO systems. *IEEE Trans. Autom. Contr*, AC-39(3):589–592, 1994.
61. S. Hara, H. Fujioka, and P.T. Kabamba. A hybrid state-space approach to sampled-data feedback control. *Linear Algebra and Its Applications*, 205-206:675–712, 1994.

62. U.K. Herne. *Methoden zur rechnergestützten Analyse und Synthese von Mehrgrößenregelsystemen in Polynommatrizendarstellung.* PhD thesis, University of Bochum, 1988.
63. R. Isermann. *Digitale Regelungssysteme. Band I: Grundlagen, deterministische Regelungen. Band II: Stochastische Regelungen, Mehrgrößenregelungen Adaptive Regelungen, Anwendungen.* Springer-Verlag, Berlin, 2 edition, 1987.
64. M.A. Jevgrafov. *Analytische Funktionen.* Nauka, Moskau, 1965. (in Russisch).
65. G. Jorke, B.P. Lampe, and N. Wengel. *Arithmetische Algorithmen der Mikrorechentechnik.* Verlag Technik, Berlin, 1989.
66. E.I. Jury. *Sampled-data control systems.* John Wiley, New York, 1958.
67. P.T. Kabamba and S. Hara. Worst-case analysis and design of sampled-data control systems. *IEEE Trans. Autom. Contr*, AC-38(9):1337–1358, 1993.
68. T. Kaczorek. *Linear control systems*, volume II - Synthesis of multivariable systems. J. Wiley, New York, 1993.
69. T. Kailath. *Linear Systems.* Prentice Hall, Englewood Cliffs, NJ, 1980.
70. R. Kalman and J.E. Bertram. A unified approach to the theory of sampling systems. *J. Franklin Inst.*, 267:405–436, 1959.
71. R. Kalman, Y.C. Ho, and K. Narendra. Controllabiltiy of linear dynamical systems. *Contributions to the Theory of Differential Equations*, 1:189–213, 1963.
72. R.E. Kalman. Mathematical description of linear dynamical systems. *SIAM J. Control*, A(1):152–192, 1963.
73. S. Karlin. *A first course in stochastic processes.* Academic Press, New York, 1966.
74. V.J. Katkovnik and R.A. Polucektov. *Discrete multidimensional control.* Nauka, Moscow, 1966. (in Russian).
75. J.P. Keller and B.D.O. Anderson. H_∞-Optimierung abgetasteter Regelsysteme. *Automatisierungstechnik*, 40(4):114–123, 1993.
76. U. Keuchel. *Methoden zur rechnergestützten Analyse und Synthese von Mehrgrößensystemen in Polynommatrizendarstellung.* PhD thesis, University of Bochum, 1988.
77. P.P. Khargonekar and N. Sivarshankar. $\mathcal{H}_2$-optimal control for sampled-data systems. *Systems & Control Letters*, 18:627–631, 1992.
78. U. Korn and H.-H. Wilfert. *Mehrgrößenregelungen.* Verlag Technik, Berlin, 1982.
79. V. Kučera. *Discrete Linear Control. The Polynomial Approach.* Academia, Prague, 1979.
80. V. Kučera. *Analysis and Design of Discrete Linear Control Systems.* Prentice Hall, London, 1991.
81. B.C. Kuo and D.W. Peterson. Optimal discretization of continuous-data control systems. *Automatica*, 9(1):125–129, 1973.
82. H. Kwakernaak. Minimax frequency domain performance and robustness optimisation of linear feedback systems. *IEEE Trans. Autom. Contr*, AC-30(10):994–1004, 1985.
83. H. Kwakernaak. The polynomial approach to $\mathcal{H}_\infty$ regulation. In E. Mosca and L. Pandolfi, editors, *$\mathcal{H}_\infty$ control theory*, volume 1496 of *Lecture Notes in Mathematics*, pages 141–221. Springer-Verlag, London, 1990.
84. H. Kwakernaak and R. Sivan. *Linear Optimal Control Systems.* Wiley-Interscience, New York, 1972.

85. S. Lall and C. Beck. Model reduction of complex systems in the linear-fractional framework. In *Proc. IEEE Int. Symp. on Computer Aided Control System Design (CACSD'99)*, pages 34–39, Kohala Coast, Island of Hawai'i, Hawai'i, USA, 1999.
86. B.P. Lampe. Strukturelle Instabilität in linearen Systemen – Frequenzgangsmethoden auf dem Prüfstand der Mathematik. In *Mitteilungen der Mathematischen Gesellschaft in Hamburg*, volume XVIII, pages 9–26, Hamburg, Germany, 1999.
87. B.P. Lampe, G. Jorke, and N. Wengel. *Algorithmen der Mikrorechentechnik.* Verlag Technik, Berlin, 1984.
88. B.P. Lampe, M.A. Obraztsov, and E.N. Rosenwasser. $\mathcal{H}_2$-norm computation for stable linear continuous-time periodic systems. *Archives of Control Sciences*, 14(2):147–160, 2004.
89. B.P. Lampe, M.A. Obraztsov, and E.N. Rosenwasser. Statistical analysis of stable FDLCP systems by parametric transfer matrices. *Int. J. Control*, 78(10):747–761, Jul 2005.
90. B.P. Lampe and U. Richter. Digital controller design by parametric transfer functions - comparison with other methods. In *Proc. 3. Int. Symp. Methods Models Autom. Robotics*, volume 1, pages 325–328, Miedzyzdroje, Poland, 1996.
91. B.P. Lampe and U. Richter. Experimental investigation of parametric frequency response. In *Proc. 4. Int. Symp. Methods Models Autom. Robotics*, pages 341–344, Miedzyzdroje, Poland, 1997.
92. B.P. Lampe and E.N. Rosenwasser. Design of hybrid analog-digital systems by parametric transfer functions. In *Proc. 32nd CDC*, pages 3897–3898, San Antonio, TX, 1993.
93. B.P. Lampe and E.N. Rosenwasser. Application of parametric frequency response to identification of sampled-data systems. In *Proc. 2. Int. Symp. Methods Models Autom. Robotics*, volume 1, pages 295–298, Miedzyzdroje, Poland, 1995.
94. B.P. Lampe and E.N. Rosenwasser. Best digital approximation of continuous controllers and filters in $\mathcal{H}_2$. In *Proc. 41st KoREMA*, volume 2, pages 65–69, Opatija, Croatia, 1996.
95. B.P. Lampe and E.N. Rosenwasser. Best digital approximation of continuous controllers and filters in H_2. *AUTOMATIKA*, 38(3–4):123–127, 1997.
96. B.P. Lampe and E.N. Rosenwasser. Parametric transfer functions for sampled-data systems with time-delayed controllers. In *Proc. 36th IEEE Conf. Decision Contr*, pages 1609–1614, San Diego, CA, 1997.
97. B.P. Lampe and E.N. Rosenwasser. Sampled-data systems: The L_2-induced operator norm. In *Proc. 4. Int. Symp. Methods Models Autom. Robotics*, pages 205–207, Miedzyzdroje, Poland, 1997.
98. B.P. Lampe and E.N. Rosenwasser. Statistical analysis and H_2-norm of finite dimensional linear time-periodic systems. In *Proc. IFAC Workshop on Periodic Control Systems*, pages 9–14, Como, Italy, Aug. 2001.
99. B.P. Lampe and E.N. Rosenwasser. Forward and backward models for anomalous linear discrete-time systems. In *Proc. 9th IEEE Symp. Methods Models Autom. Robotics*, pages 369–373, Miedzyzdroje, Poland, Aug 2003.
100. B.P. Lampe and E.N. Rosenwasser. Operational description and statistical analysis of linear periodic systems on the unbounded interval $-\infty < t < \infty$. *European J. Control*, 9(5):508–521, 2003.

101. B.P. Lampe and E.N. Rosenwasser. Closed formulae for the $\mathcal{L}_2$-norm of linear continuous-time periodic systems. In *Proc. IFAC Workshop on Periodic Control Systems*, pages 231–236, Yokohama, Japan, Sep 2004.
102. B.P. Lampe and E.N. Rosenwasser. Unterordnung und Dominanz rationaler Matrizen. *Automatisierungstechnik*, 53(9):434–444, 2005.
103. F.H. Lange. *Signale und Systeme*, volume 1–3. Verlag Technik, Berlin, 1971.
104. V.B. Larin, K.I. Naumenko, and V.N. Suntsov. *Spectral methods for design of linear systems with feedback.* Naukova Dumka, Kiev, 1971. (in Russian).
105. B. Lennartson and T. Söderström. Investigation of the intersample variance in sampled-data control. *Int. J. Control*, 50:1587–1602, 1989.
106. B. Lennartson, T. Söderström, and Sun Zeng-Qi. Intersample behavior as measured by continuous-time quadratic criteria. *Int. J. Control*, 49:2077–2083, 1989.
107. O. Lingärde and B. Lennartson. Frequency analysis for continuous-time systems under multirate sampled-data control. In *Proc. 13th IFAC Triennial World Congr.*, volume 2a–10, 5, pages 349–354, San Francisco, USA, 1996.
108. L. Ljung. *System Identification – Theory for the User.* Prentice-Hall, Englewood Cliffs, NJ, 1987.
109. D.G. Luenberger. Dynamic equations in descriptor form. *IEEE Trans. Autom. Contr*, AC-22(3):312–321, 1977.
110. J. Lunze. *Robust multivariable feedback control.* Akademie-Verlag, Berlin, 1988.
111. J. Lunze. *Regelungstechnik 2 - Mehrgrößensysteme, Digitale Regelung.* Springer-Verlag, Berlin, Heidelberg, ..., 1997.
112. N.N. Lusin. Matrix theory for studying differential equations. *Avtomatika i Telemechanika*, 5:4–66, 1940. (in Russian).
113. N.N. Lusin. Matrizentheorie zum Studium von Differentialgleichungen. *Avtomatika i Telemechanika*, 5:4–66, 1940.
114. J.M. Maciejowski. *Multivariable feedback design.* Addison-Wesley, Wokingham, England a.o., 1989.
115. J.M. Maciejowski. *Predictive control - with constraints.* Pearson Education Lim., Harlow, England, 2002.
116. A.G. Madievski and B.D.O. Anderson. A lifting technique for sampled-data controller reduction for closed-loop transfer function consideration. In *Proc. 32nd IEEE Conf. Decision Contr*, pages 2929–2930, San Antonio, TX, 1993.
117. K.F. Man, K.S. Tang, and S. Kwong. *Genetic algorithms.* Springer-Verlag, London Berlin Heidelberg, 1999.
118. S.G. Michlin. *Vorlesungen über lineare Integralgleichungen.* Dt. Verlag d. Wissenschaften, Berlin, 1962.
119. B.C. Moore. Principal component analysis in linear systems: Controllability, observability and model reduction. *IEEE Trans. Autom. Contr*, AC-26(1):17–32, 1981.
120. R. Müller. *Entwurf von Mehrgrößenreglern durch Frequenzgang-Approximation.* PhD thesis, University of Dortmund, 1996.
121. A.V. Nebylov. *Warranting of accuracy of control.* Nauka, Moscow, 1998. (in Russian).
122. A.V. Nebylov. *Measuring parameters of a plane near the sea surface.* Saint Petersburg State University Academic Press, St. Petersburg, 2000. (in Russian).
123. K. Ogata. *Modern control engineering.* Prentice-Hall, Upper Saddle River, NJ 07458, 2002.

124. V.G. Pak and V.N. Fomin. *Linear quadratic optimal control problem under known disturbance I. Abstract linear quadratic problem under known disturbance.* Preprint VINITI, N2063-B97, St. Petersburg, 1997. (in Russian).
125. K. Parks and J.J. Bongiorno. Modern Wiener-Hopf design of optimal controllers – Part II: The multivariable case. *IEEE Trans. Autom. Contr*, AC-34(6):619–626, 1989.
126. R.V. Patel. Computation of minimal-orders state-space realisations and observability indices using orthonormal transformations. In R.V. Patel, A.J. Laub, and Van Dooren P.M., editors, *Numerical linear Algebra techniques for systems and control*, pages 195–212. IEEE Press, New York, 1994.
127. T.P. Perry, G.M.H. Leung, and B.A. Francis. Performance analysis of sampled-data control systems. *Automatica*, 27(4):699–704, 1991.
128. U. Petersohn, H. Unger, and Wardenga W. Beschreibung von Multirate-Systemen mittels Matrixkalkül. *AEÜ*, 48(1):34–41, 1994.
129. J.P. Petrov. *Design of optimal control systems under incompletely known input disturbances.* University press, Leningrad, 1987. (in Russian).
130. K.Y. Polyakov, E.N. Rosenwasser, and B.P. Lampe. DirectSD - a toolbox for direct design of sampled-data systems. In *Proc. IEEE Intern. Symp. CACSD'99*, pages 357–362, Kohala Coast, Island of Hawai'i, Hawai'i, USA, 1999.
131. K.Y. Polyakov, E.N. Rosenwasser, and B.P. Lampe. Quasipolynomial low-order digital controller design using genetic algorithms. In *Proc. 9th IEEE Mediterranian Conf. on Control and Automation*, pages WM1–B5, Dubrovnik, Croatia, June 2001.
132. K.Y. Polyakov, E.N. Rosenwasser, and B.P. Lampe. `DirectSDM` - a toolbox for polynomial design of multivariable sampled-data systems. In *Proc. IEEE Int. Symp. Computer Aided Control Systems Design*, pages 95–100, Taipei, Taiwan, Sep 2004.
133. V.M. Popov. *Hyperstability of control systems.* Springer-Verlag, Berlin, 1973.
134. I.I. Priwalow. *Einführung in die Funktionentheorie.* 3. Aufl., B.G. Teubner, Leipzig, 1967.
135. R. Rabenstein. *Diskrete Simulation linearer mehrdimensionaler Systeme.* PhD thesis, University of Erlangen-Nürnberg, 1991.
136. J.R. Ragazzini and G.F. Franklin. *Sampled-data control systems.* McGraw-Hill, New York, 1958.
137. J.R. Ragazzini and L.A. Zadeh. The analysis of sampled-data systems. *AIEE Trans.*, 71:225–234, 1952.
138. J. Raisch. *Mehrgrößenregelung im Frequenzbereich.* R. Oldenbourg Verlag, München, 1994.
139. K.S. Rattan. Digitalization of existing control systems. *IEEE Trans. Autom. Contr*, AC-29:282–285, 1984.
140. K.S. Rattan. Compensating for computational delay in digital equivalent of continuous control systems. *IEEE Trans. Autom. Contr*, AC-34:895–899, 1989.
141. K. Reinschke. *Lineare Regelungs- und Steuerungstheorie.* Springer-Verlag, Berlin, 2006.
142. G. Roppenecker. Fortschr.-Ber. VDI-Z. In *Vollständige modale Synthese linearer Systeme und ihre Anwendung zum Entwurf strukturbeschränkter Zustandsrückführungen*, number 59 in 8. VDI-Verlag, Düsseldorf, 1983.
143. E.N. Rosenwasser. *Lyapunov-Indizes in der linearen Regelungstheorie.* Nauka, Moskau, 1977. (in Russisch).

144. E.N. Rosenwasser, P.G. Fedorov, and B.P. Lampe. Construction of MFD-representation of real rational transfer matrices on basis of normalisation procedure. In *Int. Conf. on Computer Methods for Control Systems*, pages 39–42, Szczecin, Poland, December 1997.
145. E.N. Rosenwasser, P.G. Fedorov, and B.P. Lampe. Construction of state-space model with minimal dimension for multivariable system on basis of transfer matrix normalization procedure. In *Proc. 5. Int. Symp. Methods Models Autom. Robotics*, volume 1, pages 235–238, Miedzyzdroje, Poland, 1998.
146. E.N. Rosenwasser and B.P. Lampe. *Digitale Regelung in kontinuierlicher Zeit - Analyse und Entwurf im Frequenzbereich.* B.G. Teubner, Stuttgart, 1997.
147. E.N. Rosenwasser and B.P. Lampe. *Algebraische Methoden zur Theorie der Mehrgrößen-Abtastsysteme.* Universitätsverlag, Rostock, 2000. ISBN 3-86009-195-6.
148. E.N. Rosenwasser and B.P. Lampe. *Computer Controlled Systems - Analysis and Design with Process-orientated models.* Springer-Verlag, London Berlin Heidelberg, 2000.
149. E.N. Rosenwasser, K.Y. Polyakov, and B.P. Lampe. Entwurf optimaler Kursregler mit Hilfe von Parametrischen Übertragungsfunktionen. *Automatisierungstechnik*, 44(10):487–495, 1996.
150. E.N. Rosenwasser, K.Y. Polyakov, and B.P. Lampe. Frequency domain method for $\mathcal{H}_2$–optimization of time-delayed sampled-data systems. *Automatica*, 33(7):1387–1392, 1997.
151. E.N. Rosenwasser, K.Y. Polyakov, and B.P. Lampe. Optimal discrete filtering for time-delayed systems with respect to mean-square continuous-time error criterion. *Int. J. Adapt. Control Signal Process.*, 12:389–406, 1998.
152. E.N. Rosenwasser, K.Y. Polyakov, and B.P. Lampe. Application of Laplace transformation for digital redesign of continuous control systems. *IEEE Trans. Automat. Contr*, 4(4):883–886, April 1999.
153. E.N. Rosenwasser, K.Y. Polyakov, and B.P. Lampe. Comments on "A technique for optimal digital redesign of analog controllers". *IEEE Trans. Control Systems Technology*, 7(5):633–635, September 1999.
154. W.J. Rugh. *Linear system theory.* Prentice-Hall, Englewood Cliffs, NJ, 1993.
155. V.O. Rybinskii and B.P. Lampe. Accuracy estimation for digital control systems at incomplete information about stochastic input disturbances. In B.P. Lampe, editor, *Maritime Systeme und Prozesse*, pages 43–52. Universitätsdruckerei, Rostock, 2001.
156. V.O. Rybinskii, B.P. Lampe, and E.N. Rosenwasser. Design of digital ship motion control with guaranteed performance. In *Proc. 49. Int. Wiss. Kolloquium*, volume 1, pages 381–386, Ilmenau, Germany, 2004.
157. M. Saeki. Method of solving a polynomial equation for an $\mathcal{H}_\infty$ optimal control problem. *IEEE Trans. Autom. Contr*, AC-34:166–168, 1989.
158. M. Sågfors. *Optimal Sampled-Data and Multirate Control.* PhD thesis, Faculty of Chemical Engineering, Åbo Akademi University, Finland, 1998.
159. L. Schwartz. *Methodes mathematiques pour les sciences physiques.* Hermann 115, Paris VI, Boul. Saint-Germain, 1961.
160. H. Schwarz. *Optimale Regelung und Filterung - Zeitdiskrete Regelungssysteme.* Akademie-Verlag, Berlin, 1981.
161. L.S. Shieh, B.B. Decrocq, and J.L. Zhang. Optimal digital redesign of cascaded analogue controllers. *Optimal Control Appl. Methods*, 12:205–219, 1991.

162. L.S. Shieh, J.L. Zhang, and J.W. Sunkel. A new approach to the digital redesign of continuous-time controllers. *Control Theory Adv. Techn.*, 8:37–57, 1992.
163. I.Z. Shtokalo. Generalisation of symbolic method principal formula onto linear differential equations with variable coefficients. *Dokl. Akad. Nauk SSR*, 42:9–10, 1945. (in Russian).
164. S. Skogestad and I. Postlethwaite. *Multivariable feedback control: Analysis and design.* Wiley, Chichester, 2nd edition, 2005.
165. L.M. Skvorzov. Transformation algorithm for mathematical models of multidimensional control systems. *Izv. Akad. Nauk, Control theory and systems*, 2:17–23, 1997.
166. V.B. Sommer, B.P. Lampe, and E.N. Rosenwasser. Experimental investigations of analog-digital control systems by frequency methods. *Automation and Remote Control*, 55(Part 2):912–920, 1994.
167. E.D. Sontag. *Mathematical control theory – deterministic finite dimensional systems.* Springer-Verlag, New York, 1998.
168. D.S. Stearns. *Digitale Verarbeitung analoger Signale.* R. Oldenbourg Verlag, München, 1988.
169. R.F. Stengel. *Stochastic optimal control. Theory and application.* J. Wiley & Sons, Inc., New York, 1986.
170. Y. Tagawa and R. Tagawa. A computer aided technique to derive the class of realizable transfer function matrices of a control system for a prescribed order controller. In *Proc. IEEE Int. Symp. on Computer Aided Control System Design (CACSD'99)*, pages 321–327, Kohala Coast, Island of Hawai'i, Hawai'i, USA, 1999.
171. E.C. Titchmarsh. *The theory of functions.* Oxford science publ. University Press, Oxford, 2 edition, 1997. Reprint.
172. H.T. Toivonen. Sampled-data control of continuous-time systems with an $\mathcal{H}_\infty$-optimality criterion. *Automatica*, 28(1):45–54, 1992.
173. H.T. Toivonen. Worst-case sampling for sampled-data H_∞ design. In *Proc. 32nd IEEE Conf. Decision Contr*, pages 337–342, San Antonio, TX, 1993.
174. H. Tolle. *Mehrgrößenregelkreissynthese*, volume 1, 2. R. Oldenbourg Verlag, München, 1983, 1985.
175. J. Tou. *Digital and Sampled-Data Control Systems.* McGraw-Hill, New York, 1959.
176. H.L. Trentelmann and A.A. Stoorvogel. Sampled-data and discrete-time $\mathcal{H}_2$–optimal control. In *Proc. 32nd Conf. Dec. Contr.*, pages 331–336, San Antonio, TX, 1993.
177. J.S. Tsypkin. *Sampling systems theory.* Pergamon Press, New York, 1964.
178. R. Unbehauen. *Systemtheorie*, volume 2. R. Oldenbourg Verlag, München, 7 edition, 1998.
179. H. Unger, U. Petersohn, and S. Lindow. Zur Beschreibung hybrider Multiraten-Systeme mittels Matrixkalküls. *FREQUENZ*, 1997. (einger.).
180. K.G. Valeyev. Application of Laplace transform for analysis of linear systems. In *Proc. Intern. Conf. on Nonlin. Oscill.*, volume I, pages 126–132, Kiev, 1970. (in Russian).
181. B. van der Pol and H. Bremmer. *Operational calculus based on the two-sided Laplace integral.* University Press, Cambridge, 1959.
182. A. Varga. On stabilization methods of descriptor systems. *Syst. Contr. Lett.*, 24:133–138, 1995.

183. M. Vidyasagar. *Control system synthesis.* MIT Press, Cambridge, MA, 1994.
184. L.N. Volgin. *Optimal discrete control of dynamic systems.* Nauka, Moscow, 1986. (in Russian).
185. S. Volovodov, B.P. Lampe, and E.N. Rosenwasser. Application of method of integral equations for analysis of complex periodic behaviors in Chua's circuits. In *Proc. 1st IEEE Int. Conf. Control Oscill. Chaos*, pages 125–128, St. Petersburg, Russia, August 1997.
186. J. Wernstedt. *Experimentelle Prozeßanalyse.* Verlag Technik, Berlin, 1989.
187. E.T. Whittaker and G.N. Watson. *A course of modern analysis.* University Press, Cambridge, 4 edition, 1927.
188. J.H. Wilkinson. *The algebraic eigenvalue problem.* Clarendon Press, Oxford, 1965.
189. W.A. Wolovich. *Linear Multivariable Systems.* Springer-Verlag, New York, 1974.
190. W.A. Wolovich. *Automatic control systems.* Harcourt Brace, 1994.
191. W.M. Wonham. *Linear multivariable control - A geometric appraoch.* Springer-Verlag, New York Berlin ..., 3 edition, 1985.
192. R.A. Yackel, B.C. Kuo, and G. Singh. Digital redesign of continuous systems by matching of states at multiple sampling periods. *Automatica*, 10:105–111, 1974.
193. D.V. Yakubovich. Algorithm for supplementing a rectangular polynomial matrix to a quadratic matrix with given determinant. *Kybernetika i Vychisl.*, 23:85–89, 1984.
194. Y. Yamamoto. A function space approach to sampled-data systems and tracking problems. *IEEE Trans. Autom. Contr*, AC-39(4):703–713, 1994.
195. Y. Yamamoto and P. Khargonekar. Frequency response of sampled-data systems. *IEEE Trans. Autom. Contr*, AC-41(2):161–176, 1996.
196. D.C. Youla, H.A. Jabr, and J.J. Bongiorno (Jr.). Modern Wiener-Hopf design of optimal controllers. Part II: The multivariable case. *IEEE Trans. Autom. Contr*, AC-21(3):319–338, 1976.
197. L.A. Zadeh. Circuit analysis of linear varying-parameter networks. *J. Appl. Phys.*, 21(6):1171–1177, 1950.
198. L.A. Zadeh. Frequency analysis of variable networks. *Proc. IRE*, 39(March):291–299, 1950.
199. L.A. Zadeh. Stability of linear varying-parameter systems. *J. Appl. Phys.*, 22(4):202–204, 1951.
200. C. Zhang and J. Zhang. H_2-performance of continuous periodically time-varying controllers. *Syst. Contr. Lett*, 32:209–221, 1997.
201. P. Zhang, S.X. Ding, G.Z. Wang, and D.H. Zhou. Fault detection in multirate sampled-data systems with time-delays. In *Proc. 15th IFAC Triennial World Congr.*, volume Fault detection, supervision and safety of technical processes, page REG2179, Barcelona, 2002.
202. J. Zhou. *Harmonic analysis of linear continuous-time periodic systems.* PhD thesis, Kyoto University, 2001.
203. J. Zhou, T. Hagiwara, and M. Araki. Trace formulas for the H_2 norm of linear continuous-time periodic systems. In *Prepr. IFAC Workshop on Periodic Control Systems*, pages 3–8, Como, Italy, 2001.
204. J. Zhou, T. Hagiwara, and M. Araki. Trace formula of linear continuous-time periodic systems via the harmonic Lyapunov equation. *Int. J. Control*, 76(5):488–500, 2003.

205. K. Zhou and J.C. Doyle. *Essentials of robust control.* Prentice-Hall Intern., Upper Saddle River, NJ, 1998.
206. K. Zhou, J.C. Doyle, and K. Glover. *Robust and optimal control.* Prentice-Hall, Englewood Cliffs, NJ, 1996.
207. J.Z. Zypkin. *Sampling systems theory.* Pergamon Press, New York, 1964.

Index

Printed in the United States
58445LVS00001B/28-33